"This is a passionate, convincing portrayal of our planet in crisis. . . . Tobias reveals points of light in the global darkness, and proposes a philosophy and ethics that could turn the tide against the approaching apocalypse."
—Joseph Kulin, executive publisher, *Parabola* magazine

"This eye-opening book should be required reading for policy makers. It is impeccably researched and chock full of vivid images drawn from the author's firsthand experiences. . . . Solid, fascinating, disturbing, *World War III*, best of all, offers a blueprint for our spiritual and material survival. For your children's sake, read this book!"
—Ingrid E. Newkirk, national director, People for the Ethical Treatment of Animals (PETA)

"This is not science fiction. This is not a diatribe to frighten you. This is not an arid book of academia. This is a profound, prophetic, poetic call to arms. . . . *World War III* must not happen. Michael Tobias shows us how that is possible."
—William Shatner

"*World War III* is the most dramatic and up-to-date book chronicling the collision of population, development, and the environment. Not since Paul Ehrlich wrote *The Population Bomb* in 1968, which sold two million copies, has a book portrayed the war humans are having with their habitat. Michael Tobias has researched his book using the most authoritative sources . . . [his] writing style is fluid, articulate, and incisive."
—Robert W. Gillespie, president, Population Communication

"With shamanistic wisdom, author Michael Tobias moves the reader to care and to see a larger picture than we might have before. . . . We are all participants, for better or worse, in this global problematique that Michael so accurately identifies as *World War III*. We must now fully participate in this war in order to end it. With the authority of firsthand experience and the scholarship of factual documentation, Michael shows us the way."
—Dr. Michael W. Fox, vice president, The Humane Society of the United States

# WORLD WAR III

## Selected Works by the Author

### Books

*Rage & Reason*

*A Naked Man*

*Environmental Meditation*

*A Vision of Nature: Traces of the Original World*

*Life Force: The World of Jainism*

*Fatal Exposure*

*Voice of the Planet*

*After Eden: History, Ecology & Conscience*

*Believe* (with William Shatner)

*Mountain People* (edited)

*The Mountain Spirit* (edited with Harold Drasdo)

*Deep Ecology* (edited)

*Deva*

### Film/Television

*Voice of the Planet* (10-hour dramatic miniseries)

*Kazantzakis*

*Black Tide*

*Antarctica: The Last Continent*

*Ahimsa: Nonviolence*

*A Parliament of Souls* (26-part series)

*Cloudwalker*

*Animal Rights*

*Ozone Crisis*

*Science Notes* (32-part series)

*The Power Game* (4-hour series)

*A Day In The Life of Ireland*

# WORLD WAR III

## POPULATION AND THE
## BIOSPHERE AT THE END
## OF THE MILLENNIUM

# MICHAEL TOBIAS

BEAR & COMPANY
PUBLISHING
SANTA FE, NEW MEXICO

LIBRARY OF CONGRESS CATALOGING-IN-PUBLICATION DATA

Tobias, Michael.
    World War III : population and the biosphere at the end of the
millennium  /  Michael Tobias.
      p.    cm.
    Includes bibliographical references and index.
    ISBN                                        1-879181-18-5  :
    1. Overpopulation.   2. Overpopulation—Environmental aspects.
    3. Population policy.   4. Man—Influence on nature.
    5. Environmental degradation—Social aspects.   I. Title.
    HB871.T63   1994
    363.9'1—dc20                                        94-5019
                                                          CIP

Bear & Company, Inc.
Santa Fe, NM 87504-2860

Cover Design: Robert Aulicino © 1994

Cover Illustration: Richard Elmer © 1994

Interior Design: Marilyn Hager

Typography: Buffalo Publications

Author Photo: Jane Gray Morrison

Editing: Kay Hagan and Sonja Moore

Printed by R.R. Donnelley on 100% recycled, acid-free paper

1  3  5  7  9  8  6  4  2

*For Jane Gray Morrison,*
*my soulmate, editorial conscience, and inspiration,*
*who labored over this book with unstinting dedication.*

*with Love and Gratitude.*

# Contents

# Acknowledgments

I want first to thank Mr. John Warren and fellow trustees of the Bixby Foundation for their generous support of this project. Without them, it could not have happened.

Mrs. Marilyn Brant Stuart, president and executive director of Population Education Committee, has (with the help of her many board members), worked assiduously and selflessly in assisting me, as has Mr. Robert W. Gillespie, the brilliant president of Population Communication. I owe both of them my extreme gratitude.

A special thanks goes to the directors and various staff of the United Nations Population Fund—in China, India, Indonesia, Kenya, and Mali—who kindly assisted me in some of my journeys, particularly Stirling Scruggs and Kate Bourne, J.S. Parsons, Tevia Abrams, Dr. Sheila Macrae, Charles Olenja, Paul Matogo, and Isaiah Adeleke Ebo.

In addition, I am indebted to the following individuals, organizations, NGOs (nongovernment organizations), and government agencies:

China: Madame Peng Peiyun, minister of the State Family Planning Commission; Huang Baoshan, director, International Relations, State Family Planning Commission; Professor Qu Geping, head of China's Environmental Protection Agency; Li Bing, senior staff member and translator, NEPA; Professor Wu Cang Ping, People's University, Institute of Population Research; Dr. Tian Xueyuan, director, Population Research Institute of the Academy of Social Sciences; Mr. Shi Yuanming, State Family Planning Commission, Department of Foreign Affairs; Liu An Ru, chairman of Taicang County Family Planning Commission; Zhang Jinglin, deputy director, Ministry of Public Health; Lin Yin Ya, deputy director, Shanghai Municipal Family Planning Commission; Sun Rang-Chun and Mrs. Wang, Shanghai Family Planning Commission; Dr. Qian Xinzhong, former minister of the State Family Planning Commission and Minister of Health; the Population

Research Institute of the Academy of Social Sciences; Beijing North District MCH Clinic; Dr. H. Yuan Tien, Population Research Institute, Peking University; Zeng Yi, deputy director, Institute of Population Research-Peking University; Dr. Ru Xiao-mei and Guo Lidong, Department of Foreign Affairs, Ministry of Public Health; Family Planning Commissioner and Associates, Xiao Shi San Li Chung, Fang Shan District.

Indonesia: Dr. Emil Salim and the Ministry of Environment of which he is minister; Dr. Russell Betts, World Wide Fund For Nature; Dr. Ratna Tjaja, bureau chief, Contraceptive Services, and Dr. Sardin Pabbadja, Deputy for Programme Operations, of BKKBN; Mr. Rozy Munir, Nahdatul Ulama; A. Hadi Pramono, editor, Tropical Biodiversity, Indonesian Foundation for the Advancement of Biological Sciences; Dr. Does Sampoerno, Pathfinder International; Professor Dr. Ismail Suny; Jack Reynolds, technical assistance coordinator, KB Private Sector Family Planning Project/A.I.D.; Dr. M. Habib Chirzin, chairman, Central Board of Muhammadiyah; Dr. H. Sugiat Ahmadsumadi, Skm., Regional House of People's Representatives, Jakarta; Dra. Sumarni MPA, BKKBN Chief, Program Operations; Dr. J.R.E. Harger, programme specialist, Regional Office for Science and Technology for Southeast Asia, UNESCO; Dr. H.A. Sanoedi; Dr. Malicah and associates, Aisyiyah Muhammadiyah; Dra. Sumarni; Ibrahim Ali Abul Zahab, Food and Agriculture Organization of the U.N.; M.S. Zulkarnaen and the WALHI Environmental Forum; Dr. Firman Lubis and the Yayasan Kusuma Buana.

India: Kamal Nath, Minister of Environment and Forests; Ms. Maneka Gandhi, former Minister of Environment and Forests; Mrs. Usha Vohra, Secretary, Department of Family Welfare for India; Mrs. S. Bhatnagar, director, National Institute of Health and Family Welfare; Professor Ashish Bose of Nehru University; Mrs. Avabai Wadia, president, Family Planning Association of India; John J. Dumm, director, Office of Population, Health and Nutrition, A.I.D., New Delhi; Dr. Mrs. M.R. Chandrakapure, Director of Health Services for Maharashtra State, Bombay; D. Singhai, deputy secretary, Rural Development Ministry; K. Gopalakrishnan, general manager, PSI—Population

Services International; Dr. V.A. Pai Panandiker, director, Centre for Policy Research; Dr. Prema Ramachandran, deputy director, Indian Council of Medical Research; Dr. Pathak and associates, International Institute for Population Sciences, Bombay; Dr. P.S. George and Phil Roy, directors, Centre for Development Studies, Tiruvananthapuram; Dr. C. V. Pratapan, Director of Health Services and Family Welfare, Tiruvananthapuram; Dr. Jayasree, Medical Officer in Charge, Primary Health Centre, Puthenthope, Kerala; Professor B. Padmanabhmurty and associates, School of Environmental Sciences, Nehru University; Razia Ahmed, president, Animal Rights Mission, New Delhi; Ayesha Bilimoria, associate editor of *Sanctuary* Magazine, Bombay; Himanshu Malhotra, MM TV News, Ministry of Environment; and finally, our good friend and brother, R. P. Jain.

Kenya: Ambassador S.B.A. Bullut, director, National Council for Population and Development, Ministry of Home Affairs; Dr. Kanyi, Chania Clinic; G.Z. Mzenge, executive director, Family Planning Association of Kenya; Mrs. Kalini Mworia, International Planned Parenthood Federation-London; Narendra and Prati Shah, and Chetan Shah; Elizabeth Obel-Lawson of the World Wide Fund For Nature Regional Office, Eastern Africa; the director and staff of Chogoria Hospital; the midwives of the village of Gaatia; Varchand and Tara Shah; Mr. Daudi Nturibi; Leonida Atieno Ongari, assistant project officer, Minnesota International Health Volunteers, Dagoretti Community Health Services, Nairobi; Alice N. Githae; and Ndungu A. Kahihu, The Kenya Scouts Association.

Mali: The staff librarian of the Ministry of Environment; and John Anderson and his associates, particularly Carl-Eric Guertin of the Entraide Universitaire Mondiale du Canada (EUMC); Dr. Doucoure Arkia Diallo, Chef de Division Sante Familiale et Communautaire; Mr. and Mrs. Franck-Schosseler; the Cooperative des Femmes Pour l'Education, la Sante Familiale et L'Assainissement.

United States: Nick Stonnington, whose initial encouragement proved critical to my writing this book—thank you, Nick! The California Department of Fish & Game, Natural Heritage Division; Joan Babbott, former executive director, Planned Parenthood of Los Angeles;

Melinda Cordero, Promotoras Group; Shanti Conly of Population Action International; Peter Morrison of RAND; Dr. Stephen Morse, Rockefeller University; Deborah Jensen, California Policy Seminar; Jones & Stokes Associates; California Nature Conservancy; Agricultural Statistics Branch, California Department of Food and Agriculture; Leon Spoggy, agricultural commissioner, Los Angeles County; Oscar Mendiola, Los Angeles County Mental Health Division; the Southern California Association of Governments; John Moore, Research Division, Air Resources Board of Southern California; Suzanne Devorak, president, International Federation of Social Workers; Mrs. Hong, Research Division, Los Angeles County Health and Epidemiology; Darlene McGriff and Nancy Veera, California Department of Fish & Game; Dr. Mary Evelyn, Chairman of Religions, Bucknell University; S. Woodson-Bryant, research/public affairs, Southern California Gas Co.; H. Kilburn, Southern California Air Resources Board; Heal the Bay Group; Paul Mills and David Abbey, Department of Public Health, Loma Linda University; William Linn, senior project scientist, Environmental Health Service, Rancho Los Amigos Medical Center, Downey, CA; Paul Papanek, chief, Toxics Epidemiology Program, County of Los Angeles Department of Health Services; Cal Trans Research Department; Ray Lund, director, Biomedical and Environmental Sciences at UCLA; Professor John Letty, UC Riverside Soil Sciences; Southern Cal Edison Statistics Branch; Jeff Girsh, environmental consultant, Denver; Fred O'Donnell, head, FAA for Southern California; Dr. Tom Merrick, the World Bank; Dr. Maurice Van Arsdol, chairman of the Population Education Committee, for use of the Population Research Library at the University of Southern California; and Dr. Joan Kaufman, Harvard University.

In addition, I owe a debt of thanks to the economist, Dr. Atul Shah of Bristol University, to Avram Davis, and to the Jain communities of Kenya, Canada, England, India, and the U.S., as well as to Dr. Schwan Tunhidone, director, Wildlife Conservation Division, Royal Forestry Department, Thailand, and the monks of Thyanboche Monastery in Nepal and Taktsang Monastery in Bhutan.

For my research in Antarctica, I am grateful to the scientists and personnel at Base Esperanza, Palmer, Jubany, and Arktowsky.

A few of the key sources included: the *World Bank Working Papers*; Peter Warshall's *Mali—Biological Diversity Assessment*; the Kenya Wildlife Service, *Policy Framework and Development Program, 1991-1996*; *Mistaking Plantations For Indonesia's Tropical Forest: Indonesia's Pulp and Paper Industry, Communities, and Environment*, by WALHI and YLBHI; Dr. Qu Geping's book, *Environmental Management In China*; the *World Resources Report 1992/93*; Charlie Pye-Smith's book *In Search of Wild India*; Ramesh and Rajesh Bedi's book *Indian Wildlife*; the many writings of E.O. Wilson, particularly his recent *The Diversity of Life*; the works of Paul and Anne Ehrlich, including *The Population Explosion*; the World Wide Fund for Nature/National Development Planning Bureau-Jakarta *Biodiversity Action Plan*; *In Our Own Hands*, by Deborah B. Jensen, Margaret Torn, and John Harte; various UNFPA and A.I.D. working papers; and the *Los Angeles Times* newspaper.

A special, all-enduring hug goes to photographer, poet, filmmaker, and explorer Robert Radin for reasons only he can appreciate; and to Claude, Carol, Janet, and Marina for holding down our respective forts.

I wish to thank my literary agent, Julie Castiglia.

Let me extend a heartfelt appreciation to the publishers, Barbara and Gerry Clow—inspired earthlings of this, and many prior lives; to Diane and Jody Winters, and Marilyn Hager—president, chief publicist, and art designer (respectively)—as well as to the entire dedicated and genuine staff of Bear & Company—true gems of Santa Fe, and of book publishing in general.

And finally, much love and profound thanks to Betty, William, Marc, Jean, Don, and Alvaro; Trooper, Mac, Charlie, Josie, Stanley, Feather, Hey-Boy, Heidi, and Toto, the family without whom little would be possible.

# Prologue

From the first glimpses, one is almost certain to be qualmish of demographics, or—for purposes of mental hygiene or financial gain—to somehow manipulate this uneasiness or distrust in order to shy away from the truth of overpopulation and what it means for all of us. Yet the numeric verities are not easily obscured or eluded. Leafing through countless tomes of statistics, one's impatience and skepticism turn to real alarm. The sheer weight of numbers known vaguely to be human beings, the burden of their illimitable data, and the substantial meaning of our impact, escalate in the psyche at a speed approaching four human births every second, over fourteen thousand per hour, nearly one hundred million per year, begging for a quality of life that is increasingly impossible to satisfy. As other species quietly, rapidly, fade away and the imbalance of human numbers continues to escalate, a hint of substantial personal peril surfaces, as if one's life were suddenly held hostage by cosmic forces totally out of our control. Yet unlike, say, an earthquake, we alone are to answer for this situation. Only we can ameliorate, and hopefully reverse, the debacle of our unrelenting proliferation.

A humorous perspective may lift this threat, though only momentarily. In one part of India, for example, family planners complain bitterly that the men do not know how to use condoms. And in the U.S. recently, there was a furor over a company which, basing its argument on "patriotic" feelings, applied for a condom trademark labeled, "Old Glory". The company prevailed.

But in the most alarming, existential sense, contact with an overpopulated human world easily dispatches reflective individuals toward unrelieved darkness, mired and subsumed in quantities, zeros, and decimal points beyond their ken. No single capability can circumvent this unimaginably teeming future. Overpopulation, from the perspective of the individual, can be thoroughly debilitating. Panic, escapism, homophobia, wrath, persistent forlorn, and bitter fatalism are all patterns associated with today's unprecedented densification of human beings. Unlike social insects, the overcrowding of most animal species results in often aggressive behavior and a negative psychological profile. There is nothing remotely aesthetic about an infinity of humankind. Nor do such statistical aggregates make for the satisfactions of anonymity. Rather, this is the numbing day-in, day-out dilemma of family planning. Those in the social services and family health fields apply a somewhat different set of psycholinguistic rules to their practice; a nomenclature that is all but shorn, necessarily, of self-defeating nihilism. Such words and phrases as compassion, and caring, career challenge, professional opportunity, or—optimistically speaking—humanity's final chance, come immediately to mind.

The humdrum surface of family planning is imprinted with every living miracle, every child, but also totally conceals the lurking horrors of human impact. The term "family planning" and its politically correct intentions conceal something like one hundred million sex acts a day. This lascivious appetite, ordained by evolutionary impulses, would, within a matter of minutes, necessarily smother the most jaded connoisseur. The officialdom of family planning gives no sense of the worldly lust, the inventive positions, polyglottal gasps, redundant upward-rising ever-unslaked passions, the silent enduring, and too frequent death throes that accompany the comprehensive sex act from beginning to end. Nor does the bureaucracy enshrouding population control give any clue to the crucial realm of love, the family and community dynamics, religious scruples, the ancient or subliminal rituals, courtship, or aggression that universally signal sexuality and its offspring.

But then, beyond the few moments of carnal pleasure, comes the biological truth: The universal contractions, the chasm of darkness out of which the human miracle tastes the honeyed light and fresh air. In many developing countries, there is no theater of technology assisting birth, but rather a single midwife, whose tools are often no more than a stone for cutting umbilical cords and cow dung for stopping bleeding. A third of all deaths among teenage women in the Third World are likely to occur during these merciless moments. Among those who survive childbirth, nature honors the unwary blush of young mother-hood—or conversely, in some countries, contests that innocence and reason for pride—with the dismal realizations of poverty and the very possibility, perhaps, of a child who will be scorned, oppressed, possibly even murdered if it should be a female.

By foot, or in mobile vans or boats, come the health care workers, usually too late to do more than ensure the status quo or provide nutritional advice, immunize the child, hand out rehydration supplies, or counsel on the benefits of breast-feeding, spacing, and an IUD insertion. The timing, supplies, level of counseling, and degree of contact vary, of course. One can scarcely pretend to sustain an interest in all the specifics, which rage and wend according to continent, country, region, culture, language, district, city, town, village, sub-village, family, and individual.

But even the casual observer will be struck by the unthinkably acute whorl of pain directly beneath and to all sides of humanity's fertility: the collision course, long in the making, between human beings and the biosphere. I speak of nothing less than war in this regard. For purposes of absolute clarity I call it *World War III*. By this descriptive finality I do not mean it in the speculative sense—the war-game scenarios that have occupied the minds of countless academic and government cadres, puzzles played out in the leisure of futurism, or the sterility of think tanks. This war is already upon us and has been for some time. It is like a terminal cancer that corrodes the body from within, too often admitting no signs of itself on the surface of the person, or only at the very end. This is why the overwhelming proportion of all medical expenditures are spent during a patient's last month of life, because that is, typically, when people experience their worst agony and rapidly succumb to the final hours of doom. Usually, nothing can save such patients. The illness has gone on too long. All that can be done is to massively attack the mature disease with radiation, or remove a limb, an organ, large sections of tissue, or undertake a transplant. But usually, science can only prepare the victims for death by dulling their senses and thus reducing pain to the extent possible.

The world is currently in the throes of a mature disease, brought on by both the best and the worst of humanity. In either case—whether best or worst—the totality of our infliction upon the Earth is akin to a war that we have waged unceasingly. It boasts of all the analogies normally associated with conventional warfare: double envelopment, vast assemblages of soldiers wreaking havoc, scorched earth strategies, aerial sorties, temporary ceasefires, paramedic corps and battlefield triage, wavering negotiations, overcrowded sanctuaries, unflinching dictators, patriotic gore, sadism, cultural madness, rapine and plunder, suicide, and near total genocide. These hostilities and absolutes are taking place at this very moment everywhere, in every country, upon virtually every square kilometer of the planet—even where no guns are fired, no fires visible. Even where alleged wilderness remains, and the economy is booming, the atmosphere above is contaminated, or the water, or the soil, its resident wildlife and the genetic viability of their populations increasingly cornered by advancing armies on the outer perimeters. But unlike the previous two world wars, this is the final war. We must not imagine that there will be ticker-tape parades for the victors. In 1987, when the world contained nearly 700 million people fewer than in 1994, the World Commission on Environment and Development described the state of "environmental decline" as a "threat second only to nuclear war." Yet, the demographics of our future indicate, possibly, a tripling or quadrupling of our numbers, with all the attendant ecological stress.

The above-mentioned demographics inherent to family planning, all the obtuse quantitative calculations, census data, and fertility surveys, define a veritable war against the planet when they are coupled with realities of human impact, be they deliberate or naive. I am referring to such causes as housing starts and road building, activities which arrogate and degrade land previously inhabited, traversed, or relied upon by far more creatures than the incoming human migrants. This sphere of conflict crescendos with the massive felling of forests and the horrifying operations of slaughterhouses and factory farms, the economics of poaching and complex disruptions by the tourism industry, the short-term motivations of capital goods production and commodities speculation, the intensity of agriculture that, like all forms of construction, defiles previously natural land, the putting up of fences and pipelines that check otherwise habitual movements of animal populations, the course of vulnerable supertankers and chemical or radioactive-emitting and transporting traffic flow, the depletion of stratospheric ozone, the melting of glaciers, the impoverishment and desecration of nearly every organic and inorganic resource and natural sink by our extraction and consumption of energy, the countless regimes of manufacturing and waste discharge, the pathologies of defense and offense, local, tribal, ethnic, regional and national conflicts, the feverish building of whole towns, and the accretion of megacities. This approximately 12,000-year-old human pattern defines the hunter, scavenger, tamperer, bulldozer, destroyer that is man and woman. They are born, suckle, dream, and want only to live. They have a soul, whatever that might mean. They avoid pain, seek out pleasure. These basic needs are biologically fixed. All animals and plants share in the gusto of life. Every population is genetically primed by these universal volitions.

And yet for humanity, whose brains and footprints are larger than those of the largest dinosaurs, we are plagued by our very presence in the natural world. Some are plagued more than others, of course, the "some" referring at present to probably three-to-four billion individuals who are poor, oppressed, in ill-health, or hungry; the "others" designating those who have thus far managed to maintain what they consider to be a dignified and pleasurable sojourn in this lifetime. Rich or poor, there is no escaping the awareness and worry of a planetary struggle, nor the appalling spasm of global extinctions that we have deliberately unleashed.

Humanity is furiously engaged in World War III at this time. Our ever-increasing numbers are bearing down on Earth's finite resilience. We are using nature as ammunition for our many struggles amongst ourselves (economic, political, cultural), while directly assaulting and out-maneuvering her every flank. We have defined the enemy and penetrated to its command headquarters, if you will, by razing or impaling upon a mighty saw the planet's greatest reservoir of biological diversity, the rain forests. In this offensive, humanity seems bent on taking no prisoners while holding the entire world captive for another day.

This is a new kind of war, global in scope yet remarkably concentrated wherever there is booty to be had. The spoils refer to all those desirable effects, the guarantees of a way of life to which more and more people would like to become accustomed, including such things as value-added trade surpluses, secured supply lines of fossil fuels, refrigerators, skyscrapers, and intellectual property rights. For most of the world's population, however, clean water and medical supplies, electricity, and nutritious food would suffice for now. Our weapons may be as sophisticated as those used against Iraq in the Persian Gulf War, or as immemorially simple as a plow scratching a bit of hard turf in the highlands of Ethiopia. While the conflict in the Persian Gulf imposed myriad levels of stress on the environment (decimation of marine dugongs, coral reefs, countless species of fish, and diverse desert habitat, as well as the extrusion of chemical effluents into the upper atmosphere), the Ethiopian farmer imposes an inobvious but nonetheless war-like burden on the land, altogether justifiable yet tragically linked to the incremental destruction of the nonhuman world.

Not surprisingly, the fervor for this war is carried on with our most impressive instincts. What we call civilization, progress, economic prosperity, even, remarkably, human empathy, the "enemy" (our biological surroundings) more than likely views differently. Economic prosperity in China is systematically destroying Chinese environment, from snow leopards in Tibet to estuaries on the Yellow Sea. The same sorry syndrome, to varying degrees, is being perpetrated throughout every country in the world, where the differences between city and village are vanishing in the shadows of expansive and destructive development. The end-loser of all this human "progress" is nature. The runaway pressures on the biosphere of nearly six billion human bodies are converting an original, untainted world into a meaningless "resource." As we extinguish those limited resources, leveling whole ecosystems in our quest to feed, clothe, shelter, inoculate, and educate the nearly two hundred fifty thousand newborns every day, we rely upon increasing concentrations of the machinery, and the monetary or political profits, of war. The efficiency of exploitation is fundamental to human aggression, and always has been. This is a war fraught with great nerve and compulsion because our species knows that it is engulfed in conflict, has the means, possibly, to reverse the aggression, but cannot think to do so.

The greed of World War III goes well beyond political ego or personal gain. It encompasses the patriotism of religious and cultural norms, the lifestyles, the sense of history, which many hold sacred. That sacredness is itself steeped in battle, making it difficult today to isolate clarity or fashion unambiguous convictions. Living with contradiction has become the hallmark of this war, whose psychological fallout shadows everyone. And because no region, no mind, no soul is spared the ravages of this conflict, there is virtually no "clean room" left on the planet, no place where one can go to breathe unburdened. In other words, it is now next to impossible to measure in any comprehensive terms the alternatives to war, or even to take an accurate compass reading of human nature, though many well-intentioned thinkers, observers, scholars, artists, scientists, and family planners are trying to do so.

In spite of our denial, inescapable self-interest, and the inferno all around us, we have known, at least since the time of Plato, that overpopulation is biologically and morally disastrous; that nature meets imbalance with brutality. Destabilization may have nothing to do with numbers, per se, but with excess. Morality, like the Ten Commandments or the Golden Rule (Matthew 7:12 and Luke 6:31), is a crucial human notion that has arisen, I believe, to combat excess. It is a beleaguered ideal, an even more ungainly tool. But it serves, if anything, to guide humanity toward balance in nature.

Morality disintegrates on the battlefield, which is where we are now; having habituated to all the wisdom and virtue of the ages, only to fall back, unmollified, on old habits. Yet there are sparks of conscience, piecemeal traces of hope in every locale and country. Between countries there are conventions, treaties, alliances, and policies that seek to stop, or slow, the war. Most importantly, there are legions of well-intended individuals—heroic, conscientious objectors. I am referring to that seemingly mundane collective of several million family planners who for much of this century have been trying urgently and humanely to curb the rate and pace of human fertility. These are people who, to varying degrees, understand the scope of this war and are determined to make a difference. One is reduced to such platitudes ("to make a difference") when confronted by such overwhelming odds. But the truth remains: it is possible to mend wounds, to change minds, to minimize suffering. Our awareness makes it our responsibility to do so. We cannot presume to make the crisis, this advanced illness, go away, or even begin to appreciate the actual danger we are in, unless we have first thoroughly diagnosed its causes and consequences. Then, like surgeons, we will have a more confident sense of where we are hemorrhaging, and whatever antidotes, policies, and resolutions we may summon will have at least some measure of credibility. Family planning has prevented the births of several hundred million humans in the twentieth century. While the projected impact of that additional contingent might be negligible, against the existing five and a half billion people, it is, nevertheless, a nation larger than the United States that has not come into being. And that is significant. Furthermore, it is now well understood that in simply satisfying the demand for contraceptives by those individuals presently lacking access to them (an estimated 125 million couples in the developing world), a substantial, non-Draconian hurdle will have been overcome. Free or nearly free access to contraceptives, guided by strong political leadership, is proving to be the key to slowing down overpopulation even more than education, which in some countries like Bangladesh and parts of Africa has proven to be of secondary importance. Nevertheless, education, consciousness raising, is key to long-term stabilization of human ethics.

Overpopulation has nothing necessarily to do with the density of people, or even the extent of their number, though the two facets seem inevitably linked to a predictable outcome. More certainly, overpopulation is a synonym for the environmental impact of our species, the result of human behavior that stems from economic goals and the ill-fitting power of individuals over nature and one another. Ecologically sound behavior, even when multiplied by large numbers of people, need not be destructive, at least in theory. Nobody knows the upper ceiling of biotic viability, or "carrying capacity" as it is called, under conditions of high human consumption. At what point does an ecosystem, a collection of ecosystems, wither and expire? The deleterious effect of radiation on a population of white pines or human blood cells is no longer a mystery. But what about the global extinction of frogs, birds, apples, and domestic animal varieties? What happens when the Garden of Eden; with tens of millions of species, is replaced by a few somber spurts of weed, short-lived flies, and wilting cheatgrass, scattered atop the dour alkaline flats of untoward destiny? Our relative inexperience in this arena is to be regarded with profound relief and adumbration. For the day we should haplessly exceed the Earth's carrying capacity will likely admit to little documentation, possibly no consciousness of any kind. Living within the rubrics of carrying capacity requires a certain mode of agreed-to behavior, as yet to be achieved. But with ample mental and physical nurturance, cultural incentives, and a politics of humanitarian social engineering, there is reason to be optimistic. And despite an erratic record and oddly unspectacular techniques, family planning has a crucial role to play in fostering the preconditions of that sustainability.

In places like East Los Angeles, Shanghai, Bangkok, the villages of Kerala and western Java, the outbacks of Bangladesh, and the sub-Sahara, I have spent time with specialists—some trained, some less trained—who are quietly working with mothers and fathers, students and educators, village elders, community, tribal or religious leaders, local and national governments, and with numerous NGOs (nongovernmental organizations) to engender two-child families, quality health services, and appropriate ecological development. The troubled chronicle of these many diverse endeavors is often inspiring. We have no other choice but to continue working to limit human numbers as compassionately, and with as much perspective, as possible. We are not obligated to finish the work, simply to act as if we could. The reasons for this global exercise in sexual and egotistical restraint are obvious by now: every consumer adds kindling to the fire that is raging across the planet, destroying all that thereupon dwell.

Yet, while small families and family planning are fundamental to the planet's future (or family *welfare*, as some countries call it), human behavior itself is out of control, and threatens to override any and all strides in the population domain. Will 12 billion soldiers do more or less damage than 6 billion? Will hand-to-hand combat escalate, or will the destruction be consolidated by a new generation of weapons, terrorism, and scarcity of primary resources, like water? What kind of evolved damage will more numbers inflict? This consumptive frenzy is fueled by laws, tax breaks, national economic priorities, and government regulations. Economic growth is everywhere judged to be a good thing. For nearly three hundred years material gain has been deemed the best form of birth control. More importantly, the accumulation of wealth has been presumed the natural evolution of humankind. Yet those few remaining "pre-industrial" communities and the records we have of earlier ones, as well as hundreds-of-millions of existing mountain, jungle, desert, arctic, and coastal indigenous peoples, more often than not reject personal advantage when obtained at the expense of others. Human dignity, artistic enterprise, the joys of parenthood and community need not be preconditioned by acquisitiveness and the total separation of human life from that of the surrounding environment. Technological cultures would assure us that there is no point in indulging historical, ecological or ethnographic nostalgia, preferring to ignore the fact that the majority of human beings alive at the end of this millennium subsist in essentially nontechnological, rural surroundings. There are countless technical solutions to technical problems. There are few, if any, technical antidotes to nontechnical problems like overpopulation and overconsumption.

In the face of several billion have-nots plunging ever more deeply into poverty, hunger, and disease, Western politicians and cultural isolationists have stridently declared, "The American way of life is not negotiable." This credo, a defense mechanism against the incoming tide of population global pressure, has devastated women, children, and most plant and animal species, hundreds of billions of individuals with individual souls, by a total behavioral insistence that destroys, separates, and self-justifies. The motto hinges upon a prevailing doctrine that (to quote President Clinton) "The only way wealthy countries can grow richer is if there is global economic growth and we can increase trade."[1] The destructive belief system holds that wealth is good and that the more it costs to achieve riches, the more those riches are worth. It is a disastrous, psychologically bankrupt conception of humanity that insists upon the inequitable distribution of land nobody should own in the first place, and the tainting of all biological forces with synthesized chemicals and damaging technology. This is not to say that *all* technology is destructive, and therein lies hope for certain technical solutions. But by generally having attempted to supplant nature with human technology, we have in fact unleashed a more dangerous jungle than ever existed prior to primitive man.

Every living being is a miracle deserving of reverence. These myriad souls—women and children and men of all ages and characteristics—are only the most obvious victims of the war. There are others, less visible but no less noteworthy. I am not referring to worms in deep sea vents, bacteria inside volcanoes, or ticks in Siberian forests necessarily, but to all future generations, for whom we will be but ghosts. Both developed and developing countries are allied in their war against the future, a fact often missed. A Western consumer has a far greater vested interest in the continued extraction of global resources—and resulting biodiversity loss—than his counterpart in developing countries, though the Western individual is rarely, if ever, in touch with the causal chain of ecological desecration of which he is a glaring part. Most, if not all, of this Western impact has no other justification than the perpetuation of his pleasure. Whereas for those multitudes of poor and malnourished in developing nations, consumption has an altogether different definition, namely, survival. Billions of people are presently stripping trees and gathering biomass for fuelwood, killing animals, blowing up coral reefs to more efficiently get at the fish, poaching, encroaching on national parks and sanctuaries (often their former homes from which they may have been evicted), having to denude their half-acre or so of farmland, stranded in resource-exhausted hinterlands, or migrating jobless to the congested, narrow margins of stifling slums.

Harrowingly, both types of consumers are killing the world, which does not provide the human heart, or policy makers, any easy choices or roadmaps. But there are selective choices to be made, and there is still time to make them. Perhaps not time to save 20 percent, or even, eventually, half of all species on Earth, but time to save at least the other half.

I want to believe that the human population has the strength, the enthusiasm, the reasoning facility, and the conscience to alter its course. It may even be that, by biological intuition, democratic impulses long ago anticipated the ungainliness and disarray of the human presence with an eye toward correcting it. There are countless kernels of hope, but if we do not seize them, then we might well muddle onward, unable to give up the fight, addicted to ossified political boundaries out of kilter with countervailing ecological priorities, hopelessly caught up in the cry of battle, poisoned by greed and the hoarding of personal assets, undeterred by the death of nature all around us. In that case, failing to formulate a new international order of empathy, of clear-headed biological policies and imperatives, to check our numbers and to seriously modify our laws and behavior, we will continue to fail as a species. Not contributing to the Earth but destroying it. Not offering to the pool of life but plundering it. Ironically, this colossal crime against the world and possibly the cosmos will be masked by ever more numbers of people, a mark of success in traditional biological terms.

Uninhibited human fertility is surely no triumph. Depending on which projections one is inclined to believe, and in the absence of any species-specific self-destruct mechanisms, *Homo sapiens* will have increased to between 11 and 12 billion sometime in the middle of the twenty-first century. According to higher projections, we may attain to 15 billion. In Africa, for example, if the number of children per couple does not come down from over six to just over four during the remainder of the 1990s, the difference by 2035 will amount to 1 billion additional people. That is some measure of the time bomb inherent in the so-called Total Fertility Rate (TFR), the number of live children that a woman has during her reproductive years. And if we should somehow triple or quadruple our present numbers, then the existing best estimates will have been proven to be in error (that would not be surprising) and there will be probably nothing much to stop continued growth long into the future. The long-term biological outlook might then be best described as no bang, but a slow and agonizing whimper, with consequences better left for future environmental archaeologists to unsort.

This volume endeavors to analyze these battlegrounds by focusing upon the major regions of the planet where particularly heavy demographic pressures are conflicting with dense plant and animal communities. I first set the stage by examining the primeval backdrop and the living truth of balanced populations, most notably those in the Antarctic, as well as among certain human tribal groups of limited but nonetheless important diagnostic value for the future. I then paint a picture of destabilizing influences and trace their history back to the origins and ephemeral nature of biological populations.

I lay greatest weight and emphasis on my analysis of China, India, Indonesia, Africa, and the United States, the five largest population hubs in the world. I view these critical ecological battlefronts through the eyes of various demographers, historians, theologians, scientists, medical specialists, family planners, politicians, and the locals themselves—farmers, cottage industry laborers, children, mothers, and teachers. The majority of the book is rooted in my contemplation of these five bioregions. Their respective dilemmas are massively suggestive of the forces and global crash occurring everywhere, but more perilous in the degree to which those pressures have already taken a toll. As bad as it is now (total population in 1994 of approximately 3.3 billion), China, India, Indonesia, all of Africa, and the United States together are looking at possibly as much as 60 percent of the human race, or 7 billion people, sometime in the middle of the twenty-first century. At the same time, these five regions as yet harbor between them the greatest biodiversity on Earth. Destructive patterns have been set in motion. The clock is ticking. A few years are all the time left to save that precious heritage.

In my chapter on China, I examine the pace of economic development and its environmental fallout. Against this context, I recount the political lessons from China's ambiguous history of family planning efforts and project different possible scenarios for its future. While the nation has already made unprecedented strides toward reducing its numbers—with little encouragement from the rest of the world—the coming generations, boosted by wild spending, enormous growth, and increasing socioeconomic disparities on top of a new baby boom, offer few reassurances that the country will be able to take control of its demographic time bomb.

Similarly, my chapter on India yields a picture of intense ecological destruction. Despite Indian family planning having been initiated in earnest nearly forty years ago, the country continues to see runaway fertility. At the same time, India's human rights violations—including an abusive regime against females and an unmonitored, persecuted child labor class that ranks in sheer size as one of the largest nations, by itself, in the world—offer little hope of simple reconciliation or child and maternal health care antidotes. India's several hundred million residents who are below the so-called absolute poverty line have no choice but to strip bare the remaining forests, while the fast emerging Indian middle class is beginning to achieve a modicum of economic security—which it has been denied throughout history—but at the unfortunate price of totally undermining all remaining semi-wild habitat throughout the country.

Indonesia presents a less dire scenario, though the long-term ecological trends are terrifying. While family planning efforts have been rigorous to date, the built-in fertility momentum, coupled with the country's unapologetic need to secure an economic future for what promises to be 360 million Indonesians in the middle of the twenty-first century, has targeted one of the last enormous tracts of rain forest in the world for exploitation. This blueprint for blatant development is veiled, unfortunately, by terms that would indicate otherwise.

In my chapter on Africa, I take the reader throughout Kenya and hundreds of miles across the sub-Saharan country of Mali, as well as examining trends and data from a dozen other African nations. In Kenya, I focus upon the 1993 *Demographic Health Survey*, and its optimistic findings, and the Agency for International Development program in the town of Chogoria, where the first substantial evidence of fertility decline has emerged. I examine how, in a country of rapidly dwindling resources, escalating poverty, inflation, and ecological mayhem, the fact of a Chogoria has come about, what will be needed to sustain its previous successes, and whether any African projections may be extrapolated from the data. In Mali, I look at the cumulative impacts of drought—the result, in part, of global greenhouse gases— and the environmental war that have been sweeping over the Sahel intensely in the 1970s and 1980s. In profiling the devastating decline of most ecosystems in West Africa at the hands of one species, it becomes especially clear why population control—with its implacable links to food insufficiency and ecological degradation—is the number one issue to contend with if Africa is ever to hope for anything like remediation. In examining the operational strategies and projections of the Kenya Wildlife Service—symptomatic of the challenge confronting all remaining African wildlife—the damage to biodiversity from ecotourism comes heartbreakingly into perspective. Present in Nairobi for the official release of the *Kenya Demographic and Health Survey*, I was in a position to survey responses to it and penetrate the web of ambiguities surrounding its dramatic findings, while drawing certain conclusions about the future of family planning and ecology throughout Africa.

I then proceed to question the widely shared perception that the Western, developed countries can fend for themselves, unaffected by the same forces and global links that are rapidly eroding the political and ecological stability of the rest of the world. Indeed, ecological ruination in the rapidly industrializing nations, from Mexico and Brazil to Poland and South Korea, is sounding a global wake-up call; and is seen even under those rarified conditions of near zero population growth, whether in Japan, Italy, or the Netherlands. I examine the highlights of this global biodiversity crash, the logical outgrowth of continuing unmediated war. The crash follows from the cumulative results of runaway urbanization, energy expenditures, agricultural expansion, the loss of genetic viability among countless populations, the total appropriation of forests and soil and marine vitality, and the massive tide of animal exploitation. Corresponding to these assaults is an explosion of ecological viruses, of hunger and disease and the sheer weight of human misery.

This picture of population explosion and global war-fare will only get worse if we do not act to curtail our numbers and our behavior in dramatic and cohesive ways by the year 2000. Many have looked toward the United States for political, economic, and behavioral antidotes. Millions of immigrants still dream of the good life in America. But in the United States, with a focus on Los Angeles, I profile the true meaning of human impact, dispelling any illusions that a mere 260 million Americans are somehow magically immune to the biological dynamics of population growth, or have any compelling or obvious answers for the rest of the world. Throughout the U.S., I chronicle what ecologist Paul Ehrlich has termed the IPAT equation—that human impact is equivalent to the size of its population, multiplied by the level of its affluence and the scope, range, and quality of its technology. But America's most dire population bomb is underway in Southern California (the Santa Barbara to Ensenada megacity), a magnet of wealth and disarray—an ironic illusion for much of the world—that has adversely altered every living system in the region. Condors, released back to the wild, are flying into electrical wires and frying, while two dozen human murders in a single weekend in Los Angeles is not uncommon.

Beyond the profile of Los Angeles, America's behavior will be placed in global perspective so that the true dimensions of human population can be assessed, not merely by the numbers, which are entirely relative, but by consumptive impact and the corresponding need for urgent restraints.

In chapter 7, "Demographic Madness," I consider the demographic future according to current best-guess projections by the World Bank, the United Nations, and other organizations. Much is made of the so-called "stabilization" phase in the middle of the twenty-first century, whence "classic demographic transition" should effectively bring to a halt our species' continued numeric growth. This assumption is based only sporadically on readable precedent. More and more demographers have dismissed demographic transition as no longer relevant. Much of it is mere guesswork, the lodging of logical hope based upon imputed behavioral characteristics of humankind that may, or may not, hold true. In fact, as I shall reveal, future projections harbor enormous discrepancy and ambiguity. It has proven impossible to statistically account for a woman's full reproductive life, or to predict the vagaries of population momentum, particularly in view of the rapidly growing numbers of young males and females. Nor is the prediction of new fertility technology or pro-life legislation a facile matter. There are causes for limited optimism, but even more reason to be concerned: greenhouse conditions for fertility, intimating not stabilization but continued growth; ever-escalating wars against nature; and a global population by the year 2100 potentially surpassing 20 billion. Such futuristic numbers are woven of mathematical paradox—even metaphysics—that evokes a new nausea, the possibility that we will be unable to effect planetary remediation.

In my final chapter entitled *Global Truce,* to effectively counteract this dismal picture, unambiguously born out in most parts of the world, I consider the alternative future, one still within our grasp. Analyzing the actual costs/benefits of compassion, vegetarianism, and prudent altruism, of living ethically within a global commons and making appropriate and decisive trade-offs, I try to offer the hope that is more than justified by the many technical, political, cultural, community, and individual solutions being implemented or discussed throughout the world. These encompass a wide dissemination of ecological virtues that will engender the psychological and community preconditions for a philosophical shift in lifestyles, regardless of the culture; a "new human nature," which I would like to summarize by way of a heartfelt "credo" comprising twelve ethical suggestions:

1) vegetarianism;

2) a universal Total Fertility Rate (TFR) of 1, or zero. The goal would be, essentially, to skip a generation of youth as much as possible. If a couple should desire a second child, then let them adopt. The world's orphans are in far greater need than the unconceived. Part of that equation must therefore include a new generation of laws and tax incentives making for much easier and more attractive adoption;

3) a host of other sweeping democratic tax reforms that would work toward engendering a global steady-state ecomony, one

characterized not by free trade, but a condition in which host countries, all industry and small business, stock exchanges, trading partners, private entrepreneurialism, donor mechanisms, and private banking and fiduciary institutions were obliged to behave in strict accord with their bioregional carrying capacity ("living within their natural regenerative and absorptive capacities," as economist Herman Daly has described the process). This would, among other things, require a complete revisioning of national accounts thinking so as to incorporate natural asset depreciations in the reformulation of GNP and taxable income;

4) a revolutionary end to all economic growth, and the beginning of true human development. According to this hypothesis for healing, recession and resource scarcities, a decline in housing starts and all heavy industry, and vastly higher prices for energy and those goods incorporating nonrenewables are all positives; whatever serves a scaling back of human activity will only effect a rechannelling of otherwise destructive, escapist, and largely delusional exploitation into more benign and biologically acceptable forms of physical expenditure;

5) the recognition of ecological concerns as the number one issue of international and subsequent national security;

6) the full revelation of those national security threats should, in turn, generate a much more urgent understanding of the implacable links between demographic pressure, warfare, and ecological degradation. Military spending and priorities would necessarily shift, hence serving to liberate enormous funds for domestic and international family planning, and far more consistent and humane health care, equal opportunity for women and indigenous people, enriched educational opportunities for the young, and cruelty-free standards of nutrition for everyone;

7) a biblical amendment to the U.S. Constitution countering the right to bear arms, that would give all individual animals and vegetative habitat the right to live free of human ex-

ploitation of any kind. This would incite a radical proliferation of noninvasive new ideas, turn-abouts and de-escalations
in human technology and culture, from the emergence of
new sustainable agro-forestry techniques and the zealous
maintenance of biodiversity—of seed types and viabile genetic
communities—to the reappraisal of all biological patents and
biogenetic engineering. It would require a chlorophyll-friendly
reinvention of urban space, and the dramatic expansion of
biomes, habitats, and species and sub-species as candidates
for listing as endangered, rare, or threatened. It would create
new mechanisms for increasing the area of designated
wilderness and greenbelts. This legislation would put an end
to those compromises, incompatibilities, legal inconsistencies (such as between States) and endless structural deferrals
characteristic of most government agencies mandated with
protecting the environment. Such an amendment would also
challenge the massive and unrelenting biomedical research
establishment, and all those industries, special interest groups,
and universities that have arrogantly persisted in destroying
animals for profit or scientific greed;

8) an international mobilization in search of sustainable (alternative) energy strategies and applications;

9) the end to all artificial subsidies, except in those instances
where a "Green" process, product, organic family farm, or conversion required a governmental "visible hand" to socially
engineer and ensure its acceptance or renewal;

10) the creation of a Global EPA (Environmental Protection
Agency) and sound policing instruments;

11) an international legal body that would determine interregional immigration policies based upon the calculated un-
burdening of biodiversity hotspots. Annualized employment
and retraining programs, the creative use of migrant labor
pools, and ecological—as opposed to political—boundaries
would become primarily components of the equations used
to stabilize refugee populations and put them to work as a force
for nature;

12) a massive reawakening of the many principles, priorities, moral choices, and practical methods of nonviolence, as practiced by those individuals, communities, and spiritual and artistic traditions that have historically recognized the equality and right to happiness of all sentient beings.

None of these transformations will be possible, however, if the root causes of their desperate necessity—human behavior and demographic pressure—are not viewed with the appropriate sobriety.

These strategic developments are all viewed as behavioral requirements for any true resuscitation of the natural environment, in which I include a natural human population, where children are loved and cared for, and their parents may find joy and dignity in their lives. Realistic curbs on overpopulation are only so good as our corresponding contemplation of, and behavior toward nature (an equation that encompasses our treatment of one another). Both are battle fronts; both family planning and ecological sustainability factor equally in our scaling back humanity's unprecedented aggression against the Creation. We are no longer speaking about "utopia," per se, but about ensuring a biological future. What is best may be beyond humanity's reach, at present. But what is good may be possible.

The pattern of these two endeavors I offer, in the end, as a master plan for a global truce, a way to cut short this melancholy war that is destroying Earth and all her precious progeny.

# WORLD
# WAR
# III

# The Balance of Nature: Antarctica

From several miles away across seething ocean waters, peering into the magenta dawn, I focused my binoculars upon a seemingly unreal gathering of mosquitoes frozen to a tabular iceberg the size of a Hopi mesa that was heading north at forty knots. It was a radiant December morning and I was traveling to a dozen military and scientific stations in the Antarctic on a polar-class naval vessel with Argentina's then minister of defense. The purpose of this mission was to resupply some of the bases. As we neared the iceberg I realized the organisms were Adelie penguins (*Pygoscelis adeliae*), some fifty adolescents just waking up. These noble avians would spend that whole day, and the coming few years, living together, exposed to pure elements upon one berg or another, porpoising through thirty-one degree seas, diving adroitly after krill, propelled at high speed by flippers full of taut muscle. Their feet and toes are like steel. Protected by thick plumage and millions of years of adaptation, they dwell knowingly in paradise, though they are becoming anxious, I suspect. Eventually, after thousands of miles of far-flung journeys, these birds will return home to the very nests upon which they were hatched, guided by an uncanny mechanism of navigation that must rank in elegance, at the very least, with the Sistine Chapel or Antonio Vivaldi's *Four Seasons*.

I later climbed the snow-covered slopes of Monte Flora above the Argentine Base Esperanza and there communed with several hundred thousand of the penguins. There are an estimated sixty-five million of these birds in Antarctica, the only animal species on Earth that has shown no signs of any arteriosclerosis, ever. Nearly disease-free, they mate year after year with the same partners, bearing on average two hatchlings a year. So well adapted are these birds to the demands of their environment that the male Emperor penguins (*Aptenodytes forsteri*) fast for 145 days, holding an egg beneath their webbed feet, while the female is out searching for food. If the mother is delayed after the little charges are hatched, the male will secrete food from its crop wall and then regurgitate the slime to tide the little darlings over.

Adelie possess a dozen waterproof feathers per square centimeter of skin surface, in addition to two centimeters worth of subdermal fat. Consistent with their ascetical incubations, and their endurance during long windy winter nights—temperatures dropping at times below -150°F—they do not seem discomfitured in the least, coddled by the clan, and the explicit benefits of kin altruism. Birds unanimously take part in the rearing of young; love-struck widows and widowers are always searching for orphaned birds to nurture. The penguins are all emotional gab, warm hugs, and salutations, living out their ten or fifteen years in affectionate simplicity. Their rookeries are a biological chorus of aptitude and commonality of purpose.

Penguins are not entirely at ease, however. When they gather in large colonies they seem instinctively to take inventory of their number. Penguin adults understand the limits of their food supply. When there are insufficient krill to go around, or excessively inclement weather (i.e., 200-mile-an-hour winds in some parts of the continent, adding a considerable chill factor), they will favor the stronger of two chicks, deliberately starving the weaker one. It is tragic to behold. I watched, day after day, as several young hatchlings slid pitifully into quietude and then oblivion, begging all the while for nourishment that their parents would not or could not provide. In the end, the soft, limp chicks died out, I imagine serving as added insulation for the nest against the frozen earth. Add to this melancholy system the added attrition from carnivorous leopard seals by sea and large brown skuas overhead, predator birds with seven-foot wingspans that patrol the penguin rookeries, and it is easy to see how as many as 90 percent of the penguins may perish in a season. Yet, enough manage to survive so as to perpetuate their kind. And it has always been so. Paleobiological evidence suggests that the penguins have lived in their colonies on the rocky ridges of Esperanza for at least three million years.

The penguins are part of a vast biological Antarctic symphony. Some days it is closer to Mahler. At other times, sheer Mozart. But, upon reflection, I keep hearing Vivaldi. Their ancestors, the *Sphenisciformes*, have been hollering and craaacking gleefully since the time of the Eocene, fifty million years ago, when the continent pulled away from Africa and South America and mammals began to evolve there. The continent cooled by at least twenty degrees, pack ice formed, extending for hundreds of miles beyond the coastline. Every spring and summer, when the ice partially melted, it would release dense populations of algae, which in turn provided the needed nutrients for billions of Antarctic krill. I probed beneath the ice, and there saw the two-inch crustaceans feeding in the submerged hairline cracks. The krill (*Euphausia superba*) often mass in gigantic blooms. A super swarm was spotted in 1981, weighing an estimated 5 billion pounds. Most krill eggs are released in the Antarctic summer, January and February, in time for the larvae to feed on mature phytoplankton. How the krill survive in darkness beneath the pack ice all winter is as yet unknown to humans. Whales will frequently swim beneath the center of the krill blooms, blow a barrage of air bubbles upward, then plow toward the surface, their mouths agape.

Somewhere beneath this darkling seawater cornucopia glides *Pagothenia borchgrevinki*, an Antarctic fish that feeds under the fast ice and maintains within its blood the highest known levels on Earth of serum antifreeze, one more adaptive dynamic on the last continent. These waters of the Antarctic Convergence—the most voluminous ocean currents on Earth, twenty-four thousand uninterrupted miles— host an extravagance of marine organisms, especially remarkable in light of the polar frigidity that enshrouds their world. Such numbers are unapologetically a measure of population stability. Dwelling therein are approximately three hundred different species of sponge; one hundred twenty-six different fish species (including dragon, white-blooded, horse, and snail fish, eel pouts and left-eye flounders, lampreys and skates); thirty-five million seals of six species, including six hundred thousand enormous southern elephant seals (*Mirounga leonina*); eight whale species; pearly nautilus, urchins, prawns, slugs, ten-legged sea

spiders, mussels and bivalves, giant isopods, huge brown seaweeds, pro-
fusions of limpets, and squid whose size varies from two centimeters
to twenty meters.

Dismembered whales' stomachs have revealed eighteen thousand
undigestible squid beaks in Antarctic waters. Sperm whales (*Physter
macrocephalus*) routinely vomit their squid beak-replete insides. This
floating puke is fed upon by the five species of long-lived albatross. Some
seven hundred fifty thousand breeding albatross couples (light-headed,
sooty, black-browed, gray-headed, and yellow-nosed) have been known
to fly eight thousand kilometers in a week searching for the vomit.
Descendents of the now-extinct ninety-pound bony-toothed *Pseudo-
dontorn*, the albatross were the first to understand that the Earth is round.
The young are likely to circumnavigate the world several times before
returning to their nurseries.

There are fifty-two Antarctic avian species, eight of which breed
on the continent, eleven on the peninsula, the rest along the sub-
Antarctic islands, such as Marion. Measurements of bird biomass on
Marion Island some years ago revealed the presence of 32,000 tons of
fresh guano, 500 tons of dropped feathers, 350 tons of dead birds, 200
tons of eggs. Biology is multitudinous in the southern seas.

Amid tens of millions of square kilometers of high altitude aridi-
ty, Antarctica enshrines a viable system of fertility, of population
balance, which includes three hundred forty known Antarctic plant
species, even two seed-bearing angiosperms on the ice. There are eighty-
five species of moss, two hundred different lichens, in addition to the
smaller forms of lifeuntold species of bacteria, yeast, diatoms, and
foaminifera. Mites roam the interiors of seal nostrils, yellow *Xanthoria*
lichens wedge themselves into minute crannies of the remote Trans-
Antarctic Mountain walls, while cyanobacteria have been found living
inside Dry Valley rocks, heating the interiors to a toasty seventy degrees,
despite outer blizzards. This has convinced some exobiologists that
similar life must prevail within the rocks of Mars, whose climatic
variability resembles that of the last continent. One species of lichen,
(*Lecanora tephroeceta*) inhabiting Marie Byrd Land, lives off the ammonia by-product of uric acid. Only one species of urine will do, that

of *Pagodroma nivea*, a snowpetrel that just happens to breed atop that lichen. At Mount Faraway in the Theron Mountains of western Princess Martha Land, and at the remote Vostock Station—thirteen thousand feet up—six pairs of skuas have been seen. A single skua made it to the South Pole itself. Such legendary tenacity transcends the individual birds and suggests something stellar about evolutionary biology in Antarctica: trillions upon trillions of neurons sensitized to sea and salt, wind and ice, maintaining genetic equipoise and population stability in what is the driest, harshest region on the planet. Countless marine organisms in this southern ocean go into suspended animation throughout the winter, revealing yet another facet of Antarctica's perfectly adapted populations.

Not surprisingly, the sheer pageant of biological numbers in Antarctica—trillions of krill, hundreds-of-millions of other creatures—follows a pattern of triumphant speciation: a condition of population equality and sharing, of free biological socialism, that appears basic to the success of life everywhere on the planet. What is perhaps most astonishing about the various plant and animal communities in Antarctica is that they are not more numerous. Given the vast underpinnings of the food base—phytoplankton and krill—one can only admire and envy whatever built-in fertility checks have been ordained here. Certain coral reefs along Australia's Great Barrier contain as many as twenty thousand species of fish, far more than are known to exist in the entire Antarctic Ocean. Any sizable portion of tropical rain forest still standing will quickly dwarf all of Antarctica, not in mammals or even birds, but insects. Yet, the effective principles of population stability are undeviating in those other regions, as well. And though the hand of humanity has been heavy upon reefs and rain forests, tainting biological evidence which might otherwise provide a clear window on the workings of evolution, certain explicit population patterns have become perceptible.

While it is known that, on average, three hundred species go extinct every million years, the pertinacity and refinement of plant and animal evolution indicates that nearly all populations on Earth have achieved an exquisite, self-regulatory equilibrium providing them the means to preserve the vitality of their species. At least in the absence of man. In other words, they will ordinarily control their numbers. Toward that resolute goal, nature performs a constant series of numeric gymnastics, maintaining high amplitudes—where populations surge in size—and equally impressive troughs where their numbers subsequently collapse, though not all the way. Sometimes self-regulation is blatant, at other times, subtle. In the stark terms of science, numbers, per se, mean little to the planet. Eight million multicellular creatures dwell within a square inch of salt pan. In two pinches of substrate in Norway, one investigator found over ten thousand separate groups of microbes. That's not individuals, but different *types* of individuals![1] In approximately one pound of fertile soil anywhere on the planet, there are likely to be 400 billion algae, 500 million fungi, 500 million protozoa, and 10 million bacteria.[2] Eighteen thousand different species of beetle have been postulated for just one hectare of Panamanian rain forest. According to Edward O. Wilson, something like a billion billion insects are alive at any given moment, weighing more than all of humanity combined.[3] Abundant microbes thrive on the bare sands of the Namibian desert, where summer temperatures exceed 170 degrees F. Meanwhile, a single human being possesses an internal population of some 300 trillion cells, give or take 2.5 million, at any given second. We harbor some 7 million follicle mites in our eyelashes and approximately 24.1 billion bacteria in each armpit. Billions of yeasts, bacteria, and viruses crawl over our forehead, while some 100 billion cells inhabit our brain. Each of these worlds within worlds indicates a population that is perfectly poised, a mysterious and essential ecosystem.

Some populations are not so poised, however, driven by strategies whose overall design we may never want to understand. A *Polycythemia vera* cancer cell, for example, found within some luckless human beings, can generate nearly 17 trillion descendents in six weeks, an aggressive weight of twenty pounds strung together by fibrous tissues. From the perspective of the host, such metastasis implies an overpopulated cellular region, to put it mildly, though it comprises less than 6 percent of the total organism—a rate of numeric increase far less than certain economies, slums, countries, and the human race in general.

From the vantage of the cancer, whose motto is perpetual growth, however, there is nothing overpopulated about this sequence of events. Only the host knows, for sure.

Within the arena of overpopulation, there are other radical viewpoints and stabilization (i.e., death of the host) techniques, as well. Take the Norwegian and Alaskan lemmings who may multiply five hundred times their normal group size every three or four years. To put that in proper context, think of the megacity Calcutta with an added increment of street-dwellers equivalent to the total number of humans on the planet, or nearly 5.7 billion. For the lemmings this is a very temporary kind of hell, which precipitates a nonviable condition, namely, famine. Their hormones release a distress call that tells them they are too many, that they must spread out in search of new sources of sedge and seedling and shoot. Invariably, many come upon water. They try to swim to the other side, dog-paddling for up to fifteen minutes. That's the limit of their endurance. If the waves are higher than six inches, they will drown, and many do. They are true environmental refugees, a condition increasingly familiar to *Homo sapiens*. The lemmings' little corpses provide superb fertilizer for the northern grasslands, which benefits a new crop of seeds and berries, the mainstay of caribou herds, which in turn provide other predators their dinner (or trophies). Enough lemmings perpetually survive this crisis to begin the cycle of regeneration all over again, which pleases yet other predators. The first two laws of thermodynamics propel these cyclical disasters and rebirths. The Law of the Conservation of Matter/Energy suggests that the recycling of corpses is inherent to the complexity and species interdependency of the biosphere. The Entropy Law provides that excess cannot persist, that imbalance must ultimately be stabilized, one way or another.

What is true for lemmings is, of course, true for insects, be they moths, gnats, cicadas, or locusts. Population sizes of the lasiocampid moth are radically convulsed on average once in forty years, their high range an astonishing 10,000 times the low range. Ninety percent of all locusts on Earth flew to Morocco in 1988 to satisfy what appeared to scientists to be a collective hunger. Their journey over mountain, desert, and ocean was marked by what seemed to be an enormous dust cloud extending several miles in all directions. Morocco could not feed the billions of locusts. Many of the insects died, some survived. What impelled so many creatures to their predictable death? An unquestioning orientation toward homeostasis. Mass migrations, death, and resurrection are all pieces of the boom and bust puzzle.

By comparison, the United Nations Population Fund estimates that there are currently one hundred million human transnational migrants—some refugees, some temporary workers. In addition, hundreds of millions of human migrants have fled to urban centers where the effects of boom and bust can be tracked in terms of crime, crowding, pollution, and infant, child, and maternal mortality.

Unlike human urban slum dwellers, starlings, howler monkeys, and hummingbirds appear to maintain population balance by taking an inventory of their numbers every day, presumably to avoid the kinds of cyclical disasters previously mentioned. Worms and tadpoles are able to secrete growth inhibitors. Among the lynx populations of the Northwest Territories, boom and bust patterns occurred in eight-to-ten year cycles, the feline ranging in population from four thousand to eighty thousand. Sometimes, ecological overshoot can plummet the population below its threshold for recovery. Such an event can account for local extinction. On St. Matthew Island in the Bering Sea, colonizer reindeer devoured the island's plant life. Their base population of two thousand exceeded the island's carrying capacity. Within three years the herd was virtually wiped out.

Until recently, human beings were very much a part of this biological system of checks and balances, which seems to hold firm for all organisms. Four primary inclemencies kept Homo sapiens in relative population calm: high infant mortality, war, famine, and disease, all contributing to a meager life expectancy. The Black Plague (Pasteurella pestis) was transmitted from the Tibetan Silk Route to a harbor at Sicily in 1347. Rodents account for 50 percent of all mammals, and it was the friendly rat, carrier of the rat flea (Xenopsylla cheopis), that caused such demolition. At least 30 percent of the human population died out—50 percent between Iceland and India. Boom and bust need operate according to food scarcities and resulting famine. Disease, acting upon a host, or dense network of hosts, plays a similar role in the maintenance of populations. Not unlike the moths and the lemmings, Europeans witnessed a spectacular revival of their populations within a century of the Plague, exceeding their pre-Plague numbers. What is quite different about human beings (from their furry Arctic, or Antarctic cousins) is the distorted duration between our amplitudes and troughs, our booms and our busts. Through a series of seemingly unrelated technological and scientific breakthroughs, we would appear to have extricated ourselves from the normal biological rules governing all populations, engineering an enduring "grace" period that has enabled us to keep growing unhaltingly.

One could cite bipedal posture, bigger brains, extended periods of child development and maternal nurturance, stable food-gathering base camps, and a host of other paleontological distinctions that marked *Homo sapiens* as somehow destined to be different. Certainly the advent of agriculture itself was the beginning of our recent break away. More recently still, a (relatively speaking) rapid succession of mechanical inventions and social unfoldings most appreciably fueled the population boom.

By the middle of the sixth century A.D., the old Roman scratching plow was replaced by an innovative Slavic device, a heavier plow with wheels and double blades. A moldboard turned the sod. But because the new plow required a team of eight oxen, which most peasants could not afford on their own, cooperative farming came into being, setting the groundwork, as it were, for the feudal adaptation of mass labor. Eventually a three-field rotational system utilizing horses rather than oxen accelerated the productivity of medieval farms. The horse collar, horseshoes, and tandem harnessing had made such agriculture feasible.[4] Irrigation furthered this medieval agricultural revolution. By the eleventh century there were over fifty-six hundred water mills in England alone. Mill reclamation drained marshes and converted bogs to arable land throughout the Low Countries. Such technological knowhow would serve, subsequently, to exert the first curbs on the outbreak of malaria, a crucial precursor of declining deathrates.

In the meantime newly devised water-driven hammer bellows

stoked furnaces, which in turn produced immense quantities of iron, thus providing the equipment necessary to open up mining in the Alps and Scandinavia. Military armadas were converted from vulnerable ships of wood to impervious vessels of iron and steel. Energy capacity was also multiplied as the British began heating their homes with coal. All of these trends supported greater comfort, longer lives, and more people.

In the wake of the Black Plague, human deaths had coincided with a rise in certain business opportunities, not surprisingly. Such widespread mass fatalities engendered commerce like never before. One such enterprise was the rapidly expanding wool industry, which dispossessed more and more British farmers, as farm land—now without sufficient able human hands to nurture and conserve it—gave way to pasture. For every sheep put behind grazing fences, seven newly unemployed farmers and their families went without bread. Inflation bridled the destitute, compelling Elizabeth I to pass an act in 1601 forcing landowners to pay rates for the maintenance of the poor (a practice later decried by Thomas Malthus who argued that for the betterment of the world, the poor should be left to die out). The profits from wool were spectacular, encouraging overzealous dealers to write off one plot after another as sheep were permitted to overgraze. It took a few years, but English pasture land was increasingly denuded. This led to unaffordable domestic prices for wool. One alternative was to export wool. But to where?

A year earlier, in 1600, the East India Company entered into serious competition with the Portuguese and Dutch in search of such markets. With the emergence of big business in Flanders and Italy, banks began to dominate the flow of goods. Money exerted more power than land because it could be invested in such nascent technologies as bituminous coal smelting, coastal reclamation, pumps, mills, and increasingly sophisticated agricultural techniques, like the four-year rotational method employed in the Netherlands. Locals began migrating to the big cities. That's where the money was.

In 1698, one Thomas Savary obtained a patent for a steam-driven "fire engine." James Watt's firm later sold more than five hundred of these steam engines.

Other trends, seemingly unconnected to the history of steam, wool, iron, and steel, were taking place. A climatic change in eighteenth century Scandinavia allegedly compelled the Norwegian gray rat (which does not carry the plague) to find passage into western Europe, where it successfully ousted the indigenous flea-carrying black rat, thus eliminating the most incisive threat to human population.[5] In quick succession, a number of medical breakthroughs further fortified the human animal in Europe who was already busy colonizing and enslaving much of the world, including parts of North America.

As population increased, more and more labor was employed in an ever-escalating industrial base. What fully revolutionized the steam business was Oliver Evans' engine for converting high-pressure coal into usable energy. His device powered the Rocket, a train that covered twelve miles in a mere fifty-three minutes. It was the rage of England, and just in time: by 1800, nearly 14 percent of British countryside had been given up to grow feed for horses, an excessive amount of land, considering the damage already wreaked by sheep, and the amount of old growth forest consumed for construction purposes and charcoal. An alternate means of ground transport was deemed imperative and now it had arrived. While England was smashing the wheelbarrow with the steam engine, spearheading the transportation and energy sectors, it also went singlemindedly after agriculture, importing skeletons from all over Europe and Africa to use as fertilizer. Conditions for a population bomb were in place.

As the English, Dutch, French, Portuguese, and Spanish continued to spread out in search of political rank, power, and assured markets for their newly industrialized economies, the modern ecological crisis was set in motion. Captain James Cook's discovery of Antarctica was part of this outward expansionism and all that it would portend for the future. Writers such as Shelley, Wordsworth, and Dickens clearly sensed the coming catastrophe. Writing in *Hard Times*, Dickens described the "melancholy madness" and referred to "people equally like one another, who all went in and out at the same hours, with the same sound upon the same pavement to do the same work, and to whom every day was the same as yesterday and tomorrow . . . "⁶ The effects of such economic territorialism were to be felt everywhere.

Yet until very recently most human communities beyond the pale of industrialization continued to comply with the unstated rules of boom and bust, usually by chance, as in the case of the above-mentioned primal causes—high infant mortality, war, famine, and disease—though sometimes by contraceptive choice or cultural lifestyle. Mbuti Pygmies, traditional Bengalis, the cave-dwelling Tasaday of Mindanao, and Tibetans, each took culturally ordained steps to limit their numbers. They used a variety of techniques, from basic abstinence for three years following a birth, to the application of various kinds of natural secretions or animal dung to the vagina (much in the manner of modern foams), to the rites of polyandry, whereby a woman limited her fertility by marrying several husbands. In fact, several thousand traditional contraceptives are known throughout the world. Several human population groups were able to maintain, or are currently maintaining, zero population growth (ZPG), which refers to a total fertility rate (TFR) of approximately two children or less per woman. Germany, Catholic Italy and Spain are each experiencing minimal, or even slightly negative, population growth.

In the U.S., in what is today northern New Mexico's Pecos Valley, some two thousand Indians lived in relative harmony according to what now would be described as an "ecologically sustainable" lifestyle. They too achieved what can be extrapolated as a zero population growth society. The Pecos fashioned an imposing four-story quadrangular pueblo with six hundred sixty rooms and countless kivas surrounding a majestic courtyard, amid the broad expanse of the Sangre de Cristo Mountains in the mid-1400s, kind of a mini-Schonbrunn in the wilderness. For at least a century, this city, built upon a rocky ridge, thrived on local agriculture, never exceeding the semi-arid land's respectable carrying capacity.

Similarly, near what is Fresno, California, twenty thousand Yokut hunter-gatherers lived in a single city of sod and thatch huts beside what was an enormous lake three hundred years ago. It was the largest known hunter-gathering assemblage in the world, an ecologically sustainable megacity. The Yokut traded with other tribes, most notably the Piute across the High Sierra, exchanging grass, herbs, beads, and vegetables for obsidian, moccasins, clay, and fire drills. They ate wasps, skunk, deer, mussels, and clams, as well as the sweetest blackberries in North America. Their meat-eating culture was achievable only because of their relatively small number.

Farther north, along San Francisco Bay, the Ohlone Indians lived in a similarly undeviating paradise for nearly two thousand years—clad in mink, sampling the local lobster, dipping their oysters in piñon nut sauce. The life expectancy of these indigenous peoples was probably less than half that of today's average American (could they have known, or minded, the years they were missing? Definitely not) though not appreciably lower than denizens of the modern African nation, Guinea Bissau, where the average age of death still hovers just short of 40.[7]

But in spite of the zero population or negative population trends within certain countries, regions, and tribes, our species as a whole is characterized at present by a problematic global total fertility rate average of 3.2, which translated into over ninety-seven million newborns in 1992, the highest annual number for human births in history. In developing countries, where over 80 percent of all human population growth is occurring, the TFR is presently 3.9, nearly four children per couple. That can mean that towns, cities, and whole countries may double their population in as few as seventeen years (an amateurish rate of increase by comparison with lemmings or moths, but extremely serious given the unique nature of human impact on the environment). Like economic inflation, such rapid population growth diminishes the perceived value and "quality of life package" of every living organism. As competition for life-sustaining resources escalates, more and more of nature is deemed a resource. The exploitation of a resource transforms subject into object, transmogrifying beauty, nuance, uniqueness, and that which is necessarily and biologically separate from ourselves, into a mere utility. Utilitarianism has had its share of economic and ethical advocates, but in truth, it perverts the world. A child is no blessing when its generation is inadvertently mobilized in fitful opposition to its surroundings—seeking to exploit, to combat, in other words, the very "nature" that has given birth to it in the first place. Ironically, explosive human population growth was triggered by a convergence of factors that all suggested an amplification of utility and value: decreasing infant mortality, increased longevity, and greater prosperity, the result of new immunizations, enhanced energy capacity, and the Green Revolution. This strictly human ambiguity has continued to obscure the moral and biological implications of our vastly overpopulated species.

Unlike lemmings, who live and die and are reborn in a strictly defined territory—"their" territory—humanity is blessed with at least the appearance of no bounds. Most predators look to a maximally nutritious, or delicious, range of species to prey upon, and it is usually a very contained array. Only domestic and semi-domestic animals consume a larger variety of foods than their wild relatives, and this is because—as in the case of rats, raccoons, dogs, and goats—their diet has been modified by humans. Not surprisingly, human food choices have drastically narrowed as our population has increased. While monoculture can be attributed to global market forces, its fast-growing, energy-intensive, cash-crop mentality is actually a response to demographic pressure. But food acquisition, which—along with sexual partners and a breeding site—defines the total orbit of all other species, has relatively less importance among Homo sapiens. Our state of boundlessness has engendered a rare delusion of reprieve from natural codes. It has meant that we are not only global carnivores, but globally territorial. No other single population (other than those which inhabit us, or have affixed themselves to our comings and goings) has so flaunted its biological prominence.

Evidence overwhelmingly suggests that even when our numbers were not so distorted, other species which came into contact with our forebears did so at their own extreme peril. The Siberian hunters who descended upon North and South America some twelve thousand years ago coincided with the rapid extinction of 80 percent of all mammal and avian genera that had thrived during the preceding Pleistocene.

Considering the widespread remains of stone weaponry and charcoal fire pits that marked this sudden human migration, which conquered at the apparent rate of sixteen kilometers a year, such massive extinction must be more than mere coincidence.[8]

The same carnage can be traced wherever early man wandered, from New Zealand, where the earliest indigenous peoples wiped out large numbers of native avian species, to Polynesia, to the lowlands of Europe, and to Scandinavia, where man has had a negative ecological presence on the mountains there since the earliest Stone Age, according to recent studies.[9]

Furthering this view of innate human aggression, Raymond Dart quantified the high frequency of bone injuries sustained by our ancestors—from the *Australopithecines* to fourth-glacial *Homo sapiens*—typically during the hunt. Dart's description of this "blood-lust differentiator, this predaceous habit, this mark of Cain that separates man dieteticly from his anthropoidal relatives and allies him rather with the deadliest of Carnivora" suggests that in *Homo sapiens* overpopulation has more to do with ecological impact than with mere numbers. This supposition accords well with what we now know about ourselves.[10]

Indeed overpopulation is not, in and of itself, a transgression against nature. The inflated community of lemmings, in its frenzy to feed itself, does not destroy the Arctic. Quite to the contrary, its multiplicity of corpses nurtures the soil there, providing a veritable gourmet's feast for other organisms (what the 19th-century Japanese, in describing human corpses, first called "night soil"). A colony of driver-ants, made up of perhaps twenty million workers, devours everything before them, mile after mile, when they move out. According to Wilson, they are "a heavy burden for the ecosystem to bear," reducing its biomass and altering even the "proportions of species." But they have been getting away with it for many millions of years without noticeably altering the planet.[11] Similarly, in recent years the Crown of Thorns starfish near Guam have been devastating the coral reefs at the rate of one kilometer per month, a syndrome that marine biologist Dr. Richard Chesher attributes to the cumulative pressure upon the starfish from human activities. But the only known species to actually change the whole planet, to interfere with nature, is *Homo sapiens*. That interference is overwhelmingly troublesome: first, because it existed when the total human population on Earth was under one million; second, because those same aggressive convictions and habits of mind persist today, empowered by an unimaginably more deadly technology than stone hatchets and spears; third, because we are more than likely to exceed 12 billion people, a current projection that I take to be rather optimistic; and finally, because we are quickly reducing the entire planet to what biologists call a "simple ecosystem." That is one dominated (by as much as 90 percent) by very few primary species, say, red mangrove swamps, various bacteria, mites, and ants. Simple ecosystems are a total anomaly in nature. Nor is there any guarantee that a planet-wide simplicity of that kind is feasible. Recent evidence from Australia dating back to the very origins of life some 4 billion years ago, shows that even in the near beginning, microbial life was diversified, not simple.

This brings me back to the biodiverse Antarctic, where human beings, grasping after "resources," first served warning to that primeval landmass in the year 1786.[12] That is when the first sealing expedition arrived at South Georgia Island, eleven years after Captain James Cook reported sighting abundant fur seals in the region. By 1791, at least 102 ships were conducting massacres of the many seal species there. Others, the southern elephant seal *(Mirounga leonina)*, and the abundant crabeaters *(Lobodon carcinophagus)* were heavily impacted. A two-ton elephant seal once "rendered" could produce two barrels of oil. Of the 30 to 40 million Antarctic seals today, most are crabeaters, one of the more abundant mammals on Earth. Brown, fawn, and blonde-colored, they scatter out, three or four seals per square kilometer of ice, lounging much of the time like sultans, unhurried, not about to expend precious energy fleeing from the harpoon or the spear. It is believed that the crabeaters collectively consume over 60 million tons of krill each year. The vast krill populations seem to withstand such mass destruction without so much as flinching.

For some of the seals, however, the same cannot be said. One species in particular, the Kerguelen fur seal, was virtually wiped out. Today at places like South Georgia Island it is making a comeback after several recent international treaties provided policing mechanisms for its reestablishment. But, for tens of millions of years their only predator had been the killer whales (*Orcinus orca*) that hunted along the polynyas (openings in the ice), rushing and circling, pummeling the bergs with their tails, until they had managed to dislodge a seal. With the arrival of this new and ungainly species, man, the seals and the whales themselves were taken by surprise, almost overnight. The Norwegians came up with a harpoon-cannon winch technique for killing whales, which they introduced to the Antarctic in 1904, establishing the first sordid shore-based whale processing industry there. Subsequently two-thirds of all baleen whales were exterminated. Despite International Whaling Commission rulings, at least two countries continue to kill as many as sixteen minckes (*Balaaenoptera acutorostrata*) every year.

Later on, with even higher expectations, Germany's Third Reich sent warships to the southern icefields. A commander stepped from his submarine onto the ice, where he was reportedly greeted by waiting Emperor penguins whom he actually saluted with a "Sieg Heil!" Nazi aviators dropped thousands of stakes emblazoned with swastikas all over the Antarctic in an effort to legitimize their "conquest."

The Argentines went so far as to temporarily shift their nation's capital to Base Marambio on the Antarctic Peninsula and to place a pregnant woman there; she gave birth to the first supposedly Antarctic human native in 1979. When I visited the Argentine bases there were nine children spending summers there.

Antarctic Treaty nations agreed in 1961 to maintain Antarctica as an international preserve—unowned, unmilitarized, a bastion of pure research. It is a land mass that, in combination with its pack ice, extends to four times the size of the United States. There are countless regions, larger than California, where no human being will be found. However sometime in the twenty-first century it is likely that geologists and engineers and multinationals will begin to exploit this heretofore forbidding region in earnest, guided by hints of extensive deposits of iron, zinc, copper, gold, platinum and oil. In 1974 the U.S. Geological Survey estimated that the continent and adjacent seas might hold as much as 45 billion barrels of oil and 150 trillion cubic feet of natural gas—quantities roughly comparable to the existing reserves in the United States. While the 1991 Protocol to Protect Antarctica managed to overturn the mining provisions of the earlier Wellington Convention, thus guaranteeing on paper Antarctica's freedom from commercial exploitation at least until 2040, there is no way to protect Antarctica's biological integrity from humanity in general. Nuclear power is excluded from the southern ice, yet the United States breached its treaty obligations in that regard. Military might is banned from Antarctica, but it is there to be observed. It is illegal for any U.S. citizen, scientists included, to so much as touch any living being in Antarctica without a special permit. At least on paper. As global population escalates, and the consequent exploitation of resources intensifies, it is hard to imagine that Antarctica will be left alone, especially in view of its recent history.

With more than three dozen countries vying for strategic presence in this splendid wilderness, in 1993 some four thousand humans can be found at any one time among the more than sixty bases. The diminutive human populations hover precisely about those scarce ice-free nesting sites (a mere three hundred miles of possibilities, out of more than eighteen thousand miles of ice-locked coastline) where millions of Antarctic birds and mammals also congregate. The small contingent of newcomers—a species not remotely suited, biologically speaking, to the southern continent—has brought with it a host of lethal chemicals and pollutants, the indirect fallout from human demographic pressure. By that causal link I refer to the complex political, economic, cultural, and technological wellsprings of human overpopulation, whose most blatant, or least pretty, expression is its cumulative ecological damage.

McMurdo Sound, the embayment adjacent to the major U.S. base, is virtually dead, according to one senior marine biologist with the National Science Foundation.[13] Human activities in Antarctica have unleashed a steady stream of polyethylene particles, polystyrene foam, trace metals, and radioactivity (the U.S. once illegally installed a nuclear power plant at McMurdo; it leaked and had to be scrapped along with tons of contaminated soil). Bombs have been exploded on the ice, toxic waste has been buried or strewn, all manner of noxious discharges have flowed from leaking fifty-gallon drums, and an entire penguin rookery was killed by the dogs at Esperanza. At most of the bases, road building and construction has disrupted the fragile ecosystems. Dust has killed lichen species (the redwoods of Antarctica); new radio frequencies have assaulted and possibly destabilized a tranquillity that prevailed since the time of the last dinosaurs; sea-borne and air pollutants from ever increasing numbers of ecotourist ships have infiltrated the wilderness. Garbage is randomly tossed from those ships, or by land-based personnel, or burned in open pits, the smoke penetrating microbial communities and wreaking the same biological disruptions that would occur in a laboratory setting. Thousands of tourists, managers, scientists, and bureaucrats land by ship and by air and go off tromping over delicate *Tortula, Bryum algens,* and *Grimmia antarctici,* the dark, heavy clumps of moss and grass. The Chileans have already erected the first "hotel" in Antarctica with sixty beds, and they are planning others. The tourists try to play with the wildlife—children chasing birds off their nests, adults posing for photographs along side elephant seals. As late as 1964 the seals were still being slaughtered. U.S. and French servicemen actually blew up bird rookeries in the course of various construction projects, *to see penguins fly,* as one worker is alleged to have put it. Antarctic seamen have tossed hundreds-of-thousands of live penguins into ship furnaces for fuel, or have eaten them.

But an even more sinister syndrome has apparently been unleashed. Since 1970 the global greenhouse effect has begun to reduce the amount of sea ice during winter months, which in turn has adversely affected the population size of krill, as well as chinstrap and Adelie penguins, according to scientists William Fraser and Wayne Trivelpiece, both biologists at Montana State University. In fact, on King George Island the number of chinstraps have plummeted by 35 percent, the Adelies, 40 percent. At the same time the Japanese and Russians are killing five hundred thousand metric tons each year of krill, one percent of the total krill ecosystem. According to Langdon Quetin and Robin Ross, biologists at the University of California at Santa Barbara, that amount represents a significant proportion of the crustaceans, considering many are egg-carrying females.[14]

I fasted one Christmas night at an Antarctic base on the Western Peninsula where penguin steaks were being served up. (At least two vegetarian scientists in Antarctica—one from Poland, the other from India—have started greenhouses at their respective bases.) The next morning I scrambled amid garbage dumps strewn in the midst of rookeries and watched penguins struggle to extricate themselves from wire mesh. A few of the Adelie lay dead in an effluent stream produced by the station. I saw seals frolicking in oil streaks, others fleeing from military vehicles by land. At Paradise Bay an albatross nested in the charred ruins of a former station that had been burned down by a disgruntled scientist suffering from Arctic hysteria, what the Eskimos call *pibloktoq*, the total disorientation that has been known to result from long winter darkness. At yet another human Antarctic settlement, I once lay on the guano-covered rocks near a rookery and beheld the sorry spectacle of low-flying helicopters taunting the birds for sport. In their panic adult penguins were separated from their young and the stampede crushed a large number of eggs. Every day since the turn of the twentieth century similar conflicts have been played out.

High concentrations of DDE and PCB have been recorded in Wilson's petrels, in skuas, cormorants, penguins, and the Weddell seal (*Leptonychotes weddellii*). These latter marvels of creation have been discreetly observed making love under the ice. They are able to dive nearly a quarter mile down, remaining there for at least seventy-three minutes before resurfacing. How they cope with the pressure of the deep and of bodily gas exchange, or with human seaborne hazardous wastes, is as yet a mystery. Antarctic seals, like many other mammals, will absorb their own fetuses under times of stress to prevent nonviable births. It is clear that population dynamics in certain parts of Antarctica have already been skewed. Certainly in the crucial, biodiverse regions of the many bases this is so.

Such negative ecological impact in a place as otherwise remote from human population centers as Antarctica, conveys two salient facts about *Homo sapiens*, consistent with our species' history, but colossally dangerous at this point in time. First, we have demonstrated the power and willingness to lay waste whole populations, not just those that happen to be in our immediate path or conveniently placed for purposes of our dinner. Second, we degrade or ruin significant portions of habitat, thus inhibiting future generations of a particular organism or complex of organisms. By severing its basis for fertility, we deny a species the opportunity to revive itself, a condition that can only lead to its extinction. That a very small number of organisms should unleash such devastation in so short a timeframe is much like the workings of a disease pathogen, with one significiant exception: no disease on record, not even the Black Plague, has yet driven *multiple* species to extinction. The dinosaurs died out gracefully over tens of millions of years, not a mere few centuries. Today at least 83 percent of all bird species—the descendents of the dinosaurs—have become endangered due to human aggression and the complicity, knowing or unknowing, of development and economic culture.

If the destructive penchant is there, has not yet been willed or coaxed or educated out of existence, then population pressure merely intensifies it. If the penchant is not there, population pressure may or may not invent it. Pressure or no pressure, *one* human being is too many for the planet if he or she happens to initiate a nuclear war, as nearly happened in May of 1990 in Islamabad and New Delhi. Or even to sink an oil-bearing leviathan in the Antarctic Ocean, which is what happened to the Argentine naval vessel Bahia Paraiso on which I travelled four times across the Drake Passage back and forth from Antarctica, approximately two hundred years after the first sealing boat arrived in these waters. Today that sunken ship languishes off the U.S. Palmer Base where it went down, leaking tens-of-thousands of gallons of oil and diesel into the waters of the Western Peninsula. It happened during the Antarctic summer. The new hatchlings in that region were just getting their feathers. Thousands died. Seal pups perished. Skuas, feeding on the carnage, also succumbed. Despite the truism that evolution occurs in harsh places, there was no immunity to hydrocarbons in Antarctica.

The same was true in the Gulf of Alaska when the *Exxon Valdez* hemorrhaged more than ten million gallons of toxic crude into relatively pristine waters, killing well over a million animals and contaminating thousands of miles of shoreline for probably a century or more. In Alaska, I waded through ankle deep Exxon crude along a remote beach in the Aleutians, six hundred miles from Valdez, five months after the calamity.[15] I witnessed grizzly bears feeding on corpses that were oil-infested. Closer to Valdez, the salmon's navigational skills were apparently impeded by the oil. A few years later, many were unable to locate the inland streams where they were born and to which they needed to return in order to mate.

Similarly, the penguins adjoining Palmer Base, stricken with a new type of stress—the chemical pollution—didn't have a chance of surviving.

I see the ghostly ship lying there a hundred feet below the surface, the brilliant sea all but concealing a vessel on which I had traveled. I marvel at the aurora austral from the bow, thrill to the first encounter with an iceberg, feel homesick for Antarctica even from the moment I first arrive there. That fifty-million-dollar ship, with its two helicopters and hot showers and espresso bar, the finest navigational equipment money can buy, killed wantonly, all the high-tech guaranteed nothing. In the larger scheme of things perhaps a few hundred thousand birds will not be noticed, or not by humans. Nor will the true culprit ever be brought to justice—neither the man at the helm, nor the first mate, nor Argentina, which produced the ship, nor even the species, *Homo sapiens*, whose politics and economics and convoluted history would have to be subpoenaed to divine a clear definition of accountability.

Humanity is self-interested. It has always reacted extraterritorially to pressure, moving out during interglacials, claiming new territory, prizing foodstuffs, ensuring surplus, strategizing in the face of perceived threats. This is why Thomas Jefferson initiated the Louisiana Purchase at a time when population density for most of the United States was a mere ten persons per square mile—to prevent, in his words, "the exterminating havoc of one quarter of the globe" from infecting America.[16] He was referring to immigration from Europe, not Latin

America. In a similar vein, the same year Argentina delivered a pregnant woman to Antarctica, Brazil launched four hundred huge developments in the Amazon so as, in then-President Medici's words, "to give men without land a land with men."

Hunger, sex, fear, greed, wrath, and curiosity have guided our wants and our migrations. These aspects of the human world have been aggravated as our numbers—and all they imply—have increased. Had the population not escalated so, there is reason to assume that humanity would have remained in balance, much like the penguins: elegant survivors, stable, seemingly content, subject to biological normalcy without the need or desire or even the possibility of managing, or attempting to overpower, nature. This more tranquil picture does not so much reject the evidence of our carnivorous past as absorb it more gracefully, given our diminutive population size thousands of years ago. The counterargument to such nostalgia is the paradoxical possibility that with fewer humans, per capita income, and thus consumption and fossil fuel extraction, might have been even more unrestrained. Singapore is a case in point. Even poor, relatively small-sized nations, such as the Cote d'Ivoire and Cameroon, have shown enormous levels of energy consumption and $CO_2$ generation, while many oil-rich countries, (e.g., Kuwait, the UAR, Iraq, and Qatar) have virtually no wildlife protection.

The penguins and seals of Antarctica are still numerous, though that could all change instantly if ozone depletion should alter the genetic viability of phytoplankton upon which the krill, and thus the penguins and seals, depend. In February 1994, scientists at Oregon State University reported data linking the disappearance of frog species to UV, as well. Phytoplankton is known to be vulnerable to ultraviolet-B radiation. While the overall rate of CFC (chlorofluorocarbons) production has dropped by nearly 50 percent between 1988 and 1993, humanity produced some 19 billion pounds of CFCs worldwide between 1978 and 1988. The damage from that assault is still rising—literally—and will be reaching the stratosphere to wreak its cannibalistic havoc on the ozone molecule early in the twenty-first century. Scientists in Canada have shown that ozone depletion in winter is now increasing over Toronto at the astonishing rate of 5 percent per year. At that rate, in a few decades many life forms will die out. Already, small pockets of the air column over Antarctica have shown depletion exceeding 60 percent during the late spring months. One sector actually yielded a satellite image describing a greater than 90 percent ruination. Over parts of Europe and Canada, the ozone layer has thinned in winter by as much as 20 percent. In India, the number of CFC-based refrigerators is projected to reach 80 million units by the year 2010, up from 6 million in 1989.[17] China, meanwhile, with its gigantic economic boom and 1.2 billion people, is producing nearly 10 million refrigerators a year. A 1990 survey showed that half of all government officials in that country did not know what ozone or chlorofluorocarbons were. This is not surprising, however. In China some 60 percent of all workers in environmental areas have not even graduated from high school.[18]

The penguins and seals have another substantial threat, namely, global warming, which liberates ice-locked methane and $CO_2$, thus compounding the warming syndrome and threatening to melt portions of the Western Peninsula, where penguins and marine mammals live and breed. Human-induced carbon emissions will double in the next generation, from the 1993 level of 7 billion to 8 billion tons a year.[19] With a high enough $CO_2$ regime, far beyond present levels but not inconsistent with 12 billion or more human fossil fuel consumers, the Antarctic land mass might eventually be returned to its condition of sixty million years ago, prior to the great chill. Then the continent teemed not with seals and penguins but amphibians and reptiles and luxuriant native vegetation probably more akin to that of present day South America. If that were to happen, every coastal city and plain would be flooded. For that matter much of Antarctica would be flooded as well. The salinity and temperature of the oceans would change, with a corresponding impact on untold numbers of marine species.

Population dynamics are thus explicitly linked to environment. A stable environment will normally engender stable populations, and vice versa. This is known as scale, and it never varies. Human *scale*, for example, has ensured that most door handles and beds always remain approximately the size of a human hand and a human body, respectively.

The same can be said of most animals' nests and lairs. Whether the smallest known organism, the mycoplasmas virus, or the largest, a giant sequoia (*Sequoiadendron giganteum*) with as much as thirty-six thousand square feet of life, all individuals and species conform to scale, a genetic and population average, the result of boom and bust vitality. Only a disease, such as a population of cancer cells, falls out of scale, booming perilously, eventually committing suicide by killing its host. In this respect, the cancer cell has declared unthinking war upon its carrier, willing to sacrifice itself for the temporary satisfactions of explosive growth. To psychoanalyze the motives of the cancer cell is to stare face to face down the naked territorial imperative.

The case can be made that—along with the cancer cell—human beings have also declared war and are also out of scale. This pattern in our population first showed itself during the transitional Mesolithic period: the Siberian hunters descending like army ants; eight-thousand-year-old petroglyphs from South Africa depicting the earliest known battle scenes. By the reign of Sargon of Akkad (ca. 2872 B.C.) the evidence of widespread aggression is unambiguous. Chroniclers tell us that Sargon "turned (the village of) Kazalla into dust and heaps of ruins; he destroyed even the resting places of birds."[20]

The world has seen its share of perturbations, of course, that make even the fiercest of warlords seem petty and amateurish: asteroids, periodic glaciation, the fires caused by an estimated 6 million lightning blasts striking the planet's surface every second, hurricanes, and volcanos. Disasters of varying scope have moved evolution forward over a period of 3.9 billion years. In the early Cambrian period, 540 million years ago, the speed of evolution suddenly crescendoed as Earth's oxygen saturation reached 21 percent, where it is more or less to this day. With oxygen came an ozone layer, and with ozone, a veritable Big Bang of biodiversity. Two hundred million years later, at the end of the Paleozoic era, a disaster of unknown origins swept the planet, eliminating approximately 96 percent of all marine animals. Yet from the ranks of those hangers-on emerged the first fledgling dinosaurs. One hundred eighty million years later, just as rain forests were everywhere multiplying, the dinosaurs disappeared; again a few mammal species were left. Out of this succession arose the first shrews and eventually lemurs, distant ancestors of *Homo sapiens*.

Five great extinction spasms that we know of have occurred on the planet. Not one of these catastrophes has occurred because of some biologic malfunction, or disorder, let alone because of a single species. But the latter is precisely the crisis we now find ourselves confronted with. There is no turning away from it: we are each contributing to the sixth pattern of extinction on Earth, the first one, apparently, in 65 million years. This perplexing, senseless tragedy must be viewed in clear, unwavering terms. The devastation is the direct result of our total population dynamic (TPD). By that I refer to the whole social, political, cultural, and religious array of causes and consequences of World War III—a war whose symptoms have engulfed every cell, gene, and ecosystem throughout the world, and whose chemical traces are to be found on the ocean floor and in the high stratosphere. The effects of this TPD are detectable in every mammary gland and pair of testes, in the optic nerves of Argentine cattle and the mercury-infested rivers of the Amazon, in hospital beds of smog-laden Denver and across the hungry, diseased, debt-ridden, and desertified swathe of sub-Saharan Africa.

Because the planet has spent billions of years adding, not subtracting species, we can safely assume without debating its teleological whys or wherefores that biodiversity has a crucial purpose in both the short- and long-term scheme of things here on Earth. Using this criterion of planetary scale and TPD, it is clear that our species is desperately overpopulated. Those who have the means and can afford the time to read these words will no doubt suffer psychologically on account of their species' ungainliness in nature, but they will be among the least likely to experience the real human fallout, the physical ordeal of overpopulation. The true victims of that burden are poor, discouraged, hungry, illiterate, oppressed, and often sick. And there are several billion such people, perhaps half the total human population, a figure quickly growing.

*Homo sapiens* had achieved the astonishing number of 400 million at the time of the first outbreak of Plague in 1347. By 1500 we were a global swarm of 700 million! By 1850 our throngs had topped 1 billion. Thus, while our transgressive nature has not changed in thousands of years, the multibillion explosion of our actual numbers is a very recent phenomenon. It has yet to reach its amplitude. And it would be premature to speculate on the coming ecological trough. But given the persistence and scope of our malice and hedonism, it would be the worst form of stupidity, evil, and complicity to simply ignore what is happening to us, and what we in turn are doing to Mother Earth. As E.O. Wilson has rightly described, we have but one planet, and one experiment. We'd best get it right.

It is impossible to know whether the physical world could sustain 10, or 15, or 20 billion people, under any conditions of human behavior. If, for example, our descendants chose to eat fast-growing species of sustainably produced fruits and vegetables, to remain totally passive, non-consumerist, utilizing solar energy for their basic needs, not traveling, harming no living beings, confined essentially to their garden cubicles across designated desert regions of the planet where the convergence of human densities and wild biodiversity were least in conflict, there might be a case for a vastly larger human population. But how vast one cannot say. One investigator has theorized that if technology could convert all food sources to liquid, step up the efficiency of marine phyotosynthesizers, limit all energy use to that which was solely productive of food (even recycling and cannibalizing liquid human cadavers) then our species might attain the (strictly) theoretical number of 60,000 million million.[21] This merely highlights the frailties of theoretical demographics, for that quantity would translate into a planetary density of one hundred twenty people *per square meter.* At various times the U.N. Food and Agriculture Organization in Rome, and the Vatican, have suggested that the Earth could feed 40 billion people, or nearly forty Chinas. What should be clear however is that at a population size of 40 billion or even 20 billion, we will no longer be the conventionally defined species known as *Homo sapiens,* endowed with the preconditions allowing for humanity. Rather there would we some far more malicious, unpredictable, and emotionally bereft creature, stressed to the point of utter relentless insanity. That is my prediction.

And that is the awesome and terrifying lesson of population dynamics and evolution here on Earth, whether in Antarctica, or in a place like China.

At night, when I lie awake meditating on these ponderous matters, I want to believe that there is still time to counterbalance this paralyzing suzerainty of which I am a part. For my children's sake . . . for the children, there are compelling reasons for doing so. And in that platitude of long-standing respectability rests perhaps the most painful paradox of all: that many of those toddlers, however adorable, must be held back, somehow consciously curtailed from being; that the actual size of the coming generations, to whom we would devote our love and attention, our dreams and hopes and fervent prayers, must be considerably less than the present one.

# 2

# A Paradox of Souls: China

## THE ONE-CHILD POLICY

There is hope to turn the "population explosion" around . . . and around. That is the frustrating demographic ambiguity: in spite of gradually decreasing fertility rate averages in most parts of the world, the overall size of the human population is rapidly increasing. This is due in part to the ever-growing number of women of effective child-bearing age (13 to 45, approximately) whose sexuality is vulnerable in countless ways, and also to the sheer preponderance of young people worldwide who are just entering or soon to enter their sexually active years.

Curiously the awareness of and fallout from sexually transmitted diseases has had no apparent impact on this rapidly increasing population (other than some statistically controversial impact in Africa, India, and several cities in the West). Similarly, the estimated total of eight thousand murders and one hundred thousand acts of violence seen on television by the average American seventh grade student has not served to curb violence amongst youths, or anyone else. Moreover, those deaths are so negligible in demographic terms as to be equivalent to nearly zero. Indeed, the prevailing argument among cultural mavins whether in Hollywood, Paris, or Bombay, holds that repeated exposure to such violence actually perpetuates it. But the same is true of sexual matters. In the U.S., where—all things considered—basic literacy is still rather high in a comparative sense, 25 percent of all Americans will be infected with an STD (sexually transmitted disease). Information, education, and the daily raising of consciousness have had little impact on sexual behavior. In an era of freedoms, sexuality is taken for granted. Desire is enjoying a renaissance worldwide. There are more

31

pregnant teenagers, unwed mothers, and younger pregnancies than ever before. Physiologically, girls are coming into menarche earlier. From the perspective of biological supply and demand (the basis of ecology) this raises a perplexing question: what is it that the planet is calling for from our kind?

Hope eternal for a stabilized, well cared-for human population, in behalf of the planet's best interests, is the one unconditional maxim that motivates and sustains the often harsh, unappreciated world of family planning, and no more so than in China. Throughout my travels there, many family planning cadres pleaded with me to keep smiling, to manifest positive thinking. Otherwise, they said, people will lose heart and give up. Of course they're right. And there are some justifiable reasons for limited optimism. It is estimated, for example, that all of the population programs worldwide have thus far prevented some 400 million people from being born, considerably more than half of them in China. The social, medical, economic, and ecological implications of this number for the twenty-first century—if family planning assistance continues—are that 4 billion fewer consumers and despoilers are expected, a population size equivalent to the total human population of the 1970s.[1]

Moreover, the use of contraceptives by couples (contraceptive prevalence rate, or CPR) has risen dramatically throughout the world. In China according to one estimate the rate was 73 percent in 1990, though another estimate shows it at 63 percent in 1992 (CPR in the United States averages 68 percent). Throughout the rest of the developing world approximately 44 percent of all couples are protected against childbearing, on average. These gains have been made in twenty years. With increasing government support and carefully targeted social mobilization, future gains might even be more rapid and dramatic. Or they might not.

In the late 1980s a global study by the United Nations Population Fund (formerly, UNFPA—U.N. Fund for Population Activities), discerned several fundamental conclusions that could be drawn from years of family planning experience. For example, the more types of contraceptives offered, the higher the rate of acceptance by men and women ("acceptors"). Thirty-five countries were studied specifically and it was shown that those with only two types of contraceptives available (i.e., sterilization and the IUD) were less than half as successful as those where five or more options were present (the so-called "cafeteria" style). In addition, according to a more recent unpublished UNFPA document, when women had the right to choose their own form of birth control, 91 percent continued to use it, versus a persistence rate of only 28 percent among those women who were not free to choose. Such lessons have entered the mainstream of family planning policy recommendations, with varying degrees of success, in over one hundred forty countries. One particularly distressing indication of the need for enhanced contraceptive use is that a woman who may desire no more than two children will require nine to ten abortions in her lifetime to accomplish that goal, in absence of contraceptive availability.[2]

While such data and percentile points may appear far removed from the glaring abyss of ozone depletion and the blowing up of penguin rookeries, in fact they are completely relatable. From the total population dynamic (TPD), or ecological perspective, our ill graces add up. Every individual, like every vote, counts. This is a platitude of particular gravity in geographically confining nations. In China it is especially so despite the vast size of the country.

And thus, for reasons having to do with the unprecedented size of the world's current population of children, and the immense number of child-bearing females, many would argue that the decade of the 1990s is the crucial one for reversing human fertility rates before a planetary baby boom drags down our species, and condemns most other species, as well. There is little time left for concerted action, the scope of which is daunting.

This is especially true in China, where the TPD, as I have earlier described it, has presently unleashed two opposing impulses—economic gain and escalating fertility—that together defy the classic "demographic transition model." Demographic transition holds that high infant mortality and a high fertility rate will inevitably shift, with economic advantage, to low infant mortality and slightly declining fertility. As economic prosperity continues to advance, the working class is said to gain in purchasing power, to be able to spend more money on health care and less time in the home, and be inclined to have fewer children. Hence, in the third and final phase of demographic transition, both infant mortality and fertility will drop rapidly to a stable fertility replacement level. In China during the 1970s, however, infant mortality and the fertility rate dropped without a corresponding per capita rise in the country's economic performance, though the standard of living was better than in most developing countries. Now, with the newfound scent of riches, infant mortality remains extremely low but the fertility rate is rising. The standard of living will continue to increase, for some, but at the expense of tragic ecological consequences and a broad gap between those who benefit from economic good times and those who do not. In the meantime, the population will expand, guided by a nearly "hormonal sociology" that reads: times are excellent, have more babies. This is a hugely problematic scenario.[3]

"To get rich is glorious," said Deng Xiaoping in 1978, oddly echoing a sentiment of Mao Zedong thirty years earlier. In 1993 the Chinese Communist Party, about to be utterly dropped, officially endorsed profit. Such economic policy inconsistencies reflect a similar ambivalence with respect to the Chinese party line concerning fertility. Mao frequently changed his mind about the population issue. Initially he denounced birth control, arguing that the solution to more and more people was, simply put, "production." (Karl Marx had first theorized this idea, in opposition to Thomas Malthus, who more rightly perceived that in the end, the Earth's resources were frighteningly finite, no matter how energetic or inventive people might prove themselves in the future.) People, insisted Mao, were not a problem. And the more people, the more production. He was widely quoted as declaring, "of all things in the world people are the most precious." That was at a time when the life expectancy in China was thirty-five years and infant mortality raged at between two hundred to three hundred-per-thousand live births.

It is worthwhile pausing for a moment to consider deeply what a world with a human life expectancy of thirty-five would be like. This was essentially the *average* human condition prior to the twentieth century, the circumstances into which an estimated 80 billion *Homo sapiens* have been born. Collectively, they have signed their name, and it has been a signature with considerable drop shadow. Individually, a minute smattering of Beethovens and Maimonedeses have left a trail of joy and triumph. But 79.999 billion of our forebears have simply vanished without a trace, having lived short and inexplicable existences. In spite of the overwhelming pattern of early death, however, nearly all of human development, art, accomplishment, and culture has evolved and prospered under a half-life regime. To have many children was to combat this narrow life expectancy. "It is not uncommon, I have been frequently told, in the Highlands of Scotland for a mother who has borned twenty children not to have two alive," wrote Adam Smith in 1776 (*Inquiry into the Nature and Causes of the Wealth of Nations*).

What is indeed curious is that, for at least the past few thousand years, there were noted individuals who lived seventy years or more, outliving their contemporaries by two generations. That discrepancy is perpetuated in 1993, and it invites a host of ethical, technological, and cultural questions inherent to China's own evolving family planning challenges. Consider the fact that there is still one country in the world—Guinea Bissau in West Africa—where life expectancy is thirty-nine, versus a Japanese life expectancy of seventy-nine. (One of the oldest known living humans, a Tibetan, is 146, according to a recent China News Service report.) There are several countries where the life expectancy difference between men and women is as much as ten years. To die at thirty-five or thirty-nine or even forty-nine is to compress the mind—every thought, goal, dream, hope—by more than half of what is possible, assuming that one is even emotionally aware that there are those who live to a ripe old septuagenarian age. It makes no "rational" sense whatsoever to die young, if there is even the slightest chance of dying old.

China, not unlike the rest of humanity, has never deviated from its pursuit of more time: time to fulfill the meaning of life, the connection to nature, and the duty to ancestors. In 1990 average human life expectancy in China was sixty-nine years, and the mean infant mortality rate (IMR) per thousand births was 14 in the cities and between

22 and 51 in rural sectors.[4] In the 1950s Chinese mothers who had lots of children were awarded medals honoring them as heroes. Within a year or two, following declining per capita figures, evidence of much malnutrition in the hinterlands, and a desperately sagging industrial base, Mao's thinking became slightly more "sophisticated." He decided that China was heading toward economic disaster. No more medals were given out. Excess children were viewed as an impediment to efficient production. In fact, rampant pregnancy was considered criminal. In China you are not allowed to steal, to carry a gun, or to carry excessive fetuses (more than two) to term.

Communism was quick to embrace Deng's free market exhortation. Within a decade of the country's economic liberalization—throughout the 1980s—there were at least 225,000 privately run companies in the densely populated coastal provinces[5] and the GNP bounded from 120 billion to 427 billion dollars. Today the acquisition of washers and dryers, refrigerators, television sets, and a host of other household goods is soaring and production is expected to double before the year 2000. In 1980 manufactured exports were worth 9 billion dollars to China; by 1989, 37 billion dollars. Deng intends to raise China's gross domestic product (GDP) from its current $330 per person (a wildly fluctuating average, depending on the province) to $1,000 by 2000, thus tripling the size of the economy. But it's moving faster than that. Just in 1992, the country's real dollar value leapt to somewhere between 6 and 12.8 percent; no one knows for certain how much. But in 1993, economic indicators suggested a growth rate higher than 13 percent. Meanwhile, Deng's economic proposals have been added to the Chinese constitution and the notion that communism was victorious in China is being deleted, according to the Xinhua Press Agency. According to the new policy, a dollar is worth a thousand communists.

Regardless of the economic data sheet or index used, 1992–93 saw China at the apex of world economic gain for those years. (The U.S., by contrast, has grown on average at 3 percent per year for the past twelve years, until the first quarter of 1994, when growth doubled.) Coupled with its current enormous foreign trade surplus, China promises to be the economic superpower of the twenty-first century, and the largest consumer economy the world has ever seen. The country's modest motto is *xiaokang shuiping*, which means "a comfortable level of living," consistent with China's "four modernizations" of agriculture, industry, defense, and science/technology.

This fanfare of the common man resembles an avalanche of dimes and quarters. Soon it will be dollars. Every consumer will earn and spend more and more. One observes it in the sheer flow of new goods, many with foreign labels like *Playboy* and *Esprit*; in the row after row of town enterprises—joint-enterprise production facilities churning out plastics, household goods, toys, chemicals, tractors, forklifts, steel, tampons, concrete. The new onslaught has the look of freeway gridlock—not of bicycles but new automobiles, all moving slowly past one high-rise development after another. This purchasing euphoria can be deciphered inside the jugular chaos of the nascent stock exchange, and crowding frantically around the counters of any store in nearly every city and town throughout the nation's twenty-seven provinces and three municipalities. This is the first impulse, a production frenzy that will, in turn, chew up the environment in one way or another.

There is a second, related trend, an opposite one in terms of demographic predictability. Corresponding to these otherwise celebratory, consumer-driven developments are the agitations of a new runaway population boom that everyone thought, just a few years ago, had been contained. It hadn't, other than in a few urban regions like Shanghai, where migration and ecological demolition are posing their own too recognizable dilemmas.[6] It is of course true that China has made remarkable strides toward curbing its human growth rate, more than any other country in history. The total fertility rate (TFR) dropped in China from over 6.0 in 1960 to 2.5 in 1993, a 60 percent decrease. Compare that with a 47 percent decline in Mexico, 38 percent in Indonesia, and a 31 percent drop across India over the same time period. Put in a different, more spectacular context, it took the United States fifty-eight years (1842–1900) to diminish its TFR from 6.0 to 3.5. In China the same decline required all of seven years (1968–1975). Such phasing out of high fertility required twelve years in South Korea, fifteen in Thailand, and twenty-seven in Sri Lanka and Indonesia, countries rather hailed by the international family planning community, but none to the degree of China.

But China's base population is gigantic, and because the country's TFR has been stuck for a decade and is now stubbornly rising once again, like an army of living ghosts returning to haunt a nation (a perfectly comprehensible Confucionist image), there is justified concern that economic and agricultural gains will be nullified by all the new mouths to feed each day, each month, each coming year. Based upon the size of populations in those eighteen provinces with "exclusions" to the one-child policy, and incoming data from a number of researchers, it appears that since 1983 the number of couples in China having a second child has gone from 10 percent to possibly as high as 50 percent. Despite its earlier successes, China is now confronted by a colossal dilemma. According to a recent official "China Population Research (CPIRC)," study by Mr. Chen Shengli, if China's existing TFR of 2.5 continues, the country will number 2.5 billion by the year 2090; a TFR of 2.3 would mean 2.2 billion; a TFR of 2.0 would balance out at 1.5 billion. Such are the perils of decimal points.

The bombshell facing China is the unlikelihood of attaining a TFR of 2.0, particularly given that the existing 2.5 TFR is now up from 2.3, a figure that had been stable for a decade during which there was every compulsion and incentive in the world to bring the TFR farther down. But that didn't happen. In the 1990s, as the economy overheats and that TFR should most assuredly be plunging, it is continuing to edge upward. Some believe that China's TFR is now at 2.8, not 2.5. If that is true, China a century from now would number *over 3 billion people*. Whatever the decimal factor, China is re-writing the demographic textbooks through its defiance of all traditional theories. (It must be emphasized that *officially* China's TFR is now 2.0.)

There are well over 100 million married Chinese women of childbearing age in 1993. Their preferred family size appears to be two, or three, depending on the region.[7] These couples are increasing China's current growth rate of 1.4 percent, with an annual tally exceeding the total population of Australia.[8] That's a net increase of between 16 and 18 million new mouths to feed every year (some Chinese scientists are claiming 20 million), 64,000 births and 17,000 deaths every day, making for a net daily increase of nearly 47,000 people.[9] The fertility "interest" is increasing, while the "principle" grows. The total momentum, in other words, is escalating. During the 1990s, China will add to itself the equivalent of the number of people in Great Britain, France, and Italy combined, or over 160 million. The huge irony is that China's annual birthrate is now what it was in 1973, when the countrywide TFR was 4.5.

As of mid-1992, China's population was estimated at 1,165,800,000, or 22 percent of the planet. The data are derivative of the official count. But it is probably higher than that. For example, in addition to the married couples, there are, in 1993, 200 million unmarried women and girls of childbearing age. The teenagers are not statistically considered fertility risks, as they are in many other countries, despite solid Chinese information describing their vulnerability. Chinese officials insist that there are virtually no teen pregnancies in the country, nor unwed mothers. They say this is because youngsters in China are principled, moral, revere their grandparents, and so forth. They never say that Chinese teenagers are not interested in sex, which would be like saying China has transcended biology. I pointed out to officials that a recent Shanghai poll showed that one out of five women was pregnant at the time of her first marriage.[10] The officials had not heard about that poll. And maybe they hadn't. Most officials do not spend their days reading polls, tabulating the latest demographic statistics, or even worrying much about long-term trends. They are most focused on immediate problems. And this tends to be the situation everywhere in the world.

But that focus on what is proximal is also subject to imperceptible change. The demographic goals become vulnerable. The carpet can be pulled out from under. As China's economy glistens, one can expect more and more young girls to become pregnant, and to come seeking help. In the short term, even presuming a mammoth and unprecedented chastity, 100 million married eligible mothers in a country with an average fertility rate of 2.5, suggests that China will exceed 1.3 billion by 2000, and between 1.6 and 1.7 billion by 2050. Chinese ecologists have long believed that the country's maximal carrying capacity is between 600 and 700 million people.[11] At nearly double that and with 30 million newborns a year expected during the coming generation, China is living on borrowed time. (Though to talk with Chinese consumers, that is hardly the impression one obtains.[12])

Such quantitative facts have engendered a widespread premonition within China's leadership and among those in the country's many think tanks of an ecological trough lurking beneath the nation's much touted "irreversible reforms," unless the present population boom can be checked. But there are no obvious solutions. It is argued by many that the highest population China can possibly absorb would be 1.7 billion. "If we exceed that the environment would become a disaster," said the head of China's Environmental Protection Agency.

One afternoon in Beijing over tea, Madame Peng Peiyun, the warm, jovial, full-bodied force behind China's State Family Planning Commission, confided in me that further declines in the birthrate, in her opinion, will be very difficult to achieve. As she was speaking, I realized that at that very moment the birthrate was inching upwards. While 100 million residents in Szechuan Province, as well as those in five major urban areas, seem to have accepted the one child policy, the rest of the country is having two, and in some cases three and four children per couple.[13] In respect to the stated demographic goals (1.2 billion by the year 2000), such trends are disastrous.

Beijing's allocation for family planning is less than one half of one percent of the government's total budget, or $1.3 billion U.S. It is expected to double between 1990 and 1995 (the eighth five-year plan). Family planning receives an additional $4.7 billion from all other sources—donor, provincial, factory, or joint-venture contributions. That combined total averages out to about five dollars per person countrywide. In the U.S., sixteen dollars per person is the estimated cost for family planning, but that does not include the additional thousands of dollars per person born by the health care system for prenatal to postnatal care, which in China are costs born by the Ministry of Health. With a warm smile, says Madame Peng, "It is not enough."

Within a year after the Tiananmen Square massacre, most of the world was conducting business-as-usual with China, eager to exploit a receptive economy. Nearly every visitor to northern China, whether on business or pleasure, tries to get up to the Great Wall, two hours by car north of Beijing. In the winter of 1993 my wife Jane and I walked along its ramparts, bracing ourselves against a bitter cold north wind. Several hawkers clad in stuffed, surplus Chinese army jackets and sloppily killed fox fur mukluks exhibited their paltry wares; one was in the company of a tethered Bactrian camel upon whose double humps the irregular odd tourist was invited to sit and be photographed. The moribund camel embedded herself in my memory for two reasons. First, she was virtually the only living "wildlife" I was to see that winter in China, aside from some pitiful caged exotic birds in a Beijing hotel; and second, because the camel, like the Great Wall itself, was such a powerful symbol of exhausting historical tumult, the rise and fall of nomadic conquerors and great cities.

It is a tireless wall, once intended to impede the invading armies that descended from the north, while to its immediate south the Chinese themselves multiplied with inexorable zeal. The Great Wall was thus the world's first ethnic contraceptive, something like the invisible wall of state and federal troops that guards the U.S. border with Mexico. It literally kept "pure" Chinese in and non-Hans out,[14] a distinction apparently less important in an era of imported Japanese cars, Rolex watches, and homegrown McDonald's hamburgers. That archaeological relic, which glows for two thousand miles on those nights when there is a full moon, conjures a rather central paradox of our time: while nearly everyone who ever lived prior to this century has vanished, there are now far more people alive than at any other time in history. Over 22 percent of them live in China.

Chinese populations have fluctuated wildly until only very recently. Two thousand years ago, the three largest cities in the world were Rome with six hundred fifty thousand people, Loyang, China and Alexandria, Egypt, both numbering four hundred thousand. In the year A.D. 2, under the Emperor Pingdi of the Western Han Dynasty, China's population was roughly 60 million. It then decreased to as low as 21 million. In the 4th century A.D., an epidemic apparently spread across northwestern China killing, by some estimates, 95 percent of the human population.[15] By A.D. 700, the largest city in the world was Changan (Sian), a walled fortress of thirty square miles and a million people. By 1943 Sian's population had plunged to a recorded 82,589. Now its residents are breeding again. Hangchow, Chengtu, Canton, and Beijing have all achieved numeric preeminence throughout time. By 1950 however it was Shanghai that first exceeded 10 million people, along with New York and London.

Today (1993) Shanghai remains China's largest, densest city. For all that, it remains a mellow, hospitable world unto itself, and its family planning officials could not be more gracious and caring. The city's current, unofficial population may be as high as 18 million, if one includes that incoming daily avalanche of rural migrants looking for work, and other "floaters." Officially, it is 12 million. Ironically, despite the city's runaway numbers, Chinese demographers claim that the city's married female population evidences the second lowest TFR in the country, approximately 1.3. This is in the nature of statistical discrepancy—the fact is that demography is not a hard science, its data is often inconsistent, and nearly every facet of family planning easily invites levels of skepticism, uncertainty, and sometimes violent disagreement. Contention has surfaced most visibly when certain "techniques" of population control perceived to be ethically incorrect have served to enflame politically—or religiously—driven zealots, both in China and Washington.

Taicang County, just north of Shanghai, with a population of four hundred twenty thousand, has achieved the lowest TFR in the whole country at 1.1. This is the epicenter of the one-child policy, still rigorously in effect according to authorities in Beijing, but with certain unavoidable *exceptions*, namely those in some rural areas and minorities. But they are glaring "exceptions." The definitions of "urban," "town enterprise," and "rural" are currently in great flux within Chinese circles. Where there is still predominant agriculture going on, it is considered essentially rural. By that definition, "rural areas" comprise between 70 and 83 percent of the country. This implies that the one-child policy has had a very limited impact. According to one demographic report,

87.2 percent of all births in the country are rural, and as of 1986, in one national survey fewer than 18 percent of all fertile women had complied with the one-child policy. There have been countless suggestions offered by the country's experts as to how to improve compliance. Tu Ping, for example, an Associate Professor at the Institute of Population Research at Peking University, points out that in some provinces, (such as Shaanxi, one of the poorest in the country), women do not even consider using contraceptives until the birth of their first child. He argues that rather than emphasizing just one child, what is actually required is a policy of spacing, and of delaying that first birth. This has become the international model for family planning in countless other countries.[16]

For those living in the cities, where so much of China's cash flow is being generated, the one-child policy has dominated Chinese life for twenty-five years and is the basis of China's "economic miracle," the augmentation of per capita income by as much as one-third. I asked one English-speaking twenty-two-year-old family planning associate in the Nanshe district of Shanghai—the densest quarter of the entire country—whether she wasn't concerned about the psychological ramifications of an entire generation of only children. Did she not miss a younger sister, or an older brother? Not in the least, she replied. "As an only child, I feel free. My parents love me more. There is no question that being an only child gives me a greater sense of independence as a women. I am free, don't you understand?"

Taicang is one of those transitional suburban regions where young men and women are keenly exploiting their freedom. From a high-rise in the early morning, staring south toward Shanghai, there is little agriculture left. Rather, a maze of construction enveloped in a pall of smokes that conceals one hundred thirty thousand women of childbearing age, one hundred thousand couples. Taicang comes recommended on the basis of a district-wide birthrate of 9.14 children per thousand in 1991, the very best in China. Liu An Ru, the cheerful local Family Planning Commission Chairman escorted me to the city's Fertility Museum where condoms and the dangerously outmoded Chinese-made steel ring IUDs (still in use) are behind glass. There are also

photographs of deformed monsters—two-headed, protoplasmic blobs—
the apparent offspring of incestuous matches which local authorities
make much of.

Sex education begins at the age of fifteen in the twenty-eight
Taicang District schools. Secondary (middle) school enrollment is
allegedly among the highest in the world here and, I was told, 100 percent paid for by the government. (Throughout China, 97 percent of
the children get primary education, 70 percent on average go to middle school or junior high school, 30 percent go on to high school, but
thus far only 2 percent go to college, a level half that of some other
developing countries.) This town and district—both called Taicang—
embody the goal of international family planning. Get students enrolled
at the secondary level, especially those of the female gender, and your
population problems will be—if not solved—at least slowed. That is
the conventional wisdom. Taicang has taken it to heart. There is a training course for the premarital couples, as well as for grandparents who
are schooled in the latest techniques of looking after grandchildren
while the parents are off "getting rich."

What about the fertility rate for those thirty thousand unwed girls
of childbearing age, I asked Liu An Ru? No problem, he was quick to
respond. What about abortion? No specific data on abortion, he said,
other than to remind me with some pride that abortion, as well as
sterilization, was legalized in China in 1953. But then he added the
assessment that there is very little abortion in the whole county because
most people are responsible for their own contraception. (There must
be a second museum somewhere in the country with photographs on
display of aborted fetuses.) Later on, another official in the area indicated
that there "may be" 0.6 abortions for every one birth. If that's true, of
course, it implies a veritable abortion industry.[17]

The preferred mode of contraception in Taicang for 67 percent of
all "eligible" women is the IUD; condoms and pills are used by 20
percent of married couples, while sterilization accounts for 13 percent
of overall birth control. Married couples obtain pills, IUD insertions,
or sterilizations for free. High school students can get over-the-counter
pills and condoms at local retail stores. But IUD insertions are available

only after the woman has been married and had her first child. Doctors do not encourage the IUD until child number one comes along, which tends in much of China to happen immediately. Afterward, parents do not seem to want another, at least in Taicang, where six thousand Family Planning Association volunteers are out pounding the pavement every day disseminating the benefits of a one-child family, knocking on doors, and counseling parents. There is approximately one family planning association member for every twenty to twenty-five couples, the highest concentration of family planners per capita in the world. Elsewhere in China, the ratio varies from between one planner in three hundred people to one planner in twenty-seven hundred.[18]

Forty years ago Taicang's locals got their water from the river. These days, the river is black, but the people have managed to construct water purification systems and pumps. I go out to a small suburb of Taicang, the village of Ludu, along one of those joint-enterprise roads. Interspersed among fields of wheat, cotton, rice, peanuts, garlic, and sesame, are small businesses with fancy bronze gates and expansive driveways. A black Mercedes 450 SL with Japanese businessmen inside exits from one of the factories. Only a few years ago, this was an agricultural commune, farmed by Ludu's seven hundred eighty occupants. Now the farmers are in business for themselves and paying into an old age fund, which is a whole new experiment for China. A home costs between three thousand and four thousand dollars in Ludu. As yet there is no central heating in the homes, though there are occasional electric shower stalls. There is no garbage collection; instead the people make their own compost, spreading it in the fields as they have done for millennia. Ragpickers from Taicang, a few miles away, come to recycle the plastic and cans. One sees tent enclaves of such garbage along the main highway out of Shanghai.

As I am served a lunch of fresh vegetables in one of the homes (my breath is visible in the air), my pockets stuffed by generous locals with peanut candies manufactured across the street, the Mayor proudly explains how one local architect is now making eighty thousand yuan a year, nearly twenty thousand dollars. Most residents are earning two thousand yuan a year—or five hundred dollars, far higher than the national GNP level of three hundred seventy dollars. The first private automobile was purchased here in 1986. Now there are many more. According to one local, everyone "feels much better" thanks to Deng Xaioping's open door policy.

The drop in Taicang's fertility, like that in many urban regions across coastal China, occurred several years before this entrepreneurial fever infected the country. In fact a fertility plateau was attained even before the one-child policy had been announced in 1979, leading UNFPA officials to wonder why Beijing ever instituted the stringent measure in the first place.[19] The answer is a complex one whose diagnosis summons the extraordinary contradictions and difficulties that the human species as a whole must confront in the coming months and years.

The origins of the policy go back at least to 1954, when Mao began to sense that while people were precious, too many precious people could be disastrous. Various birth control conferences were convened and the Ministry of Public Health began its anti-population boom campaign in August 1956. A year later, fearful of his country's inability to feed itself, convinced in his own mind that Chinese lovemaking had reached a state of "total anarchy," Mao feared that the Chinese people were heading toward "extinction."[20] Amazingly, having attained this point of view, Mao reversed his position a year later, and adopted once again the sweet, simple notion that more people on the farm meant more food. At that time, the Secretary of the Communist Youth League was quoted as saying, "The force of 600 million liberated people is tens-of-thousands of times stronger than a nuclear explosion. Such a force is capable of creating wonders which our enemies cannot even imagine."[21] Thus commenced the fervent government mobilization of China's productive energies, a three-year misguided missile, the "Great Leap Forward," which succeeded only in undermining the country's energy and agricultural sectors—never to be repeated, promises Beijing—with resulting famine that took thirty million lives! It was a true ecological "trough," one of the worst human tragedies ever, quickly followed by a population boom, as one might predict. In fact, within just a few years, the birthrate doubled, from the low twenties per thousand, to over forty per thousand, while the death rate plunged from forty-five to approximately ten. China in the late 1950s and early 1960s had the biological look of the Arctic grasslands, home to the lemmings.[22] Premier Zhou Enlai called for a new campaign to control births. What had been, up until 1971, a three-child limit, was now altered to two children. In one year—1975—17 million women had IUDs inserted. It was an all-time global record. Nor should it be forgotten that the Chinese female's body was the nearly singular battlefield upon which this patriotic premise of population control was waged by the Beijing authorities.[23]

In 1979, presented with staggering computer demographic projections that showed more population-driven economic crisis in store for the country, Deng Xiaoping consulted with his closest aides, meditated on the future, and in April of that year when spring was in the air of Beijing, made his decision. The Office of Population Theory Research of the Beijing College of Economics had come up with its published "Population Theory" two years before, in which it was concluded, "Human reproduction, like the production of goods and services in a planned economy, must not remain in a state of anarchy."[24] It was a firm call for direct state intervention into the behavior of individuals for the good of the nation. Convinced that China's population must be restricted to 1.2 billion by the year 2000, Deng called for fertility to be reduced below the replacement level, meaning less than 2.2 children per couple. It was decided that the only way to get a TFR of 2.2 was to insist on a TFR of 1, given the many uncontrollable variables of a diverse population. This was a remarkable moment in history, if you think about it. Here was a government schooled in the propagation of all things Chinese, of China at the center of the Earth, suddenly resolving to dramatically hold back the number of Chinese. It was perhaps the truest insight into the Chinese economic personality, which so resembles that of the Japanese. The externally imposed communist ideology had been peeled back, just slightly. Underneath lay the previous capitalistic China, naked and unblushing and scared. The equation had been worked out by an inner circle of academicians on computers.

One of those scientists, Dr. Tian Xueyuan, Director of the Population Research Institute of the Chinese Academy of Social Sciences in Beijing, described for me the momentum of the final analyses. It came down to two very obvious projections tabulated far into the future: a two-child scenario versus a one-child scenario. What Tian and two colleagues discovered was that China's population in the year 2050, if computed according to the one-child norm, would equal no more than China's population in the year 2015, if driven by a two-child norm. The implications were terrifying in every respect—thirty-five years of lost economic time. What was even more unsettling was the fact that China had—and most assuredly still has today—the overwhelmingly difficult task of even attaining fertility replacement, the two-child norm, let alone the vastly more distant one-child ceiling. Tian's findings were published in the *People's Daily* in 1977. That newspaper enjoyed a circulation of 4 million readers. No one could ignore the fact that if population continued to inflate, it would only undo all of China's hard-won progress to date. Per capita income, the true value of the yuan, and the country's ability to feed itself, would be greatly compromised. That meant another famine, the stench of millions of recently dead still very much lingering in the memories of all Chinese. Nobody wanted to repeat that. Nor was China prepared to ask for help from outside donors. Communism had to go it alone. The fact that the birthrate in China had already fallen to its (then) lowest in history—from a TFR of over 6 to about 3—was not good enough. Fertility momentum was driving the country toward catastrophe.

Moreover, Chinese scientists were deeply concerned about the environmental consequences of those computer projections. Indeed, the one-child policy was an ecological strategy of unprecedented proportions and, one would have to admit, of courage based upon an old Chinese proverb that cautions, *"If we don't change our direction, we'll end up where we're headed."* Deng's policy went into effect. On September 26, 1980, an open letter was issued to the Communist party membership and cadres of the youth league urging them to set the example by having no more than one child. In September of 1982 in a report of the Twelfth National People's Party, China determined to keep its population within the limits of 1.2 billion by the end of the twentieth century. That became the magic number of family planning.

The country's leaders knew very well that a one-child policy would not be taken lightly down on the farm, where it was at odds with several thousand years of Chinese Confucianist tradition. For probably 50 percent of all farmers whose first child had been, or was to be, a girl, Beijing's new edict would eliminate their chance to bring a son into this world—a male who could help generate an income on the farm and look after the parents in their old age, the customary practice in a nation almost entirely lacking social security. (There was some protection, depending on the wealth of the commune. But, in truth, it was unpredictable, and at best austere. Under the so-called "five-guarantee scheme" of rural old age security, every one of China's more than seven hundred thousand villages must provide "grain, fabric, medicine, old-age care, and burial" for any residents requiring it, a nebulous wish list that has not ensured much confidence in those who would forsake the solidity of a son.[25])

And there is an altogether unexpected irony to the issue of children and poverty. When reporter Sheryl WuDunn returned to her ancestral village in China she met a forty-five-year-old, recently married gentleman who wanted his first child but felt pressure from his wealthier neighbors to wait until he'd made more money. In five years, said WuDunn, the village of Pan Shi (population of four hundred) had seen 80 percent of its farming occupants acquire televisions. However the poorest residents were in deep trouble. One couple had lost eleven children and their land, and were evidently being sustained by cousins on a diet consisting, primarily, of "rice porridge with salted fish and a bit of vegetables." The village itself was not apparently contributing much, if any, old age security to the poor.[26]

The one-child policy collided with a number of pre-communist Chinese mores that had governed the country since antiquity. Considering the fact that most women in agrarian societies do much, or most, of the work (a fact being exploited now in the technological sectors of the Pacific Rim), the persistent male-dominated pattern of belief that holds that the woman provides less than the male is perversely erroneous. But that distinct bias, with its corresponding false dependency on the male child, has firm roots in China and in a custom which traditionally prompted the young bride to leave her family and live with her in-laws, a condition endemic in India, as well. The habits of mind favoring males date all the way back to that amorphous, mythopoetic period in early China, where poetry, law, and the workings of the government seem to have been fused—all firmly prejudicial toward women. For example, in the Confucianist annals, *The Five Classics*, it is written, "The female was inferior by nature, she was dark as the moon and changeable as water, jealous, narrow-minded, and insinuating. She was indiscreet, unintelligent, and dominated by emotion. Her beauty was a snare for the unwary male, the ruination of the states."[27] In the fundamental ethic of the Chinese yin and yang, the feminine yin refers to all things lowly and debased while the male soars in prestige. Moreover, ancestor worship, an obligation that was incumbent upon every socioeconomic class throughout the country, whether kingly or indigent, was the duty of the eldest son.[28] It was always assumed that ancestors were present at the ceremonies, watching, waiting, judging. According to *The Classic of Filial Piety, The Doctrine of the Mean, The Great Learning,* and in various works of the revered Mencius and Li Ki, the son nourishes his parents while they are alive, mourns for them by all the appropriate rites when they are dead, and sacrifices to them throughout his own life. He has communion with the dead. The male performs his military service, pays taxes, and like the land itself, is construed by Chinese society as the major source of wealth in the country. Without the male heir, the ancestors themselves die out. The son ensures the immortality of his parents. It is the law of nature, according to Mencius and Confucius. This is why certain popular skits, comedy routines, and songs instituted under the early Chinese Communist (pro-female) family planning regimes conveyed such sentiments as " . . . Confucius and Mencius and their ilk are murderous executioners. . . . "

In the interests of a higher standard of living, education, and general harmony, the communists completely overhauled the ethical system. Whether in China, Taiwan, Hong Kong, or Singapore, by the early 1950s the woman had suddenly risen in prominence and in the equality with males now guaranteed her. Yet, in an important survey of rural reproductive trends, communism notwithstanding, the Confucian ethic was seen to be at work in at least 70 percent of all rural families having two or more children in today's China, a figure exceeding 30 million newborns throughout the 1980s. These were children with existing siblings born to couples that were questing after sons. In another survey of ten villages spread throughout ten provinces, over 45 percent of couples questioned showed a desire for three or more children, a throwback, says demographer Li Jingneng, to China's long-standing notion that filial piety was best served by perpetuating the clan; that the greatest sin against one's parents was lack of posterity.[29]

It is ironic that in much of China, the daughter is still deemed incapable of carrying on most of the manifestly religious and tradition-fostering tasks of the family, rites which cement Chinese culture and ensure continuity from one generation to the next, when in truth, her fertility is the key to everything. Yet, the Confucian ethic suggests that all primary things—the method of traditional sustenance inherent to agricultural yields, to good fortune, and peaceful immortality—are the male's domain. When the parents have died, the son has an ancestor tablet inscribed with their names. It becomes his "soul tablet" throughout life. Traditional Chinese believe each man has three souls. The first adheres to the body at death, the second is merged into the stone tablet, and the third journeys to a sort of purgatory. Throughout this process, the dead parents are watching, taking stock, laying colossal guilt wherever it is considered appropriate.

When the bride went to work, literally, for her mother-in-law (the all-powerful lao-nien), where she was expected to bring into this world male heirs, she was said to be "returning home," fulfilling her duty to the new ancestral line, which probably dated back centuries. Traditionally this new family of hers often encompassed six generations all living together. It was her sacred responsiblity, the cosmic way, to continue the lineage. Someday she too would become a lao-nien—old, despised by her daughter-in-law, finally in control of something in her life. Should she fail in her mission to give birth to a son, she was solely to blame, and it was grounds for her husband throwing her out, just as if she had stolen, been promiscuous, talked too much, or shown signs of some incurable disease, all considered to be offences against the male and punishable with banishment. Her only hope of obtaining any financial satisfaction under the law was if her own parents were already dead, or if some truly odious personality quirk on the part of her husband could be proven. That, of course, did not include the male's long-standing right to keep younger concubines, which the wife had to honor.[30] So thorough was the expected loss of a young bride to her parents that it was said, Why weed another man's garden? In other words, what point in investing in one's own daughters?[31] Of the more than 70 million estimated malnourished individuals in China, it is believed that the vast majority of them are female.[32]

Adding to this tyranny is the long-time Confucianist dogma that holds that male feelings are irrelevant to his stature or effectiveness as a family ruler, a notion first intoned in the popular Confucianist text *Record Of Rituals* from the second century B.C. The male was called the *jiayan*, "the strict one in any family." It also implied that he was free to treat his spouse unemotionally, a fact consistent with the practice of foot-binding. Filial piety, the devotion of a son to his father—the mainstay of traditional Chinese dynamics—tends to repress all emotional connections, replacing them with the cold and elaborate injunctions of duty. "Thus," writes Chinese sociologist David Ho, "they (the sons) may arrange for an elaborate funeral (including the hiring of professional mourners) for their deceased father toward whom they have little love."[33] Ho foresees the coming bankruptcy of this emotionless father-son system—a generation gap, possibly the first in Chinese history, in which the next generations prefer to express their love, their feelings, to elevate women to equals and eschew the duties imposed by the Confucianist dark-ages. That could well translate into fewer children. This is precisely what Mao and Deng had in mind.

Mao Zedon had tried to eliminate Confucianist thinking from China. Communism abolished footbinding and made it possible (as of 1950) for women to initiate divorce proceedings. The following year, more than one million women got divorced. It was just one measure of the pent-up female frustration. But while Mao and his cohorts found a receptive audience among women who were clearly tired of their religious servitude, particularly in the cities, the men were less easily converted. It proved impossible to expunge age-old religious habits, ancestor worship, belief in reincarnation, or the "woman's place" in Chinese society. While more and more women entered the labor force (36 percent by 1981), the one-child policy became explosive. There was increasing resistance to it. So pervasive were Beijing's manipulative inroads throughout communal China, where a monitoring infrastructure was adamantly in place, that many felt it was tearing the sense of Chinese identity to shreds. When officials in Wenzhou City started awarding bonuses to parents with only daughters, the initiative failed. What happened?

"Having sons is what women come into the world for," said one woman, trapped in an age-old syndrome of oppression and exhausted female compliance that neither Mao nor Deng could erase. "What's the point of it all if you don't have a son? It's what we live for."[34] Under the *Woman's Rights Protection Law* in China, women have the right to give birth to a child under quality conditions, but not with fertility freedom. The *Marriage Law of 1980* makes it the responsibility of every couple in China to practice family planning, but the real thrust and burden of the law devolves to women. The granting of equal status to women has even been viewed by some as a new form of male oppression of women, the creation of "supermoms" who now are having to work harder than ever before.

The same Confucianist sentiment is at work in China's much emulated neighbor, South Korea, with its remarkably low total fertility rate and rapidly escalating per capita income. Korea is a good example of what the Chinese would like to see occur on the Mainland, though of course in much grander style and with less ecological turmoil. While women in South Korea have been hailed as "the backbone of industrialization," Korean tradition relegates them to second class citizens, subordinate to their fathers, their husbands, and finally to their sons. Their salaries and benefits in the workplace stagnate far below that of their male counterparts.[35]

The myriad forms of bias against women are everywhere acknowledged in China, but many are now pointing out increasing evidence of change. For example, one family planning worker in Yantai City was officially quoted as saying, "Some families were not on good terms because the wife had borne a baby girl, but now family members had made up with each other."[36] This is one of the marvelous and oblique tendencies of the Chinese—to speak so succinctly (one is tempted to say, guilelessly) as to reverse thousands of years of tradition, just like that. It's a quality of mind that pervades Beijing and defines the word *expediency*.

In places like Taicang, the realities of expediency had prompted the government to first lay an appropriate groundwork for its population reforms. It did so by providing clean drinking water for Taicang's residents. It then moved on to the administering of immunizations and better health care and finally to integrated family planning. Fertility control was exercised according to a system of incentives and disincentives predicated upon the national leadership's realization that the only way to curb China's population problem was to somehow monumentally alter a history of Confucianist ethics—to break up that time-worn pattern of multiple births whose sole goal had been the perpetuation of preferably two or more males per family.

In 1979, the year the one-child policy went into effect, there were nearly 8 million induced abortions, 13.5 million IUD insertions, and 7 million sterilizations throughout China. Women of childbearing age who already had one child were encouraged to be fitted with IUDs. That was the law. Infanticide was rumored to be on the rise everywhere as a result. Many couples whose first child was a girl chose to either not report her birth or give her up for foreign adoption or possibly kill her. The data is unclear. Most believe that cases of infanticide were negligible. Others are not so sure. According to *Washington Post* correspondent Michael Weisskopf, reporting in January 1985, doctors in Inner Mongolia routinely killed second children (by horrible means) under pressure of the one-child policy. Other reports of lethal injections continue to surface.[37]

Adoptions—almost entirely of girls—more than doubled after the one-child policy. Gender-specific abortion is illegal in China, yet there seem to be few other explanations—other than infanticide—for a sex ratio indicating that at least five hundred thousand to six hundred thousand girls were officially "missing" each year in the mid-1980s.[38] Because the country manufactures nearly ten thousand ultrasound-B scanners a year, it is argued that the gender abortion theory seems somewhat plausible. But there is little data to support it from any of the provinces. In fact, the Ministry of Public Health and the State Family Planning Commission have made it illegal to even tell parents the sex of the fetus.[39]

In 1981 China's population was the first in history to exceed one billion, up from 583 million at the time of their first census in 1953. The number "one billion" sent chills throughout China's scientific community. A wave of memorandums issuing from Beijing again laid down the law: one child, that's it! Demographers and economists within China had not been prepared for a figure of one billion, but rather were looking toward 900 million. The corrected data meant that the per capita economic figures dropped instantly by 10 percent. Frightened, the country now feebly clung to a family planning vision that would hold back growth at 1.2 billion, and then shrink the country's population toward an ecologically sustainable size, with each statistical family size being approximately 1.5. This was no bland doctrine or far-off volition, but an urgent calling, much akin to the sense of a military invasion, or civil war. The leadership knew it had to take up anti-natalist arms to defend itself. By the mid-1980s, following a massive mobilization of the family planning system, there were indications throughout much of the country that TFRs had indeed dropped to 2.1. This was an ephemeral characterization, however.

Indeed, civil war aptly describes the internal difficulties. Resistance to the one-child policy was widespread enough that allowances had to be made throughout rural China almost from the beginning, exceptions to the law that have harbored a de facto two-child policy. Without this latitude, there might have been actual insurrection, the second in twenty years.[40] In 1983, 2 percent of all Chinese—21 million people—were sterilized and 80 percent of them were women. A year later, the government eased off slightly, issuing a directive that exhorted family planners to display "an attitude characterized by the heart of a dear mother, the advice of a grandmother, and the love of a sister."[41]

Madame Peng herself reiterates this sentiment. She told me that being female made it easier for her to understand other females. In her interactions with her staff, she said, "I always make sure to deal with females as if they are my sisters, mothers, and daughters." Madame Peng's "staff" comprises millions of female associates. This is one woman with a big heart.

Back in the 1980s, just as this tender leniency was purporting to infiltrate national policy, a new set of disincentives was simultaneously introduced for those couples having more than one child, thus generating conflicting signals and revealing just how difficult it is to coherently dictate personal sexual habits to over a billion people, no matter how honorable or perceptive one's intentions.[42] The impenitent mother of three children would have 80 percent of her family's income taken away under the new punitive measures. However, if the mother "willingly" submitted to sterilization, only 50 percent of her income would be withheld. Multiple offenders risked losing everything—dwelling, food rations, even their license to work. One official in the Ministry of Agriculture explained to the press that temporary coercive measures were actually a "philanthropic and wise policy"; that from the wider point of view, it was imperative that China act responsibly to curb its population explosion. Farmers, he said, were impervious to logic. They had to be forced to control their family size for their own good.[43]

Other conflicting signals from Beijing in the mid-1980s further bewildered those charged with carrying out the one-child policy—namely, a lowered age for legal marriage, and the decollectivization of the farming communes. These were blatant collision courses that immediately increased birthrates throughout the country by making it impossibly tempting for farmers to employ as much labor as they could—their own children, plural. In some areas, land was actually allocated according to the number of laborers working for a family, without delineating whether those laborers were family members or not. The contradictions cascade. Early marriage is now on the rise, though the law stipulates the age of twenty-two for men and twenty for women. Yet, in 1990, an estimated 3 million girls between fifteen and twenty-one got married.[44]

In many parts of the country, as testimony to the collapse of classic demographic transition theory, rising per capita income has actually encouraged more children, not fewer, creating a whole class of "illegals," people who can now afford to pay fines associated with having more than one child (in the case of those not otherwise exempt from the law for various reasons) assuming they are caught. Many are not caught. These internal refugees drift into so-called "floating population islands." They are modern individuals who have craftily eluded the census takers and the police, and who have resisted registering their presence and family size with any local magistrate's office, bypassing entirely the bureaucracy of population control, much like white collar tax evasion in the U.S. There could be as many as 100 million of them. Not all floaters have money; in fact many do not. This latter group comprises migrants dispossessed of their land or their rural occupations, searching throughout urban China to better their work situation, dragging their children along with them.

While the economic boom is in full swing, China is beginning to fall prey to its side effects. In many areas per capita income gains are quickly being negated by rising inflation, a source of grave concern in Beijing where credit regulations remain lax but where the memory of previous inflations, civil wars, and the toppling of governments is fresh in many minds.

For now, however, throughout greater metropolitan Shanghai, which includes Taicang, all economic indicators are bound for glory, while 1.2 million trained family planning experts, associates, and volunteers, continue to knock on doors, undaunted by the country's outrageous fertility momentum.

Every mother in the city theoretically enters into a contract with her family planning section. She agrees not to have more than two babies. Chances are she will have only one. One is the law, two is the exception. Exceptions and exclusions are confusing in China. Beijing's

State Family Planning Commission, the official agency charged with holding back the country's virtual tidal wave of newborns while medically and psychologically easing the fertility burden for women, has recognized for years that the one-child policy is fraught with imperfection; that degrees of recalcitrance persist throughout the country. Beijing has caved in, decentralizing authority, turning over the day-to-day execution of the one-child policy to local governments, which are free to exercise their own judgment in exacting incentives and disincentives. Nothing is uniform. In some areas, incentives may include monthly payments for fourteen years, or priority medical care, or schooling. In the 1980s rural couples who pledged to have just one child might be granted land or more grain. Similarly, those who failed to comply were likely to be penalized in the job market or in terms of medical care or housing. Get pregnant, loose your roof. It might even mean the cutting off of water and electricity.

But all that is changing now in capitalistic China. In Liaoning Province in 1984 and 1985, second births were authorized when the first child was a girl.[45] Many provincial and city bureaucrats ignore Beijing. Others have proven themselves relentless in their pursuit of violators. Madame Peng expressed it in a different way. "For manual work you need manpower," she told me, going on to cite this age-old platitude as the primary reason for excluding rural farmers from the one-child policy.[46] Eighteen provinces now have formal allowances for two children, she explained.

In actual fact, family planning in China is an enigma. It varies considerably according to the province, and local conditions. In economic boom times, Beijing's primary authority is its military manpower, as well as its ability to engage in international economic commitments, gaining most-favored nation status for low-tariff export of goods, and so forth. Otherwise, domestic policy in China, as in most countries, is actually a labyrinthine confederacy that is driven not so much by laws as by individuals.

One of those individuals particularly impressed himself upon me. In the last emperor's childhood home, a beautiful forested estate within Beijing looking out upon a semi-frozen lake, I sat one Saturday morning in February beneath a large Ming Dynasty landscape painting and spoke with Dr. Qian Xinzhong (pronounced Dr. Chen). It was the last day of the Year of the Cock celebrations. Outside in the cold wind, hundreds of Beijingers browsed through an outdoor vegetable market, buying and selling pears, apples, squash, broccoli, cauliflower, various mushrooms, tomatoes, garlic, onions, potatoes, lettuce, and cabbage, much of it grown in countless little greenhouses throughout the city. Comrade Qian, his white hair cropped handsomely, forehead gleaming, his bearing and bonework reminiscent of nobility, had recently published his book, *Renkouxinpian*, meaning, new look into the population problem, in which he spelled out the history and future of China's demographic situation. In 1984, when Chinese population control was at its fervent height, Qian—along with Indira Gandhi—was handed the first United Nations Population Award for his "outstanding contribution to the awareness of population questions and to their solutions." His solution—temporary in nature he repeatedly stressed—was tactful, but firm compliance with the one-child policy, though allowing for eleven special exceptions. U.N. Secretary-General Javier Perez de Cuellar cited his own "deep appreciation" of both Qian and Gandhi for having mobilized their respective countries against the greatest threat facing humanity. For its praise of Qian Xinzhong, UNFPA was bitterly criticized by many inside the Reagan Administration, which would go on to derail any American support of UNFPA a few years later based on the allegation that UNFPA had ignored Chinese human rights abuses in its pursuit of lowering fertility in the country.

There are ethical grounds for questioning Chinese authorities in certain provinces where the State Family Planning Commission's stipulated rules of behavior have been misinterpreted or abused. The very euphemisms used on occasion by Chinese family planning officials invite concern: phrases like "remedial measure" that in fact refer to coercive abortion. The Chinese reject the criticism that their policies are coercive or their statements deceptive. It all depends on one's definitions of what is right, and what is essential. Chinese family planners, Qian more than most I've met, are looking far into the future, which means, at the very least, a few decades ahead.[47]

The Chinese do not hesitate to point to abuses of a gigantic bureaucratic system which employs nearly fifty million paid and unpaid workers.[48] But they are quick to remind their critics that the U.S. engenders even greater transgressions: the laws forbid murder, yet the country has the highest murder rate of any nation. Rape, child abuse, and hard drugs are illegal, yet nonetheless rampant in the U.S. The litany of contradictions is inherent to the democratic ideal of upholding individual rights, says the Asiatic mind; the ideal suffers from free market vagaries and the hedonism of a system that encourages the person to seek success, irrespective of his connection to community or country. Thus, the onset of multinational corporate culture with loyalties to no nation; of an isolationist generation that has eschewed family and ancestors and the greater good of the people (noting the prevalence of divorce in the West). Of course such doctrinal presuppositions are all breaking down in 1993, as China models itself precisely after American capitalism. Perhaps more importantly, the Chinese argue that they do not have the same leisure to indulge all of the niceties of human rights, though they admit that Tiananmen Square was a disaster and insist it could never happen again. But they also recognize that there are simply too many people in China, they are in a state of emergency, and they owe it to themselves, and to the planet— Americans included—to reduce their number with due diligence. The different phraseologies lose clarity not only in the translation, but in the varied perspectives that politicians, social scientists, and the public at large are likely to hold.

But, as one senior UNFPA representative in Beijing told me, "the Chinese need us." His goal, and that of a sadly underfunded UNFPA in general, has been to work diplomatically behind the scenes to assist the country on a compassionate path toward population stabilization. In China this has not been easy, partly because of the many critics on the outside who may not have thought through the total picture. When China presented its deeply felt concerns about a global population crisis at the Mexico City conference in 1984, the United States response was one of marked antagonism. According to Reagan officials, fertility rates had a "neutral" impact. It was not the fault of reproductive behavior that socioeconomic conditions were miserable in some countries but of badly conceived governments. There could be no dialogue on the issue, in other words. The Reagan position was emboldened by a spineless report from the National Academy of Sciences issued by the Working Group on Population Growth and Economic Development that insisted that population growth posed no crisis. China, along with Indonesia and India, embittered and incredulous, went its own way, pursuing its own deeply considered policies with a steadiness of hand and purpose, though with no encouragement from the country club Republicans in the U.S. who instead insisted that there was no population problem, at least not on any golf course. Until the Democrats came back into power, U.S. leadership was looked upon by China and much of the Third World as sadly uninformed.

Clad elegantly for winter, proud of his one grandchild, the elderly Qian radiated a career of devotion to the ideal of population stability. Qian adopted two of his four children, the offspring of a beloved colleague who died young. In 1964 he became Minister of Public Health. During the Cultural Revolution he was prosecuted as a member of the old guard. By 1970 he had been "rehabilitated" and had resumed his post, which he kept throughout that decade, a period when those who resisted compulsory family planning were considered selfish "counter revolutionaries." From 1982 to 1984 Qian was the state family planning commissioner. Though he himself takes little credit, many see him as one of the leading architects of the one-child policy and its most clairvoyant executor.

When Qian took office, there were nearly 900 million people in his country. The rural TFR, accounting for 80 percent of the nation, was a staggering 5-plus. In the cities, three children per couple was normal. There was, according to Qian, an air of desperation in Beijing, as if the whole country was heading toward the Armageddon Mao himself had feared, some unimaginable ecological disaster induced by 3 billion Chinese. Actually, it was quite imaginable. A mere century before, such a disaster had befallen China, confirming the worst fears of Thomas Malthus' first edition of *Principles of Population*. Between civil war and famine conditions, an estimated 70 million Chinese died between 1850 and 1880, following the Taiping Rebellion of Hung Hsiu-ch'uan, who considered himself the younger brother of Christ, and whose utopian blueprint for communal life and agriculture was adopted by Mao himself. Yet by 1901, after the disastrous fallout of the Taiping Rebellion, China's population had rebounded to a seething 426 million. By Qian's time, it had more than doubled. The Cultural Revolution had totally abandoned any pretense of population control. Food production was not keeping pace, arable land was exhausted, forests were disappearing, pollution was smothering the cities, increasing numbers of malnourished rural poor were voicing discontent. These were all clear harbingers of a Chinese doomsday. At the same time, critics throughout the world were beginning to wonder what had happened to all the Giant Panda bears, and to express concern that China's uncontrolled burning of coal was contributing an undue amount of carbon emissions to the planet's upper atmosphere. Other developing countries with persistent hunger problems were concerned that China's population might unleash a competitive famine, usurping control of food aid and diverting the flow of scarce Africa-bound grain supplies from the Western powers.

Qian's Ministry set about to incite a consciousness-raising revolution. It masterminded such slogans as, "Promote vigorously the policy of one couple, one child, and control strictly and stringently two babies per couple, and strictly prevent one couple, three babies." I told him I did not think it rolled off the tongue. Of course it has a different elocution in Chinese.[49] But Qian did more than promote crafty slogans. Recognizing the inevitable dissatisfaction with the one-child policy in many rural areas, he went to the people.

"I held a discussion with farmers and their wives and posed the following offer to them," he told me. "'You are twenty-five years old now and we (Beijing) are asking you to have just one baby. But when you are fifty, you will be allowed two grandchildren. What do you think?' And they all answered, 'No problem! If in the year 2000 our children can have more freedom, then we're willing to make some sacrifice now.' I realized from that discussion that our policies must be explained clearly and simply to the people," Qian explained.

Following his deeply moving encounters with villagers throughout the Chinese heartland, Qian returned to Beijing, where he and his many colleagues set about to implement a number of (some would say self-contradictory) exclusions from the one-child rule, carried away by a sense of destiny, or compassion, or both, I imagine. But he also knew what he was doing. The "exceptions" were actually cultural incentives, spawned by a sense of social justice, fairness, and realism. Of course, it was a gamble, knowing the power of decimal points the way he did.

As of 1974 all contraceptives were made free to married couples in China. Special "contraceptive stations" for distribution were created in every province, organized and managed by the local family planning committees. Later on the Ministry of Chemical Industries and the State Pharmaceutical Administration got involved to upgrade and increase the flow of contraceptive supplies. At the same time, much in the manner of the barefooted doctor, local cadres, fortified with supplies, would go from house to house giving them out and checking to see that IUDs were working properly.[50]

Qian proposed legitimizing eleven exceptions to the one-child policy, which he described for me. Minorities (at least fifty-five official ethnic groups in China) should be allowed to have more than one child. That was only fair.[51] If the baby was handicapped, but not hereditarily so, then the couple should be able to have another. In the case of second marriages, the bride or groom who had not been married before should have the right to bear or sire a child, whether his or her mate already had had one or not. If a couple was previously judged incapable of bearing a child and as a result adopted, but then found themselves with a baby on the way, the woman would be allowed to carry the child to term. If either spouse was the one-child descendant of two generations of one-child families, then he or she could have a second. Handicapped military veterans could have two. Mountain people with heavy household work to do could have two or more. The same with fishermen in coastal areas if they could show just cause. If only one brother among many was capable of siring children, he would be allowed to have more than one. Chinese who came back to China could also have two. If a groom chose to live with his wife's family, they could have two.

In 1984 the Chinese public perceived that a major policy shift had occurred, spawned by a Central Committee Document known as "No. 7." The rumor spread that Beijing was easing up, exceptions to the rule were proliferating, and the crisis had gone away. Everyone, it seemed, was now allowed to have two or more children.[52] The Family Planning Commission was not able to put a cap on the rumor mill, or not quickly enough. China's population growth rate lurched from 11.2 to 14.8 per thousand in one year. That one hiccup in the system will translate, according to Qian, into a hundred million more Chinese in the year 2050. Such is the power of population momentum.

Professor Ma Yin Chu, President of Beijing University, had stated his own forward views back in the 1950s. They were ignored. According to Qian, had the eminent professor been listened to, there would be 200 million fewer Chinese today. Qian would not go so far as to name the true brainchild behind the one-child policy. But he stated with self-effacing animation that he and his cohorts had "successfully prevented the birth of 260 million people" in spite of many setbacks.

Qian knows his history. He is now convinced that the slightest easing up on population control policy will translate into a hundred million more newborns before you know what's hit you—a sudden cluster the size of Mexico's population, in other words. Qian introduced both practical and hard science to family planning to prevent this. It was the only way, he believed, that society would take this numbers game seriously. Coming from the medical field, he was the first to acknowledge China's need for western technology in the birth control arena. Pride was useless. What was required, he alleged, were better surgical techniques and plenty of contraceptives. Today, China has over forty contraceptive factories, more than half of them upgraded by UNFPA, and produces nearly a billion condoms, several billion pills, 22 million IUDs, and 300 million vaginal foam suppositories every year.[53] Chinese researchers have developed a new vasectomy technique that takes but five minutes and can be accomplished with blunt forceps under a lical anaesthetic. As early as the 1960s, the Chinese were the innovators of the vacuum abortion technique.

There is no question that China in 1993 has the scientific and technical acumen to accelerate the success of its family planning programs. The country's proven ability to deliver health services to most of its residents is among the most remarkable in the world. But there are recognizable areas where improvement is crucial if China is to more effectively wrestle with its fast growing population. UNFPA in Beijing estimates that by shifting from the old SSR (stainless steel ring) IUD to the far more effective Copper T 220C and 380A, as they're called, that within a decade "35.6 million induced abortions, 1.4 million spontaneous abortions, 18.4 million births, and 16,300 maternal deaths" would be prevented. In addition, the technological change alone could lower China's TFR significantly—after just two or three years, by one-tenth of one percent. These are remarkable findings.[54] All that's needed to realize them is more money, an issue of grave contention. Indeed, UNFPA's very continued presence in China, after fourteen years there, is jeopardized by U.S. Congressmen who would naively apply the same failing American population ethics to China, where there is one-third the amount of good arable land and five times as many people as in the U.S. The threat comes as a result of allegations that family planners have once again escalated their efforts throughout the Chinese countryside, echoing sentiments published in a recent book chronicling China's system of forced abortions.[55]

In broadly considering China's past and future, the recommendations of UNFPA and those of Dr. Qian together seem to point in a viable direction. To my way of viewing him, Qian embodies the full paradox of compassion tempered by unflinching realism. He knows that the nature of finitude is ruthless in a land teeming with nearly 1.2 billion people. China is only a microcosm of the whole earth, its environmental constraints emblematic of the larger problem. Any choices made in Beijing will affect nearly every plant and animal species in the country, and be felt in countless ways elsewhere in the world. Qian and his colleagues thought deeply about the one-child policy, ever mindful of the larger sphere.

If one is to speak of empathy, remember that not only individuals, but the world, too, needs empathy. Similarly, if the world is to be unburdened, individuals themselves, one by one, must also be freed so that they can walk more softly, and in fewer numbers. How do you apply yourself, given such delicate and harrowing ideals? asked Qian. He'd outlined his response. Most couples (the exceptions delineated) should concern themselves with bringing up no more than one child, a child that is loved, fawned over, disciplined, and given every advantage. Adoption, by his own example, is an additional, much needed option.

But how does one ensure a preference for an ideal family size of one child, given the trends toward two and three children? For their part, UNFPA has stressed other coefficients of the family planning future in China, that de-emphasize targets, that reject coercion of any kind, but which are urgently focused on enhancing the overall quality of family planning, as well as maternal and child health care, especially in poorly served rural areas. In addition to their emphasis on new technologies and wide-ranging contraceptive methods (at least five), they also emphasize the need for women to work outside of their homes so as to delay or diminish the mother's desire for more children. In addition, UNFPA is supporting population research in some twenty-two universities in China, developing teaching materials for young students, and integrating literacy and income generation projects with family planning. UNFPA stresses the need for psychological support of the woman and has introduced the idea of "interpersonal counseling" for those grassroots family planners who go from door to door thoughout the country. There is nothing particularly new about such recommendations. They have been part of the working vocabulary in the fertility realm for forty years. The real issue is continuing to keep that vocabulary in focus.[56]

"I myself participated in the formulation of the one-baby policy and when the policy was implemented, at that moment I felt encouraged, I was hopeful," Qian said quietly, reminiscing on a long and remarkable career. I had asked him what he was doing on the very day that the one-child policy went public in China. Did he go out for a long reflective walk through town, pick up groceries, visit his grandchild, sit around and listen to the radio? Did he open a bottle of wine and get drunk, or take his wife to a movie? What were his thoughts? And I kept trying to place this smiling grandfather, his hands in the air, wrinkled and fine, gesturing with a deeply understated conviction, beside the entire population of the United States. That was the amount of human life he and his associates would actually prevent from coming into existence. The spiritual dimension of this achievement is, like all feats of the imagination, intangible, intimated, an endless deferral whose consequences we will never know. On another, more palpable level, we can well appreciate the significance and its causes and effects, much as with the invention of the zero, long ago, in India. A negation, like nonviolence, or nonintervention, or nonproliferation. This one man, part of an idealistic team, had done something profoundly important. Mikhail Gorbachev liberated a similar number of people throughout Central Europe, a great achievement that is proving to be slightly marred by human nature. But Qian liberated the unborn, a perfect act of conscience. Conceived in ecological terms, he was like a great general who preempted a battle that was going to be waged by another 260 million consumers against the planet.

Finally, he stated in an even, though unmeasured tone, "I did envision that the country could one day make real progress in controlling growth. It was only through implementing the one-child policy that we could, in a real sense, check the population growth in China. So when it was put forward I thought it was very necessary and I was very much encouraged. Many developing countries are not able to address the problem of overpopulation squarely and therefore I cannot be too optimistic about the future of the world."

Nor was Qian all that optimistic about China, either. "Due to various blunders . . . the one-child policy will have to continue for another ten-to-twenty years because we have not reached our targets." He was referring to Beijing's goal of checking China's population at 1.2 billion by the year 2000. It is not going to happen because of what Qian saw as the unfortunate easing up on the one-child policy. Consequently, China is looking at 1.3 billion or more by the end of this decade.

Throughout time, and time and again, China has faced the abyss. For now in the early '90s, it has resurrected an "interesting" economy in the face of strange odds. When I first visited China in 1978, there was a very hard and introverted edge about her. No other people seemed quite so aware of their population and environmental predicament back then. There was tacit alarm and grave uncertainty. A year later, the one-child policy was unleashed. Since 1980, like no other country in history, this nation has learned to live with more than a billion people. This is a vast experiment. No other government, with the exception of India, can really know what China is dealing with. No one would want to. The Chinese are, by and large, enormously sympathetic, radiant people. It is easy to admire, to be enamored of, to love them, horrific events notwithstanding. But this love, so to speak, founders in the demographic den of iniquity, where the sheer colossus of China and all its numbers pose a paralyzing burden on Chinese environment, and thus, on the very nature of what it means to be Chinese.

Every smiling photogenic baby dressed in colorful silken pantaloons, bobbling across the sun-bright courtyards of the not-so-Forbidden City, or riding herd on their parents' shoulders through bicycle-laden back alleys, will steal your heart. Every scene of grandpa kneeling on a street corner to tie his grandchild's shoelaces, or families out for their leisurely late afternoon stroll together, sucking on penny candies, fawning over their kids, is poetry in motion. These age-old value-laden images conjure a familiar China that is now caught between a crossfire of biology and dreams. Whatever its family planning successes to date, China is in no position to rest on its laurels.

## CHINESE ECOLOGY AND ECONOMICS

Throughout China one hears it aired in countless guises that as the country changes from a labor- to a technology-intensive economy, increasing wealth will supercede the population crisis; that no human being who generates income is too many. The so-called human deficit will simply disappear as per capita income has risen to a comfortable level. What about rural China, I asked repeatedly, where poverty is still widespread, the economic benefits of those few major cities in the country having generated little trickle-down prosperity? Even that is changing, I was assured.

In Fan Shan County, two hours west of Beijing by a brand new freeway, I visited one particular rural township dating from the Tang Dynasty and today comprising 325 households, where this trickle down formula in fact seems to be working. The various endeavors of the entrepreneurial collective bring in 12 million yuan per year (nearly $2 million). Eighty percent of the income goes back into the collective, a very high tax burden but nobody seems to mind inasmuch as it still leaves nearly $1,200 per household per year, or $400 per person, on average. Thirty-five dollars per month is upper middle class by Chinese standards. The villagers of Fan Shan grow rice, fruit, corn, yellow beans, and wheat, in addition to cultivating several large fishing ponds. They sell their yields to markets in Beijing. They also operate machine shops, and other industries. The collective's wealth holds out other benefits for each household which thus drive the per capita income into much higher spheres. Fan Shan is probably the near future of China, at least economically.

About ten or eleven children are born each year here. During the first month of a newborn's life, the tail of a rat is posted on the door to warn people who might have a cold or other infectious disease not to come in. After thirty days the mother will take the child out and there will be a celebration. After one hundred days there is another celebration. The child will go to preschool at three, kindergarten at five. By the age of fifteen, a young boy will ordinarily be expected to go to work. Girls start manual labor at eighteen. Ten youngsters from the village have thus far gone on for university training.

What do these young people want out of life, I asked? They want cars, or trucks, or a motorcycle. They want money to build a new house, or to buy a video camera, or travel.

I went to one such new house, that of Mrs. Ma Junrose. There was a huge poster of a romantic, faraway waterfall in the livingroom, beautiful cabinets, lovely tiles on the floor, eight plants, a television, 20 to 30 watt fluorescent lights, and neo-Florentine statues in the corners. The house has cost fifty thousand yuan. Mrs. Ma and her husband paid ten thousand yuan in 1985, then continued to renovate. She has two children and dreams that the daughter will work in an enterprise factory, and hopes for the boy to study in a university.

I was served a lavish meal and given presents and business cards. The town is looking for joint ventures from the outside. They'd love someone from California to get involved in the manufacture of their little cloisonnéd steel medicine balls, that—when you shake them—make sweet music. So far, no foreigners have invested there, though some one hundred domestic joint enterprises are in place.

The Ma's spend eighty-four yuan for one ton of coal and use up ten kilograms (twelve pounds) per day for their home heating in winter. That works out to eighty-five fen per day (eight cents). The collective provides water for everyone in the village, an increasingly difficult prospect. So far, water is free to the collective. In addition, the Ma's have a few electric heaters which cost them an additional three cents per day for twenty-four hours of electricity.

Over a lunch of sautéed green peppers, vegetable dumplings, garlic, scallions, salad greens, candied walnuts, and a bitter *kung fu jyo* drink, life in Fan Shan seems very pleasant. The sun is warm here, even in February. There is no evident air pollution. As far as the eye can see, in all directions, there is order, intensive cultivation, and sophisticated irrigation. Modest homes built of brick, wood, and aluminum roofing dot the Nebraska-like landscape. There are very few trees and no animal life to speak of—only some domestic chickens, pigs, and ducks, that will be slaughtered, though Mrs. Ma says she herself is a vegetarian.

We speak of the future problems in Fan Shan—the likelihood that more and more of the young will venture into the nearby big city, that agricultural yields will decline, that inflation will begin to diminish the profits of the collective, and that pollution will spread to their town. But one of the Chinese luncheon guests says enthusiastically, "If you can tell me the problem clearly then we have 85 percent of the solution already worked out."

China is at a $1.2 trillion GNP in 1994–95. Such economic stardom is considered to be the ultimate solution to everything, whether in Beijing or Fan Shan County. The International Monetary Fund in

May of 1993 calculated that China is now the world's third largest economy, with a $1.7 trillion dollar purchasing power. Compare China's (real dollar) economic growth with its neighbors for 1993—over 13 percent in China versus India's 4 percent, Indonesia's 6 percent, Singapore's 5.5 to 6.6 percent, and Thailand's 7.2 to 8.2 percent.[57] According to one estimate (the China State Statistical Bureau) the per capita income in 1992 in Shenzhen, the special economic zone near Hong Kong, adjusted for state subsidies, was Rmb 18,000 ($2500); it was $1600 in Shanghai, $1500 in Canton, $1400 in Beijing, and $1000 in Tianjin—thus representing a substantial economic gap in relation to rural Chinese. Chinese urban affluence is now on a par with most other Southeast Asian consumption, including that of the Thai and Malaysians.[58]

Yet in 1994 more than 80 percent of the nation earns three hundred dollars per year or less, while 20 percent of the country is earning much less than that, in some remote areas as little as eighty dollars per year. At least 664 counties, or 20 percent of the nation, are below the Chinese poverty line. It means that a collective like the one I visited outside of Beijing is as much an anomaly as Beijing itself.

Yet these focal points of wealth are triggering massive emulation throughout the country. In a region like Anwang, where villagers earn on average fifteen dollars per month, less than half the per capita income of those in Fan Shan County, there is no less striving after wealth and improvement. New roads are being built, new terracing for high yielding soy and corn hewn from the mountain slopes, new sources of electricity installed so that the people can mill their own corn with machines at night and make noodles for sale by day. Cisterns are being built and inoculations administered to the livestock to improve their health and number.[59]

What on the surface would appear to be the benign (or at least acceptable) aspirations of a developing country, a national preference for a modicum of comfort—*xiaokang shuiping*—and monetary gain consistent with the economic priorities of the rest of the world, in actual fact is taking a decisive toll on Chinese ecology. Among humans, it appears inevitable that to populate is to despoil. There is no comparable situation in nature, nor has there ever been. Five and a half billion large bipedal primates with truant diets and wanton desires, equipped with agile fingers and brains exceeding, on average, fifteen hundred cubic centimeters, the ability to procreate at any time of the year, compulsively engineering oriented, and in possession of destructive technology, have engendered an untenable burden on the Earth. At least 22 percent of this ecological debt can be gleaned from the current Chinese experience. Although it will be pointed out that from the perspective of wilderness loss, that region of the planet known as China has more to lose than any other portion of the planet with the exception of the rain forest quadrants.

Professor Li Jingneng, the Director of the Institute of Population and Development Research at Nankai University in Tianjin, points out several crucial ecological problems associated with China's fast-burgeoning population/economy. China currently possesses some 100 million hectares of arable land (11 percent of the country's total land area), which places China twenty-fourth in the world for arable land availability. This is approximately 0.09 to 0.085 hectare (less than a quarter of an acre per capita), a decline of nearly 50 percent in cultivated land since 1949. This contrasts with some two acres per person in the United States. At an average density of eleven hundred people per square kilometer and growing, the country is now (in agricultural terms) at least eight times more populated than the U.S. Keep in mind that "arable land" typically refers to total or partial cultivation. It may mean enclosed grasslands that are alternately plowed. But in China arable land refers to intensively farmed soils. Loss of grasslands constitutes an additional factor of overall deterioration. There is no apparent land left in the country that can be converted for purposes of agriculture. China thus finds itself with approximately 7 percent of all farmland on Earth, but nearly 22 percent of the human population.[60] And at least 70 percent of China's population is directly dependent on the agricultural sector.

Where there is agriculture, the existence and interdependent web of most species larger than insects and rodents are vastly diminished, or eliminated entirely. The remaining 30 percent of the population depends directly on industry and urban life, which once again tends to destroy most existing wildlife. Estimates on plant and animal life in China indicate approximately thirty thousand types of bryophytes, pteridophytes, and seed plants, twenty-one hundred species of mammals, avians, reptiles, and amphibians, twenty-one hundred species of fish, and some one million invertebrate types. If these NEPA (China's National Environmental Protection Agency) figures approach even vague accuracy, then China hosts nearly 10 percent of all known fish species, 10 percent of all avians, mammals, reptiles, and amphibians, and (an unlikely) 95 percent of all known invertebrates, making it one of the most biodiverse nonrainforest nations in the world. While few all-China surveys have been carried out, and little in the way of comprehensive analysis, it is believed that 257 animal and 354 plant species are currently headed toward extinction, short of Draconian lifesaving measures. These species include the Chinese rhinoceros, the wild horse, wild elephant, mandarin duck, leaf and golden monkey, high-nosed antelope, panda, snow leopard, Chinese dolphin, and alligator. Probably nothing can save the Chinese tiger from extinction, however. The country exported over fifteen thousand cartons of tiger bone tablets and three hundred fifteen thousand bottles of tiger wine in 1992. One kilogram of pulverized tiger bone goes into the manufacture of fifty bottles of wine. A brewery in Taiwan imports two thousand kilograms of the tiger dust every year from the Mainland.[61] Nevertheless, Taiwan continues to elude world sanctions despite unambiguous CITES (Convention on International Trade in Endangered Species of Wild Flora and Fauna) regulations.

Investigations in the northeastern province of Ambdo in Tibet, where population has increased five times in the last twenty years, have shown that vast regions of grasslands (as extensive as many European countries) are being rapidly degraded or destroyed. With the incoming Chinese colonists, and Tibetans themselves who are repopulating after their tragic decimation by Chinese troops during the 1950s, nearly every major species has been listed internationally as endangered.[62] The IUCN in Switzerland (International Union for the Conservation of Nature and Natural Resources, now called the World Conservation

Union) and the World Resources Institute in Washington, D.C. report at least thirty-one animal and bird species endangered in Tibet. Wild yaks have been machine-gunned for their meat by the Chinese in the Himalayas. Other threatened species include the snow leopard, the Asiatic black bear and gray wolf, the wild Bactrian camel and black-necked crane, Cabot's tragopan and the Sichuan partridge, the Tibetan macaque and brown-chested jungle flycatcher.[63]

At the same time, hunters from abroad have been invited to Tibet to kill wildlife for fees—thirteen thousand dollars for a white-lipped deer, twenty-three thousand dollars for an Argali sheep. The Ministry of Forestry is charged with enforcing a 1988 *Wildlife Protection Act* but evidently has no authority to curb the sale of pelts. A delegation from the International Campaign for Tibet's Human Rights found baby snow leopard furs for sale in various markets, some as cheap as thirty dollars each.[64]

China, like every other region populated by humans, also exerts a prodigious commercial toll on wildlife. As reported by CITES for 1988, China exported over 2000 live primates, 89,650 cat skins, and 65,665 reptile skins. Data was not reported for other mammals or avians, probably to avoid international censure. In addition, the Chinese diet in some parts of the country includes a number of endangered species whose populations—to say nothing of the individuals—can ill-afford such attrition. So-called Man-Han banquets have been known to include bear's paw, camel's hump, white crane, even elephant's trunk, as well as the highly endangered clouded leopard.[65]

Agriculture, industry, and commerce, to say nothing of animal experimentation, factory farming, and the countless other ways in which animals and whole ecosystems are exploited, taken together suggest a colossal substitution imposed by one species upon all others. As China's population grows, the substitution continues to intensify in other ways, further adding to the human hegemony of net primary production (NPP), the amount of natural photosynthetic vegetation on the planet which has been co-opted and usurped by human intervention. The current global NPP is estimated to be between 45 and 50 percent. Every year China converts some four thousand square miles of additional land into urban or rural development—new cities, new factories, new housing developments. At least 87 percent of the country lacks any foliage whatsoever. Another one hundred twenty thousand square kilometers are now desert, twice the amount since the time of China's independence.[66] In many urban regions, such as Shenyang— whose population growth rate is now as high as 3.0 percent annually— there are a mere four square meters per person of public green lands.

The rate of deforestation in China is now at least 1.2 percent per year of the remaining stands of forest, largely on account of the need for cooking and heating fuel, where inexpensive coal is not available, as it is in places like Fan Shan County. Most remaining woods in the northeast provinces of Jeilongjiang and Jilin, and much of the forest in the southwest province of Yunnan, for example, have been hacked to oblivion. The species that have been lost as a result have not been quantified or qualified but the amount of burned board feet and destroyed hectares each year in China is biologically devastating. The forests of Guanxi Province will likely disappear by 2010 or so. Forest cover in Sichuan has been reduced from 20 percent cover in 1960 to less than 13 percent in 1994.

The forces at work in China are symptomatic of human economic growth. Norman Myers has suggested that at the beginning of the agricultural revolution some ten thousand years ago, the Earth was covered with 62 million square kilometers of forests, half the ice-free land area of the planet. Today in 1993, at most, 41 million square kilometers of forest remain. Tomorrow it will be less.

As the forests are depleted and agricultural intensity increases in a continuous effort to provide edibles for a rapidly expanding population, the finitude of China's agricultural lands becomes more and more apparent. China's grain production peaked in 1984 and has been declining since then. In the year 2000, China will need 1.36 trillion jin of grain (1 jin = 1.1023 lbs). The resource is already in shortfall and as per capita requirements continue to escalate, the ecosystems surrounding agriculture will be battered further still.[67]

Since the 1950s arable land has begun to decrease, currently at a rate of three hundred thousand hectares per year in China. At the same time, 87 million hectares of grassland are degenerated each year, another 5 million hectares becoming desert.[68] As China's massive population wrestles with its declining food yields, at least 5 billion tons of precious topsoil are washed away annually, equivalent to 40 million tons of cost-inflating fertilizer. But of course there is no true synthetic equivalency for the loss of soil. In the poorest regions of the country, like northern Shaanxi and the Haixi area of Gansu Province, virgin lands have been devastated, whole mountains denuded, as poverty accelerates the population increases. One-third of all cultivated land in China is badly eroding. The Chinese government has undertaken over three hundred greenbelts throughout the country, but for now, deforestestation, according to Li Jingneng, is one and a half times faster than the rate of reforestation, despite some very impressive strides by the Ministry of Forests. Throughout the country, mature forests will have vanished completely by 2000. The total loss of forest to date is estimated at one hundred million cubic meters.

Coupled with China's population impact on its forests (and consequently upon its already scarce amounts of rainfall) is the burden of ground-based and atmospheric pollution from coal, a regional prelude to the country's substantial contribution to the global Greenhouse Effect. Seventy-five percent of China's energy comes from coal, of which 85 percent is burned without any pollution controls. China produces and consumes more coal than any other country in the world, 1.11 billion tons in 1992—25 percent of the world's total. This is easy for China to do: the resource is relatively inexpensive, and the country is endowed with some 50 percent of all known coal reserves. Chinese open mining pits are so deep in many places that any notion of replanting or restoration is out of the question, despite China's long history of such efforts.[69]

I was told by Dr. Qu Geping, head of China's Environmental Protection Agency, that the country plans to increase coal consumption by three times in the years 2015 to 2025. In a very rough correspondence, that energy escalation is defined as a 14 percent economic increase. It will also coincide with a quadrupling of the present number of Chinese-made automobiles (to some 13 million domestically produced vehicles). Taken together, these two energy-consuming impositions will utterly undo any carbon emission controls implemented in the West. Coal accounts for many of China's pollution problems— 80 percent of all particulate matter, 15 million tons of sulphur dioxide, and a suspended particulate matter content as much as thirty times thicker than that of the U.S. during the period when U.S. standards were virtually nonexistent! The pollution fallout rate in Los Angeles per square kilometer per month during the 1950s was approximately 7.7 tons—the worst it has ever been—equivalent to the level of Mexico City in 1994. But in China in the same year, levels have surpassed 100 tons! Beijing's air is said to be thirty-five times dirtier than London's, sixteen times that of New York. In Beijing and several other Chinese cities, the rate of cancer deaths increased 145 percent between 1949 and 1979 as a direct result of coal-based industrial expansion.[70]

Alternatives to coal, such as natural gas, and alternative fuels, or solar, are nowhere on China's horizon. China's CNOOC (the National Offshore Oil Corp.) has offered seventy-three thousand square kilometers of the East China Sea to more than twenty foreign companies on a royalty basis but so far economically recoverable oil is playing a minimal part in China's energy portfolio. At best total off-shore predictions are 850 million tons, or 6 billion barrels. But with China's energy needs increasing by some 7 percent per year, the oil—which of course carries its own environmental price tags—is not going to dent the country's reliance on coal. Biogas—methane conversion—is used throughout the country for everything from cooking fuel and electricity to drying boxes of silkworm cocoons and breeding mushrooms. But it is no substitute for China's heavy reliance on nonrenewable fossil fuels. Amory Lovins believes that China might have more natural gas than oil—as well as access to fields in Siberia. Siberian gas would be cleaner, and possibly less costly than Chinese coal. But in 1994, there is no indication that China—or Russia—is making the kind of deals that would guarantee a safer environment in Central Asia.[71]

China views nuclear power as clean and safe. So far, two nuclear power plants have been installed in an effort to compensate for the many difficulties of transporting coal from the country's interior to its heavily populated coastal regions. But nuclear is expensive and is therefore not going to play much of a role in China's energy scenarios of the future. As for hydroelectric power, China's leadership is divided. Only 10 percent of the country's estimated 378 gigawatt potential (the largest in the world) has been exploited, with seventy-six thousand hydrostations and nearly nineteen thousand dams in place.[72] But, according to Dr. Qu, hydro—small or large—is not being perceived as an answer to China's energy problems. Nevertheless, the potentially disastrous Three Gorges project of the upper Yangtze River has been approved, detractors have been thrown in jail over it, and the international controversy is intense.[73]

In other ecological sectors, 16 hundred billion gallons of sewage enter China's coastal waters every year. Most of it flows untreated. In the 1950s residents of Nanjing washed their vegetables in the local rivers. By the 1970s that was impossible. The fish and shrimp had gone extinct, the water was badly polluted. By the 1980s, to quote Qu, "the water turned into unbearable black stinking liquid."[74] Many of China's cities are experiencing massive drinking water deficits, Beijing to the extent of half a million cubic meters per day. One-third of all wells in Beijing are dry. Groundwater levels throughout China are dropping, in some areas as much as one meter per year, as existing wells are over-pumped. Excessive pumping has literally lowered Shanghai. Sixty-four percent of the city of Tianjin has also subsided. As of 1988, analysis of China's rivers, ground water, air, and coastal zones yielded extremely high levels of contamination. At the same time, irrigation expansion is necessarily shrinking. Many existing irrigation channels in the country are badly choked with silt.[75]

And the problems proliferate. Some 60 percent of all Chinese factories built during the 1950s are ecologically "backward" (as they are in every developing country, and wherever communism once prevailed). During the ten years of the Cultural Revolution in the sixties and seventies, environmental issues were bypassed almost entirely in favor of growth. Throughout China an estimated 536 square kilometers are given over to waste dumps in 1994. In nine cities in recent years, at least 150 serious chemical spills have occurred.

There are 173 Chinese cities with populations exceeding three hundred fifty thousand. An effort is underway to "eliminate the difference between workers and peasants, and between the city and the countryside."[76] It is part of Qu's "synchronous development" concept. This homogenization is intended to accomplish two things: first, to increase the numbers of workers in so-called Township Enterprises, industries that can earn more profit than traditional labor-intensive agriculture; and second, to build up such enterprises out in the country for disenfranchised agricultural workers so that they will not necessarily come to the big cities looking for work but can remain in their own home regions and find alternative employment. It has been estimated that as many as 40 percent of the country's 800 million farmers will, by the end of the 1990s, have to work in other industrial sectors across rural China. That will probably mean a greatly increased environmental burden; the Chinese human population appropriating more and more vegetative cover and net primary production areas.

This is the crux of China's ecological dilemma: in stark terms, *too many people industrially or agriculturally overtaking every last corner of the country.* In Shenyang, the capital of Liaoning Province in Southern Manchuria (the Northeast), 6.5 million people have created one of the country's worst environmental mires. There is no sewage system; the central river is dead; thirty-two meters of groundwater have been lost; new industrial suburban satellite communities of between two hundred thousand and four hundred thousand are springing up in sensitive watershed areas; there is no overall protective plan; and the World Bank ranks Shenyang as one of Asia's seven most polluted cities. According to environmental consultant Jeff Girsh who, with a team of others, has been studying Shenyang in an effort to help the local government plan for future development, air and water pollution in that city make Beijing look downright desirable, by comparison.

As China's "homogenization" evolves, little Shenyangs and Shanghais, like cancer cells, will spring up throughout China. With between 160 and 200 million more people every decade wanting more and more prosperity, there would appear to be no other alternative. In the early 1990s, 24 new cities and 921 new towns are being built every year. From 1949 to 1979 Chinese housing encroached upon 709 million square meters of land, previously rural, but in just the last decade it has covered that much again. Roads are being built everywhere. But the real crisis of land conversion and deterioration in China is coming about as a result of the transfer of pollutants and population to rural industrial complexes. It is not so much a decentralizing phenomenon, say the Chinese, as one of rural urbanization, so huge is the population.[77]

Satellite data taken in the eighties over regions like Shanghai already reveal the distressing trend toward what I would call Total Human Appropriation Regions (THARs). They resemble the so-called "Dead Zones" in the ocean, where human effluents of one form or another have destroyed most or all marina biota. There are now industrial regions in China whose air is so polluted that even satellite reconnaissance fails to yield a resolved picture for months each year. These cities, such as Benxi with a population of over one million, have simply vanished beneath a sea of unrelenting air pollution.

Curbing the ill effects of urban appropriation is not easily systematized in China since each province has its own specific population agendas and growing socioeconomic disparities to contend with. Furthermore, each province is empowered to generate its own pollution-control laws, making it next to impossible to engender unified ecological standards throughout the thirty government regions. Part of the problem is China's lack of hard cash to implement expensive pollution clean-up devices—relying instead upon good will of the masses—and reasonable compromises that do not reverse sorely needed economic gains. The State Council has favored a management approach consistent with Qu's strategy at NEPA. "About half of the environmental pollution problems have been caused by poor management," Qu insists.

What *is* management, specifically? I asked. It consists of planning, coordination, supervision, and guidance, said Qu, describing the goal of his agency. Every manufacturing plant would be systematically categorized according to its level of pollution, and that includes pro-

cessing, waste disposal, and the product itself. Those industries producing unambiguously hazardous products would be shut down or the
product lines altered. Only those companies with environmental control measures in place would be allowed to start up anew. Pollution fees
would be exacted, and every municipality mandated to apply its own
local environmental legislation and corresponding financially punitive
measures against offenders.

While over two hundred environmental standards have been set
in place to help launch Qu's vision of a fruitful future, the population
bomb continues to outstrip ecological restoration. China's corporate
sector will be hard-pressed to situate industries away from population
centers, as much as the government would like to encourage it to do
so. Inevitably, the pollution burden will be divided between outlying
populations and wilderness areas themselves.[78] But the more difficult
question persists: where will the needed pollution control and management funds come from? Despite the country's breakneck economic
growth, it has not yet lifted the overall per capita income beyond $370
per year. Hence, a centralized vision of environmental legislation and
controls, and the appropriate targeting of budget line items, is the only
hope, as in most other countries. That means choosing priorities.

"We are still poor. We cannot afford to set very high standards for
environmental protection. We have to do everything according to our
practical limits," Qu quietly went on. Nevertheless, to China's great
credit and despite its poverty, for the period 1991 to 1995 approximately
100 billion yuan (roughly $17 billion, or 0.78 percent of the GNP) has
been allocated toward environmental protection. In the evolution of
U.S. budgets, such proportions of the GNP were not environmentally
expended until after per capita income had exceeded approximately
two thousand dollars. But according to Qu, much more money is
needed. The Chinese estimate that they require at least ten times their
current spending allocations to really get a handle on existing industrial
pollution and prevent new despoliation during this decade, and that's
with good management. But the payoff, theoretically, should come.
According to one study, 15 percent of China's GNP is actually lost
because of internalized costs associated with environmental damage.[79]

Such "inaccessible" and internalized costs are by no means unique to poor, developing countries, of course. Thousands of recently ascertained toxic waste dumps throughout the United States implicate a supposedly enlightened economy and yet there are few guarantees that money will ever be set aside for reclamation. Qu addresses society's lack of ecological financing in general terms: "Our policy of balanced development . . . will achieve the unification of economic benefit, social benefit and environmental benefit. It is a positive policy, which stresses prevention rather than cure."[80] Interestingly, this is precisely the system upon which Chinese medicine is predicated. China now has the highest life expectancy and one of the lowest infant mortality rates of any developing country. Its health care system, in many respects, is second to none.

While China is now regarded as the world's largest manufacturer of television sets, refrigerators, washing machines, radio cassette recorders, and electrical fans,[81] its NEPA is confident that environmental management will continue to keep pace. Such confidence is new in a country that—despite an eminent history of landscape painting and poetry—has recorded environmental degradation for at least two thousand years, since the Han Dynasty.[82] Since 1949, there have been eight aggressive five-year plans and hundreds-of-thousands of industrial construction projects launched irrespective of environmental impact. Qu points out that thirty years ago China endured a "leftist" Cultural Revolution whose slogan, remarkably, was "down with everybody and wage a nationwide war." Another slogan declared, "Near the mountains, scattered and into the caves," referring to the frenzied construction of factories and the mad dash to mobilize millions of farmers to manufacture iron and steel. Production was tantamount to the concerted national rape of China's resource base. Those who spoke out in defense of the environment were deemed to be traitors "tarnishing the reputation of socialism." Not surprisingly, the same economic downturn that caused Mao to reverse his pronatalist policy would also begin to infiltrate the leadership's thinking about ecology. Tainted fish in a Beijing market, polluted shells in Dalian Bay, encroaching desert, loss of forests, and miry water supplies impacted on the Chinese mind.

In 1972 Premier Zhou Enlai sent a delegation to the Stockholm U.N. Conference on Human Environment, indicating China's newfound concerns.

Since 1973, when China publicized its first environmental legislation the "three wastes" law—pertaining to gas, water and other residues—and the 1979 *Law on Environmental Protection* (pertaining to marine regions, lakes, rivers, and air), the leadership has sought to promote "rational" development and to convert "harm into good, relying on the masses, everybody set to work, protecting the environment and bringing benefit to the people."[83] As with the one-child policy, China's environmental management concept may be the only strategy, albeit financially underserved and imperfectly enforced, that is a realistic option. Unfortunately for now, the Chinese public is as uncertain about environmental management as it is about the one-child policy. For example, in the first opinion poll taken by NEPA, 81.1 percent of the respondents were shown to be dissatisfied with China's environment but most of them also insisted that cleanup was the government's responsibility, not the people's; 35 percent worried that "damages to the natural ecology" were impeding economic development, while 41.5 percent cited population, housing, and traffic as the world's biggest problems. Only 8.4 percent said that they would "walk out in protest" when a banquet featuring rare and endangered animals was laid before them.[84] If there is no money for adequate prevention, and the people themselves are unwilling to creatively shoulder the burden, who then will take care of things?

The overwhelming causes and consequences of China's many ecological woes were not systematically addressed until 1988, when the National Environmental Protection Agency was founded. Director Dr. Qu felt from the beginning that government, working through the people, could harmoniously balance environmental impact. It was Qu who first applied the policy of "polluter pays" in China. But he took the injunction a step further by stating that "whoever causes pollution shall be reponsible for its elimination." But Dr. Qu's most ardent vision has hinged upon his insistence that environmental quality requires enhanced management, not heavy financial inputs. Tactically speaking this is a wise move, considering China's population-driven economic ambiguities and adversities.

Having gone to Africa and seen communities that were starving amid clean air and tranquillity, Qu told me that he has vowed to steer China toward a more realistic approach, and that has meant eliminating the "pollution of poverty," as he called it. He does not accept the ultimatum that "growth must be stopped or civilization will be destroyed." To the contrary, he believes that the more highly developed and technological a culture, the more means it has to control environmental pollution. Unfortunately, that reads much like the logic of those who argue that in China the more wealth, the less severe the population crisis. But if China's extraordinary economic zeal can be taxed to help cover the costs of management, Qu's crusade may succeed.

In the meantime he insists that man must not "give up eating for fear of choking"; that "natural ecosystems are not ideal ones, (but) only through rational remaking by mankind can they be idealized. . . . Apart from economic development, environmental protection would be like water without a source, or a tree without roots."[85] In a country with nearly 1.2 billion people, the ideal of wilderness, that romance of nature, has been necessarily redefined according to human need. The impending tragedy for China is that it is one country that still possesses, according to the World Resources Institute, over 22 percent of its landmass in a wilderness state (nearly 211 million hectares), unlike India and Indonesia, which have 0.0 (zero) wilderness. (The definition of such wilderness is a parcel of land at least four thousand square kilometers in extent that is totally undefiled—no power lines, roads, railroad tracks, buildings, homes, pipelines, landing strips, no mark of humanity whatsoever.) Thus, from the point of view of wilderness habitat—with all that that implies for unimpeded animal corridors, biodiversity, moral, equitable, aesthetic, and genetic viability of nonhuman populations— China, like Antarctica, has an enormous amount yet to lose from rapid development.[86] Because its biological heritage is so complex, and its remaining wilderness so uniquely extensive, population and economic issues in China are at the forefront of global ecological concern.

Qu insists that the ideal of "back to nature" is simply not possible anymore. In this respect, he views China as a laboratory of realism. "Requirements (i.e., environmental standards) can only be raised when the economy is developed," he repeats. With irony, he adds that those countries capable of affording environmental protection do not pursue it, while those unable to afford it (i.e., China) do. It is therefore not merely the price tag, he argues, but the "philosophical system" that will effect ecological change. One need only look to China's postcommunist family planning experience to appreciate the frailty of any such philosophy.

Qu, an imperturbable optimist whose bright eyes and knowing grin carry any conversation, speaks with certainty about constructing a spiritual life for 400 million densely packed Chinese city dwellers. What can that mean, I ask, seated in his temporary lodgings five stories up in a nondescript Beijing University annex office NEPA has taken over. A surprisingly mild winter breeze rustles the bare poplars outside, the sound mingling with the enlivening chime of bicycle bells. The sun pierces the linen curtains of Qu's office. We sip tea before his huge wall of books. A city, he says, "is an ecological system composed of man, nature, and buildings. Within this system, man is the center. . . . We must coordinate the natural environment, artifical buildings, and urban spaces to facilitate man's activities so that the city is both convenient for work and for living." Qu is convinced that the "rational layout of industry" will not only ease but ultimately neutralize environmental damage.

Qu's authority has triggered a remarkable convergence of environmental protocols throughout the provinces. China has developed four hundred eighty-one natural reserves, comprising 2.5 percent of its total land area. Wastewater in many of China's drought-ridden northern cities is now recycled, while urban sewage is used in the surrounding countryside as fertilizer. Recognizing the inherent liability of coal and China's near total dependence upon it, some three hundred thousand boilers, millions of domestic stoves, and seventy thousand coal-fired boilers with low rates of thermal efficiency have all been renovated. Three hundred seventy-six chimney-control zones have been established to curb effluents and provide a steady tax stream from offenders. Over one thousand Chinese factories solely devoted to manufacturing pollution control devices are now in operation. The seriousness of recent environmental laws in China is such that one company owner was sentenced to death when it was learned that his company knowingly dumped toxic wastes in a river and killed some fish. His sentence was commuted at the last minute, but it must have literally put the fear of death into his peers.

Noise level reduction, water pollution cleanup, the enhancing of wastewater processing, the building of "clean and civilized factories" have all been moving at a steady pace. Waste gas has been incorporated into the overall energy picture and centralized heating has been integrated into new cities such as Tianjin so as to consolidate and amortize production and consumption of BTUs (British Thermal Units of energy). Also in Tianjin, 146 chimneys were disassembled, 10 million yuan and 82.3 million tons of coal saved during recent winters. Covering 859 square kilometers, the Tanggu District of Tianjin city, with its population of over four hundred thousand is one of dozens of communities centralizing heating and in all ways attempting to "control" the urban environment. In the 1980s, twelve thousand of the country's most serious polluting industries were closed or merged into more efficient manufacturing domains. In Tianjin 122 factories were shut down. In Beijing 156 factories were also ordered to cease operating. In addition, the capital saw 467 electroplating factories disbanded. Chemical plants, high energy-consumption enterprises, illegal refineries, synthetic ammonia, chromium salt, paint, and rubber enterprises were also closed down. But as yet virtually no toxic chemicals or toxic wastes are systematically regulated in China. In fact, China has even acknowledged its willingness to bury toxic wastes from other countries if the profit margin and perceived trade-off is adequate. Very few chemicals (BHC and DDT are the exceptions), have yet been banned.

Gao Dezhan, China's minister of forestry, has shown equally impressive leadership, calling for an increase in forest cover to over 15 percent (up from a current estimated 13.63 percent).[87] Guangdong Province has allegedly reforested all its barren hills; Fujian, Jiangxi, and Anhui provinces are quickly reaching the same status. Eighty percent of the newly planted vegetation is surviving throughout areas of the north. China's Green Wall (the San Bei Forest Shelterbelt System— meaning the three norths) has united tens-of-thousands of dedicated volunteers. Local communities have collectively spent over $400 million of their own money, versus the state government's $200 million investment, on reforestation and afforestation efforts. Volunteers have put in more than 500 million workdays without pay sheltering 11 million hectares of cropland with trees, thus raising crop yields between 10 and 30 percent in those regions. In addition, soil erosion has been arrested in some parts, particularly in Yulin prefecture in northern Shaanxi Province, one of the locales afflicted by the worst desertification. The Yulin people planted eight hundred thousand hectares of trees and some sixty thousand hectares of land have been reclaimed for agriculture. Tree planting in Beijing environs has been extensive as well, and this has resulted in a decrease in total suspended particulate matter by 18 percent and dust fall by 43 percent. Even the number of sandstorms off the Gobi Desert has been curtailed by 39 percent.[88]

Three thousand environmentally related court cases in recent years have been filed by local citizens from 1985 to 1993. Fifty thousand angry letters have been written regarding pollution, and eighty thousand visits have been paid by China's citizenry to various environmental protection units throughout the country. Environmental departments have been started in seventy-nine universities in China.

Taken at face value, and numerically, China's practical responses to its ecological dilemmas have been laudable, all in contrast to public polls that show a much higher lethargy factor. But the numbers, ultimately, are bound to be suspect in view of the disconcerting data pertaining to a population that keeps escalating. Seventy percent of all Chinese pollution comes from industry, it is alleged, and industry is increasing at an annual rate commensurate with population increase, or more than 10 percent. Agricultural pesticide use (159,267,000 pounds in 1992), fertilizer, herbicide, and livestock wastes, account for an enormous and probably underreported environmental burden. The "polluter pays" concept has hardly been applied anywhere in the world to the agricultural sector, and certainly not in China, where the memory of famine—and continued malnutrition—dogs the psyche. A policy of "control" meant to guarantee a "rational" approach to development is simply not predictable when the population itself is out of control. Against the inevitable economic tidal wave of opportunism in China's immediate future, "rational" is not easily translated, either. It suffers from the ambiguity of China's desire to achieve a state of well-being for its people. Where does economic well-being end, and environmental well-being begin? Or vice versa?

There are now, in 1993, nearly four hundred thousand valid trademarks in China, with record numbers having been registered in 1992. Shanghai's foreign trade increased 21.3 percent in 1992 over 1991, $6.6 billion in exports, machinery and electronic products being the country's leading hard currency earners. There is very little that can be, or ever will be, done to curb such enthusiasm. The retail business in China soared 88 percent in 1992. New industries emerged with a 20 percent overall growth rate. Some seventy thousand foreign investors poured between $5.6 and $11 billion into China in 1992—depending upon which statistics one reads—and signed contracts pledging $57.5 billion more for purposes of selling everything from McDonald's hamburgers, candy, plastics, Avon cosmetics, and Procter & Gamble shampoo, to American Standard Incorporated's seven thousand dollar whirlpools. Avon posted $50 million in sales in its first two years in China. The Chinese are a long-repressed middle class suddenly unleashed along Shanghai's teeming Nanjing Road. Consumer fever is volcanic.[89] Shipping syndicates are now moving goods between the Mainland and Taiwan. Chinese vessels put in at over 150 countries around the world. Chinese corporations enjoy joint ventures with at least thirty nations. Today, China is the eleventh largest trader in the world, having generated $165.63 billion in 1992. But its status and dollar figures are rapidly rising. Tomorrow will yield an even more euphoric profile, at least economically speaking.

Thus, despite modest gains in a few cities, consistent environmental controls in China are stymied by a frantically expanding economy that is moving as fast as the population, faster than the nation's ability to ecologically keep up. Urban sectors such as Shanghai and some outlying rural regions such as Fan Shan are seeing a corresponding growth in per capita income, but the majority of Chinese are still very poor. Within six years the government hopes to more than triple the per capita GNP. It has taken the last ten years for the economy to double. In the year 2000, 2 billion tons of coal, 400 million tons of gasoline, and 50 billion cubic meters of natural gas, will all be needed to supply the required 200 billion kilowatt-hours of a 1.3 billion population.[90] As of 1981, China was consuming five times more energy than Japan, 95 percent of which consisted of low-grade coal. Yet the energy efficiency in China remains considerably lower than in developed countries (30 percent in China versus 51 percent in the U.S. and 57 percent in Japan). Because about a third of all production capacity is still untapped, the rapid industrialization of the late nineties suggests an enormous increase in environmental degradation. According to Amory Lovins, "If China keeps burning coal and little else, the world could face an ecological disaster. It will ruin the climate for everybody."[91]

International observers have suggested ways to counter the country's coal dependency. Chinese coal is notoriously foul, usually containing 5 percent sulfur and 27 percent ash. Coal washing technology could reduce half of that sulfur content, while an integrated gasification/combined cycle would also remove a substantial portion of nitrogen oxides.[92] But, once again, that is a realm that transcends mere management: it takes money, and the national will power to allocate it.

China's new Wild West economy symbolizes the state of the world, the state of human desires and basic biologic urgings. What well over a billion Chinese do with their ancient biological heritage in the coming fifty years will coincide with the economic/ecological curve of every developing country where population is the critical factor. In developed countries where population growth is slower, economic prosperity will increasingly translate into environmental concern, erratic protection, continually declining wildlife populations, and ever more stringent laws applied to environmentally sound technologies and engineering. But such efforts will not trickle down to the poor nations of the world trying to achieve a developed standard of living. Eighty percent of the

world's population, in other words, is re-inventing the wheel beginning with the dirtiest industries. And because the developed nations
are finding themselves entrapped by global recession and, in some cases,
unimaginable national debt, their largesse and willingness to cede
anything to the poor is diminishing. Nobody is giving China free
environmental clean-up technologies. China, like every other country, needs hard dollars and seems willing to do anything to get them.
In its quest for greenbacks, China now exports more goods to the U.S.
than do Malaysia, Thailand, Indonesia, and the Philippines combined.
In all of those countries, China's impact is already lowering the value
and dollar price of manufactured exports, which translates into added
degradation of those other countries' remaining natural areas. Thus,
China's ecological impact goes global.[93]

The pulse of China's runaway economic turmoils are best sensed
in Shanghai, a city of horrendous dimensions. From atop the seventy-
story super-chic Jin Jiang Hotel, the city resembles a hazy, surreal, en-
tangled exoskeleton the size of a huge crater on the surface of Venus,
hemorrhaging bilious fluids and caustic vapors. Its minions stream, its
traffic chokes, construction zones everywhere grate with the bashing
of hammer-toting men and women salvaging pieces of steel girding from
vanquished concrete blocks, striking, unloading, digging, breaking
down old buildings by hand, re-erecting huge scaffoldings in the manner of Brueghel's memorable icon of human confusion, *The Tower of
Babel*. The renovation of Shanghai's waterfront summons images of
Tokyo or Dresden after the bombings of World War II. It is a concrete
metastasis that has erupted in all directions, pustular and fierce. The
horizons around Shanghai are a pall of congested, smoke-saturated color.

Jiangsu Province, crossed by Highway 204, which leads out of
Shanghai, is China's fastest growing area. It witnessed a 26 GDP rate
of increase in 1992. Shanghai, Taicang County to the immediate north,
the Pudong New Area, and all of Jiangsu Province are the epicenter
of the New Economy, where more than 50 percent of industrial production is deriving from joint venture and private business. The
average worker is being paid 34 cents an hour versus $3.50 in South
Korea. This explains the influx of foreign corporations wanting to

exploit the incredibly low-wage labor. China's trade surplus with the U.S. in 1992 was $18 billion, already nearly half of Japan's $43 billion. Toys, electronics, and textiles are among the biggest export items from the Shanghai region, where one in five families now plays the stock market, where the money supply grew 30 percent in 1992, where wage inflation was 20 percent, and urban inflation 11 percent. In 1992 foreign investments in China amounted to $30 billion, up thirty times since 1987.[94] China's economy, spearheaded by epicenters like Shanghai, is on its way toward achieving the per capita income of other industrializing nations on the Pacific Rim. When it does, it may well be "larger than all industrialized countries in the world combined. It would be a bit like the rise of Japan, except that China has nuclear weapons and nearly ten times the population."[95]

Some have seen Beijing's tolerance of the new economy to be a calculated policy of taxation that finances the dictatorship, just as democratic dreams were appropriated by authoritarian powers in Germany after the revolution of 1848, and again in the wake of the Kaiser Wilhelm.[96] But by the year 2000, private companies in China are expected to comprise about 25 percent of the economy, collective town enterprises about 48 percent, and Beijing no more than 27 percent. Neither in the population realm, the environmental realm, or the economic realm will Beijing continue to maintain the same level of authority over the country it wielded even as recently as June 1989, when it used force to quell peaceful prodemocracy demonstrations at Tiananmen Square. That also means that the National Environmental Protection Agency's role will become one of increasing consultancy and limited authority.

As for the role of Hong Kong in China's economic dreams, it is important to remember that the PRC has invested over $11 billion in Hong Kong as of mid-1993, and runs the second largest bank in the colony, the Bank of China. Moreover, the success of business in Hong Kong has resulted in over 3 million jobs in China itself. Given China's brief experience to date with a free market economy and considering what it knows about its own population and environmental dilemmas, it is unlikely that China will squash that golden egg.[97]

Regarding the military power of Beijing, observers throughout Asia are on the alert—particularly Japan, which recognizes that in fifty years the Japanese population will be but 5 percent of China's. "A better than average chance of at least limited maritime clashes" has been predicted for China and her neighbors.[98] And while Beijing denies it, there are reports that at least some of China's nuclear armaments in Tibet are aimed at India. The Chinese, after all, helped build Pakistan's own ballistic missile capability, a very real threat to India. China possesses over 3 million troops, in addition to six hundred thousand members of the People's Armed Police. It is boosting military spending by 15 percent, and has spooked all of Southeast Asia by basing twenty-four Sukhoi-27 long-range fighter-bombers on Hainan Island near Hong Kong. Chinese military spending is in fact believed to be double the officially reported amount—or 30 percent of the total military budget, some $12.80 per capita. Adjusted for inflation, actual spending will be about $7.4 billion—$6.40 per person per year at current expenditures.[99] That's roughly three times the amount of money allocated by Beijing for the State Family Planning Commission, the National Environmental Protection Agency and the Ministry of Forestry combined. Why? one wonders. What are the country's true priorities?

On May 21, 1992, China conducted a nuclear test equivalent to six hundred sixty thousand tons of TNT at the Lop Nor testing range in the northwest of Tibet. It was the most powerful nuclear explosion on Earth in ten years. It served to remind both Pakistan and India— neighboring nuclear powers with gigantic problems of their own—that in Asia, nothing is predictable.

Against this bewildering backdrop, several crucial questions arise, though no easy answers. What should U.S. foreign policy toward China emphasize? Economics? Ecology? Human rights? Family planning? How can these arenas be sensibly combined? One thing is clear: the U.S. must fervently support China's state family planning efforts through much more generous assistance, while encouraging U.S. foreign investment to impose the same ecological imperatives on its overseas factories, co-ventures, and products that would be required within U.S. borders.

Increasingly independent, though never separated from the truth of Beijing's military force, what can Chinese provincial leaders do to re-ignite enthusiasm for the one-child policy and ecological sustainability within their own regions? The country's visionaries—people like Dr. Qu Geping, Dr. Qian Xinzhong, Gao Dezhan, and Madame Peng Peiyun—have each provided viable blueprints for action.

# The Ecology of Pain: India

## SAMVARTAKA, CLOUD OF DOOMSDAY

In a determined effort to counterbalance China's military presence in Asia, Prime Minister Indira Gandhi began stockpiling plutonium and building up a nuclear capability during the early years of her regime.[1] India's concerns encompassed not just China, against whom she'd fought a limited, one-month war in 1962 (and would engage in border skirmishes with up to 1987) but Pakistan, her long-time adversary. In 1971, the Indo-Pakistani War (which culminated in the independence of Bangladesh, formerly East Pakistan) resulted in a death toll exceeding three million people, leaving tens-of-millions of Bengalis ever more destitute, destroying untold numbers of wildlife, and forcing at least nine million human refugees into exile in India, a nation scarcely recovered from a famine which had inflicted at least one million dead the year before.[2]

When I investigated the Indo-Pakistani War—interviewing numerous refugees in Calcutta—obtaining testimonies from eyewitnesses—a picture of widespread rape, cannibalism, and ecological mayhem emerged, the vengeful particulars of a sordid and complex disaster. Meanwhile, up in Dacca, postwar famine was taking its own additional toll. I watched members of the Anjuman-e-Mahifidul Islamic organization collecting as many as four hundred unclaimed corpses off the streets every day. In Tongee, a refugee camp outside Dacca, I volunteered to help in the handing out of foodstuffs to thousands of doomed and dying environmental victims. Newly formed Bangladesh was confronted with a population of nearly 90 million, 60 percent of whom were malnourished, many starving. I saw human corpses festering beneath scavenging dogs that dragged at the entrails. Human life expectancy was optimistically believed to be forty-three, per capita income about one hundred dollars. Meanwhile, ten thousand babies were being born every day in Bangladesh. The military dictatorship was trying to clean up the streets. They would grab "squatters," toss them into the back of trucks and drive them off to camps. Most would starve to death, like the ones at Tongee. There was a woman in the camps that I remember, young, exquisitely beautiful, dying, who smiled bountifully at me. There was radiance amid Armageddon, some impossible flirtation from the depths of melancholy, nihilism tinged with sensuality. She has remained for me a symbol of the Indian subcontinent.

Farther south, maintaining the offensive following its war with Pakistan, India then detonated its first nuclear weapon beneath the ordinarily tranquil deserts of Rajasthan in 1974, hastening the develop- ment of an infuriated Pakistan's own secret nuclear program. Throughout the 1970s and '80s, India and Pakistan remained enemies, their northern combat troops exchanging bazooka fire in the Karakoram Range. However, by May 22, 1989 it seemed as if both India and Pakistan were coming to their military senses. At that time both countries pulled back their troops from the Siachen glacier in Northern Kashmir after years of hateful conflict. Seven months later, at the Indian-Pakistani Summit meeting in Islamabad on December 30, 1989, the two prime ministers signed an agreement prohibiting attack on each other's nuclear installations, an odd concept at best but one viewed by both nations as the first step toward normalizing relations and defusing the escalating violence in India's northwestern state. In 1993, however, the Kashmiri situation appears stalled, the two countries inching back toward the tension of a few years ago.

Countless Kashmiri Muslims would like to see Pakistan invade Kashmir and rid this enchanted Himalayan sanctuary of its "Indian overlords" once and for all. Over eighty-three hundred people died in clashes over Kashmir in the three years 1991–1993. A vocal contingent of Kashmiri Muslims see their quest for liberation a just cause for a *jehad*, or holy war, such as was waged in what is Bangladesh. And in an in- terview with the press, the so-called "father of the Pakistani atomic bomb," Dr. Abdul Qadeer Khan, had stated that "if driven to the wall

there will be no option left" but the use of nuclear weapons, a sentiment underscored by Pakistan's High Commissioner to India, Riaz
Khokar, who has described Pakistan's "insecurity" as the result of his
country's "fear of India."[3] At the same time, a spokesman for the Hindu fundamentalist Bharatiya Janata Party stated, "India will talk straight
and walk straight when we have the bomb."[4] India's Congress Party
has in fact spent a great deal of money on the bomb, money certainly
better spent elsewhere.

According to Hersh's account, several months after the December
1989 nuclear peace, two hundred thousand Indian troops amassed in
Kashmir, fifty miles from Pakistan. Fearing the inevitable, Pakistani
president Ghulam Ishaq Khan and his army general Mirza Aslam Beg
were poised to unleash between six and ten nuclear weapons, which
were to be deployed on American-made F-16 fighters and dropped on
India if tensions escalated any further. "It was far more frightening than
the Cuban missile crisis," said Richard J. Kerr, a deputy director of the
CIA, who was in charge of coordinating the intelligence reporting from
Pakistan in May of that year.[5] A BJP parliament member in Delhi
warned that in the case of the next war with Pakistan, that country
would be wiped out. The war did not happen, thanks to behind-the-
scenes diplomatic shuffling between Washington and Islamabad.

In 1993 the Clinton administration placed Pakistan on a terrorist
watch list for the country's alleged supplying of arms to Muslim guer-
rillas fighting in Kashmir and to Sikh separatists in Punjab. But it goes
much deeper than that. According to Clinton's CIA director, R. James
Woolsey, an Indo-Pakistani nuclear war is "perhaps most probable." In
Congressional hearings headed by Senator John Glenn, Woolsey
remarked that "The arms race between India and Pakistan poses perhaps
the most probable prospect for future use of weapons of mass destruction, including nuclear weapons" in the world at present.[6] Like
China, Pakistan has not signed the Nuclear Non-Proliferation Treaty.
It has contracted with China for the building of an ecologically
disastrous, potentially weapons-grade, nuclear power plant 140 kilometers southeast of Islamabad in Chashma to help deflect the coun

try's eighteen hundred-megawatt shortfall in energy requirements. That's Pakistan's explanation, in any case.[7]

Looking back at the events of May 1990, CIA intelligence was convinced that Pakistani General Beg was actually prepared "to take out Delhi" according to Hersh, who believes that the two countries possess many nuclear weapons and " . . . India, frightened and besieged today by terrorist bombings and growing Hindu fundamentalism, remains always willing to fix the blame for its troubles on Pakistan."[8]

Hindu–Muslim belligerence was further fanned after Hindu zealots destroyed the mosque at Ayodhya, southeast of New Delhi, in late 1992, triggering riots and terrorist attacks that killed over twelve hundred people. There were further allegations by Hindus that the Muslims were overpopulating the country, which they are not,[9] though in Pakistan, where exhuberant fertility is deemed the best long-term revenge against India, the Muslims are certainly trying.

Since its independence in 1947, population in Pakistan has increased four-fold, from 32.5 million to more than 130 million.[10] Women are expected to procreate and raise children. That is a key perception, by males, of that gender's reason for being. Despite a female prime minister (Benazir Bhutto), the only apparent signs of change are in the increased female literacy rate—from 11.6 percent to 16 percent—and in the slightly higher legal age of marriage, and improved working conditions for females. The preferred family size norm is 4.9 children. These few college-educated Pakistani appear willing to have 4, not 4.9 children.[11] Family planners believe that strong international support of fertility control might enable Pakistan to check its population at around 335 million sometime in the middle of the twenty-first century. But because official government policy is so anti-Hindu and essentially pronatalist, it appears more likely that the country will exceed 550 million sometime in that century. A similarly anti-Indian sentiment was echoed by a former chief minister in Kashmir who stated, "We should reject the government's family planning program. This is aimed at further reducing the Muslim population in Kashmir. Every Kashmiri Muslim should have four wives to produce at least one dozen children."[12]

As Pakistan races toward a probable half-a-billion people, her arch-rival, India—the world's proving ground for birth control[13]—is currently at nearly 900 million. Approximately 11 percent, 100 million, are Muslim. The vast majority are Hindu, with small percentages of Sikh, Jain, Zoroastrian, and Christian, and a large minority of tribals, or *adivasis* (566 individual tribes). There were, as of 1980, an estimated 3743 so-called Backward Castes, 2378 Hindu subcastes, and 1086 Scheduled Castes in India, adding up to the most fragmented cultural puzzle in human history. The country speaks somewhere between 500 and 1652 dialects and languages, depending on the survey.[14] India is not one country, socially speaking; it never has been. Every ethnic and religious group typically sees strength in numbers, as they try to gain political autonomy or economic guarantees of some kind. As a result, the politics of fertility control are held hostage by a protean paranoia. And as in every other nation, these ethnic groups, pressured into becoming neighbors in the same land, have inherited stark political boundaries, some—like those in Kashmir—still internally disputed.

Roughly one-third the geographical size of China, India's human population will probably overtake that of China's by about 2050; already it is twice the density of China's. Bound together by the besieged Himalayas, these two countries—India and China—will comprise as much as one-third of the mid-twenty-first century's human population. In fact, many see India hitting 2.5 billion by the end of that century. This fact of human numbers is the most critical perspective that can be adopted when thinking about India, a country whose demographic and ecological instability poses tragic consequences for the planet.

Regardless of its many boundaries, socioeconomic distinctions, dizzying number of languages, customs, and gods, the Indian subcontinent as a whole knows only one war; it is being fought inadvertently, though at times deliberately, against nature, a masterpiece of nature that is crying out.[15]

In the beginning, those cries were less audible, the pain less evident. The forests were enchanted, an ideal, the abode of Lord Siva, also known as the Wild God, Rudra, the Ultimate Yogi. They were a source of inspiration, a sanctuary for worship, absolution, and repose.

Because Siva was believed to protect all wanderers in even the harshest wilderness, Indians long ago traveled fearlessly, building an ancient network of roads through nearly every forest in the country. In this time of innocence, the Wild God inadvertently invited domestication, the emergence of cities. Even twenty-five-hundred-years ago, there were as many as forty thousand villages in the country, connected to one another by the roads. People routinely wandered into nearby woods seeking communion with monks and sages who dwelled under the tropical and temperate canopies. It was inevitable that the forests should come under intense devotion, scrutiny, and finally, pressure. Indians learned how best to exploit the woods. Ancient dietary manuals hailed the fruit as being the essential foodstuff, recommended green trees for fashioning farming utensils, advised upon the vast pharmaecopia to be hewn from the tropics. The cultivation of orchards was considered in and of itself a form of meditation.[16]

Because the size of the human population was still relatively modest, whatever damage was wrought could be reasonably absorbed by the natural environment. From this (as yet *sacred*) pragmatism arose what has been described as the foremost tradition in India, namely, *Pancavati*, meaning "five groves." The word surfaces in virtually every known Indian language and refers to five kinds of sacred trees: the Indian *banyan*, associated with bearing human progeny; the *peepal*, abode of Lord Visnu, every part of it ready-made for utilization; the *asoka*, which eases suffering and provides relief for certain gynecological problems, as well as *ghee* (fat) for numerous food dishes; the *bela*, said to possess the most curative fruit in the land, more nutritious, even, than mango; and the *harada*, a sure cure for every ailment. Subsidiary to these five were a profusion of other plants and shrubs and trees, all part of the spiritual, vegetative hegemony. Communities throughout India tended to substitute and introduce local species into the sacred grove . . . coconut, banana, basil, myrobalan, jackfruit, sal, neem, keekar, babool, sami, tamal, sandal, blackberry, and so on. All were sanctified, all subject to consumption, but never total consumption. The clearing of forests was forbidden, a sin (*adharma*) against the body of God. The original Sanskrit term for forest was *aranya*, meaning literally, "no war."[17]

At least one Indian philosopher of the sixth century B.C. was purported to have known nearly eight hundred seventy-five thousand species. (A century later, Aristotle and his compatriots were aware of fewer than five hundred.[18]) This profound knowledge of Indian biology was advanced in order to safeguard nature, not to foster its compromise. Yet, familiarity with the sacred fueled all manner of ecological violation. The inevitability of this contradiction was implicit in what writer V.S. Naipaul has labeled "a wounded civilization." That which was spiritual became profitable, particularly as population increased. The bounty of nature inspired human excess. Ironically, and fittingly, the analogy of simultaneous creation and destruction is inherent to the personality of many of India's ancient deities, a clear commentary on the human psyche.

Much has been written about poverty in India, mostly by non-Indians who cannot fathom a better term for it. But it is an invented term, culturally relative and much-maligned, in need of at least one qualification. I would suggest that some forms of poverty—of managing "less with less," to use California urban designer Sim Van Der Ryn's phrase—are not without certain affirmative, even critical qualities that date back to India's beginnings. Westerners are fond of the phrase, "Nobody wants to be poor": a maxim that seeks to justify persistent, if not blind, consumerism; a myopic and self-serving understanding of nature that would morally uphold human colonialism on the basis of a misreading of Darwinian evolution.

Thousands of years ago most of our ancestors had not yet caught this fever of consumption. Our needs, curiosities, and pleasures were met without having to degrade and alter nature's rhythms. There is a clear threshold of violation that is reached either by an individual's excessive behavior—as when one person burns down an entire forest— or by the inevitable mass destruction that results when a population as a whole exceeds in its myriad demands the carrying capacity of its territory. The formula by which we can tell when we've destroyed something is not hidden from us, nor does it imply any kind of science. By degrees, destruction has been clearly marked throughout time and anybody could always read it and understand it.

Typically, so-called ecological overshoot resulted when a population found itself marooned atop an exhausted food base or plot of terrain. Like other species in nature, whether starfish or wildebeest, that population then moved on to the next file or tract of land. When the new niche was exhausted, our ancestors shifted to yet another locale. Locales, until recent times, were seemingly expendable by human standards, because regions were always large enough to absorb injury. And when the ecological stress upon whole regions was noticed, our ancestors simply migrated, searching out new pales and dominions. Thus emerged a style of life known in northeastern India as *jhuming*, or "shifting cultivation." Every year in India, 5 million hectares are processed by this ten-thousand-year-old method of migratory agriculture. Once the human tribe moved on, the burned out fields were left fallow, or at least that was the ecological recompense prior to a population boom.

In the 1940s, anthropologist Furer-Haimendorf visited Nishi villages in what is today India's northeastern-most and least populated state, Arunachal Pradesh. The tribal people lived on mountain ridges in the jungle, descending into the valleys to catch game. Population was (even relative to today) totally sparse. " . . . feuds and revenge killings then undoubtedly acted as a check on the growth of the population. This check has now been removed," writes Furer-Haimendorf, "and the availability of medical services has limited the effect of epidemics, which on occasions took a heavy toll. An increase in the Nishi population is hence inevitable."[19] In the late twentieth century the building of motorable roads has encouraged the abandonment of former remote ridgeside communities and the reestablishing of population centers in the broad valleys. Plows and bullocks have been introduced for tilling the land, which has suddenly become valuable. Intensive utilization for rice-growing purposes has meant the "transformation of the natural environment" and the steady decline in fallow periods. Moreover, "with the increasing population growth, land is gradually becoming an object of conflict . . . whereas previously men used to quarrel about women and mithan (cattle) most disputes now arise from competition for land."[20]

Environmental conflict in India occurs on a vast scale, and in areas that the rest of the world rarely hears much, if anything about. In the State of Bihar, for example, local impoverished farmers from dozens of villages have been fighting absentee landlords over a flood plain known as the *diara* along the Ganges. This silted region of some 32,390 square kilometers spread over five districts provides the sole source of food for tens-of-thousands of locals who would otherwise probably starve, or poach all remaining wildlife, or migrate to cities and vanish in the slums. Landlords have responded by hiring mercenary gangs to extract high rents from these farmers. But it's ridiculous: the farmers have no money. They live from day to day. They are untouchables. The dispute over land and money in this region has already claimed at least ten thousand lives in the decade of the eighties.[21]

There are some 65 million Indian indigenous peoples like those of Arunachal Pradesh, and hundreds-of-millions of untouchables like those in Bihar. They have high birthrates and their infant mortality rates are declining. In essence, they are a population boom within a population boom, much like China's minorities. Attempts by the government to translocate the tribals out of parklands have been ineffective, by and large. More at issue is the tribal way of life that—under conditions of overpopulation—is tragically in opposition to essential forest ecology. It means that more livestock will munch on more leaves, more people will cut more wood for basic necessities, and there will be less and less primary forest for native species.

Where there is no competition—the condition of humanity many thousands of years ago, and vestigial in a few remaining outposts—humankind knows how to be benign and still live well. It is not wishful thinking to want to be like those people. Indeed, the whole realm of sustainability, what many today think of as a kind of scientific method essential to our survival, derives simply from the wisdom of some of our ancestors. Not all of them, however. It is known for a fact that many traditional, so to speak "land-based," groups were environmental terrorists in their time. But it is equally true that countless peoples have been, and presently are, "poor" yet leading relatively happy, productive, and sustainable lives.

The test of sustainability has been proved—and contested—in numerous ways. For example, recent data from Belize indicates that the harvesting of medicinal plants from tropical forest plots can yield more dollar value in the regional market than agricultural produce or lumber from those same plots.[22] But two caveats quickly emerge: first, there is a limit to the regional market for medicinal plants. The moment that market is necessarily expanded, new destructive forces descend on the source of supply.[23] Second, as the larger market economy imposes certain conditions of cost and extraction, the increased temptation to exploit the land more intensively for profit forces harvesters to confront difficult ecological choices that may conflict with the dynamics of the group decision-making process, in the case of a commons, or with personal greed, or insecurity, in the case of private holdings. Among the fishermen in the Indian southern state of Kerala, these various factors have been devastating. Short-term profit motivations and an indiscriminate harvesting of the marine ecosystem, have led to a syndrome whereby "it actually pays to bring ruin to the commons" in the minds of Kerala's new breed of capitalists.[24] In addition to this community and personal dilemma, even where otherwise subtle biodiverse products in India have witnessed a new economic appreciation, exploitation by local populations has been totally unsustainable.[25]

There are still a few purely regional economies scattered throughout the world. Such harvesters and farmers are people living directly off the land—but isolated, unconnected to any larger human systems of exploitation. In India, until recently at any rate, one of the last of these was on the Andaman and Nicobar islands, where the Jarawa and Onge people had managed to remain fairly inaccessible to most outsiders, at least until the late 1970s. From what can now be ascertained, their small portion of rainforest was essentially untouched. Among such communities, there was a sufficiently complete break in human linkages to afford a micro-population the opportunity to sustain itself in a pure state, as in a clean room laboratory. We know enough about them, in principle, to endeavor to apply the reasoning of their preservation to larger population areas, though with a host of difficulties. There are few natural clean rooms left and they are vanishing rapidly. In the last fifteen years colonists from the Indian Mainland, searching for jobs, dreaming of paradise, have virtually driven the Jarawa; and Onge to extinction. As of mid-November 1992, there were 80 Onge and 31 Jarawa; the emigre population had swollen to over 277,000, up from 30,971 in 1951, an annual growth rate during the past decade of 4.7 percent, even double that for the rest of the country.

India's immediate neighbor to the northeast, Bhutan, an ancient Buddhist nation with a per capita income on a par with India's, and a population 650 times smaller, is poor and miserable by most official world statistics. The "International Human Suffering Index," for example, equates Bhutan with sub-Saharan Burkina Faso.[26] When U.S. Ambassador to India John Kenneth Galbraith visited Bhutan in the late 1960s, he had to do so by yak and it took nearly a week to reach the capital. There were no roads, no landing strips, no phones, virtually no electricity, hardly a wheel to be found in the country. When I went to Bhutan in the mid-1970s, the whole country resembled one alpine ski resort—without the ski lifts, tourists, tour buses, or knowledge of such things. Bhutan may be the only country in the world where sustainable development on a national level is underway; where greed has been curbed, the commons revered, and a nationwide ecological sensibility embraced and codified. "Small is Beautiful" is the young King Jigme Wangchuck's motto. Ecotourism has been limited to no more than five thousand visitors a year (up from twelve hundred when I first went) though in 1992 only three thousand foreigners visited the country, at a cost of two hundred fifty dollars per person per day. Most (though not all) large energy and development projects have been derailed before they ever got started. Over 60 percent of the country is still completely cloaked in primary forest. The largest city, Thimpu, contains a mere twenty thousand people. Whatever the explanation for the nation's wisdom—a tiny and benign Buddhist population of 1.3 million, a religious aversion to killing anything, an invigorating climate, a magnificent Himalayan isolation—one thing is instantly clear to anyone who visits the country: Bhutan may be poor by Western standards, but it is one of the richest of nations with respect to quality of life and purity of spirit. However, all is not paradisiacal in Bhutan either. Recent population pressures from non-Drukpa (non-Bhutanese) immigrants, primarily Indian and Nepalese, along the southern border with India, have resulted in armed clashes and increasing threats to the country's stability, posing the threat of ethnic civil war. One more clean room is disappearing just as we begin to know about it.

India's rural poverty, which accounts for nearly 77 percent of the country or well over half-a-million villages, is reminiscent of Bhutan's own negligible per capita economy. More than half of all Indians are below the *garibi rekha,* or poverty line.[27] But the spirit of India, no matter what else one alleges about the country, is as mesmerizing as it is troubling. I have been returning there for twenty years, enamored of its people, landscapes, and passions. Such feelings render any supposed insights into India's population crisis oddly suspect, and I will be the first to admit it. While there can be no romancing the extreme direness of much that is Indian, it would be an equally insensitive and ignorant blunder to overlook India's beguiling capacity and mind-boggling range of traditions. India represents a very clearly discriminating population, which, in spite of high illiteracy, harbors capabilities grossly underestimated by almost everyone, Indians included. But the contradictions abound, and they assault one's perception. One former minister in the government likens contemporary India to Nazi Germany, with respect to its violence—violence in the form of oppression of women, of the environment, and of all animal life.

Poverty in India is defined by outsiders. On the inside it is something other than mere poverty, but pain that dictates and separates the haves from the have-nots.[28] Because so much of Indian *nature* is in pain, there are few human residents who can elude entirely, or at minimum fail to recognize, the thralldom of suffering. But that pan-Indian recognition itself has not proven particularly susceptible to improvement, or even empathy, under the law as most rural women in the country might tell you. As one Agency for International Development (AID) official in New Delhi put it, "India is in a time warp." The Hindi word most apt for this sentiment is *kal*, which means both yesterday and tomorrow, and tends to summarize the Indian approach to getting anything done. Perhaps the most noteworthy example of a social anachronism— one which many Indians condemn, but just as many uphold—is the continuing designation and prejudice attached to an untouchable class (*harijans* or *shudras*, also known as OBCs—"Other Backward Classes)." They still predominate throughout the country (60 percent of the populace, as noted by the Indian Mandel Commission). Such people typically are forced to live on the margins of village life; they must walk extra miles to find water because the local village will not allow them to share it. The untouchables are barred from all jobs other than what they were "born to do," and from marrying out of their realm, and are rarely ever represented in government. Such class designations reflect an ancient hierarchy, sociopathic and insensitive, that is part religion, part expediency. Like the gypsies and orphans of the country— also condemned—the untouchables are considered the dregs of India, and yet they are its life's blood.[29]

India may have begun much like the Andaman Islands, or Bhutan, as a sort of spiritual clean room. In the hallowed late afternoon quiet of a white marble temple I have sat peacefully listening, watching, attuned to a life force that insinuates the most delicate nuances amid flurry. That silence within chaos is a uniquely Indian oxymoron, an introspective bustle. One feels its dramatic energies in the innocent, blessed glances of the working people who cannot, will not, hold back a smile at the slightest provocation, in spite of adversity. One experiences it among the indomitable children who have gaily converged upon me wherever I have wandered; and at the entrance to nearly any doorstoop where there is assurance of being invited in. The country intimates contagious affection and is easily capable of bringing out the best in a person.

India also summons the past. Twenty-five-hundred-years ago the country must have been extraordinarily graceful, majestic, intoxicating. But as with today, it must also have been seething. How else to explain the fact that two of India's premier religious leaders—the twenty-fourth Jain Tirthankara, Mahavira, and Siddhartha Buddha—both born during the sixth century B.C. into royal families, were to renounce all of their possessions, become celibate, and spend their lives wandering throughout the countryside analyzing the sources of despair and seeking ways to remedy a rampant problem through a comprehensive philosophy of nonviolence (ahimsa)? The conclusions they reached (what the ancients knew to be enlightenment) were nothing short of family planning, ecological sustainability, and a deep, abiding reverence for all living beings, consistently transmuted into action. Whether Jain, Buddhist, or Hindu, the soul was believed to transmigrate upon death into other living incarnations on its journey toward moksha, or nirvana, or atma—liberation, non-being, the world soul. This was to become the goal of religious life with one's karma, the feedback of one's actions, stipulating the conditions under which a person would return in the next life. A ceaseless series of reincarnations (samsara) ultimately led to release from this painful world. But in the meantime, each rebirth took the form of some other organism—a cricket, an elephant—and thus, the human soul was intimately related to all other life forms. The Hindu God Vishnu, the preserver, was identified with the eagle, the boar, the fish, and the turtle; the swan was associated with Brahma, the creator; the Brahminy kite with Durga, goddess of fortune. Even the princely rat figures as the mount of Ganesh, the god of wisdom whose head is that of an elephant.[30]

By the third century B.C., following Alexander the Great's retreat[31] from India and Chandragupta's consolidation of the Mauryan Empire, South Asia is believed to have contained some 50 million people. Population pressure was already being felt, the gods seemingly in competition. Then, one thousand years ago, the Sword of Islam struck across northern India, beginning with the raids of Mahmud of Ghazni. A succession of Turko–Afghan Muslim invaders proceeded to demolish one Hindu temple after another. Kanauj, one of the largest cities in the world at that time, was decimated, fifty thousand faithful Hindus allegedly slain in a single day. The God Siva's much counted-on magic lingam (symbolic penis), made of iron and suspended in a "womb-house," did nothing to stop the massacre.[32]

By the time the East India Company first dropped anchor at Surat in 1607, Mughal rule had fully transformed a country of spirit into one of coveted matter. Emperor Jahangir declared that the British appeared to produce nothing of interest and were refused entry. After all, in Surat's teeming bazaars one could purchase everything from leopard skins and peacock feathers to albino elephants and slave girls. Five years later, however, the East India Company returned to Surat with a bold display of cannon power, which worked its effect. Beginning with this military foothold, England began cultivating India, at first as a voracious consumer of Indian goods such as calico, indigo, and saltpeter. Soon Bombay, then Orissa and Bengal opened their arms to the British, granting trading concessions free of duties, and more importantly, the right to purchase land for the building of factories.[33]

That same Mughal culture that received the British had already escalated the indigenous pace of human appropriation throughout the country. Jahangir, who kept a nature diary, boasted of having personally killed nearly fourteen thousand animals between 1580 and 1616. The emperor Aurangzeb employed his entire armies—when they weren't killing people—in the flushing out of game.[34] In a National Museum of New Delhi miniature,[35] a paradise scene from eighteenth-century India depicts a lotus pond with crocodile and ducks, peacocks in the trees, and a tiger couple at the water's edge. The scene is punctuated by the central fact of four hunters—scantily clad females with sword, and bow and arrow, and rifle. They have not the slightest compassion in their hearts. One of the women, seated in a tranquil yogic position, takes aim with her gun, seconds away from killing the tigers. The whites of the womens' eyes shine with menacing calm. Yet this was the same nature accorded inviolate status in India's most revered texts.[36]

But the Mughal impact on nature was amateurish next to the incoming British (and to a slightly lesser extent in the south, the Portuguese) who devastated whole species and cleared the forests in order to grow tea and coffee and a host of other "concocted" export products. However one might profile the range of indigenous exploitation that had gone on for thousands of years, earlier mayhem now all appeared rather sustainable, by comparison with the new foreign plunder.

In Kerala, for example, virtually the entire coast was cleared of its natural vegetation and replaced with coconuts, once the Portuguese discovered the superior quality of coir over wood for cordage, caulking, and ship rigging. The wide diversity of animals that inhabited the coastal thickets was exterminated, while birdlife that could, migrated inland. In addition to the astounding loss of wildlife, the soil suffered increased erosion as a result of the superficial root structure specific to the coconut tree.[37] Throughout the Western Ghats—that range of southern hills that parallels India's west coast for sixteen hundred kilometers from the mouth of the river Tapti in Maharashtra to the southernmost tip of Tamil Nadu—populations of the Malabar giant squirrel, lion-tailed macaque, and Nilgiri langur were threatened as their forest homes were converted to tea and cardamom plantations by the foreign conquerors. Even hundreds of years ago, the fragmentation of ecosystems began to sever one genetic group from another.

Today the jigsaw puzzle of India's biological heritage is impossible to recognize. More than 90 percent of the pieces are missing and what is there is surrounded by roads, fences, and human scavengers, men and women born into a world where they themselves are instant ecological victims.

On top of India's existing labyrinth of roads, a proliferation of railways and factories were introduced to accommodate Britain's greed. Following the Indian Mutiny of 1857, the British Crown took over the running of the country from the East India Company; which it managed like an aggressive corporation. By the 1860s, according to a local conservationist, the British were shooting any animal that moved. Local Indian hunters who had not previously thought to kill off all the wildlife, now did so systematically for profit, given Britain's insatiable demand for feathers, skins, exotic foods, and forest products.[38] British forestry meant maximum extraction for the British; neither India's environment nor Indians were part of that equation.

To examine the old archival pen-and-ink drawings featuring the bored young British hunters in India with their local guides standing over the corpses of tigers is to glimpse the nauseating fate of empire, what Oswald Spengler politely summarized as "the decline of the West." In 1911 King George V along with twelve thousand helpers came to

India and Nepal to hunt. The party rode into Chitwan on six hundred elephants and proceeded to kill thirty-nine tigers, twelve rhinoceroses, and four bears. In the early 1900s, according to author Charlie Pye-Smith, there were twenty thousand *shikaris*, or hunters, just in the Punjab alone. It was not uncommon for some spoiled, indolent lord or another to shoot down four thousand birds at one go. Preservation laws were enacted only on paper.[39] No such regulations were about to curb the British passion for hunting, especially since profit had entered into it. Adding her own show of support, in 1961, a year before India went to war with China, Queen Elizabeth took part in a tiger hunt at Ranthambhore, today a beleaguered national park. A photograph shows the Queen standing with other "dignitaries" beside the corpse of a large tiger her party had bagged.[40]

Whatever contradictory gains were made by those few British who cared about wildlife, India's independence and a population of nearly 400 million overwhelmed the existing sanctuaries with a people's desperate hunger for land. A new *National Parks Act* was initiated in 1955, during the same period when the Indian cheetah went extinct, yet as of 1965 only twenty-six hundred square miles of the country had been granted sanctuary status. More land than that had been preserved a century before in the United States, where the population was a small fraction of India's. What this meant was that by the time India—under Indira Gandhi—had truly taken its biodiversity loss seriously, with the commencement of Project Tiger in 1973, the damage was nearly irreversible.[41]

Indian wildlife and habitat loss poses heartbreaking choices for a country headed toward doubling, possibly tripling, its population size in the twenty-first century. Approximately 4 percent of India is protected (on a par with the global average). That land is divided into more than seventy national parks and three hundred seventy wildlife sanctuaries. However every one of these allegedly protected areas is falling prey to the surrounding human tumult. As a result nearly every ecosystem and a huge number of species are in trouble, though opinions differ greatly as to the actual total number of species in India.[42]

Along with their primeval habitats, India's tigers and rhinos, turtles and crocodiles, elephants and even India's national bird, the peacock, are all being driven to *ecological virtual reality,* to mere fantasy and memory, in other words, by poachers and the countless hunters working on behalf of breeders and international zoological collections, but mostly by the very encroachment of human beings trying to survive.

At the same time the cow and the buffalo, reverence for which Mohandas Gandhi described as the core of any vegetarian Hinduism, are today being massively butchered for their steaks under cruel and deplorable conditions, along with goats, chickens, sheep, camels, and nearly every other domestic creature.

This animal peril is the direct outcome of the ever-decreasing availability of space and natural resources. Consider that Indian industry is now producing over one trillion pounds of iron, steel, cement, and fertilizer each year, an avalanche of extraction. Since 1951 India's energy use has increased thirty-seven times per capita. In 1989 what 850 million Indians purchased in more abundant quantities than anything else were soap, nonbiodegradable detergents, and batteries, as well as cotton and jute, paper and plywood, limestone and rubber.[43] The fact that 80 percent of all Indians are living in varying conditions of impoverishment, with little or no access to safe water, sanitation, or quality medical attention, has meant that whatever consumption does occur takes place in a highly concentrated manner.

Adding to the desperate character of that equation is the fact that the primary gains of the Green Revolution are over in India. Indian agriculture, based principally on cereal crops such as maize, wheat, rice, bajra, and a host of coarse grains known as jowar, is now rapidly in decline, yet the requirements for foodgrains are going to increase from 146 to an estimated 238 million tons per year.[44] It has been shown that 87 percent of all Indian farmers are suffering from degraded land.[45] As of 1985 half of India's 3.5 million square kilometers, equivalent to 329 million hectares or approximately 1.35 percent of the Earth's terrestrial share, was already being deemed "wasteland," and as much as 260 million hectares additionally moisture-stressed.[46] What this all means is that food production is no longer guaranteed or self-sufficient, while the marginalization of small plots and the growing inequities of food distribution have exacerbated a growing nutritional illiteracy. Per capita availability of pulses, grams of protein, and edible oils has all been decreasing steadily. Not surprisingly, 85 percent of all Indians have substandard diets. The National Institute of Health and Family Welfare states that 65 percent of all children in India are malnourished. Considering that more than half the country is under the age of fourteen, and that another 160 to 180 million newborns will have to be fed, somehow, by the end of the 1990s (at an untenable cost to the country of 16,000 rupees per infant), this data is absolutely devastating.[47]

This glimpse of widespread famine on the horizon is further strained by the fact that there is a widely recognized sex bias with respect to food consumption within the Indian family. Women inevitably receive much less food than men, even accounting for weight differentials. This nutritional oppression impacts mental capacity and individual motivation, further aggravating the existing crisis of illiteracy among women in the country, 75 percent of whom cannot read a newspaper. Illiteracy varies from state to state across India. In 1921 less than 2 percent of all Indian women were literate. More than seventy years later the all-country average is still less than 25 percent, and almost double that for men.

These trends—and others—have utterly invalidated earlier predictions about the country with respect to food consumption. For example, the U.N. Food and Agriculture Organization published a study in 1982 suggesting that Indian agriculture could feed 2.6 billion people. But that was before a subsequent investigation showed that at least a third of the country's arable land was "completely unproductive." Later it was argued that the Indo-Gangetic plains by themselves could easily feed 1.8 billion people (the World Bank stabilization figure) assuming "the Indian diet remains essentially vegetarian as it is today."[48] That too has proved to rest upon sadly erroneous assumptions. According to a recent anthropological survey of India headed by Dr. Kumar Suresh Singh, it was found that 88 percent of India's communities— of all Indians—eat meat, which can mean anything from field rats and baby crocodiles to civet cats and jackals.[49] So much for the sacredness of life, the reverence for cows, and the sadly deluded Western ideal of a pacifist India. In addition, at least one-sixth of the Gangetic plains are now periodically devastated by floods as a result of watershed destruction in the Himalayas where sown areas are frequently three times the recommended amount, population density per hectare may exceed ten persons and ten cattle and nearly half of all springs have gone dry. Eight thousand square miles of Indian arable land are lost in the form of top-soil erosion every year: this is six billion tons, double that of the U.S.[50] (But remember that India has one-third the territory of the U.S.) At the same time all of the associated problems of intensive irrigation have come back to haunt the country. Virtually no new irrigation expansion is possible in India and more than a third of existing channels are clogged with silt, are leaking, or have added to the existing alkaline wastelands. In addition, two of the country's largest states, Uttar Pradesh and Bihar—the breadbelts of the Gangetic plains—find more than half of their residents well below the poverty line. This does not argue for the kind of sound environmental tendering that would ensure fallow periods, forest cycles, proper irrigation, and agricultural restraint—all of the key elements of the aforementioned FAO report.[51]

The Ganges, as well as India's thirteen other major rivers, is severely polluted. Nearly a billion and a half liters of untreated sewage are cast into the 1580-mile-long "Ganga" every day, though the Ganga Action Plan has been trying to intercept this effluent. Eighty percent of all hospital patients in India are victims of environmental pollution, mostly water-borne.[52] Many victims catch their viral diarrhea, typhoid, and infectious hepatitis by drinking holy water (charanamrita) at the countless temples, where disastrously high levels of fecal coliform bacteria—levels eighteen times higher than the Bureau of Indian Standards—have been consistently recorded.[53] Nearly 60 percent of all Indians are without access to safe water, 93 percent of the populace in Bihar. Water tables in some parts of the country have fallen by a hundred feet. In Haryana and Punjab, key agricultural states, aquifers are nearly exhausted.

Many other Indians live in atrociously polluted cities, or in the vicinity of unregulated heavy industry. New Delhi is considered to have possibly the worst air pollution in the world for many months each year. The sheer ubiquity of the two-wheelers (called "fart-farts" locally), which contribute 65 percent of air pollution in the city, makes them nearly impossible to control for emissions.[54] An astounding half of all adult residents in fifty villages surrounding the State of Orissa's Talcher Thermal Power Station have tuberculosis. Ash from the power station contains nearly 63 percent silica.[55] Along with TB, the villagers of Talcher are suffering from a host of additional infections. Women are particularly vulnerable, as they may spend 80 percent of their lives indoors, exposed to the equivalent of a constant low inversion of smoke and other chemical pollutants. Among school children in Talcher, 60 percent have tuberculosis.

Along with declining agricultural yields other woes are benighted by India's swarming numbers and swelling pollution regimes. As many as 115 million Indian children who should be in school are not. Rather they are working under adverse industrial conditions with little respite offered by the government. The 1986 *Child Labor (Prohibition and Regulation) Act* offers them no protection from hazardous conditions, or from the frequent beatings and sexual abuse of their employers. Nor are children allowed to form unions to protect themselves, under a 1926 law. The situation represents a colossal area of human rights abuse, which some foreign parliamentarians have tried to stop through the power of boycotting imports of Indian products made by children. But nothing has changed.

India's child labor force is equivalent to one of the ten largest national populations in the world. The work conditions have engendered a painful breeding ground of serious health maladies from chronic bronchitis to stunted growth. The children routinely handle toxic chemicals like potassium chlorate and zinc oxides, starting at three in the morning and going until evening. Girl workers outnumber the boys three to one. There is no such thing as holidays for them and no government policing. Children from the age of three years are employed making matches and fireworks and carpets because of their alleged nimble fingers. Older children labor in stone quarries in Kerala, slate factories in Andhra and Madhya Pradesh, in deep mines in Meghalaya; they work in road project trenches, manage handlooms in Kashmir, or spend their young lives handrolling Indian cigarettes (Bidi) in nearly every town and city. They work electroplating and hand presses as well as glass furnaces that put out seven hundred degrees of heat. Children are, along with women, India's most exposed and oppressed class. Environmental epidemiology among them is next to impossible to establish, though the diseases are expressed in India's tragically high infant and child mortality statistics.[56]

The oppression of people and of biological habitat follows identical patterns. All human ecological exploitation backfires. Where people are directly dependent on the land, this self-perpetuating cycle is most apparent. The majority of India's natural pathways, and 80 percent of the human population, are ecologically degraded to varying degrees. Two prevailing aspects of overpopulation come immediately to mind. First is the theory that the minimum space needed by any animal to survive is approximately the square (in body size) of its length.[57] Second, according to the so-called "area effect," an increase or decrease in space by 90 percent either increases or decreases the number of potential species (or presumably, existing ones) by half.[58] That, too, is terribly optimistic.

Both of these biological determinants of what, over time, should be described as a minimalist ecosystem are greatly compromised by a large population wreaking havoc with available space. What I have called the THAR effect, or Total Human Appropriation of a Region, results from industry, agriculture, irrigation, and the countless other urban, suburban, or rural ways by which humans overintrude. But there is no greater, no more visible area mal-effect, no more devastating loss to India than that of her forests, the substance and background of so much of the country's religious landscape, the source of water catchment and subsequent food self-sufficiency, and of fuel and fodder supply, the basis of employment for tens of millions, the roof overhead, the floor underfoot, the realm of most terrestrial biodiversity. All are fast disappearing.

My first glimpse of this deforested war zone was from the air. I flew on across the country one remarkably clear day, from Pakistan to

Myanmar, following just south of the Ganges, scanning with my binoculars. During four hours over Indian air space, I saw all of two-minutes-worth of forest, and even those few stands were probably degraded. While the "official" forest cover hovers around 19 percent, the former minister of environment and forests, Maneka Gandhi (widow of the late Sanjay Gandhi, daughter-in-law of the late prime minister Indira Gandhi, and one of India's truly radiant and empathetic beacons) indicated otherwise to me.[59] At her home one evening she explained how she had studied the satellite images and discovered that, at best, India retained 9 percent, not 19 percent, of its forest. And even that 9 percent figure came bolstered by the presence on satellite of sugarcane crops. Satellites do not distinguish between forest and mature sugarcane stands, which are quite high. Furthermore, the sugarcane, unlike the forest, admits to virtually no biological community, while inviting the heavy burden of soil-destructive pesticides.

India currently begets 20 million newborns every year. With a total fertility rate (TFR) of 4.1, an average life expectancy of sixty, a GNP per capita of one hundred sixty dollars, and a total population-doubling time of thirty-four years, this population is poised to totally decimate India's few remaining forests, whether the existing cover is 9 percent or 19 percent, in very few remaining years.

However, according to minister of environment and forests Kamal Nath, a dashing sort of political figure blessed by a sense of urgency that lends enormous zeal to whatever he says, India is the only country in the world that has actually enjoyed a net accretion of forests. The apparent contradiction, an extreme one, is symptomatic of all facts and figures in India. When Nath's experts informed him that two million hectares were afforested in 1992 he demanded proof. They handed him satellite images, different ones than Maneka Gandhi had seen, presumably, and added that every new seedling had been entered into a computer. Nath wanted to see what the seedling survival rate was so he appointed some NGOs (non-governmental organizations) to make a random check of at least 10 percent of the alleged replanting. According to Nath, the survival rate was excellent.

But Maneka Gandhi is skeptical. She says that the government has on occasion admitted that it has a 70 percent survival *failure* rate. The planters are paid four rupees per seedling, whether it survives or not. The faster they work, the more money they make. Survival of seedlings is not their first order of interest; digging holes and slapping down seedlings is what counts. In New Delhi, says Gandhi, there are eight thousand government planters; which is more than the total number of forest planters in the rest of India, yet seventy-nine hundred of those eight thousand planters, she insists, do not really know how to plant a tree. She should know. She wrote the training manual herself after having tested the staff. The top senior planters had come to her house to plant a very hardy weed called *Lantana*. After about a week it died. She had it dug up to do a sort of weed postmortem. It turned out the gardeners had planted the *Lantana* inside the plastic, and had made similar mistakes throughout the city, even planting trees upside down! If they can get away with it in the nation's capital, consider what goes on in the hinterlands, she said.

According to Mrs. Gandhi, in India there is little accountability. The law states you cannot cut down a tree unless it's dead.[60] So people set fire to trees or debark them and then when they have died they are "legally" cut down and taken away. "If you look at the trees anywhere in India," she says, "you'll note that they have very few branches."

Mrs. Gandhi's home, protected by round-the-clock soldiers carrying machine guns, and shaded by quite a few trees, adjoins one of the most polluted roads in New Delhi. "You'll get brain damage if you stand for ten minutes at the crossing," she exclaims, referring to the poor traffic cops who must withstand the constant chemical barrage. Along the roadway, she had her staff plant large neems, sacred in India, which allegedly give off the highest amount of oxygen of any tree species. (The government is currently developing contraceptive creams for women from the neem.) But the arbres all died along Mrs. Gandhi's road because in the mornings the slum dwellers would come and break off the branches for a variety of uses. "I can keep planting trees and they keep reducing them to toothbrushes," she says with a weary smile.

India's tree-planting efforts are "a total sham," Mrs. Gandhi reiterates, pointing out that her former Ministry of Environment and Forests is allocated less money than any other ministry in India. Until 1981 the Ministry itself was set within a beautiful grove of forest; then the Asian Games committee convinced the Government to build huge high-rise structures and tear down all the trees. Gandhi replanted with ten thousand seedlings that are now coming up. She supervised the planting. They are her legacy. Two years after coming into office, she says, corrupt officials threatened by her views, her attempt to legislate a host of new animal rights legislation, her concerns over unregulated industrial pollution, and her efforts to revamp forestry programs, managed to get her booted out of office. She believes the election was rigged.

India does not utilize its remaining forests as a revenue source—as they do in places like Malaysia and Indonesia—but as a community resource, says Kamal Nath. (Malaysia takes in an annual five billion dollars in income.) Nath points out that he, too, could reap such profits, if he chose to do so. Rather, he says, he's trying to preserve his forests, which is why India sought financial encouragement from international lending agencies. "I told the President of the World Bank, I said, 'Do you understand this, I wouldn't need you. But we don't cut our forests. We are doing no lumbering.' This is what the world doesn't appreciate. We have no forest contractors in our country. Our total requirement of timber products is 18 million tons per year. We import it. Despite having a bad balance of payments position, we imported 40 billion rupees worth (in 1992). I didn't have to do that. But I did it to protect our own forests. Not only as a community resource, but as a responsible country we understand what forests mean. But forests are our national resource. It's not a global resource, as many countries would like it to be."[61]

This "think globally, act locally" concept raises certain, seemingly insoluble ethical questions. Economists have long wondered how to morally address trade-off possibilities between different consuming groups. But if "consumer sovereignty" is deemed politically obligatory— as it usually is in peace time, and as a community resource suggests— then, as economist A.K. Bhattacharyya has described, "the gains or losses suffered by any population set x 1 cannot be set off against the losses or gains of another population set x 2." This is why the North/South dialectic remains utterly unresolved. One man's torment is another's pleasure, but both are politically equal under international law, particularly the GATT, which has seldom acknowledged environmental considerations as valid terms of denial or bias between trading partners.[62]

The 18 million tons per year that Nath refers to as India's total timber requirements are only the beginning, however. And this is the terrible dilemma confronting politicians like Kamal Nath. Pressure on the forests of India comes not so much from "consumption" as from subsistence. In the West, it is consumption that kills. In India it is the lack of consumables, a paucity exacerbated every day, as nearly fifty thousand newborns come into existence. And every day, at least 400 million Indians carry wood out of the forests in bundles on their heads to provide fuel, food, housing, fodder, (and toothbrushes), as well as sheer income for those children. The law allows for one large bundle per day, as much as you can carry. And this translates into about 190 million tons of fuelwood per year, at India's present population. Nath is quick to point out, as he did at the Rio Summit, that the government must allow its people the opportunity to consume its forests, sustainably it is hoped. "God forbid that India should ever take to industrialization after the manner of the West. . . . If an entire nation . . . took to similar economic exploitation, it would strip the world bare like locusts!" says Nath.

Indians are doing a pretty good job of it on India, certainly. They cannot afford oil.[63] If Indians could afford oil—or if the forests were somehow rendered off-limits to Indians (an impossibility)—then, says Nath, the price of a barrel of oil would jump to a hundred dollars worldwide. It is this latter point that Nath hammered home at Rio when he insisted that the Northern economic bloc recognize that its continued franchise would be undone if the South should fail to conserve its own resources. "And hence, it is in the interest as much of the North as of the South that there not be any parsimony in participatory mechanisms designed to give expression to this recognition of the principle of shared responsibility," he argued. To this end Nath, like nearly every other minister of environment in the developing world, is not simply talking about management, but about money, which Nath feels is owed to India. "We are seeking recognition of the principle that since historically it is the exploitation of the resources of the South that has fueled environment-damaging development in the North, a small fraction of the prosperity so secured must, as a matter of obligation, be devoted to survival into the future."[64]

The language of "obligations" is an unwelcome one to Western ears. But such intent is more fiercely sabotaged by the cruel truths of self-preservation in India. Trees are what the politicians have to give to the people—not money. One headload of wood per person per day. By subsidizing the poor in this way, the government believes it is slowing down the pace of total destruction, but is helpless to prevent it. In Varanasi, where there are no forests to be plucked, an electric crematorium has now been built because people cannot afford the eight hundred to one thousand rupee cost for a wood pyre on which to dispose of their dead according to Hindu custom. It used to be that the same funeral pyre cost fifty rupees.[65] This is but one example of the impact increasing population is having on a declining resource base. In India land is divided into three categories: revenue land, smaller private holdings called *patta*, and existing forest. All are destined for exploitation. As for revenue land (government nonforest lands that account for a vast majority of India), anyone can homestead (squat) there. When they have squatted long enough, it becomes by law their private *patta* at which point they can burn it down if they should so desire. You see the squatters everywhere in India. There are several million of them in Calcutta and Bombay, Delhi and Madras. By law, squatters' rights even supersede forest rights. Several cases have been challenged in the courts. The squatters always win. India is a tenacious democracy. But as India is discovering, democracy and ecology are at odds when the population overshoots its carrying capacity. China learned the same lesson from a different political point of view.

The Indian squatters are free to consume the forest (other than forest sanctuaries—though even these last few hundred refuges are succumbing to population pressures). Throughout the country, most animal species have already been evicted to *nowhere*, as will be described later on, while India's seminomadic masses clear one new area after another. Unlike leaf-cutter ants, however, these human confluences take the whole trees. While chimney smoke is China's most lasting symbol, the smoke from small outdoor cooking fires, whether on city streets, along roadsides far away in the country, on mountain ridges, in tea plantations, or in jungle swamps, is what most indelibly characterizes India's war zone, its unrelenting contribution to the global Greenhouse Effect.

The so-called Tree Patta Scheme introduced by the central government in several states, provides legal rights to the landless poor, allowing each family approximately one acre on which they may collect loppings, fruit, flowers, leaves, and deadwood. They may also keep bees. It is a law oriented toward saving human lives and trying—against unholy odds—to save the trees. But it is not working well enough.[66] Estimates of "adequate density," board feet, hectares, primary forest, and secondary forest, are at bewildering variance. Differing definitions of what constitutes a "forest" invite multiple interpretations. According to one government figure, 2.5 million acres of trees are lost every year.[67] But the World Resources Institute in 1992–93 stated that India appears to be losing a larger percentage of its forest every year than any other country in the world—over 3 percent of its total remaining stands, which would translate into fifteen thousand square kilometers annually. By comparison, Brazil is losing 13,820 square kilometers a year, or 0.4 percent of remaining forest; Ecuador is losing 2.4 percent, Indonesia 0.9 percent, or 10,000 square kilometers.[68] At this rate, simple calculations suggest that within another ten to fifteen years the forests of India will be gone, truly gone. Given its climate, and the monsoon floods from the Himalayas, India is on its way to becoming the next Sahara. Ten thousand years ago the Sahara was also tropical, like some remaining parts of India. Hippos waded through lush riverine canyon lands. Millions of birds nested in forested areas. Primates and big cats luxuriated in salubrious nests and lairs, as they still do in a few remaining parts of East Africa. Over time, meteorologic vagaries displaced North Africa's curtain of rain more toward Europe, exposing the Sahara and eventually extinguishing much of its diversity. Now a similar fate, strictly human induced, strips India bare. In parts of the country, deforestation has reduced annual rainfall from 150 to 75 inches.[69]

The forests, it appears, will be lost. But that's not all. The precious grasslands are also losing their battle against humankind. India has 11 percent, or eleven hundred of the world's species of grasses, but they are being rapidly eliminated by the largest cattle population in the world and by human development, or reduced to dry steppe, a mix of grass and thorn scrub.[70] The country hosts 50 percent of the world's buffalo, 15 percent of all cows, 15 percent of all goats, and 4 percent of all sheep. Population pressure among these animals in the twentieth century has reduced necessary grazing space by ten times, leaving less than 12 million hectares of suitable pasture. As a result of malnourished cattle, milk yields in India are among the lowest in the world—173 kilograms per cow versus 3,710 kilograms in Denmark.[71] And thus, the ecological backfire delimits the very calcium content in the bones of children. One sees cases of rickets throughout India. It is the perfect irony: a country with vast numbers of milk-producing bovine destroys its feeding base and thus depletes the quality and amount of the milk. The Gangetic Plains has lost nearly all of its grasslands. When grasslands are managed properly, they can grow back quickly. Not so forests and shrub, which are being increasingly denuded by India's goat and sheep and free grazing cattle populations, which—lacking sufficient grassland— extend their grazing behavior toward the shrinking forests, where they would not ordinarily venture because of tigers and lack of grass.[72] The biodiversity of the grasslands has been completely unsettled, with many open plains–dwelling species losing out to livestock overpopulation. In many regions of India the wolf, the Bengal and lesser florican, chinkara, partridge, and quail are disappearing, as are the great Indian bustard and blackbuck.[73]

Coastal regions are also mired in the population vortex throughout India. For example, where there are people, there is pathogen-bearing excrement, lots of it. I spent one sunny morning surveying beach feces along a four-kilometer tract in Versova, a northern suburb of Bombay, where a thousand or so squatters live in shacks on the beach, butting up against half-a-million dollar highrise condomiums. Every squatter leaves his or her fecal matter in an individual clump on the sand. There is not a single outhouse in the region for these people. And no one, apparently, has thought to dig a communal hole that would thus ease the burden on an otherwise magnificent coastal area.

In Goa Indira Gandhi made it illegal to build hotels within five hundred meters of the high-tide mark. Under the ministry of Maneka Gandhi, this stipulation was rigorously enforced. Not so under the successive administration. The tourist industry is developing over seventy hotels right on the beach, which means the probable extinction of the sea turtles in that region and the total alteration of all natural habitat. On the southeastern coast the endangered Ridley turtle is killed for its blood, an alleged cure for asthma. Near Calcutta rare freshwater turtles are captured and eaten alive.[74] Turtles, like frogs, are extremely sensitive to changes in their environment. When the turtles die out, one can be reasonably sure that the preconditions for natural fertility supporting a multitude of other species have been skewed, perhaps irretrievably.

India's coastal waters receive some twenty thousand tons of pesticides and fungicides and thirty-one thousand tons of detergents every year. Like the turtles, the country's coastal frogs (which would consume many of the pests, and could be targeted at crops to help minimize bioaccumulation of toxins in the food chain) are being exterminated, their legs sautéed and served on the tables of European three-, four-, and five-star restaurants.

Most of the country's coastal mangroves are also almost gone. A mere seven hundred hectares are left in all of Kerala—the size of a small ranch in New Mexico. In the Gulf of Kachchh (also spelled Kutch), twelve thousand tons of cuttings are taken each year by the locals for camel fodder, boat building, firewood, even fishing poles. And, according to local scientists, the people evidence no idea of the damage they are inflicting.[75] In addition, India's coral reefs—like those of neighboring Sri Lanka—are almost gone. According to studies by India's National Institute of Oceanography, forty thousand tons of coral are mined just from the Gulf of Mannar alone, at India's southern tip. From the Lakshadweep in the west to the Nicobar Islands in the east, the situation is the same: uncurtailed devastation.

Adding to the country's other water-related woes, India finds itself with fifty-one large hydroelectric projects languishing in various stages of incompletion, some since the time of Independence. Though the World Bank finally pulled out of the enormous Narmada Valley Development Project begun in 1987, the enterprise continues to muddle along—thirty large dams, one hundred thirty-five medium-sized dams, three thousand small dams—the largest such project in the world, larger even than China's own ecological boondoggle, the Three Gorges Dams Project. Hundreds-of-thousands of local people with nowhere to go will be forced to go there. Two hundred fifty villages will be submerged, along with a hundred thousand acres, a third of it scarce forest lands. Additional loss of biodiversity. For what? More power lines and irrigation ditches. The environmental protestors shout, *"Koi nahin hatega—Bandh nahin banega"* (Nobody will move—the dam will not be built!). Unfortunately the first of the huge dams has already been started in Gujarat. It is as high as the Golden Gate Bridge and half-a-mile long. Officials estimate it will irrigate nearly seven thousand square miles, supply 3.5 billion liters of drinking water per day for the locals, and produce 1,450 megawatts of hydroelectric power at a cost of roughly 7 billion dollars. Ironically, with more than 1,554 existing large dams in the country, Indians have access to only half the water they enjoyed at the time of independence.[76]

In the midst of all this country-wide demolition, perhaps no other environmental battle has captured the attention of Indian ecologists as has the fight to save the tiger, a species at risk of extinction, whose plight symbolizes that of every other exotic creature in the country. By 1972 there were an estimated 1,827 Indian tigers left, down from nearly fifty thousand at the turn of the century. At the twentieth anniversary of Project Tiger (held in New Delhi during February 1993) scientists argued with some pride that there were now four thousand Indian tigers.[77] Maneka Gandhi disagrees with that self-congratulatory estimate. "The entire figure of this Tiger Conference that just took place is rubbish," she told me days after the august assemblage. "It is based upon data manufactured by a certain so-called tiger specialist who multiplied a base population twenty years ago of three hundred by the number of presumed couples, their likely number of three offspring each two years, year-by-year, substracting 10 percent for poaching, subtracting 10 percent for other calamities, and thus arrived at a number of four thousand. He's a demographer. Nobody has a clue as to the number of tigers we have and I'm willing to bet it's not over a thousand." She says that police are constantly sweeping down upon various poaching rings. The government knows who many of them are. Pelts are confiscated. A single diamond merchant in Bombay was found with three hundred tiger nails in which he was preparing to insert diamonds and emeralds for export jewelry. The smuggling of tiger parts, particularly bones and skin, is big business in India. Tiger bones sell for $50 to $120 per pound. Yet nobody ever goes to jail because the police are so eager to take bribes instead.

Dr. Karan Singh who initiated Project Tiger insists that the tiger preserves must be expanded from nineteen to thirty areas, an unlikely possibility given the alarming trend toward denotification (sanctioned destruction).[78] What this means is that the state or national government has been officially chipping away at those existing reserves, either through a variety of commercial contracts, or often equally destructive extractive allowances for locals (a form of recompensation for those who were displaced—social justice). At least 1,175 square kilometers of tiger preserve have been lost just in the last six years to various forms of encroachment and denotification. If anything the nineteen protected areas will continue to dissipate, without a corresponding compensation in terms of either new parks or newly acquired land.

Consider the following examples. In just one state, Himachal Pradesh, thirty notified wildlife sanctuaries and two major national parks are now "imperiled." Wildlife authorities are said to "feel helpless in providing full protection to some of the rare species. . . . "[79] In other parks the government has stopped just short of declaring wildlife "industrialized." Poachers have reduced the estimated number of tigers at Dudhwa National Park from 104 to 21. There have been confirmed reports of poachers using helicopters to spot the tigers within the Corbett Tiger Reserve in Uttar Pradesh. One professional tiger hunter from Bangladesh claims to have killed over 56 of the rare animals.[80] With the ousting of the local villagers from what is now Ranthambhore National Park, there has been much resentment and encroachment by those dispossessed. The locals were not involved in the creation of the park and, in many cases, have taken to reentering the 115-square-kilometer buffer zone surrounding it in order to extract wood, badly needed for fuel and fodder. Eventually the buffer zone will be gone and the park will be encroached upon. The locals are barred from the park itself, but not tourists, and this perceived injustice has invited poaching as a revenge against the authorities.

At the Melghat reserve in Maharashtra, pressure from seventeen thousand villagers forced the government to give up 25 percent of the reserve in favor of small timber production.[81] The chief minister of the State was described in the press as "eager to win the support" of the people.

Thumbing their noses at the central government in nearby Delhi as well as at their state's own *Forest Conservation Act*, the chief ministers of Rajasthan have handed out some three hundred licenses for mines within Sariska National Park. For a century prior, maharajahs slaughtered tigers and hoofed animals there. By 1955 it had been deemed a sanctuary, though much of the forest had already been extinguished, and with it many of the resident cats, chinkara (gazelles), and nilgai (blue bulls). But most recently the park has proved a windfall to those who would prefer to see it mined of its dolomite and limestone and marble. When a local environmental group sued in the Supreme Court, there was a temporary stay on the park's destruction. The environmental leader was beaten up by the miners and then, adding insult to injury, was fined five hundred rupees in court for daring to interfere with what was proving to be a profitable venture for the state of Rajasthan, namely, $1.5 million a year in revenues from mining the park.[82]

One of the national ministers, according to Gandhi, is a businessman whose family enjoys a virtual country-wide monopoly on the tall power transformers. That same individual, she alleges, actually used his formidable leverage in government to have eighteen new coal-based power plants built, three of them in wildlife sanctuaries. He evidently sent a team of his own experts in to give a statement that there was not a single crocodile left in the Dholpur sanctuary in Rajasthan, where in fact there had been a tremendous comeback of the crocodile population. But his own experts said they saw no such revival and the installation of the power plant subsequently went forward.

There are some fifteen hundred villages in the existing nineteen tiger parks. The government is increasingly saying it cannot stop the deluge of denotification requests. Thus there is a nuclear plant proposal in Nagarjunasagar, a hydroelectric project that could destroy Manas, and a dolomite mining project already underway at the Buxa tiger preserve, like that in Rajasthan. In 1991 Darlaghat sanctuary in Himachal Pradesh was denotified for a cement plant. India is running roughshod over the entire country in the name of survival.

The question has become clear: will the government simply denotify (sanction the destruction of) more and more protected areas as adjoining village populations increase?[83] Meanwhile biologists and systematists throughout the world argue over strategy: whether to encourage governments to create and safeguard large protected areas or numerous small patchwork sanctuaries; how many animal migration corridors to implement; how large the parks themselves, and the buffer zones, should be.

When an all-India bumper crop of sugar cane diminished profits for harvesters in the Lakhimpur–Kheri district of Uttar Pradesh in 1980, the farmers began pressing authorities for an alternative form of income, killing the man-eating tigers that had roamed into the fields outside of Dudhwa National Park, which could then be stuffed and sold as trophies. Dozens of farmers had been killed by the tigers and meager compensation paid by the government. Further compounding the dilemma, tribal peoples forced out of the national park had taken to meting out revenge on the tigers, whom they blamed for all of their woes. The tiger conservationist Billy Arjun Singh argued vehemently for "reformation of the marauders," (referring to the tigers, not the farmers or tribals). But this was an arduous process at best, given the all-around pressure on the tiger's habitat—poaching by humans of the tiger's own prey species, rampant felling of trees within the park, and the alteration of the local herbivorous food chain, upon which the tiger ultimately depended, as a result of the unlawful grazing of cattle within the preserve. In the end, lacking even basic tranquilizing guns, the forest officers themselves participated in killing the so-called man-eaters.[84] The tigers never had a chance.

A similar conflict prevails throughout the Sundarbans, a vast region of swamps between India and Bangladesh. Once the dense human population killed off most of the tiger's normal prey—such as the gaur and the rhino—people themselves were preyed upon by a few of the big cats. In addition, human pressures on the local sundar tree, after which the surrounding mangroves are named, have resulted in further depletion of the overall ecology. The sundar is used for firewood and for building; thousands of boats transport bundles of wood every day through the tiger's habitat. With over 15 million Bengali migrants having swarmed into India (Bangladesh being India's Mexico, so to speak), the pressure on the tigers is overwhelming—millions of people versus a few hundred tigers. And still, locals, who could avoid confrontation with the animals if they merely took some care, blame the tigers for their own problems. (In fact, only 25 percent of the tigers are known to have ever developed a taste for human flesh.)

As bad as it is for the tiger in India, it is far worse for the Asiatic lion, an animal revered in India's ancient Hindu text, the *Atharva Veda*, but which began going noticeably extinct throughout India around 1814. The lion's last strongholds are the Gir and Girnar forests of Gujarat. When Lord Curzon was invited by the Nawab of Junagadh, on whose lands the lions lived, to go hunting for them, Curzon declined. When he learned there were only a dozen Asiatic lions left in the world, he agitated for their protection. But when a local famine killed off the lions' remaining prey, the lions in desperation went after nearby livestock, as well a few herders. This complicated the issue.

In 1966 six hundred square kilometres of the Gir forest were declared a lion sanctuary, but the human population pressures were intense by then. By 1971 nearly five thousand local maladharis (cattle herding families) and seventeen thousand head of cattle, in addition to another forty-eight thousand cattle from other areas, came to Gir for grazing during the monsoon. An estimated 90 percent of the lions' food was appropriated. By 1974 there were 167 lions. Though the herders and their cattle were relocated away from the sanctuary as of 1979, farmers are currently allowed to cultivate nearly 40 percent of the protected Gir valleys for cash crops. At the same time cattle continue to enter the sanctuary illegally while local harijans steal meat from fresh lion kills. The pressures on the felines are doomsdayish.[85] Nature has not provisioned for nearly a billion carnivorous humans to live agreeably side by side with a few big cats.

To reiterate: the same biological contests now affect every species in the country. The elephant is the most visible case in point. There are an estimated twenty thousand wild elephants left in India. Their migration paths have been destroyed or crisscrossed with electric wires to try to prevent their movement across farmlands. Skirting the southern boundary of the Mudumalai sanctuary in Kerala, eleven thousand cows and buffalo graze almost entirely on the park forest, incrementally destroying the corridor that had been, at great effort, preserved for elephants.[86] More than half of the elephants are found in the northeast of the country, and this is precisely where forest clearance for human development is occurring most aggressively. While the elephant is listed as a Schedule I endangered species in India, there are some local villagers who have complained that the forests in at least one northeast state, Assam, are overpopulated with pachyderms, citing five thousand elephants, of which one thousand are domesticated. A few of the four thousand wild elephants stand accused of having killed twelve people since 1990. In addition the government's ban on *mela shikar*, the traditional Assamese elephant-catching system disallowed since 1978, has put another one thousand persons out of work. A public complaint has been asking for elephants to once again become fair game for hunters.

A much less visible but nonetheless endangered Indian species is the musk deer, of which a mere two thousand are left. Its pod of musk now commands as much as fifteen thousand rupees, five years' salary for many Indians. A kilogram of the musk is worth well over fifty thousand dollars on the international market, making the deer's future less than likely. Musk deer farms could theoretically prevent this species from going extinct.

Even more endangered are the last few magnificent hangul stags of Kashmir whose sanctuary, the Dachigam Wildlife Preserve founded in 1951 on the outskirts of Srinagar, appears doomed as the result of population pressure in myriad guises. By 1971 the World Wildlife Fund's census revealed a mere 153 animals left.

Meat-eating poachers (mostly Muslim) in Manipur, between Myanmar and Nagaland, have virtually wiped out one of three remaining thamin races in the world—a rare circular-antlered deer known locally as sangai, the "deer that looks back at you." As of 1979 aerial surveys showed a mere thirty left.

It required six permanently established, armed antipoaching patrols to save the barasinga deer from extinction, and that was inside the Dudhwa National Park in Uttar Pradesh. Such patrols are few. The tragic truth is, India cannot patrol itself, anymore than a rich country like the United States has managed to keep illegal aliens from crossing its huge southwestern border.[87]

In Assam where the only Indian ape (the monogamous, affectionate and jovial hoolock gibbon) is to be found—along with the golden langur, the clouded leopard, the pygmy hog, wild buffalo, and rhinoceros—poachers have been known to kill twenty-one rhinos in a single month, while locals have been killing the hoolocks.[88]

In Kerala (which has the highest population density of any state in the country, the most recent estimates being some 750 per square kilometer) at Periyar Wildlife Sanctuary, two hundred forest guards and wardens have proved insufficient to keep poachers away. Elephants are slaughtered for their tusks (the large tuskers having been already wiped out, the younger ones are in great peril); the slender loris for its eyes; and lion-tailed macaques for their skins and their meat, believed by some locals to increase a man's potency—as if that were a problem in India. The monkey in the wild "speaks" certain words that some insist are actually Hindi words. Another local monkey, the Nilgiri langur, is hunted for its blood and organs, which are said to cure human ailments.[89]

In the high Ladakhi Himalayas, only one hundred Przevalski black-necked cranes and one hundred fifty Tibetan gazelles remain, as well as a meager few snow leopards, Tibetan lynx, and Pallas's cats. On occasion the Indian military is called out to protect these animals. But out along the Andaman, Nicobar, and Sentinel islands, far from the military, poachers have had a heyday, selling in open marketplaces endangered turtle eggs and rare meats of wild boar and chital, the giant monitor lizards. Nonindigenous hunters deploy large arrays of traps— much like gillnets spread out over hundreds of yards. Many of the local creatures are endangered and listed under Schedule I of the *Wildlife (Protection) Act.* Of the islands' 240 species and subspecies of birds, the endemic ones are mostly rare or already gone.

The precise outcome of these trends may be subject to some speculation. But it is only a question of time before Indians discover that they have brutally reduced their once bountiful jewel of creation to a sterile wasteland of simple ecosystems—weeds and desert—bereft of all but human diversity. Such is the imponderable volatility of the country's demographics and the sheer, untempered survival needs, bluntly sought, by so many people.

The prognosis for all animals in the country is grim. When I asked Maneka Gandhi what hope, if any, there might be, she was angrily resolved in her response. "None!" she repeated. Her face had the look of one who had wiped away quite a few tears in the past and had now come to accept—almost with a coldness—certain inevitabilities about India and the way of the world.[90] Though having recently lost her post as Minister of Environment and Forests, in many respects Maneka Gandhi had lost none of her power. Instead of moving whole mountains, her goal now was to save individual lives, to spare as many of the billions of doomed domestic creatures as possible. In this respect her goals correspond with those of all conservation biology: to save as many other species, and individuals, as possible, recognizing that human nature will not permit true egalitarian coexistence with other life forms. We seem relentlessly programmed against such harmony.

Mrs. Gandhi now runs the Sanjay Gandhi Animal Welfare Center in New Delhi. She writes a column and lectures occasionally but spends most of her time fighting to stop an epidemic of animal cruelty throughout her country. India is by no means unique in this realm, but the animal rights movement hardly has even a foothold in Asia, as yet.

Animal rights' advocates in India are viewed as the lunatic fringe," says Gandhi. "I don't have anything to be defensive about. I don't respond to critics. I am a very good politician in that I look after people as best I can. I don't believe in animals *or* people. I believe in animals *and* people. Or people and animals. Love doesn't stop. I'm not pessimistic. I'm realistic. In a country like this we shouldn't kid ourselves. But at least one can create an awareness. Then you'll have less people throwing stones. It's a drop in the ocean. But it's something."

Perhaps the first and only moment of national consciousness regarding animal rights occurred during Maneka Gandhi's term as minister. New Delhi filmmaker Himanshu Malhotra's half-hour chronicle of the horrors of the Deonar abattoir in Bombay, a film that was cosponsored by Gandhi, aired one night over India's national network, Doordarshan. This was one of the first powerful glimpses of animal suffering ever televised in the country.

What the viewer witnessed was the arrival of cattle trucks at the slaughterhouse. The animals had been tied together and loaded on top of each other, four abreast. By the time they arrived to be killed, their bones were broken, they were starving and terrified. Many were diseased. Some had already suffocated or died of heart attacks. Acid was poured into their nostrils and tobacco rubbed into their eyes to make them get out of the trucks. Most could not, would not move. So men dragged them by their tails.

Ironically, the only "law" that seems to be observed in most of the slaughterhouses in the country is one concerning the slaughterer's self-pride: he must wait until the animal stands up before killing it. It's a macho thing. The slaughterers persuade the animals to stand by forcing electric or molten iron prods up their anuses. Frantically, the twelve-hundred-pound-animals lurch up one last time, as a group of young people converge upon them with rusty hatchets and saws. They chop and slice and it can take a half hour before the convulsing animals are out of their misery.

The filmmaker Malhotra told me of his initial hopes that maybe at long last something would change in India. The film elicited some twenty thousand local protests for a day. Incoming cattle trucks were blockaded. The price of meat soared from thirty to sixty rupees per pound. But then, just as quickly, the furor died down. The Municipal Commissioner in Bombay insisted he was too "tied up" to deal with the problems and people accepted that.[91]

In New Delhi little children "trained" to be "brave" are led into the Idgah abbatoir where it is believed that as many as fifteen thousand animals are killed every day. Anybody is invited to enter a slaughterhouse and pay one rupee to kill an animal or as many animals as that person desires—any animal other than a human, it should be pointed out. I once visited such a slaughterhouse in the middle of the night. Young Indian boys, no older than ten, were there with long blades in a concrete bunker slaughtering dozens of sheep as if it were some kind of sport. The experience was as close as I can imagine to going insane, or wanting to murder someone.

There is no enforcement of laws, no stun guns, no interest whatsoever in alleviating pain and suffering. Just brazen, sloppy, horrible killing. Much of the meat is shipped to Arab countries. There are over five thousand butchershops that engage in such slaughter in New Delhi alone, over one million throughout the country. The law stipulates that cows cannot be killed until they are sixteen years old. But few are the experts in India who know how to tell when a cow is sixteen years old—or eight years old for that matter.

Over three hundred thousand buffalo are slaughtered in this and other equally detestable ways every day, as well as countless other smaller domestic meat-bearing creatures. Aside from the human emotional, ethical outrage, there are other costs associated with the slaughter. For every goat bred to be slaughtered, the country actually loses some sixty thousand rupees because the goats wander around public lands chewing up all the vegetation. It is estimated that 60 percent of Indian wasteland is the result of goats that have been bred for meat. The city of New Delhi has taken out a third loan for sixty-three million dollars recently to attempt to replant grass and shrubs in the surrounding Aravali hills, but no corresponding effort to stop breeding goats for slaughter has occurred.

Authorities blame it all on the Muslims' lust for meat. But there are between six and seven hundred Hindus eating meat for every Muslim. Much of the slaughtered pig, goat, and buffalo meat is going to fast-food chains throughout India. At the Muslim religious slaughters, such as the Bakhr Id festival thousands—hundreds of thousands—of camels, goats, sheep, and bullocks (to name but a few of the animal types) are killed, having been starved and left without water for days prior. Why feed them when you know you're going to kill them?[92] One Islamic animal rights leader came to visit me in Delhi. A woman in her late thirties, she was shaking, in obvious mental anguish. She needed outside help, she begged. The situation according to her was so desperate that she was prepared to martyr herself. How? I asked her. By going to Mecca during the annual *Hajj*, or pilgrimage, and protesting the routine slaughter there of tens-of-thousands of animals. Then what? I asked. Then they will kill me, she said.

In Tamil Nadu, Kerala, and West Bengal, cattle droving goes on throughout the night. Young herders run the animals, beating them unmercifully to speed up the journey. The cows are already half-starved by the time they arrive at the killing grounds, where their heads are bashed in with pipes. That's how it's done in the south.

Millions of old cows too emaciated to interest the butchers wander the streets starving to death. This is what India's supposed worship of the bovine comes down to. Desperate, the animals are forced against their nature to "steal" fruit, vegetables, anything they can find in the bazaars, or the garbage dumps, or along the dusty street surfaces. Vendors possess bottles of acid and splash the animals in the face to keep them away.

A huge population of bullocks are used for pulling carts. They are the tourist's notion of a quaint rural India. I once met a woman writing one of those "India On Five Dollars A Day"-type books and she told me that she loved the country because of all the "wonderful" cows.

There were no cows on the streets in Manhattan, where she came from. Seventy percent of these doomed cart-pullers rapidly develop open painful sores that often become cancerous. The animals die typically within three years, or are subjected to the slaughter routine first. No one has yet thought to design a more humane (let alone, economical) yoke, which is the cause of the open sores. Mrs. Gandhi has put out a call for such a design.

The horror stories are endless, of course. Every silk sari in India results from some fifty thousand silk worms being boiled alive within their cocoons. Crocodiles are lassoed, strung up, and stabbed in the neck repeatedly until they grow quiet. Snakes are grabbed in the wild and their heads nailed to a tree where they are skinned alive. The fashionable Karakul hat is made of lamb's wool. The ewe (the mother) is forced to have a premature birth. How? She is struck with an iron rod or steel pipe or hammer, as many times as it takes, to induce the birth. The newborn is then skinned alive so as to ensure soft fur. Civet cats are caged in a tiny container and beaten every few days because when they are in pain they secrete musk. The poachers continue inflicting the pain as long as the animal lives. In India they test talcum powder and lipstick by thrusting them down the throats of squirrel monkeys. When the monkey dies the "researchers" figure the dose was too much. In Goa, according to several animal rights activists in the country, the Catholic Church continues to sanction a tradition in which children tie up pigs to trees and then devour them while the pigs are yet alive (a practice not altogether different from a long-standing tradition of horrible animal sacrifice in Florida that was upheld by the U.S. Supreme Court as an expression of religious freedom). I hear the laughter and scorn of butchers into the night—an inconsolable cry coming from India's heartland.

The misperception of a vegetarian India is merely one of many myths. With the exception of the eight-to-ten million Jains, the one million Bishnoi, and scattered members of other religions in the country, there is absolutely no consistent respect, let alone reverence, for animals in India—or not anymore, consistent with the fact that only 12 percent of the country is vegetarian. In the realm of medicine and science,

India employs intensive duplicative experiments on animals, if for no other reason than because they are so cheap to come by. Five hundred monkeys are killed for every single monkey in any other country. The antivivisection board, started by Mrs. Gandhi and incorporating the earlier Animal Welfare Board, was discontinued once she was out of office. There appears to be no monitoring or regulation of animal experiments or of the use of animals in commercial research.

While the existing *Prevention of Cruelty to Animals Act* of 1960 is a very strong piece of legislation, it remains completely unread and unheard of by most people, including the police. Global treaties like CITES (the Convention on the International Trade of Rare & Endangered Species) makes little difference in India. Selling ivory is technically forbidden but it can still be sold under the law according to a provision allowing for the closeout of earlier inventory. Lacking any substantial body of organized field operatives, the government is virtually powerless to detect or curtail the sale of endangered species' parts even if it were serious about doing so, which it is not, according to Mrs. Gandhi.

Peacocks drop a feather a day. Peacocks are the national bird and cannot be harmed. When Gandhi became minister she was skeptical that the law was actually protecting them, given that the country was exporting a million feathers a month. At that rate, no village merchant is waiting for one peacock to drop its one feather and then sending a little boy to go fetch it every day. He's going to kill the bird, which is exactly what was happening, making India the only country in the world that kills its national bird for export. Gandhi thus banned all exports, but the government is now planning to renew monthly quotas, evidently.

Even "man's best friend" is not immune to this carnage—quite to the contrary. Authorities go after the communal (stray) dogs, the hungry ones, with sweet cajolery, then break their legs with metal tongs, drag them by the hundreds into a tiny room and starve them to death, or douse them with water and begin the slow process of mass electrocution that typically can take half a day before the writhing animals expire. There is sadistic pleasure taken in these little holocausts.

Sometimes the slaughter occurs right out on the street. Nobody seems to mind. The animals are simply clubbed to death. It's all part of the government's supposed rabies control program. Yet the government itself has stated that at least 50 percent of all rabies in the cities of India come from pet dog bites, not from communal dogs. But there has been no effort to persuade dog owners to give their pets antirabies vaccines, a practice that has resulted in zero rabies cases in Spain and Hong Kong, for example.

Nor has India undertaken sterilization campaigns for its animals, though it has been quick to do so for its women. In the U.S., canine sterilization costs forty dollars per animal. In India, according to Mrs. Gandhi, the cost is *two cents*. If the city of Calcutta were to spend the 1.2 million rupees each year allocated for the indiscriminate killing of dogs, on their sterilization and the WHO antirabies vaccine instead, the problem would be solved. From an economic standpoint, the cost for the fourteen required injections for those humans who have contracted rabies is ten times the amount that would be spent by just one Indian city to curb the problem to begin with.

As I sat with Mrs. Gandhi listening to her many tales of tragedy, she dabbed a handkerchief to the flare-up of mange on her lip that she had caught from one of the many communal dogs she had rescued and was nursing back to health at her home, and she smiled. At that moment, I recognized hope for India.

"It comes up in fits and spurts," she theorized. "There are thousands of animal rights' groups throughout the country but there are no funds. We don't know what to do. When you don't have money you forget about it. You can be full of zeal when you're eighteen years old. But then by the time you get married and you're twenty-two and your husband says to hell with it, you shouldn't waste your time picking up stray cats, that's the end of it. There are plenty of people moaning about the rhinocerous and the tiger because they know there's nothing they can do about it. They're going to go extinct. But I want to show the people the common crow, the dog, the cow, the camel, the goat, the water buffalo, that lives right here and is suffering. Wherever you are you must make a difference. That's all one can do."

After enough exposure to these unimaginable cruelties in human nature, an inescapable conclusion surfaces, or at least it does in my own mind: the vague, incalculable hunch that human rights have got to be *overcome*, somehow conquered. What are our rights—to rape, plunder, kill, destroy, declare war upon, all other species? Those are the human rights that have to be abolished if the planet is to survive. Mrs. Gandhi—and many others like her throughout India—wants to see hundreds of animal hospitals established, vast new reforestation programs that are realistic, a ban on the exportation of meat, an end to cruel circuses and zoos, the establishment of training centers for wild animal vets, as well as for members of parliament, where they could see what was really going on in their country. Just to make them aware, to encourage new laws, to step up enforcement of those laws. To train far more antipoaching units and to allocate considerably more money from the annual budget toward environmental protection and restoration.

Short of an ethical revolution in India that coincides with serious efforts to immediately curb the population boom, the country will remain in a state of war. One of the clearest examples of how demographic pressure impacts the habitat and animal life comes from the region surrounding the northern town of Dehra Dun where Project Tiger was launched. Two hundred twenty thousand new settlers to the region cleared twenty-seven square miles of forest. India unleashes 1.5 million new settlers every month and that number is growing. Forest clearance for development means destruction of animal life. But the government of India sees development as its only hope to transcend human poverty, even while recognizing the increasing poverty of the environment. The *National Land Use Policy* of 1988 states as its two priorities: "To meet the consumption needs of a growing population by increasing productivity of the integrated land resource in the country" and "To prevent any further deterioration of the land resource by appropriate preventive measures."[93] But no one has yet divined a way to accomplish this hat trick.[94] Already, the countless development battles have left India devoid of any wilderness, according to the World Resources Institute definition of the term—four thousand square kilometers untouched by humankind.[95] There is only one remaining five hundred square kilometer section of the country that can be described as a wilderness—the Palpur Kund Sanctuary in the State of Madhya Pradesh.

A few years ago when the government of India wanted to try to expand the possible territory for the Gir lions, to their bewilderment they discovered precisely what zero wilderness means: there was no other place for the lions to go in the whole country. Palpur Kund was unacceptable because there were some tigers already in residence. Tigers and lions have only ever been known to get along under captive breeding situations.[96]

Just as the animals and their diverse habitats are slipping away, so too the human population plunges ever deeper into self-inflicted turmoil. This is a reciprocal promise, little appreciated in India until just recently. The per capita gap between the needs of a swelling population and the actual availability of clothing, housing, health, water supply, sanitation, electrical generation, and transportation facilities, is rapidly escalating. Of the 3,245 towns and cities throughout the country covering nearly seven million acres, only twenty-one, or 0.6 percent, have even partial sewage treatment facilities.[97] According to the National Institute of Health and Family Welfare, just to maintain the status quo in India the following additional requirements must be met in 1995 to keep pace with the runaway population: as many as 1.2 million additional hospital beds, 190 million meters of cloth, a minimum of 3 million new dwellings, 10 million tons of food, a staggering 127,000 schools (others have computed as many as 150 schools per day needed), 373,000 teachers, and 4 million jobs. This, of course, does not indicate the extent to which the country is already underserved in each of these domains. Researchers H. Simon and B. Sharma conclude, "If this situation remains unchecked, there will be socioeconomic chaos and confusion. People will be caught, at both micro and macro levels, in the vicious circle of poverty, underdevelopment and the wasteful demographic patterns of high fertility associated with high morbidity and mortality."[98]

Infant mortality is already between 80 and 95 per thousand, on average. Child mortality (from ages one through five) is a staggering 170–185 per thousand. Most of these latter deaths are believed to be avoidable. What kills the children is typically a combination of respiratory infections, diarrhea, malnutrition, measles, and whooping cough, and persistent, unattended fever.[99] Rural deaths are twice those of urban deaths. However virtually no epidemiological studies have yet been conducted among the more than 50 million urban slum dwellers, a figure expected to double before the end of the twentieth century when 35 percent of the country will be urban. (I did hear from several Indian scientists that the slum dwellers appear to be evolving remarkable immunities to the diseases that are killing everybody else. I did not see any data to support this theory however.) The annual medical breakdown in India is hard to absorb. Tens-of-millions of new cases of TB every year (more than half the world's total, in India), 3.3 million new cases of leprosy, 14 million new individuals stricken with filariasis (blood parasites), 1.7 new malaria cases, 1.4 additional water-borne diseases, and so on.[100] The first known HIV-positive case appeared in India in 1988. According to the World Health Organization, there are now an estimated five hundred thousand HIV-positive individuals—and nearly 2 million prostitutes. Some regard India as the next Klondike for the hungry virus and its predatory mutations.

And so it goes—facts, data, trends, calculations, statistics. A picture of human turmoil and ecological entropy. Of internal unrest not appreciably different than the agony in Somalia and the Sudan, Bosnia and Liberia, except on an infinitely vaster scale and with only an occasional terrorist bomb blast, assassination, or near nuclear exchange. But this is no civil war, and the distinction is an important one. What's happening in India as in China involves the total appropriation of nature, with profoundly negative human consequences. Nobody in particular is at war, though there are persistent ethnic outbreaks of killing—some well-known—as in the Punjab—others little mentioned in the world press. Stand upon any street corner, or in any field of India; chances are you will not be aware of this remarkable conflict, nor at all convinced of its myriad dimensions, though one cannot fail to perceive a certain battle-weariness in the air. Ecological plunder has entered the mindscape, blurring distinctions, losing focus altogether on the highlights.

To this day there are rat temples in Calcutta where the little creatures are treated like visiting royalty. In Rajasthan hundreds of desert Bishnoi, a subsect of the Hindus, have martyred themselves protecting the desert's trees and gazelles, while the Jains—whose entire religious and social orientation has consistently embraced compassion for all living beings—have founded *panjorapors*, sanctuaries for animals throughout many parts of the country. (See chapter 8 for a thorough discussion of Jainism and its importance.) Put in context however these scarce living traditions of love and decency and common sense are mere remnants, emotional throwbacks to an earlier age when there were literally millions of lay people who followed the spiritual gurus, or teachers.

But considering the whole of India—of Asia—what could have gone wrong? How to account for such pervasive cruelty and sadness? Simply stated, the abovementioned "literally millions." You see, the God Krishna himself allegedly had 16,108 wives.[101] Nobody knows how many children, or grandchildren.

## SEXUAL PLUNDER

India's transition long ago from a spiritual to a market economy has resulted quite systematically in the rape of nature and the negation of any "clean room." Furthermore, the socially upheld custom of intense utilization has extended far beyond the forests, the mangroves, the grasslands, the mountains, the coral reefs and coastal waters, to the bodies of women. Exploitation of India's vast biodiversity has coincided with an ongoing denial and ignorance of the human female's cultural plight. The psychopathology of these connections hints at the country's complex ecological and population crisis, a fact of no little significance to family planners who are confronted by stubborn historical biases favoring the male. Today's family planning slogan is, "*Beti ho ya Bete. Bacche Do He Achhe*" (Boy or Girl. Two is Best). Yet in India (as in China) the working slogan appears to be "Two Sons Is Best."

As early as the seventh century B.C., the *Dharmasastras*—The Rules of Right Conduct first codified by the famed lawyer Manu *("The Laws of Manu")*—placed women under the control of fathers, husbands, and then their own sons.[102] A girl was to be married prior to puberty—before the likelihood of her being raped, in other words. (In more recent times, because the amount of the dowry is lower for younger daughters, early marriages have resulted in India having the highest rate of teenage pregnancies in the world.[103]) The marriage rituals traditionally took two weeks. From the first nuptial arrangements, the mother-in-law was poised to rule the daughter-in-law. This was her only domain of power in her entire life. And thus, victimized over enough generations, women themselves served this male-dominated system in the guise of fertility taskmasters passing down from one generation to the next the male-engineered abuse.

During the marriage ceremony, both past and present, the spice turmeric *(haldi,* a member of the ginger family) was crushed and its decorative yellow powder rubbed on the couple's bodies to ensure that there would be children. As they walked around the sacred fire the groom's loin cloth and the bride's sari were also dyed in turmeric. Once the wedding ceremonies were concluded, the couple bathed in turmeric-laced water.[104] After her marriage (usually to a man outside her village) the young woman joined her husband's clan, moving away from her parents and friends, who might never even see her again. She was thus an economic write-off, as far as her parents were concerned.

Parents always wanted to marry a daughter off as young as possible, to cut down on their investment and dowry costs. She had to be a virgin, totally uneducated, passive, at the mercy of whatever man would have her. It was then the young bride's responsibility to bring males into the world. During her pregnancy she was told to eat little so that the baby wouldn't grow so large in the womb that it couldn't come out. And she was warned against green vegetables, which were believed to cause miscarriages. She was also forbidden from drinking milk because it was understood that milk makes the "baby stick to the uterus during delivery." When the daughter was ready, the *dai,* a village midwife, would push on the mother's stomach during labor, unaware that this could burst the uterus. Cow dung was applied to stop the bleeding. The umbilical cord was cut with a stone.[105] Once she had delivered a sufficient number of male children, the mother might secretly opt for some form of birth control for which, in ancient times, there were numerous prescriptions. It was said that "the woman who will drink for three days . . . a decoction of the Kallambha-plant and the feet of jungle flies will never have children;" or, "the woman who will eat every day for a fortnight forty Mashas of molasses . . . which is three years old will remain barren for the rest of her life." Animal intestines were used as condoms. Pulpless lemon, certain leaves, the lactic acid from fermenting acacia tips, and honey, were all placed in the vagina to impede sperm.[106] However, the Hindu, in general, was not eager to hinder birth inasmuch as that would eradicate the future evolution of a long line of ancestors looking toward reincarnation. Thus, in the *Bhagavad Gita*, Lord Krishna tells Arjuna, " . . . I am born from age to age."[107] Women came by necessity to rely upon the salvation afforded them in the next life, because this one was liable to be desperate.

One of India's most famed, beloved, and controversial demographers, Ashish Bose, puts it this way: "This country loves children, they simply love them. But probability theory works like this: people do not want large families, they want two sons. But probably, in trying to get two sons, you'll get four daughters first. In Kerala they love girls. In tribal societies they love girls. In Mizeram and Nagaland they love girls."[108] But among Hindus—the vast majority of Indians— girls traditionally were considered something of a disaster. "They say Indians want sons because they want to go to heaven and the Hindu religion says this and this. Nonsense," says Bose. "It is on account of the sheer strategy of survival." When Bose did field work in Rajasthan during a drought, he says he was reminded forcefully of this truth. The locals were afraid to send their daughters long distances to fetch water. Nor could girls settle the constant disputes and fights, or protect the family in case of natural disasters where physical strength was required. "Unless you have two or three sons you can't survive," Bose remarked. "The government can't help you. Only after the disaster will they do anything. Not before. If a fruit hawker gets sick, he'll have no food if even for one day he can't work. But if he has sons, they can shine shoes, sell cigarettes, sit on the street and get money, do something. Girls can't do that. In a country where there is no social security at all, no unemployment insurance, no old age pensions, nothing for the poor people, then what happens? They have to rely on the solidarity of the family. They must have sons. Or your brother's sons, or your nephews." Male preference is additionally predicated on the belief that unless your son lights your funeral pyre at death, you will not go to heaven.

Whether twenty-five hundred years ago, or tomorrow, conditions have not changed dramatically with respect to the woman's place in Indian society. Once honored, even revered, in the sacred literature, in fact, the mother was more usually desecrated and deemed filthy throughout her life. If a newborn was unlucky enough to be born female in India, particularly in rural India, the mother-in-law was likely to meet the exhausted and typically anemic mother with a sentiment that suggested, "Nothing was born." After the birth, the mother was considered polluted and she would have to take great pains to refrain from entering the kitchen or touching anything that others might come into contact with (whether her offspring was male or female). If the household were a mere hut, this would not be an easy injunction to follow. After forty days, a Brahmin priest was called in to purify the home and toss out "contaminated" materials. The child was then named and purity restored.

It was considered a sign of status for the husband that he should be wealthy enough to keep his wife locked away indoors. Thus evolved the various systems of purdah among both Hindus and Islamic societies. Muslims expanded the borders of purdah to the chador, the practice of placing a veil over the woman's face.[109] The girl would never inherit land nor would she have much involvement, if any, in public life. Her concerns were strictly those of the fields and the interior of her hut. The Hindu bride's one and only sacrament was the marriage, which in principle gave her religious rights. But those rights were questionable inasmuch as she was not allowed to take part in any religious ceremony without her husband being present. "A woman," said Manu, "is never fit for independence."[110] The ancient *Laws of Manu* have prescribed her suppression, and her total dependence on the men in her life, from childhood to old age. She must be unquestioning, a sex object, a vessel, obedient to her husband at all times, without a voice. She must bear her husband all the sons he commands. He may be a monster, unchaste, unfaithful, but she is to remain his devoted sex slave.

For thousands of years Hindu widows were not allowed to remarry, or even associate with men. This was true even if a woman had never even married but only been engaged at the time of her fiance's death.[111] Spiritual recompense, it was alleged, would come in the next life for women.

Several million widows every year continue to endure torment in India. *Sati*—the burning alive of the widow—refers to a "devoted wife." In Rajasthan it was acceptable at one time among the Rajput husbands, if defeated or disgraced, bankrupt or depressed, to kill their entire families before killing themselves. To this day *sati* (banned by the British in 1859) is occasionally practiced, though there is uncertainty whether the women are murdered sometimes by a whole community (their death made to look like voluntary suicide), or have been drugged and persuaded, or have simply found the funeral pyre less unendurable than widowhood.

The slow starving to death of girls (female mortality being ten times higher than male mortality among children as late as 1973) merely reflects the Indian consensus among men that the female is useless, and shamed a family by her very being.[112] In a survey of the abortion clinics of Bombay, only one male fetus was found to have been terminated out of all the thousands of female fetuses. Twenty-two million girls were "missing" from the 1991 census. In addition, though it is now illegal for doctors to use ultrasound and chorionic villus sampling to determine the sex of the unborn, many do it as a means of assisting the parents in their preferential gender abortions.[113]

Concurrent with these trends, and perhaps at the root of many of them, the Indian male's infantile fantasy has always viewed young women as objects for sexual gratification. Data on this has begun to surface from many quarters in the country. According to psychoanalyst Sudhir Kakar, there is "widespread sexual misery" among all classes in India, a country with no equivalent word or expression for "orgasm" in many Indian languages.[114] "In this culture, a strong woman terrifies men. Helplessness has to exist. . . . A woman can be raped. It's very violent. The image of hurting a woman does excite the audience," says Udayan Patel, a psychoanalyst of films in Bombay. When the 1991 hit *Rihaee* (meaning liberation) premiered in India—depicting a bold adultress who decided to keep a child by her lover and dared to inform her husband, whom she also loved—controversy exploded. Indian men were so threatened that many tried to keep women from getting into see the film.[115]

Part of the male's desire to rule over helpless women was achieved through child marriage, which has been part of the Indian experience for many centuries. The excuse often given was that this protected the girls from being captured by invaders.[116] Eventually, when Islam became established, Hindu parents who could not sell off their daughter to a husband frequently sold her to a temple where she became what was called a *devadasis*, or prostitute for the pleasure of the priests. The sale provided poor families with a means of survival. Today, the sales are not to temples but to cathouses throughout India and beyond, as far away as Dacca and even Ho Chi Minh City. Indian girls as young as seven and eight make up the bulk of these sex transactions. One-third of all prostitutes in Bombay, Delhi, and Madras are now HIV positive, an additional aspect of the oppression of the female in India.[117] Along Falkland Road in Bombay, Dr. I.S. Gilada, head of the Indian Health Organization, concurs with the World Health Organization in declaring the epidemic out of control. Among his many endeavors, he wanders the streets in the red light districts handing out free condoms. So far, the government has confirmed fewer than three hundred deaths. But it is estimated that three million Indians will carry the virus by 1996 and by the early twentieth century India will exceed Africa (where an estimated 6 million currently have the disease) in numbers afflicted.

Child pregnancies make for devastating physical problems. A young girl's pelvic regions are unformed and often unprepared to carry a fetus to term. High infant and maternal mortality result.[118]

Similarly, there is a much higher incidence of fetal death during childbirth as well as neonatal deaths during the first month. Child marriage was not prohibited in the Indian Parliament until 1929, and it still persists throughout India. At least 75 percent of all marriages include a female spouse who is a minor. You see the girls carrying their young, often begging for food, not as a ploy or an organized pastime but out of desperation. In India 25 percent of the 12 million girls born each year die by age fifteen. These innocent, elegant sprites take one's breath away, and by indelible means burn their lovely images, one by one, into your heart. All the more so because they are exhausted—gentleness surrounded by inconceivable hatred, by Hindu spirits that have labeled them as unclean, defiled, the cause of all problems in the country. Their only hope is the birth of at least two sons in marriage. If a female eventually embarks on a spiritual path, she is likely to become an even greater outcast, deemed crazy—only her children can give her spiritual salvation. But all too often they are her curse, especially if she is one of many unfortunate widows with only daughters as a legacy. The prodigious fertility symbolism that pervades Hindu culture mocks such a woman's actual predicament. If she is bold or modern enough to remarry, she "knows" that according to the *Laws of Manu* she may be reborn in a jackal's vagina.[119]

The miasma of cults, sects, and superstition confuses sexuality in India, scorning while adoring the pretty female, disgracing her if she perpetuates herself, awarding her only a lifetime of labor and submission if she should be so lucky as to have brought into this world a darling son. Against this backdrop the Hindus' religious scruples are annoyingly pricked by the vague chatter of the government's so-called "population crisis." But what can it mean to one who is following his *dharma*, or destiny? And for whom abstinence is only an academic, meritorious concept for later in life when the ascetic stages of spiritual development are applauded?

Ashish Bose recounted the time that he had received fancy posters, from an international donor organization, describing overpopulation, that is, a world containing over five billion people. He showed these expensively printed items to some villagers who responded, "Ah, very good!" Bose explained that five is a small number. The villagers can understand it. A billion they don't understand. Remember that the majority of all Indian women, and more than half the men, are illiterate, though the government is trying hard to change that. It is not a question of a specific language—Hindi, Gujarati, English, or what have you. Whatever they can count with their hands is manageable, is good. So five is a good number. Five children is good. Even five rupees is good. Up to twenty is a good number. People who cannot count beyond twenty are never going to comprehend 5 *billion*, but if you tell them that there are 20 billion people in India, they will say, Ah, very good!

The great and lasting irony, then, has been India's self-proclaimed worship of nature, and of the female, whose countless deifications continue to exert an emotional, aesthetic, and spiritual hold on the country, as well as its international reputation.[120] Little good such adulation has ever done real women. In the southern State of Karnataka, young girls are still kept sexually enslaved in the service of the goddess Yellamma. Because parents feel compelled to pay their daughter's dowry at marriage (in essence, to sell their daughter), she is considered an economic liability, a disaster. Despite being illegal since 1961, the dowry system remains steadfast. A woman who poured scalding soup down two of her newborn daughters' throats and buried them under piles of cow dung—one of the most widely practiced methods of infanticide in southern India—described the daunting effect of a ten thousand rupee ($350) dowry. A catastrophe for the family. A survey in the southern state of Tamil Nadu by the Community Services Guild of Madras found that more than half the 1250 women questioned had killed their day-old daughters presumably because they too could not imagine being able to pay dowries later on. Thirty-eight percent said they'd do it again if they should happen to give birth to another female. Throughout Tamil Nadu infanticide is occurring—whether among farmers, tribals, or wealthier, educated city folk.[121] In response the regional government has begun leaving cradles in primary health centers so that parents who might otherwise murder their daughters, can instead anonymously and conveniently drop them off to be raised by the state.[122] This is a country not of laws, but expedients.

Given this extent of sexual, psychological, and cross-genderal dysfunctionalism throughout much of India, the challenge to family planners seeking to curb ingrained habits has proven especially onerous. There have been some gains, particularly in the south of India. But where the majority of Indians live, forty years of family planning has still not managed to arrest a population momentum that will soon see India becoming the most heavily populated, ecologically bankrupt nation in human history—a country constantly on the edge of widespread religious conflict, its forever shuffling leadership in possession of a nuclear arsenal. There is still reason to be hopeful, nevertheless; to envision the strategies and precise mechanisms that might combat this bleak state of affairs, to marshal sufficient political will and social consciousness so as to hold back the avalanche of future generations from entirely overwhelming an already precariously stressed country. It will require nothing less than the government's admission that the nation is in an ecological state of war—an admission that must be much more than a mere admission. Only then is India likely to begin dispatching the countervailing forces necessary to get the job done.

## THE FAMILY WELFARE CONUNDRUM

An estimated 1.5 million slum dwellers reside in India's capital, New Delhi, most congregated in inner cities of tent-like clusters known as *jhuggis* or *jhopadpatti*. The people are known as *jhopadi*, hutmen, and can be seen at Yamuna Pushta, Vasant Kunj, and below Vikas Marg, well-known parts of town. These are the established slums. Many of the male denizens keep sporadic, low-paying jobs of some sort. They are the lucky ones. But there are countless other, transitory enclaves of humanity, beside the river, by the railroad tracks, before high-rises, in open fields, alongside every road, where a truly dreadful level of slum life can be observed. In some cases, as would be expected, the squatters' tastes and personality dictate their topographical preferences. There are tens-of-thousands of these less settled newcomers to the Delhi area. They set up their belongings in a dust cloud, a pile fashioned from whatever assemblage of nicely recycled scrap may be available, lean-tos of cardboard and discarded garbage, the odd steel sheet or portion of mattress, no different than the homeless in the United States. Because New Delhi still boasts more patches of green than possibly any other metropolitan region in the country, the razor sharp edge of desolation in the city seems, at least superficially, just slightly lifted.

One afternoon I visited with a few families who had staked their claims alongside an open trench adjoining a congested roadway in the middle of the old city. They were from Rajasthan, the desert state several hundred miles due west of Delhi. For an hour I spoke with Nanibai, an attractive young woman, married by her account at fifteen, with three children clinging to her side. Savan, seven years old, liked to tease Unam, his five-year-old sister. And there was naked Sham, three years old. Nanibai says she's lived in Delhi forty or fifty years. But how old are you? I asked. About twenty or thirty, she said.

They lived with other Rajputs in half a dozen huts side by side. Many of the women made little decorative stuffed horses, in cottage-type industry known generically in India as papa and achar after the fried crackers and pickles that are sold along every road in the country. Nanibai and her husband slept on the floor of their shelter, the three children on the one bed. The hut, partitioned into the bedroom and "dining" area, was 5 feet by 8 feet by 9 feet. An old black and white TV set sat next to the bed on a used carton. A frayed electrical wire was jerry-rigged to the outside. Nanibai's husband—who was out drumming at some wedding for the afternoon—paid an outrageous 100 rupees a month for the illegal service to street-wise pirates who tap into city electricity. That 100 rupees (about $3) was 40 percent of their monthly income!

The biggest problem, said Nanibai, was the lack of a toilet. In Rajasthan, one could just go anywhere. Here, in the city, it was not so easy. This cluster of Rajputs—migrants from the desert where they could find no work, no food, no future, though evidently plenty of toilet space—pooled their money in order to have the right to obtain relatively clean water from a tap a block away. That was a luxury. Nanibai also complained about the local schools. She herself had never been to school in her life and she said she'd heard the teachers were not interested in poor children. Unless you had money and influence, you would not be educated.

"I dream of owning a palace," she said ruefully. "What would it look like?" I asked. "The Taj Mahal?" "No. I just want a toilet and a refrigerator," she replied.

As for Savan, he explained how he'd like to play drums in the manner of his dad. Is that all? I asked him. "I want to do many things," he applied, hoping that I might even help him define the range of possibilities.

I did not ask his mother whether she used contraception. I did not feel right about doing so.

The traffic flow was bumper-to-bumper, the air pollution as stifling as the noise. Another young Rajput mother joined us with her own four children, one of whom she was breastfeeding. We all kidded around for a little while. I gave them all some money, then took a few carefully studied photographs so they would not feel I was simply giving them handouts. They were grateful. I asked Savan if he would one day teach me to play drums and he smiled.

Was life any better for them here in Delhi than in Rajasthan? A similar question has been posed about the over two billion people in the developing world who have migrated into cities. There are no answers, obviously. Each family, each individual within the family, tastes a slightly different aspect of life. The comparisons fall to pieces. More toilets there, more water here. I had spent time throughout Rajasthan a few years ago during a forty-eight-month drought that left millions of cattle dead and compelled several other million human victims to flee from the Thar desert. Life held no panacea there, either. Yet, in spite of the turmoil and mass migrations into the cities, Rajasthan remains the fastest growing state in India. Contraceptive prevalence rate (CPR) is only 3 percent. The 97 percent of women who decline to use birth control are not ill-informed about family planning. According to Ashish Bose, women there just don't see the need for it before they have two children. Which means two sons. Bose asked a

Rajasthani woman, "How is it that every person is saying we want two sons?" She answered, "What a foolish question you have asked. Why do you have two eyes?" Throughout India, boys are considered a gift of god; they bring their kismet, their own destiny into this world and both Hindus and Muslims believe that. Every man is predestined to receive what is already written in stone for him. Lending a practical aspect to this predestination, one Rajasthani father of four was heard to say, "I need five children. One to look after my goats. One to look after my cows. Another to tend to my sheep. One to help me on the field. And one to help my wife at home."[123] With such a population boom, and the associated ills, a big city like New Delhi seems attractive to the Rajputs, at least until they have made the arduous journey and find themselves in the slums. "When you're there there's no there," said Gertrude Stein.

I have seen a dozen differing figures pertaining to the percentage of literate women in Rajasthan, just since 1991. The average rate among those figures is 11.4 percent.[124] Similarly the average official figure for enrollment of girls in middle schools is 16 percent. The average IMR, or infant morality rate, is 103 per thousand. However, in Rajasthani districts like Jaolor and Barmer, which have an alleged female literacy rate of 6.2, Bose believes it is closer to zero, acknowledging that many who are allegedly "literate" are really able to sign their name, and that's all.[125] There is one bright spot in all this, which I mention in some sad jest: public smoking has been banned in Rajasthan.

When Rajiv Gandhi came into office, he was at once confronted with a failing family welfare program. That's what it's called in India. The phrase "family planning" was abolished after an earlier administration's unhappy experience with compulsory sterilizations. Gandhi called called upon demographer Ashish Bose for help and Bose was quick to enlighten the prime minister. Sir, your officials are misleading you, Bose began, recounting his earlier conversation. What they are telling you, Bose said, is that in a few states and union territories the program is doing very well—these regions of Kerala, Goa, Chandigarh, the Andaman Islands, and Tamil Nadu. And they also admit that in five or six states the program is not doing quite so well—states like Bihar,

Uttar Pradesh, and Rajasthan. Bose put it all in brutal perspective for the prime minister by reminding him that 40 percent of India's population resides in just four states, the "not quite so well" northern Hindi-belt, which is plagued by the highest infant mortality (well over 100 per 1,000), the highest TFR (5.4), the youngest marriages (16-17), the fastest growth rates (2.27 percent), the lowest life expectancies (49 for women, 52 for men), and the worst contraceptive prevalence rates in the whole country (25 percent versus an all-India average of 43.3 percent). And not surprisingly the preferred family norms were between five and six children. Birthrates were inert at 35 per 1000. Those four states—Bihar, Madhya Pradesh, Rajasthan, and Uttar Pradesh—Bose nicknamed the BIMARU states. *Bimaru* means "sick" in Hindi, or "not so well." Bose, who laughs a lot, thus disarming most solemnity, loves to fashion such acronyms. To put Bimaru in global perspective, consider that one of the four, Uttar Pradesh, benumbed by these demographic facts of life, is by itself one of the largest countries in the world, with a population of nearly 150 million.

The United Nations Population Fund (UNFPA) in India sees the Bimaru states as the final showdown for India. So does the Agency for International Development (AID). Both donors have been in India for many years trying to bypass a labyrinthine bureaucracy that does not take kindly to Western advice or controls. If you give money to India, attach no strings. Donors dictate policy nearly everywhere in the world, but not in New Delhi. UNFPA (with World Bank involvement) has been trying to raise female literacy standards in Rajasthan by financially assisting families so that their daughters could get a primary and secondary school education.[126] In Haryana it took three years for UNFPA to wrangle a contract with the central government. AID has spent four years trying, in essence, to give away $325 million to India. But because both donor agencies want certain assurances that their money will be utilized as effectively as possible, they have asked that Western experts be part of the implementing teams. This has been far more difficult to secure than expected.

"We have to be mature on all sides and let the process happen. There are no morals to be drawn," the UNFPA representative in New

Delhi, Mr. Tevia Abrams, told me. And an AID official added, in reference to India's frustrating bureaucratic underbrush, "We will out-India India," meaning that AID is prepared to withstand nearly all obstacles in order to get the job done. AID's ten-year program, known as IFPS, Innovations In Family Planning Services, was designed to be a collaborative Indo-U.S. effort to bring in fresh ideas, to try one last time to save the Bimaru states from total demographic disaster—a prognosis based upon data suggesting total runaway fertility, the fact that throughout the northern states a very high percentage of males are in primary schools but only 55 percent of the females, and the many ecological barriers northern India faces. The women of the Bimaru states constitute a population of nearly 340 million, a constituency larger than all but two countries in the world, China and India itself.

The GOI (Government of India) prefers total control. As one Ministry of Family Welfare official conveyed, "India doesn't need foreign cars so why does it need foreign contraceptives?" Indian women have been using the short-lived 200 B Copper T IUD rather than the 380 A Copper T which lasts for eight to twelve years (the one both AID and Indian Council of Medicine recommends). Usha Vohra, secretary of Family Welfare has rejected the better IUD because, she says, Indian women only use IUDs for two years and then reject them because of inevitable infections. She says that the twenty-three thousand auxiliary nurse-midwives (A&Ms) throughout the country are not yet properly trained to detect such infections though it is the Ministry of Family Welfare's goal to foster that competency. AID wants to help the government by setting up such training in all ten of the medical schools throughout the Bimaru region. The GOI is interested, but on its own terms.

In addition, the AID program would focus on emphasizing longer spacing between pregnancies, on women's literacy, on agricultural and milk co-ops, and on motivation and education. NGOs would play an important role in degovernmentalizing the program so as to make it more personal. Massive statewide social marketing would be part of the picture. AID would contract with an Indian firm to manufacture quality and inexpensive pills and condoms. But AID will not release any of the $325 million until targets are agreed-upon and tangible progress achieved. As one official put it to me, this is "India's last chance to really reform its program." During the 1980s, there were five lost years during which AID and GOI argued over contracts, resulting in the deobligation of $33 million—in other words, AID took back the money. India was looking for a magic bullet.[127]

India is once again looking for that free magic bullet and it hopes that AID's $325 million giveaway is at least 15 percent of the solution; ($325 million equals 15 percent of India's overall family welfare capital projections over the coming decade). For speed of implementation, AID wants to disperse the funds directly to the NGOs. GOI has said No, the money must go through the government finance office, which many outsiders view as the definition of a morass. It's yet to be seen whether the project will really take effect.

I spoke with the representative of a condom distributor in New Delhi with whom AID would contract as part of the IFPS project. An earnest and forthcoming chap, Mr. Gopalakrishnan was general marketing manager for Population Services International (PSI), whose upstairs offices were in a charming residential area in Delhi's South Bloc region. A herd of cows was foraging out front on the afternoon I arrived. Following the Bhopal disaster, Gopalakrishnan told me, the government of India dissociated itself from Union Carbide, which happened to be the distributor for 50 percent of all condoms in India. This severely diminished the supply to the Bimaru states. Other companies, like PSI, were invited to take up the slack. When the government discerned that only 5 percent of eligible couples throughout the country were protected by condoms, it decided to make a variety of brands available at a government-subsidized price. India's condoms are the least expensive (and the flimsiest) in the world. World Health Organization specifications are slowly taking effect—thicker condom, better lubricant, better foil, better quality control. Female condoms are now going through field trials throughout the country. PSI distributes the government's Nirodh, their own brand Masti, and their own oral contraceptive for women, Pearl.

Condom use is low in India, despite there being two billion condoms in the country's inventory at any given hour, because the focus has always been on terminal methods. Moreover—astonishingly—many Indian men seem to get befuddled (for a variety of reasons) in the presence of a condom. It was learned, for example, that the Madras branch of the Family Planning Association of India buried tens-of-thousands of unused condoms in the ground. Why? Little demand. Why little demand? Because many people claimed they did not know how to use them. The local Family Planning unit, rather than risking embarrassment at their lack of success, chose to bury them. According to one scientific consultant in the area, even "bank executives" were in the dark as to their use. Because the quality of Indian condom manufacturing is so inconsistent, there tends to be much tearing. Incipient cracks in the latex are broken through when semen—which travels at 40-to-90 kilometers per hour during ejaculation (and not just in India)—hits the tip. Thus, say sex consultants in India, there is an uneasiness associated with their use. [128]

What are Indian couples using for contraception? Seventy-five percent are going in for sterilization, and nine out of ten of those are tubectomies. [129] The problem with sterilization in India, as elsewhere, is that people normally take the procedure *after* the demographic damage has been done, or as they say in Delhi, after the "horses have gotten out of the barn." Furthermore, sterilization figures are dropping. In 1990 slightly more than 4 million Indians were sterilized—3.8 million women had their fallopian tubes ligated or removed and 249,000 men had vasectomies, a drop of nearly 300,000 from just the previous year. [130] That's why there is a lot of rethinking going on now throughout the country. If India is ever to begin to approach fertility replacement—a TFR of 2.4 in other words (down from its country-wide rate of 4.1)—at least 60 percent of all couples will have to be protected by some form of contraception. However even that 60 percent is now suspect as a reliable barometer of declining fertility. In the wealthy State of Punjab, a contraceptive prevalence rate of 68 percent has produced a TFR exceeding three children. To make matters worse, Indian officials are concerned that even the current 43 percent average contraceptive prevalence rate is slipping. In fact the number of couples protected during the prime reproductive age group—fifteen to twenty-nine—is a mere 16 percent, from whence it has not deviated for years. As far as abortion is concerned, Indians—unlike Central Europeans—have never perceived medical termination of pregnancy as a form of contraception. Abortion is legal up to twenty weeks in India, but as few as 10 percent of rural clinics are set up to provide for one. (Ironically, in the United States that figure is 11 percent).

The debate over priorities rages on between spacing, IUDs, terminal methods, and the condom. As of 1993, it looks as if 5 percent of all Indian couples are protected (more or less) by condoms, about 7 percent by short-term IUD (with a 2-percent IUD failure rate per year), and a very measly 2 percent by oral pills. That totals up to about 14 percent. The best wisdom indicates that these figures will have to be doubled, while maintaining a high sterilization rate. However, in a 1988 national survey, half of all couples were shown to have *never* used contraception or had any contact with family planning personnel and only 20 percent of women under the age of twenty-five reported using contraceptives. Different polls are like different branches of the same government that never speak to one another. When the polls are judged side by side, they overlap, or distort. The same is true of family planning data in general. Usually the real figure is worse than expected. What it all means is that the fertility rate among younger women— the backbone of population increase—is actually rising, based on the overall population momentum, both in rural and in urban regions.[131]

But that's not the only bad news. Apart from rapid and spectacular strides needed in the contraceptive prevalence realm, it is fairly well understood by demographers that India will have to get its birth rate down to twenty-one per thousand from a country-wide average of thirty-one. In addition, the death rate will have to plummet to about nine per thousand (some argue six, even five), to provide the needed assurances to couples that they can rest easy, so to speak. Mr. Gopalakrishnan PSI believes this is all possible perhaps by the year 2018, not by any magic bullets or miracles, just persistent dedication on the part of family planners. Of course by that time India will just about have overtaken

China's population. And as one prominent family planner in India put it, "We don't have time to wait for trends."

"All we can do is to try and avoid more damage," Gopalakrishnan told me.

One serious impediment among many to this twenty-five-year vision is the ongoing Hindu–Muslim crisis, particularly in the northern areas. The reason is that virtually no politician in the north dares speak of family planning there. If a Hindu should be perceived to be denouncing Islamic population growth, he is labeled a racist, as wanting to inhibit Muslims from achieving greater parity with the Hindu majority. And if a Muslim should expostulate on the subject, he is more likely if anything to enflame pronatalist sentiments among his constituents, a minority. But there is an even deeper meaning to the reticence among those in power—a serious lack of political will that has crippled India's confidence in fighting the battle. That is the troubling legacy of Sanjay Gandhi who, in some respects, was a visionary along Chinese lines.

The roots of Sanjay's vision date back to the beginning of the twentieth century. In 1892, 25 percent of the Indian budget was going toward paying off British pensioners, not toward helping India. A series of economic devaluations of the rupee against the British pound, in addition to famine and the introduction from China of plague-carrying rats in Bombay harbor, created the worst economic depression in India's recent history, just at the moment that Britain and the West were rapidly ascending.[132] Between 1895 and 1905 medical, economic, and ecological conditions were dire enough in India that the country's population actually began to decline from excess mortality in absolute numbers to below the level of the first census, which had been taken in 1872.

Until 1921 this decrease continued despite a birth rate per thousand of 48.1. That's because the death rate was 47.2. Malaria, TB, polio, diphtheria, tetanus, and ailments associated with malnutrition and the breakdown of the immune system were the major killers. Among children, bacterial pneumonia, measles, and diarrhea were, and continue to be, the major causes of deaths. But consider how amazing the numbers are. Nature, somehow aware that there were 260 million carnivorous *Homo sapiens* packed in at 83 per square kilometer in a finite space, ordained a graceful, almost imperceptible attrition, a net loss of one per thousand. This was a clear window on how biological determinants prefer to operate in a vacuum (prior to human medical engineering). If those forces had continued, that declining population rate would have manifested an India with fewer than 200 million in 1993. Per capita income, potentially, would be on a par with South Korea's. The matrix of human pain would have been hugely softened, freeing up other energies that could be applied toward conservation and the amelioration of nonhuman pain. (This is not a guaranteed equation, of course: South Korea is an ecological quagmire. Other positive forces have yet to be liberated there.) Having subverted nature's manner of resolving demographic pressure however, India's density in the early 1990s averages 216 per square kilometer (176 for rural, 30,002 for urban). Population increase is 2.22 percent annually, approximately 18 million and counting, more than the whole population of Sri Lanka.[133]

What happened? In the decade following the 1921 census, everything changed for India. By 1931 small medical breakthroughs and slightly improved social conditions had rapidly pushed the death rate down to 36.3 from 47.2 per thousand, the single largest advance in India's demographic history. Family planners, knowing what such technological successes actually implied, were already ahead of the statistics. The first Indian birth control clinic was begun in Bombay in 1925 by Dr. R.D. Karve. And in the 1930s, birth control crusader Margaret Sanger's message was heard. But Mohandas Gandhi refused to lend his support to her, thinking birth control unnatural. He stressed abstinence and the rhythm method, ignoring the fact that very few women in India in his day could count beyond ten or twenty. Between 1921 and 1941, 66 million people were added to the subcontinent. By 1951 another 44 million for a total of 361 million people. The Health Ministers, meanwhile, all strove toward lowering the death rate and encouraging longevity. These were humane endeavors, after all, aided by the increasing sophistication of science and the outreach of pharmaceutical multinationals.[134]

By the early 1950s the efforts of the nongovernmental Family Planning Association of India, established in 1949, were seen to be too few too late, for already the population growth rate had surmounted 2 percent, up from 1.5 percent during the previous decade. Independent India recognized that smaller families contributed to a healthier economy and to healthier women. The first five-year plan in India sought to stabilize population so as to foster the national economy. "Economics" was, and continues to be, at the basis of family planning policies. Nehru recognized that the economic well-being of each individual would define the country. He was to say, there is not a population problem, there are 400 million such problems.

In the early 1960s when the National Family Planning Program was launched, its goal was to reduce the country's birth rate to twenty-five per thousand by 1975. That year came around and the country's population had swelled to 620 million. At that level of increase, it was projected that India would reach a population of 2.275 billion by the year 2025.[135]

When Indira Gandhi assumed power in 1966, she placed family planning under the Ministry of Health and by 1969 announced that family planning was "to be given the highest priority and to occupy a pivotal place in the economic development of the country." The government promoted IUDs, then passed in 1971 the *Medical Termination of Pregnancy Act* (legalizing abortion without a battle eighteen months before *Roe v. Wade*), as well as a law forbidding females from marrying before the age of eighteen (a provision difficult to enforce). By the mid-70s, male sterilization had become the vogue, following the success of several vasectomy camps in the states of Kerala and Maharashtra. For thirty days from late November to December in 1970, a mass vasectomy camp called a Family Planning Festival was held at Cochin in Kerala. The medical personnel performed 62,913 vasectomies and 505 tubectomies, exceeding by 400 percent all previous sterilizations. It was hailed as a breakthrough, the result of preparations in the press and a broad coalition of concerned writers, organizers, medical institutions, and government officials. Family planning cadres gathered the data on eligible couples, publicity units went into the field throughout the Ernakulam District, one hundred medical officers with staff and equipment were assembled, the compound was made rainproof and decorated festively with garlands. Free canteens, counters for issuing incentive money, an entertainment auditorium with twenty-four hour-per-day variety shows, and free condoms were all part of the appeal. In a terribly ironic way, there must have been something slightly prurient about it all.[136]

A year later, more than one thousand vasectomy camps were held throughout Gujarat state and 221,933 vasectomies performed in a two-month period. Circus-type clowns and jokers drumming tom-toms and walking on tall bamboo stilts advertised the camps from village to village. The success of these efforts has been called an important landmark in family planning history.[137]

When the mass vasectomy camps were tried in Uttar Pradesh however, they resulted in four tetanus deaths, enough to prompt abandonment of the camps. Subsequently the governments of Haryana, Rajasthan, and Uttar Pradesh began enforcing sterilization practices through other means.

Financial assistance from the central government to the states was soon linked to sterilization performance. Legislative discussions regarding compulsory sterilization took place. (It should be pointed out however, that compulsion is basic to most health programs. As social scientist Robert Gillespie, president of the Population Communication organization, has pointed out to me, "there is no such thing as a voluntary program." The control of most diseases, whether yellow fever or smallpox, has always made for required identification certificates, immunizations, even quarantine.)

Then the pace of fertility control suddenly escalated. Indira Gandhi declared a state of emergency as a means of consolidating her authority when the Allahabad High Court tried to overturn her election in June 1975. She imposed censorship on the media and promulgated a Twenty-Point Program to improve the country. Her son, Sanjay Gandhi, was given the responsibility of focusing New Delhi's

authority on three aspects of the initiative, namely, population control, adult education, and afforestation, a remarkably informed combination of disaster counter-strategies. In October of that year, union
health minister Dr. Karan Singh sent a message to the prime minister
which stated, "The problem is now so serious that there seems to be
no alternative but to think in terms of introduction of some element
of compulsion in the larger national interest."[138] The measures,
deemed necessarily "drastic," drove sterilizations up from 2.6 million
in 1975–76 to 8.1 million the following year. Sanjay Gandhi and his
mother sincerely believed that there was no other choice. India was
in a state of crisis, and Indira Ghandi's emergency declaration was meant
not only to preserve her own regime but to underscore the severity of
the population explosion.

The findings of the 1978 Shah Commission inquiry revealed that
Uttar Pradesh police and the Provincial Armed Constabulary had
been given specific targets to achieve (the number of people brought
in, not for questioning, but for sterilization). Homes, whole neighborhoods, were surrounded in the middle of the night by hundreds of police
who would literally drag men and women out of their beds into the
waiting clutches of surgeons. Those who resisted or protested were
arrested.[139]

Fearing civil war, Indira Gandhi called for national elections in
March of 1977. Not surprisingly she and her party were disgraced. In
Uttar Pradesh, the most populous state of India, the ruling party could
not win even one seat in Parliament out of the eighty-five seats in that
state. Such was the fury of India's illiterate masses. The incoming Janata
Party did not want family planning even mentioned. Contraceptive
prevalence throughout the country fell to below 6 percent, thus
engendering the explosive growth being witnessed today.

By extraordinary powers of persuasion—a testimony to the Nehru
dynasty's hold on the country—Mrs. Gandhi managed to return to
power in 1980. This was also a measure of how badly managed the
government was under the Janata Party. At first Mrs. Gandhi down-
pedaled family planning. But when the 1981 census showed that the
population was rising at 2.2 percent per year, and that there were at

least 17 million more Indians than expected, she again pushed for population control, though noncoercively and with some tact. But the government was not keen to listen and, according to investigators V. Panandiker and P. Umashankar, family planning in India has been the "victim of electoral politics" ever since. For the next decade family planning stayed in hiding. And it was during this uncertain time that female tubectomies replaced male vasectomies as the primary choice of contraception in the country. Today, rather than confront the lingering ill-feelings regarding family planning across the Bimaru states, many of India's parliamentarians are more eager to visit China, Indonesia, Singapore, and Thailand to learn about these issues. And this is one of the reasons that the runaway populations of Bimaru are going to be difficult to check, no matter how much money the government, with assistance from AID, UNFPA, and the World Bank, throw at the situation.[140]

In spite of India's rejection of compulsory family planning, the truth remains that the additional 6 million sterilizations under Sanjay Gandhi probably prevented 2 million births. Multiply that number times likely offspring, over several generations, and one is into a European-size nation. In fact, if India's population growth had been checked at 1.5 percent a year since Independence (the original hope) versus today's 2.2 percent growth rate, there would have been some 200 million fewer Indians, with all of the corresponding increases in the quality of life.[141]

This message is well understood in India. When UNFPA sponsored a book of children's drawings, the subject being environment and population, a resounding cry of awareness rose up from young people in villages all over the country. Awareness was pared down to basics: not enough food, water, space, trees, or animals. Sangeeta, eight years old, writes, "Helpless before the assault of people, maybe the trees wish they could run away." V.P. Raghunath Rao, twelve years of age, writes, "Less population more happiness. Let's stop destroying the environment."[142]

Only Draconian measures seem capable now of altogether combating India's difficulties, yet these are the last sorts of measures any politician wanting to remain in office would advocate in India. The country has thus boxed itself into a democratic and economic corner. One profound example of this is the dowry system. If the dowry could effectively be outlawed, punishable by the full force of the existing laws, population growth would decline. It might not change demographic fundamentalism, particularly in poor rural areas, but it would certainly impact urban parents. Knowing they had nothing to fear from the financial burden of a dowry, they would be less concerned about supplanting a daughter with an additional male child. However the dowry system is more widespread than ever before, as are dowry deaths and wife-burnings. No person has ever been convicted under the dowry law, even though the government knows the system is responsible for thousands of cases of infanticide every year, and the perpetuation of a disabling economic and cultural bias favoring males.[143]

India's overall response to this complex bottleneck of fertility uncontrol has been characterized by some as a policy of resignation, of de facto destiny, of good intentions subsumed in mere rhetoric. Or as one former joint secretary in the Union Health Ministry has said, "The tragedy is that prolific breeding is not a disease. Children are not an epidemic. An outbreak of population doesn't hurt anybody immediately. With the result that no one really cares."[144]

Many people do care, of course. The government is up to its neck in population-fueled crises and crusaders. A barrage of control strategies has been tried out, endless think tanks convened, more than a few careers made and ruined on various micro and macro policy initiatives. A specialist once called in, invented a bead necklace for women that would help them remember the rhythm method. Green for dangerous days, red for good days. But children got hold of the beads, mixed them up, and half of all the women got pregnant.

Several years ago the application of government incentives emerged as India's new hope for the future.[145] In some deferred bonus schemes on tea estates, women would receive their accumulated bank earnings once their childbearing years were finished, a kind of fertility ransom. Those polled throughout the country liked the idea of incentives but said they wanted immediate rewards for fertility compliance, not offerings many years down the road. They wanted cash, land, livestock, or free baby-sitters. In Gujarat, women vowing to cease all childbearing were awarded complimentary bank accounts and monthly stipends. A host of other remunerative schemes has been tried and discussed, all for those couples who elect to get sterilized after the birth of their second child. Private companies have also given incentives to their employees.

But the incentive schemes are vulnerable to corruption. Because the central and state governments were giving out money to both doctors and clients, there was a built-in temptation for both parties to lie. In 1988 at the first conference of the Central Council of Health and Family Welfare in New Delhi, it was stated that the previous year had been a milestone for family welfare, with an "all-time high record" of 20.5 million fresh acceptors into the program. But as investigative journalist Raj Chengappa discovered, some 75 percent of all fresh acceptors were over the age of thirty and had already had more than four children.[146] In addition some of the new acceptors were said to have obtained repeat vasectomies because, of course, they wanted the 180 rupees prize money and the surgeons were willing accomplices (any doctor should be able to easily distinguish a client who has had a vasectomy, and thus avoid a repeat performance).

To "solve" the population bomb, some have argued in favor of a national campaign to reintroduce breast-feeding. Others have focused on the overall unemployment rate among females, which is 80 percent. The Integrated Rural Development Programme budget has allotted 30 percent of its financing to women but little has actually happened. Many see the hope for India with the rising middle class (a throwback to the classic demographic transition theory). A few have once again suggested "playing God, but soberly" (broadly interpretive). Others insist on no more targets but simply maternal and child health care. Or getting more NGOs into the act (there are already an estimated ten thousand separate nongovernmental organizations in India), the privatizing of birth control, and removal of the sterilization emphasis. All of these components have been called upon in different states and union territories, at various times, and to varying degrees. Many look toward that magical moment when the sun pierces through the storm and the convergence of multiple factors effects a transformation.

Despite certain dramatic successes—such as an estimated 90 million unintended births prevented in the last forty years[147]—a fertility pall hovers over the country. The population rate has not gone down in twenty years. Nine out of ten babies in villages are still delivered by typically untrained *dais*. Refrigerators frequently break down, or do so temporarily as the electricity fluctuates. And when that happens, vaccines for polio or diphtheria-tetanus are ruined, though they may still be mistakenly administered. Meanwhile, the auxiliary nurse midwives or A&Ms are the least trained and most neglected and underpaid of the whole health system, yet they are the ones who are supposed to immunize children, attend deliveries, and provide postnatal care. Medical training for pregnancy related needs and problems (or for knowing how to detect repeat sterilization fraud) is inadequate. According to Bose, as many as 90 percent of all alleged "new" sterilizations in Uttar Pradesh may be falsely reported. If such levels of medical or bureaucratic incompetency seem unlikely, consider that in one recent study, according to Conly and Camp, "over half of government health staff failed to correctly identify when in the menstrual cycle a woman is most at risk of pregnancy."[148] I asked a leading official of the Indian Council of Medical Research about this rather astonishing allegation. The official's response was, "Why is it so important that doctors know such things about the cycle? After all, we are not advocating the rhythm method. It would be idiotic to offer a woman over the age of twenty-five anything other than sterilization, when she still has thirty years of childbearing ahead of her." What about the cafeteria-style approach, I asked, referring to a multiple of contraceptive choices, the tried and true test of any successful family planning program worldwide? "Not in India," said the official. "Here, it needs to be 100 percent sterilization by the age of twenty-five. Only in the poor northern sector where the infant mortality rate is ninety per thousand or more should you insert the IUD after the second child. Once that child looks like it's going to live, then you counsel that woman to be sterilized as well."

This emphasis on sterilization, Sanjay Gandhi's own prescription for India's future, hinges upon the presumption that young couples will be in favor of it, which so far they have not been. Too many other psychological, religious, and cultural factors particularly acute during an average person's most fertile years (his or her twenties) have preempted this otherwise singularly effective approach. The burden of government-proposed contraceptive progress has thus shifted to other wider ranging methods, like two to three years spacing between the first and second pregnancy (best attained by a breast-feeding regime), and state of the art long-term contraceptives, such as Norplant, which was recently sanctioned by the Indian courts after being held up by feminists—ironically, on grounds that it would curtail a woman's freedom of choice. But the fact remains, however ingenious the new technologies, such as Norplant, India's huge population of sexual neophytes has thus far indicated little interest in any kind of birth control whatsoever.

How, then, is India to break through this numbing veil of conflict? The country is buffeted on the one side by seamless introversion, scientific self-assurance, and implacable custom; on the other by political paralysis and a rash of differing perceptions, superstitions, priorities, fears, and inflexibilities from village to village, from state to state.

In late February 1993, I met with the secretary of Family Welfare, Mrs. Usha Vohra, and her two key staffers to try to solicit the government of India's most current thinking and initiatives in this arena.

While many had warned me about Usha Vohra ("she eats nails for breakfast" was how one well-known social scientist in the country described her), I found her to be gracious, articulate, and by all appearances deeply committed to helping India find a way out of its population labyrinth.

A stately woman with a stolid and penetrating gaze and draped that day in a striking sari, Mrs. Vohra was formerly with the Trade Fair Authority and charged with boosting exports. Now she was boosting family planning as she detailed for me the country's new assault plans. In late 1991, Prime Minister P.V. Narasimha Rao had taken what were reported to be decisive steps to overhaul India's mired family planning programs. That task now devolved to Mrs. Vohra. She laid them out as if preparing to land at Normandy—a sweeping vision of battle, which, to listen to her, suggested credible maneuvers. In India, however, the "enemy" is always underestimated.

Mrs. Vohra reiterated the government's belief that population control is one of the most important priorities for the country. The government has provided community health centers at the 3,000- to 5,000-person level throughout the country, or one medical clinic for every four villages. This translates to 130,000 sub-centers and 23,000 primary health centers. Six hundred thousand workers and health volunteers service this infrastructure, often having to walk five kilometers in a day going door-to-door to meet with mothers and counsel on the advantages of birth control. In 1989 when Mrs. Vohra began, the infant mortality for all of India was 91 per thousand live births. Of course it varied from a high in Orissa of 126, to a low in parts of Kerala of 17. In 1993 however the country-wide average had declined an incredible eleven points in just four years. How? By a strengthening of the immunization program, according to Mrs. Vohra. The government's goal is to bring the IMF down to sixty by the end of the decade.

However that could pose other problems reminiscent of the 1930s. According to Dr. Prema Ramachandran, deputy director of the Indian Council of Medical Research, there is as yet no evidence linking a declining infant mortality rate to a hoped-for decline in the total fertility rate. What has been shown in many countries is a coincidental drop in TFR once maternal and child health care was improved. But then India has followed no rules. And if Usha Vohra's target of an IMR of sixty should be achieved without a corresponding increase in contraceptive prevalence, then India's net reproductive rate will soar even beyond the much-speculated 2 billion plus stabilization figure sometime in the next century.

Getting the IMR to sixty—part of the Child Survival and Safe Motherhood project—will require providing most of India's rural populations with safe drinking water. That poses nearly insurmountable ecological and economic problems the country has not yet managed to figure out. Other aspects of the government's program should be less difficult to achieve. For example, the country plans to escalate its war against anemia and typhoid and is currently strengthening its technology in anticipation of more complicated deliveries and high-risk cases.

In addition more than 135,000 auxiliary nurse midwives are being trained throughout the country. At the same time, the *dais*, or traditional birth attendants, are also being retrained, as money permits, and supplied with newly provisioned kits that basically supplant the Stone Age with twentieth-century appurtenances.[149] Such practical education programs coincide with the widespread recognition in India's medical circles that ante- and postnatal care is crucial to the whole fertility process, though with 80 percent of all births taking place outside of hospitals such screening and follow-up have been almost entirely absent from India's experience.[150] Further compounding the existing rural fertility dilemma is the fact that only about 25 percent of graduating physicians end up in rural areas where the majority of Indians live.

I met with Dr. Chandrakapura, head of family welfare for the State of Maharashtra, which contains 10 percent of India's population, and where a new program is underway that hopes to set a precedent in this domain. From her office atop the St. George Hospital complex in Bombay one windy and incense-rich afternoon, she laid out the details. Despite her budget having been stalled for several years at 570 million rupees, or seven rupees per person (twenty cents), and her state desperately requiring some two hundred more primary health centers, as well as mobile units with paramedics for remote tribal areas, she has undertaken some impressive developments.

In the Nasik district, she had launched a new Health Card program that ensures long-term care for the whole family. The program started as a medical test case, but family welfare logically "piggy-backed" onto the initiative. She took eight primary health centers and developed a system of profiling the health of each family member in the area. The auxiliary nurse midwives are dispatched to each household once a month where they maintain continuing medical surveillance, examining each member of the family and checking off a list of dozens of health and fertility parameters. All the children are immunized in the district. Nothing so comprehensive has ever been tried in India.

Whether it will prove to be a model for other parts of the country is budget dependent. "The cards are expensive to print," Dr. Chandrakapura told me. "One rupee, eighty paisee per card, every five years, on account of the plastic laminate. At two hundred thousand people that's ten million rupees, or two million per year just for that one district."

It comes out to chicken scratch in a country like Singapore or Switzerland. But here a few cents is the crucial determining factor in whether an infant's chance of survival is increased or not. I never saw it presented in such focused and merciless perspective. It was calculated in the mid-1980s that the government of India was spending a mere twelve cents per person on basic health care. And yet, multiply that times 900 million and it surely adds up for a poor country.

But the money is there, if the sense of urgency is there. India numbers among its imperatives the developing of nuclear weapons, the cleaning up of the Ganges, subsidies for irrigation, and the building of hugely expensive hydroelectric plants to fuel more local industry. In a nation with thirty competing regional governments, ethnic turmoil, and a vanishing resource base, New Delhi's choice of national priorities has not been made any easier. Outside Dr. Chandrakapura's offices in Bombay, at least, the essential budget requirements seem crystal clear. The city's population is drifting toward an estimated 17 million in the next couple of years, though—as in Shanghai—it is impossible to know for sure. Average density is estimated to be 3726 per square kilometer. The glittering lights at night of the Colaba Causeway, the spectacular bays, the bustle of Nariman Point, conceal the handicapped, rotting city itself, which is set like a sparkling jewel amid garbage. Not even the kites and vultures and crows, which consume an estimated eighteen tons of that garbage every day, or the human ragpickers who recycle some of it, can hold back the millions of pounds of untreated, undisposed of wastes that fester and mutate in the streets and alleys, the beaches and every clearing. Over half the residents live in slums and another few million are without toilets or water taps. The poorest of the poor inhabit *jhopadpatti* tent cities like Chunabharti,

Mahim, and Siom. The oldest one, Dharavi, is said to be the largest slum in the world.[151]

With so vast an impoverished population, it is understandable how the use of monetary incentives as a means of stimulating popular interest in birth control has long played a part in the government of India's family welfare strategy, whether cash disbursements or a preferred loan for a water buffalo. Back in New Delhi Secretary Usha Vohra explained how birth attendants are now receiving ten rupees for every delivery they conduct, assuming they follow the prescribed checklist, which includes bringing the pregnant women to an auxiliary nurse midwife for tetanus/toxoid immunization early enough and making sure that there are two to three antenatal checkups.

As an incentive, an acceptor currently receives nine rupees and the doctor six rupees for each IUD insertion. A vasectomy is worth a reward of 180 rupees, a tubal ligation 250 rupees. Government employees (1 percent of the total work force) are granted various benefits if either spouse volunteers for sterilization. The money for such enticements comes out of both state and central government budgets. GOI allocated ten billion rupees to family welfare in 1993, or $350 million, a 15 percent increase over 1992. In 1994, the budget is expected to increase by another 12.7 percent.

But it's nowhere near enough money, says Mrs. Vohra echoing Madame Peng Peiyun's concerns in Beijing. And so she is counting on the many NGOs to get involved. She wants to make family welfare a "people's program, neighbor to neighbor. Every department of government, every chamber of commerce, every individual. Only then," she states forcefully, "do we stand a chance of compressing the time necessary to accomplish what needs to be done, to civilize properly," as she puts it.

NGOs are deeply involved in India's environmental and fertility dilemmas, perhaps none more effectively than the National Family Planning Association of India, whose longtime president, Mrs. Avabai Wadia, met with me in Bombay a few days before all the terrorist bombings in the winter of 1993. A buoyant, girlish visionary who has seen it all, Mrs. Wadia believes, with Einstein, that God is in the details. "If you try to take in the big picture," she said, "you get into paralysis by analysis. We know nearly everything. But it's the doing that counts." And she pointed out with justifiable pride that at Independence, the life expectancy in India was thirty-two. In 1994, it is sixty.

With eighty thousand members in some 3164 local voluntary groups across the country, Mrs. Wadia has for many years been spreading literacy, installing water and sanitation facilities, and developing income-generating schemes for women. In 1991 FPAI's endeavors were encouraging. In the District of Malur in Karnataka state, for example, 150 newlyweds from 138 villages agreed to delay their first pregnancy by two years, while 500 boys and girls pledged to postpone their marriage by two years beyond the legal age. Over ten thousand men and women attained literacy and nearly thirty-five thousand boys and girls (near parity) were shepherded into schools and Balwadis (preschools). Nearly thirty-six hundred women were provided jobs and 8 million rupees generated. Savings accounts were opened by 8330 men and women. Smokeless stoves were introduced into nineteen hundred households, while eighteen thousand people were provided with safe drinking water systems. Tens-of-thousands of tree saplings and fruit gardens were planted, and nearly sixteen thousand new couples accepted family planning. Such efforts are inherent to the struggle that engulfs a country as rife with numbers as India. It breeds philosophy, a sense of perspective, a sense of hope that refuses to cave in to the equally galling premonition that whatever incremental gains may be won, the built-in population momentum and its ecological devastation are nearly irreversible. For people on the front lines, like Mrs. Wadia, irreversibilty is not what matters. Gains are what count; gains and sensitive, sensible damage control.[152]

As we spoke, Mrs. Wadia radiated optimism in her impassioned descriptions of small groups of illiterate hopefuls clustered together around a kerosene lamp in a hut during the monsoon rains learning from a teacher who was herself just in the process of becoming literate, one step ahead of her peers. To know that another 6087 women were initiated into the world of reading and writing, of the dignity and equipoise and freedom attached to those skills, was no small matter.

And these were precisely the kinds of endeavors that Mrs. Vohra up in Delhi was referring to when she spoke of a people's program.

Imparting literacy is in fact the cornerstone of Mrs. Vohra's whole mission. The Ministry of Family Welfare calls it the Total Literacy Program. "Where there is literacy there has been a decided impact on lowering family numbers," Mrs. Vohra said, echoing Mrs. Wadia's own beliefs.[153]

From a somewhat more satirical, but nonetheless affirmative perspective, Professor Ashish Bose also wrestled with demons in the night—the big picture—yet ultimately came back around to the details, and to the possibilities for unleashing an enormous reserve of energy and skill in the Indian people themselves. A few days before I visited Bose at his home in New Delhi, he had opened the Indian National Family Planning Conference in the State of Orissa by referring to three factors responsible, in his opinion, for India's family planning failure. First, foreign *kubudi*, which means "bad advice" or "bad intelligence" in Sanskrit. He was referring to financial incentives. "Why do you give money? Tomorrow the villagers will tell you, I'm brushing my teeth so you give me money! If you start giving money, the World Bank is not rich enough to bail us out, not 160 million couples in the reproductive age group." The second factor involves IEC—or information, education, and communication. Bose challenges all such raising of consciousness. "You know what I call IEC—incompetence, extravagance, and corruption," he says. "They give a tape recorder that never works, a TV, which only the village headman will look at, and they put some posters on the walls that nobody can read, they show some lousy films, which nobody understands. It's all a waste of money. Under conditions of mass illiteracy, what education can you give them? I have yet to meet a foreign expert who knows how to communicate with illiterate people. In Rajasthan, the female literacy rate is only 9 percent. Now what communication strategy will you have? Have a small family? You know what the villagers say? 'A mad dog has bitten you. You people from Delhi are all mad, why do you come and tell us all these things.' In Bihar, the women tell us, 'Are you crazy, this is the age of marriage, these girls will be raped. Do you think this is Europe, is it Sweden? Get lost.'"

The third fact was what Bose described as the total lack of medical ethics in India; the complete absence of postoperative care and of medical malpractice insurance. Everyone would have to be sued in this country he said. In India when the patient dies from incompetent treatment, the survivors are lucky to get five thousand rupees, $140. It was the same at Bhopal. One doctor boasted to Bose how he had operated on five hundred women in one day. And he wanted an award from the government of India for doing so. In a public lecture Bose suggested that rather than an award, that doctor should be arrested. "Our women are being treated like cattle by the health delivery system," he declared, painting horrible pictures of women lying on the floor in cold December months without being looked after following a sterilization procedure. They were given their incentive money and told to get lost. Bose has studied what happens to that incentive money. Invariably, he told me, the husband ends up spending it on drink or gambling. It's too small a sum for any substantive nutrition, for any education of their children, even for buying a nice shirt.

But even that would be all right, in a free country, if the sterilizations were really occurring. But according to Bose, nine out of ten of them are frauds, the whole program is bogus, sustained by foreign ideas.

I asked him what he thought about the new initiatives in the Ministry of Family Welfare, new investigative commissions, new ideals.

"If you want to shelve an issue, appoint a commission," he said in quick retort. "We can write beautiful English. I have no doubt that our population policy will be a very well-written document. But so what? Who needs that policy. The World Bank, or our people?"

Bose labels the entire Family Welfare program by one of his characteristic acronyms, COMIC: Contraceptive technology that is unsuited to the culture, Monetary incentives that are utterly fraudulent, and Information and Communication that are a disaster.

"If you can't bring drinking water to people, you have no right to ask them to enact birth control for two children," says Bose.

His solution to all these grievances is something akin to a literacy corps of engineers. Like Usha Vohra, Bose sees India's hope to lie in a people's movement, not in the dank corridors of Delhi's bureaucracy. He calls for a mobilization of every retired military man, government officer, teacher, and unemployed graduate. In just one district that he cited, there are one hundred thousand retired military men. And he pointed out that India is producing thousands of graduates in English literature who are useless to the country. Mobilize all of them to impart reading and writing skills to the masses and there would be a true renaissance in English literature. Indian women who more and more want contraceptives but must keep their wishes hidden from their husbands, would all be able to read the instructions on the boxes. It may not be Shakespeare or Tagore, but it's crucial reading nevertheless. He believes 100 percent literacy could be achieved within three years.

Bose is convinced that education is the only way. Not laws, not dire warnings. "We have doubled the population and are still surviving," he reminds me. "If you say to the Hindu that the situation is so bad, that there is no water, no food, no land, et cetera, then he will simply say, 'OK, only God can save us, Hari Krishna, Hari Rama.'"

But, in fact, with the rate of growth as of 1993 continuing to increase, and with 500 million new Indians in just forty years since Nehru, nobody is scared, he goes on. (And as he earlier described, most people in the country can't imagine 500 million, let alone a billion.)

"The house is not burning. They understand philosophy. They understand that your children must be better than you. So you must educate them. It's the Japanese approach. It's the Hindu approach. My chauffeur has only two children. They are not chauffeurs. They are working in companies. They are graduates. Upward mobility. So by saying that India's population will be more than China's in the year 2025, what a ridiculous statement to make! Who is worried about 2025. You ask a villager, he wants to know whether tomorrow he will get two meals. He's not worried about India's population in 2025."

Bose calls upon the model of religious management as one possibility for understanding India's potential for mobilization, such as the country's Ramakrishna Missions. "It is the easiest thing to get fifty thousand people to take a bath in a puddle. They feed all those people in one day without any disorder. They can teach the government how to manage large numbers. They are marvelous managers. The government has not tapped the people's power." There are, for example, an estimated two hundred fifty thousand traditional healers and herbalists in the country who are already providing an estimated 80 percent of all medical services, a fact apparently little dealt with by the Central Government.[154] Of course, how does one "deal" with that fact?

During the Rajiv Gandhi administration, there was an attempt on paper at least, to mobilize two million villagers in a campaign to spread literacy and the two child family norm. Krishna Kumar, the man behind the Kerala sterilization camps (they were called festivals) of the early 1970s, was hired to make it happen. But soon thereafter, Kumar was transferred to another job and that was the end of it.[155] However, in the Ahmednagar district of Maharashtra, Drs. Rajnikant and Mabel Arole were able to pioneer their nongovernmental Comprehensive Rural Health Project in the town of Jamkhed by relying upon the villagers themselves. The Aroles realized that until they involved the local people, rather than "official" outsiders, there could be no long-term success. So they trained an illiterate tribal woman of the region in basic health care, and soon other village health workers came to be trained as well. The project spread to over sixty-five villages. Antenatal care and nutrition classes for mothers were given, early detection of diseases, the handing out of contraceptives, prevention of blindness by providing vitamin A pills, and lessons in water purification were all part of the program. It became so famous that committed doctors from other parts of the country started showing up in Ahmednagar. Ironically many of these professionals had to take refresher courses from the illiterates so as to become familiar with realistic rural needs and treatments.[156]

Ultimately, avows Ashish Bose, India needs more of these village-based mobilizations. It's the only way. But, he confesses, India also needs a little luck. According to Hindu scriptures, without luck one is lost. "Unfortunately," Bose says with a grin, "Indians rely on that luck more than anything else."

## THE LESSONS OF TAMIL NADU AND KERALA

While demographers and family planning experts throughout India have for several years hailed the two southernmost states, Tamil Nadu and Kerala, as proof that India has what it takes to reverse its population crisis, most NGOs and international donors, and even the Central Government itself, are well aware that these two states are not easily "replicable." Kerala, with a population equivalent to that of California (just over 30 million) but one-third California's geographical size, is the only state in the country where women outnumber men, 1032 for every 1000; where the literacy rate is higher for women than their male counterparts, 93 percent versus 92 percent. Infant mortality in Kerala varies from 17 to 33 per thousand, while the birth rate itself is between 19 and 23 per thousand. These are the best figures in India, by far. In a 1988 survey of contraceptive use—including traditional "natural" methods—the State of Kerala showed the highest usage among couples, at 80 percent. More and more the thinking throughout India is that to firmly establish a two-child family norm in the country, nothing short of the "Kerala method"—an 80 percent contraceptive prevalence rate (CPR) will do.

While the figures are also encouraging for Tamil Nadu—a 22.4 per thousand birthrate, and an average marriage age for women of 21—the state's literacy rate is at 55 percent (high by Indian standards, but not compared with its neighbor, Kerala), and its infant mortality is at 57 per thousand. This latter figure is extraordinary considering that it was at 93 in 1980. Tamil Nadu is a poor state, with little economic potential for women, at present. Kerala is also plagued by what many consider to be a depressed economy. Both states suffer from badly mired environments. Yet, in addition to Goa, a union territory negligible in size and population, and thus a bad model from which to extrapolate, these two southern regions offer abiding hope that India possesses the power of stabilization, under the right circumstances.

Some Indian demographers have viewed the achievements of Tamil Nadu as a more realistic example to strive toward in implementing similar, hoped-for curbs on the runaway Bimaru populations. According to T.V. Anthony, the former chief secretary and chairman of the State Planning Commission in Tamil Nadu for several years, Tamil Nadu offers the most realistic model of improvement for the rest of the country. Anthony attributes the beginnings of Tamil Nadu's fertility success to an early reformer, Periyar Ramaswamy who, long before family planning, was emphasizing late marriage so as to avoid even three births. Today in Tamil Nadu, the "ideal span for reproduction" that officials refer to are the years twenty-two to twenty-seven, along with an "ideal birth weight" of three kilos and a minimum weight gain of eight kilos by the mother during pregnancy. Such standards have necessarily entailed a three-year spacing between first and second pregnancies in order not only to increase the mother's health but to alleviate pressure upon the family's purse strings so as to encourage the parents to send that first child to kindergarten.

In the mid-'80s, Tamil Nadu under chief minister M.G. Ramchendran initiated a remarkably logical—yet in all of India untried—program, the Mid-Day Meal Scheme, to help defray protein deficiency among students who made it to kindergarten. Since then, every day a hot meal of rice and vegetable curry has been given to nearly a million children in sixty-seven thousand centers throughout the state. In addition some one hundred thousand expectant mothers are given health food packets every afternoon. Two hundred thousand otherwise unemployed and poor women are recruited within their villages to pass out the food in the centers and schools and to reach the slum children. It is not an inexpensive program—at a cost approaching $75 million dollars—or approximately 25 percent of the entire family planning budget for the whole country, paid for not by Delhi but by Tamil Nadu. While there is no hard evidence linking the meal to a lower birthrate, the program does reflect unique political willpower on the part of the state government, a sense of urgency that has not yet caught on in most other Indian states. The impact of this one activity seems to have been extraordinary. Aside from improved health, female literacy has risen significantly, primarily because parents now want their children in school where they are likely to get their best meal of the day.

Auxiliary nurse midwives, so-called lady health visitors, doctors, and *dais* were also encouraged throughout the decade of the eighties in Tamil Nadu to persuade at least 60 percent of all mothers who had had one or two children already, to accept an IUD. Eighty percent of all mothers who had already had a third child were canvased to accept sterilization. As a result of massive training programs, IUD acceptor levels rose eight hundred percent in just six years so that the mean age of tubectomy acceptors fell from 30.1 in 1980 to 28.4 in 1990, while the mean age mean for IUD acceptors fell from 27.4 to 25.8 in the same period, comparable to Kerala's figures. Taken together, these many considerations reduced the Tamil Nadu "span of fertility" from eighteen years (as in most of India) to seven years. Parents were counseled on the wisdom of having children while still young enough, wealthy enough, and healthy enough to enjoy and nurture them. Appealing to the parents' vanity, the message has been spread that too many children age the parents prematurely.

Every department in the local government has gotten involved in Tamil Nadu, from the Marriage Register and Revenue Department to the Social Welfare, Health, Education, and Information and Publicity Departments. Inspired by the chief minister's own passionate commitment to family planning, Tamil Nadu has seen a declining birth rate of about one per thousand every year after two to three years of preparation. T.V. Anthony points out that in the 1960s, India moved from being a P.L. 480 dependent nation (referring to its receipt of U.S. food aid) to being rather self-sufficient in the agricultural sector, by selecting two districts in every state that offered the potential for radical growth. Inputs from the central government were provided that assured the flow of high yielding rice and wheat strains, fertilizer, pesticides, irrigation support, and instruction. The experiment was successful and other districts in many of the states throughout the country replicated the system. Anthony believes that Tamil Nadu's success in the family planning arena is similarly replicable in the Bimaru states, given sufficient educational investments that are tailored to local customs and sensibilities, and realistic targets set for each well-studied factor (TFR, IMR, age of marriage, and literacy). Anthony believes that with but three years of preparation, the subsequent decline in birthrate of one point per thousand per year could be expected, with the success spreading throughout the country within three to four years.

What Anthony calls for is nothing less than a whole new way of life, one that promotes the postponing of a first child until the woman is twenty-seven, and pushing up the legal age of marriage for women to at least twenty-one. He advocates deemphasizing contraception and sterilization in favor of noninvasive planning and choice, ethical and cultural shifts in thinking, in other words. His prescriptions seem appropriate for Tamil Nadu, which has already begun the transition toward population stability. If those same stabilizing ingredients can be specially targeted toward towns adjoining ecologically sensitive parts of the state, then truly Tamil Nadu will be worth emulating.[157]

In seeking to extend the lessons of Tamil Nadu to the rest of India, certainly two relevant, if not key, components include finding the money for the mid-day meal and for expectant mothers' health packages. Given current fiscal priorities, the central government appears in no position to undertake such financing. It will thus be left up to all the individual states and territories throughout the country to recognize that healthy, educated children and parents will make a critical difference to the economy and the ecological well-being of the region. Whether environmental safeguards can somehow be incorporated into this scenario remains questionable, given India's experience thus far, and the fact that virtually no country in the world—whether heavily populated or not—has yet managed to increase its GNP without imposing considerable ecological damage.[158] Tamil Nadu's western neighbor, Kerala, is a case in point.

Kerala got a jumpstart on female literacy over the rest of India, and the developing world for that matter, because its leadership in the early nineteenth century under the maharajas, was adamant about providing all citizens an education and health care. Keralan society was essentially matriarchal. Women could inherit property. Girls in Kerala were never susceptible to pressure from their parents to get married. Furthermore, because the wars and religious conflicts of northern India have had little, if any, impact on this distant state, there was no history of fertility insecurity. Other than certain British atrocities carried out against local women, the state could claim a relatively tranquil history. All of these aspects of Kerala's past have predisposed its present population toward zero population growth.

But it would not be an easy task for the Bimaru states to overnight adopt Kerala's cultural disposition.[159] Nor could central and northern India easily envision Kerala's unique political tapestry. A party pluralism in Kerala has meant that virtually every community in that state is duly represented. Untouchables are not disenfranchised as elsewhere in the country. Muslims and Christians are represented in the Indian Union Muslim League and Kerala Congress; the Ezhava and Nair Hindus have their Socialist Republican Party and National Democratic Party; there is also the Marxist-led Left Democratic Front, as well as the Congress-led United Democratic Front. Curiously, because Kerala is the densest and most educated population in the country, it shows the least violence, and harbors the least crime. The average size of a village in Kerala is 17,500. In the rest of India, village size is between five hundred and one thousand.[160] On December 6, 1992, when riots erupted in northern India over the destruction of the Ram Temple, Kerala witnessed no ripple effects.

In spite of Kerala's political and social isolation from the rest of India however, her demographic successes—like those of Tamil Nadu's—have not yet solved persistent and serious ecological degradation consistent with the rest of the country. As I journeyed along Kerala's magnificent coastline, visiting various villages in the jungle in the company of a local physician and two government family planning officials, these problems took on sharp definition. Many factors are involved in Kerala's TPD—Total Population Dynamic—the intense, human appropriation of natural vegetative cover and biological primary production.

As of 1974 fish (from among the fifty-four or so exploited species) and prawn harvests began to fall in Kerala from both overfishing and industrial pollution. Economic overconsumption resulted from a simple idea, the mechanization of boats. Once it was shown to be possible, this shift from traditional boats induced merchants from other business sectors to exploit fishing for profit. Because the coast was deemed a "commons," access was free and unlicensed. Between 1966 and 1985 mechanized trawlers (registered under hard-to-trace corporate names, because the government at least officially required the boats to be sold only to traditional fishermen, known as artisanal) grew in number from a few hundred to twenty-eight hundred. Total catch for Kerala plunged in just nine years by nearly 50 percent. If the goal were sustainability, traditions, presumably, should have been preserved against the inevitable onslaught of mechanization, through subsidies and by strictly enforced quotas.[161] As the profit motive expanded, fueled by a new international demand for processed prawns, the trawlers indiscriminately clear-cut whole shoals and sea-bottom species. Prawns, once perceived as the poor man's protein and priced at 240 rupees per ton in the early 1960s, were valued at more than 14,000 rupees per ton by 1985. Profits, after cost of diesel, labor, and the boats themselves, were high for at least a third of those engaged in the trade. But for the commoners, that vast majority of fishermen in the state, not to mention for the marine ecosystem, the overfishing has been disastrous. As has been pointed out, "For capitalists, given their short-term perspective and under the given conditions of investment, the ratio of profits from indiscriminate harvesting of the commons to the profits from regulated and sustainable harvesting are large."[162]

While overfishing has wrought havoc in the coastal waters, declining profits and lack of available land have been at the heart of Kerala's agricultural stagnation and unemployment. The unemployed in Kerala exceed 18 percent, the highest level in the country. Every year a new Keralan labor force finds fewer jobs. Declining per capita acreage inhibits self-employment schemes. Much of the problem stems, paradoxically, from the fact that 55 percent of the population has various secondary school and college degrees. They're overqualified for most available menial tasks. Thus, in what must be judged a most perverse twist on Indian aspirations, that state which has best managed to educate and dignify its people finds that it can least support them, or do so without raping the environment. Education has also meant that Kerala's unions are pervasive and well-organized, and this has had a chilling effect on potential outside investors who realize that they will be unable to set up sweatshops in so literate a region. Adding a demographic twist to an existing paradox, some have observed that the reason women marry later in Kerala (thus ensuring fewer children) is that they are not interested in marrying unemployed, and possibly unemployable, males. It is a bizarre situation.

Does it then imply that a depressed economy and high literacy are the real preconditions of low fertility rate?[163] How are all these unemployed graduates surviving? The answer relates to yet another, unreplicable aspect of Kerala's low fertility profile, namely, temporary employment for Keralan males in the Persian Gulf states. Men frequently go away for years at a time, sending portions of their income back to support their families in Kerala, and that often translates into an entire cadre of young men and women just entering the work force, who can afford to remain stubbornly unemployed.

And for those many males who remain in Kerala—those who are not away sending money home to their families—the state government has taken a lenient attitude to income-generating encroachments on forests. This partial compensation for hard economic times has been described as "a policy that has in effect served as a powerful incentive to spread the encroachment mania far and wide and to denude the forest lands of the state continuously without rhyme or reason."[164] Even the Nilgiri Biosphere Reserve, the last large complex of any ecological integrity in the south, encompassing parts of Tamil Nadu, Karnataka, and Kerala, is now under siege by such encroachment. Fifteen hundred elephants inhabit the complex, along with tigers, gaur, sambar, muntjac, and chital. But forest-depleting livestock are increasing at astronomic rates along with the human population, and the local electricity board is trying to launch yet more hydro projects to entice settlers to move to the area. It is doubtful that the Reserve's wildlife can withstand this economically motivated human migration much longer.[165]

In reality, the state government's laissez faire policy is nothing new. In one form or another it has prevailed throughout Kerala's history. The Portuguese introduced fast-growing exotics centuries ago, utterly transforming large tracts of natural forest into coconut, cashew, teak, pepper, and rubber estates. That economic expedient diminished

Kerala's forests by the amount currently "missing" from Brazil, or 14 percent. Figures suggest that 87 percent of the Cardamom Hills in the Idukki District were forested in 1905, but by 1973 forest regions covered only 31 percent of that district. Conversions elsewhere in Kerala were even more extreme. The Wynad forest has been reduced from 44 percent to 9 percent; the Southern Highlands went from 53 percent to 8 percent. By 1988 over 7 million hectares of Kerala had been brought under coconut cultivation, and another 3.4 million hectares had been converted for fast rubber growth. The high price of both coconut and rubber make them unaffordable for most Keralans, who import cheap palm oil for cooking.[166]

While natural vegetation covered 44 percent of the state at the turn of the century, today there is less than 7 percent left.[167] And though the forests are rapidly disappearing, agricultural expansion has nearly ceased in Kerala since 1960. There is no more land. Increased cropping intensity has not compensated for a declining food base but merely added to the burden of pollutants.[168]

At the same time, the introduction of so many new species like cardamom, the state's chief foreign export earner, has occasioned terrain instability and a pernicious load of silt in the coastal zone, where 95 percent of remaining marine food sources are to be found. Though Kerala experiences two monsoons each year, only since 1980 has the region ever suffered such devastating floods, the result of watershed destruction. The nutrient-poor silt deposits from topsoil erosion have added an additional level of stress to the marine environment and a significant decline in overall marine activity has been tracked.

Kerala's unique backwaters have also become inundated with dissolved synthetic fertilizers, giving rise to increasing eutrophication and exotic weeds, and further destroying the coastal zone down river. Add to that the untreated cubic kilometers-worth of sewage, industrial wastes, and solid wastes from 30 million people.

Widespread ecological compromise in Kerala, a (relatively speaking) demographic panacea, poses very serious conceptual and ethical considerations. I say "relatively speaking" because 30 million human carnivores with short-term profit motives densely packed into a small state with a history of environmental malpractice are not likely to suddenly revert to ecological stewardship, not unless they can figure out how to stabilize the economy for all of the people in a benign way, an unlikely occurrence given Kerala's extremely high density of population. Thus, the rest of India and the developing world might well try to emulate the basic fertility replacement patterns of Kerala. But unless a corresponding effort is made to redefine what's good in life and to wean populations away from the crude reliance on GNP indicators toward ecologically sustainable, non-violent business practices and ethics, then zero population growth by itself will not be the answer. India invented the answer thousands of years ago—*ahimsa*, nonviolence. It is India's only hope. As Martin Luther King, Jr. put it, this is not a question of violence or nonviolence, but of nonviolence or nonexistence.

Many Keralans know this for themselves. The SPS (Samaj Parivartana Samudaya) and KSSP (Kerala Sastra Sahitya Parishad) organizations have fought to save the coastal waters, the rivers, and the Western Ghats. They have organized nonviolent protest marches and worked with villagers to restore their native wildlands. The KSSP prevented the Silent Valley Dam project from going forward and has launched environmental impact assessments throughout the region. But the pace of destruction continues, only slightly hindered. So one is confronted with a perplexing scenario: if not Kerala—which has managed, after all, to achieve an extraordinarily educated populace, whose largest towns boast of more bookstores per square block than most capitals in the West, a region which has proved the possibilities of NRR:1, namely, of natural fertility replacement—if not the Keralan Method, then what are the best means of achieving population control that are tied to even partial ecological amelioration?

There are programs working in at least two areas: ecology and population. On the one side, there are the ecologists. With over thirty-four hundred environmental NGOs throughout India, countless small-scale restoration efforts suggest that India possesses the know-how and the will to fight for ecological balance. The Gujarat Energy Development Agency, for example, has used controlled *chulha* stoves that reduce firewood consumption over traditional wood-burning stoves by 40 percent. In addition, biomass grown on fallow lands in Gujarat has then been converted in wood gasifiers as an alternative, renewable fuel. In cities throughout India, new recycling techniques are being advocated. In Bombay one company has begun spraying a microbial mixture of bacterial culture in a slurry of cow dung solution onto garbage, transforming it into manure and selling it as organic fertilizer. (One downside to this otherwise innovative scheme is that local ragpickers could be put out of business.)

Elsewhere throughout India ingenious solutions to water management are to be found—in Tamil Nadu, Ladakh, the Thar desert of Rajathan. Dr. B.D. Sharma, a highly respected Gandhian, has been going around promoting village "republics," smaller ecological units of population within greenbelts where there is spiritual and practical unity to be attained. These true Gandhian fraternities would reject the Western paradigm of technocracy by which the Third World has been entrapped, seeking ecologically sustainable ways of life. The ideal has realized change in countless villages. For example, in the impoverished deforested town of Ralegan Sidhi, along the Western Ghats of Maharashtra between Bombay and Pune, local men and women recruited under the concept of *shramdan* (volunteer labor, a Gandhian precept) dug percolation ponds to bring back scarce water, planted trees and fashioned hillside terraces to hold the soil. The terraces turned green. The villagers (illiterate by New Delhi's standards) then installed biogas digesters for converting organic wastes into methane cooking fuel. In less than twenty years, the village went from meeting only a third of its food needs to producing 100 percent, with surplus stocks that could be sold at profit in adjoining villages. Per capita incomes rose from a mere $10 to $80, still half the mean national average, but a miracle within the context of the region.[169] In the village of Sukhomajri near the Shivalik Hills, soil conservationists worked out a scheme with the local people resulting in an earthen water-harvesting dam and an elaborate system of irrigation that ensured equitable distribution of water while checking siltation. At the same time they developed "social fencing" to curb free-cattle grazing and thus increased the grassland yields. The combination of these efforts over a period of a few years tripled crop production in the village. Similar conservation and rural development schemes exist in every state of the country, from continuing *chipko* (tree-hugging) activities in the State of Himachal Pradesh, to Jain activism throughout the country.

On the population side, I have been impressed by various committed and sensible central and state government projects as well as by the innovation and deep concerns of so many of the country's demographers, clinicians, and village officials. But the formulating of specific economic policies that effectively induce population control while ensuring ecological sustainability has proven almost impossible at every level of bureaucratic, social, and consumer demand. The circumstances and needs vary from Kerala to Tamil Nadu, from Ladakh to Rajasthan, from Maharashtra to West Bengal, and often the energies and focus of population control and environmental restoration would seem to have nothing in common. The State of Haryana, for example, has the second highest per capita income in the country, yet it is very backward in some respects—the status of its women considered to be among the worst in the country, its forests almost completely denuded. A comprehensive financial package, targeted toward women, afforestation, and species protection, would be necessary to uplift Haryana. The money is there. The will, apparently, is not yet there, despite the state's proximity to New Delhi and efforts by UNFPA and the World Bank to help Haryana. Such contradictions abound throughout India. One cannot understand them through mere statistical analysis. Every village demands its own unique solutions and requires unusually creative thinking from any political administration.

It would surely be desirable to transpose Kerala's extreme fertility successes to other regions of the country, but not the state's devastating industrial and environmental shortcomings. The ecological strategies that have worked for the desperately poor residents of Ralegan Sidhi may not be at all applicable to an intensely literate, unionized state like Kerala.

Tamil Nadu, her triumphs less pronounced thus far than Kerala's—not having enjoyed Kerala's nineteenth-century predisposition toward literacy and matriarchy—may nonetheless be the key to understanding how India might best focus on both mass family planning and ecology mobilizations. The techniques for achieving these goals may be as uncomplicated and elegant a solution as providing free lunches to children at school (though this by itself, is unlikely to solve the population problem). But such elegance will only mock Tamil Nadu if it is unable to control female infanticide within many of its villages, or check the state's massive animal abuse in slaughterhouses, or curtail destruction of its beaches, turtles, and elephants.

The jury is still out as to whether family welfare and environmentalism in India can, or will, pave the way for sexual caution, a widespread cafeteria approach to contraceptive methods, and the propagation of a true reverence for existing life, enshrined in the legal codes but sadly absent from any substantive monitoring or policing infrastructures. Two hundred fifty million females in India are currently of childbearing age. Of the fifteen largest states in India, the ten most populous ones have witnessed a disastrous and (at least officially) inexplicable drop in the rate of contraceptive prevalence and sterilization from 1991 to 1993. The drop in IUD insertions was 7.6 percent, in oral contraceptive use, 14.1 percent, and in all other contraceptives, 3.5 percent. The sharpest falls occurred in Assam, Bihar, Haryana, and—most discouragingly—in Tamil Nadu itself. The report, undertaken by the evaluation and intelligence division of the family welfare department over the months of April through November 1992, also noted sharp declines (16.3 percent) in the distribution of commercially available contraceptives even though government subsidies meant that the condoms were free.[170]

The world's population is increasing at three children plus every second, approaching three hundred thousand a day. In the decade of the nineties nearly a billion people will be added to the planet (some say precisely 958 million, though there is no such thing as precision in demographics). Whether 958 million or 1 billion, it is believed that India will contribute about 200 million to that figure, 20 percent of whom will be below even India's generous definition of the poverty line.

Many Indian demographers look to China, but fear the political repercussions of coercion. They look to Thailand, which has cut its growth rate from 3.2 percent in the mid-1970s to 1.6 percent in 1993, through massive raising of consciousness and distribution of literature and condoms—even at traffic jams (of which there are plenty)! And they look to Indonesia, which has also seen a steady decline to under 2 percent annual growth from 3 percent two decades ago, though not without its own severe environmental problems as will be discussed in the next chapter.

In comparing India's situation with so-called RICs—Rapidly Industrializing Countries, such as Brazil or Mexico or Sri Lanka—Indian scientists acknowledge that they are as much as thirty years behind these other nations in provinces of income, education, and health. While India's TFR was 4.4 in 1985, South Korea and China had done much better than that twenty years before. Even neighboring Sri Lanka was doing better than India as of 1971. India currently provides a mere 44 rupees per eligible couple. That means seventy-five cents per person— and only if that person is married. The consensus now is that India's family welfare program is trapped. And that the population explosion is yielding diminishing returns even for its meager per capita investments.[171] In fact many argue that unless a lot more money is invested in Indian family welfare—massive amounts—then even the current three hundred fifty million dollars a year is money much wasted.

In the 1980s the Indian family welfare slogan was, "*Two or three children are enough.*" In the 1990s, it reads, "*Next child not now, after two never.*" Many argue that the future slogan of family planning should be simply "*No more,*" that morality and population control need to be separated, that voluntary sterilization programs must be forcefully and urgently enhanced if India's internal war is ever to be peacefully resolved.

The greatest living sources of paradox in India are decreasing mortality and persistent poverty, two demographic facts of life which translate into increasingly more numbers—men, women, and children who will be forced, from the exigencies of sheer survival to rampage across every remaining hectare of India like driver-ants. That situation to date has condemned the country to a most unfortunate syndrome of contradictions: that which is most loved—nature, women and children—tends to be most abused.

· The World Bank projects population stabilization in India at 1.852 billion during the last quarter of the twenty-first century, though it will hit 1.35 billion in the year 2025. The government of India has cited other projections, believing that the country's population could reach between 1.6 and 1.8 billion by the year 2035. Such discrepancies highlight the difficulties through which demographers must tread.[172] But as Ashish Bose reminded me, "Unless we go beyond the cold calculus of births and deaths and feel the heartbeat of the people, demography will remain a dismal science of population, dominated by doomsday predictions based on mechanical projections, which can now be done in a matter of minutes on the computer. Statistical competence is not enough to understand the population problem. . . . We have produced a whole lot of technical demographers who cannot go beyond decimal points."[173]

What the numbers do tell us, however, is the cumulative weight of impact, about which science and detective work yields very real, very tangible problems, decimal points and all. At the turn of the century, educational costs alone for India's growing population will be roughly $50 billion a year, a large share of the country's resources. In the very best case scenario a huge percentage of the country will still be illiterate.

The urban population of India will exceed 325 million by the year 2000. Even if allocations for sanitation and water should manage to be doubled, they will at best maintain the status quo, which is desperately inadequate today, and the basis of much disease, frustration, infant mortality, and population growth. Actual real dollar costs associated with disease in India have not been calculated.

By the year 2000, India will need to be investing approximately $20 billion every year in the housing sector, just to keep up with the increase in newborns. This, according to P.D. Malgavkar, is beyond the means of the nation.[174] And even if that money were allocated, there would still be hundreds of millions of untouchables living in substandard shelters, leaf huts, and urban slums of cardboard and dried buffalo dung. The so-called "quality of life package" for India's future, including environmental protection and restoration, access to safe drinking water for humans and other species, adequate nutrition, clothing, housing, education, and health care, all appear similarly to be beyond India's means, if one wishes to perceive it that way.

On the other hand, what else really matters to a nation if not such basics of the social contract? What else is a government for? India's annual GNP exceeds $300 billion now, her long-term public debt 21 percent of that, or just over seven dollars per capita.[175] The country at present spends 17 percent of its annual budget on defense, but a mere 2 percent for health, 3 percent for education, and essentially nothing for social welfare or housing.[176]

Considering India's overwhelming dilemmas, certain fiscal priorities are begging the kind of reappraisal that would leave ample economic elbow room to preserve what's left of the country and ensure a dignified and creative future for its people—only if the voters and the politicians themselves are committed to doing so. Otherwise one of the oldest and most noble of human civilizations, along with the planet's most remarkable diversity of biological habitats and creatures great and small, will vanish painfully, like so many other human cultures and ecosystems during the past hundred thousand years of humanity's lineage.

# 4

# Nature Held Hostage: Indonesia

## THE INEVITABILITY OF THE MASSES

Sometime around four o'clock each rain-sullen morning during winter, in the stillness of sprawling Jakarta, the call to the faithful rasps and twines, as thunder rumbles gently across the distant hills. The muezzin's cries are no longer broadcast from atop minarets, but by citywide megaphones. Sultry, grating, and inflictive—and from which there is no escaping—these prayers punctuate my sleep with the tension of Islam, fill me with apprehension, and echo beyond Java to all of Indonesia's nearly seventeen thousand islands. Overhead in the heavy darkness, the fan turns lazily, momentarily disorienting the mosquitoes.

Across three thousand miles Indonesia begins to rise. If you turn your ear toward the enormous industry of this rapidly developing world you hear the shuffling of the work day: of weavers taking up their bamboo baskets, of embroiderers and bricklayers, of tilers and fishermen and garment makers, of those engaged in the preparation of snacks, or the manufacture of brassieres, or rattan wickerworks and kerosine stoves. One detects the clatter of women walking to beauty parlor posts, rope makers bicycling to their shops, brush makers gathering their materials, sandal fabricators pulling their leather. By dim light, the workmen who decorate copper lamps can be seen huddling over an early morning breakfast of rice and tempeh. With poles and nets the caretakers of fish nurseries begin their myriad tasks, while processors of soybeans, pastry, and cakes take up their positions. Others can be heard scraping various husks for animal fodder, or harvesting, peeling, washing, cutting, cooking, and fermenting the cassava. Some are busy recycling rope, others pounding millet. Train stations are bustling. Taxi drivers are already at work. The middle class—those earning as much as six hundred forty dollars a year—are pouring coffee for themselves, reading the morning's *Jakarta Post*, while the storm-soaked streets steam in the unearthly cornelian glow of a clearing dawn.

The air is thick and mucilaginous, and through it softly strides a labor pool out into the paddies and fields, more than 20 million farmers, upon whom 170 million others depend. They toil, principally, for rice, maize, sweet potatoes, soybeans, ground nuts, coffee, tea, tobacco, sugar, clove, pepper, and nutmeg, cassava, coconut, and corn, in a constant struggle with the loss of topsoil and the corresponding deterioration of watersheds. The land is redolent with agricultural odor that dates back several thousands of years. But, an estimated 10 percent nutritional deficiency now enshrouds all this erstwhile effort.

Beyond the farms, small-scale industries proliferate in ten thousand directions. Rubber tappers, who collectively produce more natural rubber than any other country in the world; extractors of palm oil; manufacturers of various copra products; and the omnipresent animal jobbers who "produce" nearly 600 million chickens, 25 million ducks, 11 million cows, 3 million buffalo, 11 million goats, 6 million sheep, 7 million pigs, and 700,000 horses every year for commercial and dietary use (slaughter, by any other name). Well over 2 billion pounds of meat are now consumed annually by Indonesians.[1]

In addition, Indonesia exports over 11,000 primates, 88,000 live parrots, and over 3 million reptile skins in a given year. But the term export hardly conveys the truth of the activity—the hundreds-of-thousands of other animals killed in the process of obtaining live "specimens"; the fact that the young are frequently left to die without their parents. As with any other country whose economic indicators look applaudingly at desecration and ecological disaster on the basis of subsequent contracts and jobs targeted at restitution, export suggests progress and prosperity. In the case of animals and animal by-products, export conceals the savage entrapment and horrid conditions of shipment—the animals subjected to torture and trauma, to wire cartons, asphyxiating cases, gunnysacks, and overstuffed crates that lack food, water, or enough space even for the animals to move, over a period that can last weeks, sometimes months, or even years. Such export removes the resident creature from its rainforest paradise, in the case of Indonesia, to the dark basements of an animal research laboratory, some taxpayer-supported university where the animals are doomed; or to zoos where they are humiliated, frustrated, and condemned to unholy, forlorn lives; or to decrepit, diseased holding cells prior to distribution to pet stores and private owners, or other forms of hell. Or, finally, to department stores as leather belts, briefcases, shoes, and furs. Such is the woeful itinerary against which population policies are ultimately contending. For it is in this wide guise, however obscure, that animal suffering infiltrates the country at the hands of a burgeoning human contingent. While there are the odd vegetarian dishes, Indonesian food predominantly reeks of meat. And there are rumors that filter down. Of orangutans (meaning "man of the forest") being grabbed from the wild and shipped to Taiwan on oil tankers. Some four thousand of these herbivorous anthropoidal apes (*Pongo pygmaeus*) are believed to be stranded in Taipei—too large and ungainly for the delicate Chinese who prepurchased them, then who cast the animals out onto the streets where some of the orangs have been found, either dead or begging for food. A few have been shipped back to Indonesia for attempted reintroduction into the forests, but most perish from respiratory diseases.

I am inclined to single out such incidents not randomly, but in an effort to begin to grasp the broad swathe of devastation. The last mangrove preserve along Jakarta's coast, a meager 15.4-hectare area, has now been turned over to developers by the minister of Forestry, in spite of the minister of Environment's protest. This was home to long-tailed and Javanese black monkeys as well as large numbers of migratory bird species. Now it will be another strip of hotels and golf courses. Elsewhere in the city, a few blocks from the World Wide Fund for Nature offices (the WWF, known in North America as the World Wildlife Fund), tens-of-thousands of rare, wild-caught birds are kept in squalid cages, exposed to the heat of day to be sold to passers-by. Jakarta has become a magnet for poachers, a lawless epicenter for those wishing to profit from the agony of other species. The WWF representatives know the poachers by name but are hesitant to interfere in domestic legal matters such as these when many greater priorities call for their fragile negotiations with the government—the saving of whole habitats, for example. Furthermore, if the WWF or any other well-intending NGO, or animal liberationists were to buy up all the desperate creatures, they would be confronted by three emphatic difficulties: First, their lifesaving expenditures would only subsidize the illegal trade, thus encouraging the zealous poaching of yet more birds. Second, many of the birds have been captured from distant parts of Indonesia. If released around Jakarta, they might not find the specific habitat and edibles necessary to keep them alive, and are thus probably doomed no matter what happens. And finally, neither the WWF nor the government of Indonesia is financially able—all things considered—to police so vast a country, or rescue all the endless victims of human greed and ignorance, or relocate them to safe havens. Phrased in the popular jargon, Indonesia reflects a condition of necessary triage much like that depicted in William Styron's *Sophie's Choice*, and Thomas Keneally's *Schindler's List*, whereby the needy are too many, the means of amelioration too limited. The country, like the world, is simply too large and too cruel.

Even *identifying* safe havens is becoming problematic. As far as away as Indonesia's remote Aru Islands southeast of Maluku, there are inexplicable mass deaths of pearl oysters occurring every day. As of late 1993, 60 percent have gone extinct as a result of an unknown illness, or pollution, or overconsumption by locals. People are hungry. Oil tankers are unmonitored, spills untreated.

One hears of Indonesian fishermen bombing coral reefs, even within park boundaries, to get at the fish; of military generals who have been given whole islands as perks and have then destroyed lagoons to build hotels.

Out among the mountains, miners are busy digging and dynamiting for sulphur, nickel, manganese, phosphate, gold, tin, silver, bauxite, and coal, their effluent streams percolating into the country's groundwater and sending enormous loads of toxic waste downriver.

Beyond the paddies, amid primeval forests, smoke is rising. What used to be designated with admiring distaste as "the jungle" is now more formally described as "biodiversity." But there is nothing polite about what's happening in those woods. There the morning sounds of awakening are very different from those of the city: power saw and lumber mill noises mingling gigantically with the plaintive, melancholy worship in the predawn hours, and the bewildered cries of myriad animals torn from their homes. So goes the agony of discord and development.

With sunrise comes the full vision of Jakarta, a megacity reminiscent of the slums of India and the rapid concrete and metal ascent of Shanghai. A city of 11 million, Jakarta's population is growing by 10 percent a year, an effect of migration, high fertility, and the promise of an economic boom. For 25 million dollars, anyone can start a bank in the country now, a recent liberalization policy of President Suharto's benevolent patriarchy. Dozens of gleaming high-rise financial institutions and over two hundred private banks currently enliven the city's major thoroughfares, conduits for the international investment now bolstering much of the country's commercial sector. In 1993, 227 new industrial projects were unveiled and "exports" projected to exceed $18 billion. Iron sponge, electric machines, plywood, pulp, textiles, and a medley of other value-added forest-based products are now shipped around the world, while factories churn forth alkali benzene, polyethylene, methyl ethyl peroxide, and a host of other chemicals quite lethal if uncontrolled—a dizzying array of human activities in search of foreign exchange.

A multiplicity of priorities assails Indonesia, but there is no confusion: Indonesia represents enormous growth in terms of an energy-suppressed population. One hundred ninety million consumers in the country demand an ever-improved quality of life. The demands will be met somehow, with or without the cooperation of the developed countries. At the same time, Indonesians are systematically attempting to reduce their birthrate in order to increase per capita income. Most residents can understand that fewer children at the dinner table mean more food to go around. But despite this conventional wisdom, and many years of serious family planning, the economics have not yet been reconciled: the country finds itself ecologically cornered. With a nearly 2 percent population increase per year, Indonesian human numbers are expected to double by the year 2030, assuming even that existing levels of birth control and family planning do not diminish (and there is some cause for alarm in that regard). As it is, the country's population has virtually doubled since the 1961 census, when the recorded figure was 97,018,928.

Despite this doubling of population, it could have been much worse, for Indonesia has managed to increase contraceptive prevalance from 10 percent use in 1971 to nearly 50 in 1993, thus preventing the births of over 12 million people. Viewed even more broadly, if the number of users of some form of contraception continues to increase at least until 2005 as it has since 1971, 57 million fewer Indonesians will exist than might have, the equivalent of the population of Thailand or France.[2] In 1970, Indonesia's population was 118 million, with 78 million located on Java and Bali. The 1992 population was over 185 million, with a contraceptive prevalence rate (CPR) of 49.7 percent, and an average life expectancy of sixty-one years.[3]

To support that doubling of inhabitants, and based upon *projections* for resources needed to sustain a future quality of life package, so-called, already more virgin tropical rainforest overall has been claimed than in any other country, with the exception of Brazil. And this implies the calculated annihilation of perhaps millions of species unless alternative technologies and economies can fashion a system of warranted utilization congruent with a wide range of local responses and rural poverty.

"Only if the poor can enjoy a better life will they be able to give their contribution to the solution of global environmental issues," said Indonesian president Suharto. Seven out of ten people in developing countries depend on forests for their heating and cooking; for houses, furniture, fences, tools, and household implements; for any foreign currency. According to the Indonesian Department of Forestry, those basic demands have resulted in the loss of more than 500,000 hectares of forest each year throughout the country, while only 40,500 are annually reforested. Officially to date, 7.5 million hectares—or 9 billion trees—have been cleared to meet local needs, and this figure does not include the equally enormous attrition from private industry.[4] In addition, thirty million hectares of allegedly protected forests are, in fact, wide open to illegal logging and pioneer agriculture.[5]

Indonesia is the fourth largest country in the world. By comparison with China and India, its population is still "manageable," the country rather spacious, and its environmental and family planning leadership very much aware of their options. Yet, a converging tragedy has already been plotted on countless charts in Jakarta, though darkness and the demise of nature are on few Indonesian lips. As elsewhere, the colossal debasement of the natural world has been rephrased in the lingo and currency of hope, feasibility, and furtherance. Because poverty has been identified as the source of all destruction, its alleviation has logically been deemed central to rainforest preservation, longer human life expectancy, a diminishing fertility rate, and a decrease in infant, child, and maternal mortality. Wealth, symbolized by those enormous banks along Jakarta's swank, tree-lined avenues, is unambiguously viewed, with the same hysteria as in the West, as the solution to everything.

Ironically, in the midst of this unabashed quest for dollars, Indonesia's leaders are quick to seize upon the vulnerability of the Western economic model. They point out that developed countries pollute, consume, and destroy far more than developing ones, citing Austria, for example, which exploits 3.3 cubic meters per hectare of timber compared to Indonesia's 0.27. Moreover, Austria's total forest area is slightly less than 3.7 million hectares while that of Indonesia is somewhere around 100 to 109 million; these numeric differences separating the temperate and the tropical forest countries invite ethical and policy comparisons based upon population and country size, GNP, and—most importantly—the varying levels of biodiversity. Indonesian government officials, while not unaware that biodiversity is vastly more multitudinous in a rainforest than a temperate forest (possibly by a million-fold), are adamant about their right and their need to exploit their own forests. Unlike India's official position, however, Indonesia is not focused upon trees as a community resource but as the only guarantee of sorely desired international exchange. The economics, driven by Indonesia's population bomb, are fickle. No matter what policy of monetary gain the country pursues, regardless of the ethical and just positions it adopts to justify its financial aspirations, *nature* will suffer. Human nature itself recognizes that suffering, and well understands it has manufactured a social and political language of false certitudes so as to obfuscate and deny the truth, or to go on in spite of truth, or to satisfy itself with half-truths. Living in false consciousness, reconciled to compromise, Indonesians, like everyone else, have harbored a global lie and termed it "survival."

The lie comes from knowing better, even if "better"—the absolute safeguarding of even a small piece of wilderness—has proved nearly impossible to achieve. Within this context, a knowing Indonesia is attempting to fashion a compromise that is at least *a little better.*

India, China, and Indonesia together account for nearly 40 percent of the world's population. While India and China have already suffered untold ecological pillaging, their populations rising seemingly out of control, Indonesia still has time to implement lifesaving measures in the two crucial antiwar sectors, family planning and conservation. Because Indonesia has developed an economic sweet tooth, has a huge population of young people, and more biodiversity to lose than nearly any other country, its struggles require urgent, international empathy and assistance.

Upon moving past customs at Jakarta's rather chic airport (done over in tribal architecture), the first thing to meet one's eye is a sign that reads *Ruang Ibu Dan Bayi* meaning "Nursury," the room situated next to baggage claim. The implications go beyond a mere sign— whether male or female, children are adored and looked after in Indonesia. Among the Bata people of North Sumatra, following a three-day wedding shindig, the chief blesses the bride and enthusiastically implores her to bear sixteen baby girls and seventeen baby boys.[6] Elsewhere across Indonesia, the pace of reproduction has been toned down considerably in the past twenty years, thanks to an aggressive national family planning program and a predisposition among countless rural, and some urban, communities to honor the state's authority. The government is seeking to achieve a total fertility rate (TFR) of 2.1. At present, however, the figure seems inert at 3.02 by many estimates—at least one leading NGO (the Yayasan Kusuma Buana) places it between 3.2 and 3.4—in either case a considerable improvement over the Bata priest's recommended TFR of 33, and dramatic evidence of progress over Indonesia's countrywide TFR of 5.6 since 1960. Even more optimistically, the Agency for International Development (AID), the largest donor to Indonesia's family planning program, believes that the country-wide TFR is currently 2.8. The methodologies, sampling techniques, and local research by which these diverging assessments are derived are not necessarily conflicting; only the results are. A spread between 2.8 and 3.4 is, of course, the difference between tens-of-millions of people over a single generation.[7]

A gap in consistent data is evident. Nevertheless, to put the country's indisputable fertility achievement in proper context, remember that Indonesia is 85 percent Muslim, making it the largest Islamic nation in the world.[8] (Except for Turkey's TFR of 3.31, most of the other forty Islamic countries show TFRs between 4 and 7.7, the latter figure from the Republic of Yemen. Many countries, like Saudi Arabia, Pakistan and Iran, exceed a TFR of 6.) The more than one billion Muslims are the fastest reproducing religious group in the world.

How, then, has Indonesia accomplished so much toward stabilizing its population already? Some have argued that the achievement is the result of a "dramatic shift toward self-choice marriages" and nuptials at much later ages, in addition to the increasing trend toward normalizing remarriage and divorce.[9]

But a more pervasive explanation pertains to Indonesian *ulamas'* (religious leaders) interpretations of Muhammed's teachings. Many social scientists discount the influence of religion on contemporary family planning, whether in Catholic or Islamic countries.[10] But without religious accommodation in Indonesia, much of the government's considerable efforts would have been in vain.[11] Indonesian Islam favors birth control and was thus immediately receptive to the government's own family planning programs. Many Muslims insist that family planning is inherent to the *Qur'an* itself, once it is properly understood.

But that understanding is largely an Indonesian phenomenon, shared by few other Islamic countries. Consider that in Egypt a growing fundamentalist sector among its 60 million residents has denounced birth control as a Zionist plot, despite President Hosni Mubarak's constant reminders that the nation's resources are almost exhausted because of unchecked population growth. Mullahs in Iran have claimed that family planning was a Western plot, though that sentiment is slowly changing. In the Sudan, birth control is considered shameful by many, who believe that the best the poor can do for the world is to offer as many children as possible to Allah for his care.

But the fact remains, if it is so interpreted, that Allah, the Prophet Mohammed, and the many collections of the *Hadith* detailing statements by the Companions of the Prophet, all confirmed permission to use birth control, both pre- and postconception. For Allah, birth control is not even a religious matter, only a lesser, worldly matter, about which, according to some commentators, he was pragmatic, far-sighted, and liberal, having already emphasized that the religious destiny of the universe would not be affected by worldly matters. "Go ahead," said Allah, with respect to birth control. "Nature will in any case take its course."[12]

According to most Islamic religious scholars, the four original *Imams* (Abu Hanifa, Shafe'i, Malik, and Ibn-e-Hanbal) accepted contraception. The formation of the embryo and the sexual act were given absolute legal parity (the male did not have greater standing than the female), masturbation was not frowned upon, and according to historian B.F. Musallam, "Agreement on the sanction of contraception was so general that the permissive view became a culture cliché." Long before Europeans were openly discussing the fine points of birth control, Islamic jurists were doing so, supporting the use of intravaginal suppositories and tampons.[13] Moreover, Islam prohibits intercourse while a woman is nursing, recommends suckling for two years, even accepts coitus interruptus *(azl)* as a form of contraception. Islam wants a large number of followers but unlike conventional Catholicism is not willing to propagate those minions at the expense of a quality existence.

Thus, economics has always entered into Islamic feelings about family size.

Why then have most Islamic nations failed to achieve a low fertility rate? The answer appears to be the dogged devotion to two particularly simplistic statements in the *Qur'an*, namely: "Marry and beget children," and "Marry so that you multiply because I will make display of you in front of the other nations on the Day of Judgment even I will make a display of those prematurely born *(alsiqt)*."[14] Indonesian Muslims have managed to see beyond these facile provocations. They have not, however, granted the woman the free right to have an abortion.

There is no discussion of abortion in the *Qur'an* itself. Conservatives (fundamentalists) prefer, in absence of precise guidance, to err on the side of caution and thus prohibit abortion, except in cases of doctor-certified mother's peril.[15] Among Muslim nations, only Tunisia, Turkey, and Bangladesh have legalized abortion. But many Indonesian Muslims would like to see legalization at least openly discussed in their country. Though the government denies it, Indonesians are aware that millions of "illegal" abortions are being carried out, often resulting in maternal deaths.[16]

One morning I sat with Mr. Abdurahman Wahid, chairman for the Nahdlatul Ulamas, one of the largest Islamic organizations in the world, based in Jakarta. At least 25 percent of all Indonesians follow this middle-aged gentleman. With an acutely sensitive interpretation of human affairs, Mr. Wahid began by clearing the air. He wanted to let me know that not all Muslims fit the Western parody of them, then proceeded to quite unexpectedly praise the Jews and Israel. He then explained how Nahdlatul's most fervent effort was the raising of consciousness regarding birth control, and the subsequent need to preserve the natural environment. But, he argued, the government itself was faltering, having skirted the issue of permanent contraception "for fear it will be seen as too anti-Islamic. I know my people and how they react," he said. "If the scientists can give a level of confidence about recanalization, reversible sterilization is OK with the Muslims." The official government family planning agency, BKKBN (the Indonesia National Family Planning Coordinating Board, launched in 1970) has not advocated it because "they have no guts," said Mr. Wahid. And he went on, "Let's open up the debate on abortion."

Islamic scholars, using relevant *Hadith*, or, historical commentaries on the *Qur'an*, have stated that so-called "ensoulment" occurs one hundred twenty days after conception. Abortion prior to that time is not considered *wa'd*, or murder, though not all *ulamas* agree on this point. If the mother's health is imperiled, abortion is actually mandatory under Islamic religious codes, however. Interestingly, despite much that is made of the male's oppression of women in traditional Muslim societies, mutual consent for sterilization is also mandatory inasmuch as procreation is considered a "right" of both spouses. Husband and wife must discuss the matter, in other words. The problem in Indonesia is that the most conservative elements of Islam have scared the government into relying upon shorter-term contraceptives and refusing to even open up the abortion issue for debate. Pronatalist Muslims have argued that "He who does not marry for fear of having a large family has no trust in God," as well as stating that "If they are poor, God of His bounty will enrich them." [17]

The government of Indonesia has been able to verify a 35 percent confidence level for reversible sterilization (the so-called recanalization technique) at seven official centers in the country. But most Muslim leaders want to see a 51 percent assurance rate before they will sanction it. Do the leaders speak for the people? Not in the rural areas, according to former UNFPA country representative Jay Parsons who explained to me that the authority of Islam in the countryside does not usually extend beyond the front doors of a mosque, especially for women.

I later met with several members of the other large Islamic organization, the Muhammadiyah (Pusat Dakwah Muhammidiyah). Begun in 1912, Muhammadiyah accepted family planning in 1968. Its members advocate spacing between births, consistent with the *Qur'an*, and maintain eighteen hospitals and one hundred twenty maternity clinics throughout the country. They teach Dakwah, their population communication program, to 4 million students in twelve thousand schools and thirty-five hundred kindergartens. They also have schools for nurses and graduate fifteen hundred every year. Among the organization's leaders, I heard a marked insistence that a total fertility rate of seven children was fine if the family could afford it. Not so, according to Dr. Maftuchah Yusuf, the head of Aisyiyah, Muhammadiyah's female branch and an outspoken female professor of Islamic philosophy and environment, gravely concerned about deforestation and the population bomb. She is pessimistic about the government's targets and the possibilities of reducing environmental destruction. Using the *Qur'an* to promote birth control and environmental purity, Dr. Yusuf tells everyone she meets that "Allah demands clean water and a clean soul." She is deeply committed to reducing the birthrate. Grassroots activism is languishing, in her estimation, particularly in the rural areas of the country where she and her elegantly white-robed female students and midwives have launched sex education for thirteen-year-olds beginning in 1990, something the male side of Muhammadiyah would not do. The males speak of reproductive health, not sex education, because "sex" connotes something in their minds that transgresses Islamic laws of modesty. The debate between Islamic men and women on this subject strikes of psycholinguistics more than it does family planning.

According to Mr. Wahid, Indonesians "cannot afford" too much traditional Islamic law. In other words, conservative inhibitions are no longer practical in a world where human population has exceeded environmental carrying capacity.

The problem Indonesia faces by refusing long-term methods of contraception is the high rate of discontinuance of short-term methods, a trend that does not bode well considering the government's recognized need to greatly accelerate the number of new contraceptive acceptors in the coming years. According to one recent study, "Even in the Asian nations where contraceptive acceptance is increasing, high rates of contraceptive discontinuation continue to plague program planners."[18] Demographers studying Indonesia have pointed out that—consistent with the psychology of contraceptive freedom elsewhere in the world—a woman's ability to chose her method is crucial to the likelihood of her maintaining contraceptive vigilance.[19]

One positive, medium-term trend is that Norplant has been embraced more enthusiastically in Indonesia than in any other country. The reason for this cultural endorsement predates Islam's arrival there. Among the Balinese and West Javanese, it was considered beautiful for a woman to insert gold or silver needles under her skin. Or for a man to have stainless steel inserted in the palm of the hand. These insertions, known as *susuk*, were also believed to ward off illness, to maintain strength. Sometimes the metal was inserted in the chin, the breast, or the face. The Population Council did not know about *susuk* when they developed Norplant, but it is not at all surprising that the contraceptive insert was so easily adopted by Indonesians. Fortunately, the five-year life of Norplant has not been deemed "long-term" by the conservative Muslims.

Promoting two official slogans ("Two is enough, male or female, it's the same," and "Small, happy, and prosperous family"), the BKKBN employs eighty-one thousand workers, five hundred thousand village volunteers, operates seventy-six thousand contraceptive distribution centers (VCDC) across twenty-seven provinces, and, in addition, coordinates the activities of three hundred thousand family planning acceptor groups and tens of thousands of other volunteers.[20] For all this, compared with China's 50 million family planning volunteers, the BKKBN is vastly understaffed. Yet, for twenty years the agency has distinguished itself through ingenious publicity, rigorous family planning administration, and impressive results.

A variety of family planning "rewards" (they don't use the word "incentive") have been offered to those contraceptive acceptors who have demonstrated initiative and persistence over a period of years. Rewards have included livestock, televisions, or irrigation pumps, visits to President Suharto's palace, hybrid coconuts, babysitters, and all expenses paid pilgrimages to Mecca.[21] Recipients are made to feel like super-patriots and their pictures invariably get into the newspapers. Direct cash allotments have not been part of Indonesia's rewards programs, though village lotteries for family planning acceptors, as well as discount cards, have been initiated.[22] Youths as young as seven are encouraged to pledge and sign that they will not marry until twenty, will postpone the birth of their first child until four years after marriage, postpone the birth of their second child another four years, and then stop having kids forever after.

With the exception of China, no other country has been so aggressive or successful at reaching the subvillage level. Furthermore, in Indonesia family planning makes the initial inroad and other integrated development projects follow suit, piggybacking to get to the micromanagement level—and that is very distinctive about the country. As Dr. Sardin Pabbadja, BKKBN's immensely likable deputy director for Program Operations explained, family planning maps for every village in the country have been fashioned. Each household is color-coded on those maps for age and health profile, all the female's monthly cycles, and the status of her contraceptive use. Dr. Pabbadja put such a map up on the wall and described his agency's approach to increasing contraceptive use and maternal and child care. Every second week of the month, local family planning staffs meet to go over the maps and see what needs to be done, where supplies or motivations are lacking, where roads have been washed out, where children are sick, where local infrastructure has grown weak. They then dispatch workers from door-to-door.

On paper, the system is superb, compromised only by Indonesia's rugged terrain and endless island chains. Movement from village to village is frequently difficult, curtailing the predictable flow of monthly contraceptive supplies. This is one reason why long-term contraceptives are so crucial in Indonesia, and why the Islamic resistance to them (on grounds that any permanent check to children, or a mother's free right to have children, is morally wrong) has served to escalate Indonesia's population bomb, and corresponding ecological crisis. In fact, according to a confidential World Bank report in 1991, the shift to longer-lasting contraceptive forms seems to be the key to making family planning work in Indonesia. How does a government best manipulate the economy to foster long-term contraception? That is the issue, and it is not a simple one.[23]

Most births in Indonesia occur, unassisted by doctors, in rural areas where 83 percent of the population live, and where the rural literacy rate has actually fallen from 39.1 percent in 1971 to 28 percent in 1980, a level comparable with rural India.[24] The traditional village midwife, or *dukun,* delivers more than 50 percent of all babies in the country, often under conditions of superstition and an absence of all hygiene, also as in India. As a result, Indonesia's maternal mortality rate is extremely high—450 per hundred thousand, or seventy-five times that of other developed countries. There is great variability throughout the country with regard to availability of basic health and antenatal care. On many of the outer islands like East Nusa Tenggara and South Sulawesi, the use of family planning services is 80 percent below the national average. In parts of eastern Java, there is a high incidence of child malnutrition. Countrywide, depending on which data you accept, infant mortality is somewhere between fifty-eight and seventy per thousand, midway between the IMRs of India and China. Among many of the outer islands a large percentage of children with uneducated parents receive no diphtheria, tetanus, or polio immunizations. Indonesia's total percentage of expenditures on health and education is considerably lower than the share of gross domestic product (GDP) in other Asian Pacific countries, such as Thailand and Taiwan.

Just as infant mortality has always been linked to high fertility, one might expect that high maternal mortality would similarly be linked to low fertility; that the logic of death would intervene to check a mother's enthusiasm for more children. To a certain extent, this is true. What prevents lower fertility is the lack of available contraceptive supplies, particularly in winter months when the rains and flooding are insistent. The week I first was in Indonesia, sixty thousand families were evacuated in central Java, three hundred thousand people in east Java, as the denuded watersheds vented runaway rivers of silt and mud and high water. During these monsoon periods BKKBN workers are unable to reach many of the sixty-seven thousand villages throughout the country. And these are the same periods when millions of women get pregnant. The government is working to correct this situation by training and dispatching one auxiliary nurse-midwife, or *bidan,* to each village.[25] The hope is that the "A&Ms" will settle down and marry in their respective new communities, thus ensuring the presence of a local expert for many years who might radically transform the condition of poor medical services.[26]

At the same time, income-generating schemes for women have been undertaken throughout the country. On the outskirts of Bandung, a city of more than two million at the base of a volcano a few hours southeast of Jakarta, I visited one such enterprise. Local women had their own bank accounts, community loans from the government, and a system of production (in this case, cassava by-products) that brought in enough money in many cases to give an otherwise impoverised, totally dependent woman a sense of freedom and dignity, and a job out of the home. It is believed that once the mother is working, her life will improve and she will have less inclination to want more children.[27]

In spite of this well-organized and broadly defined mission to curb future births, Indonesia's population is growing rapidly because of the large existing number of young newlyweds. In wrestling with this paradox of numbers, the BKKBN has increasingly sought help from outside, nongovernmental consultants, such as Dr. Firman Lubis, the executive director of the Yayasan Kusuma Buana NGO, which maintains six clinics in the poorer parts of Jakarta, oases of options for women.

There are visionaries in family planning throughout the world whose commitment and warmth is nothing short of heroic. Dr. Lubis is one of them. Born in the capitol's wealthy Mentang area—a shaded, luxuriant neighborhood near the American Embassy set amid tropical gardens—the middle-aged doctor has given himself over to poor people, particularly women and children. Dr. Lubis used to consult for BKKBN until, in his words, it became too huge a bureaucracy. He felt restricted, wanting to tackle Indonesia's population dilemma directly and independently. To dramatically lessen the number of people on the planet is an aspiration not normally ascribed to a single individual's capabilities. It is one thing to provide medical care as a doctor, to alleviate suffering among individuals, but to aspire to demonstrably change the world through these actions—to lower the birthrate precisely so as to counteract the ecological fallout from a population boom—is an act of dogged courage in the face of futility. In 1981 Dr. Lubis was the first person to introduce Norplant on a large scale anywhere in the world. Over seven hundred thousand Indonesian women now use the implant. By comparison, there are slightly more than five hundred thousand users in the U.S.[28] But Dr. Lubis's efforts go far beyond Norplant. And he has focused his efforts on certain districts, refusing to be daunted by either the pervasiveness of poverty or the density of population.

In many parts of the country, government birth control policies have always relied upon ethnic predisposition to follow authority. In Bali that system is called *banjar*, in central and east Java it is the so-called *hamlet* system. These are community action hierarchies that tie individual responsibility to that of social good. In its earliest development, the BKKBN integrated family planning into the *banjar* system to ensure cooperation from local headmen. According to Jay Parsons, because traditional Javanese attach near "mystical reverence" to their leaders, the common people always have believed that politicians would do right by them, and were thus zealously altruistic, eager to assist family planners however they could.[29] At the same time, local government based upon consensus, or *musyawarah* in the local Bahasian language, is well established in Java—so much so, says Parsons, that the individual voice is heard at all levels of local and national government, something like the recent reemergence of town hall meetings in the U.S. But this inbred mechanism for grassroots mobilization is not found among the Sundanese; hence Jakarta is the most difficult region in the country for controlling fertility. In his research, Dr. Lubis found that urban family planning was lagging throughout Indonesia but mostly in Jakarta, which is a mix of Sundanese and other cultures. The Sundanese are West Javans, with an entirely different political orientation on the local level.

The Sundanese people comprise nearly 20 percent of the Javanese population, or nearly 30 million. In fact, non-Javanese ethnic groups throughout Indonesia account for nearly two-thirds of the total population and inhabit 90 percent of the land. This means that future authority-driven fertility control may be exceedingly difficult to achieve in the country, unless consciousness raising can socially engineer an actual cultural change.

The island of Java, on whose northwestern end Jakarta is located, constitutes 7 percent of the land of Indonesia and contains the majority of the population, or 110 million. Its extreme density of 835 persons per square kilometer in a region the size of Wisconsin has resulted in the deterioration of every major watershed on the island. Java is losing 3 percent of its productive capacity and nearly 737 million metric tons of soil each year, just in the highlands.[30] Jakarta is where Indonesia's economy is fueled and policies of far-reaching ecological importance determined. But volatile immigration and exodus to and from the city are increasingly stressing and defining the decision-making process. Urban overpopulation in Indonesia's few major cities is creating a self-fulfilling prophecy of economic exploitation throughout the whole country. As one Malaysian NGO put it, the differences between the economically advantaged consumers of the "North," and the disadvantaged consumers of the "South" can be charted within every major city of the developing world as well, where there are enormous political and economic disparities between the "haves" and the "have-nots." The rich elite of the South are shaping national policies according to their exclusive needs, irrespective of the far vaster requirements of their poorer countrymen and national environments, just as in the North.

Because the BKKBN has had less impact on curbing overpopulation within the cities, being focused instead upon rural Indonesia, family planners like Dr. Lubis are able to make a difference with potentially far-reaching ecological implications. Dr. Lubis does not hold with the BKKBN that Jakarta has achieved fertility replacement, or a TFR of 2.1 to 2.3, but rather believes it to be between 3.4 and 3.5. His considerably higher estimate is consistent with the sheer number of children everywhere to be seen in a region like Pesangan Baru, where the density is sixty thousand people per square kilometer, the average number of children between three and four per couple, and average family income about sixty dollars per month. These are not the poor, who can be seen living along the canals and railroad tracks and have even more children, but the lower middle class. The government's position that Jakarta has achieved the two-child norm may, in fact, only apply to rich neighborhoods, an internalized demographic discrepancy similarly to be noted in the cities of Brazil, Iran, and even the United States.

Here, in Pesangan Baru, where one of the six clinics of Dr. Lubis are to be found, the children flock around him with deep affection. He delivered many of them into this world. The kids run up and down the narrow dirt alleyways in a festive mood. It's early afternoon on a Wednesday. Nobody seems to be in school.

Lubis started his clinics in 1981. Those who can afford to pay for services are charged a minimum fee commensurate with their income. Those who cannot afford anything are served for free. Dr. Lubis's teams are traditional pediatricians and family planners who go into the homes. Moreover, the clinics themselves are like homes, spotlessly clean, among the most intimate and inviting I have seen anywhere in the world, much like a tropical version of the northern California "home birthing" scene.

"We cannot depend on the government," says Dr. Lubis. "We are doing what the government is afraid to do. It is our task to help women. It is a matter of guts. We believe abortion is not a contraception but a backstop. We educate people not to have it. But we do what we have to do." And to that end, Dr. Lubis's six clinics in Jakarta are all medically licensed for various surgeries, like sterilization.

While we wandered together through the Pesangan Baru, he explained that less than 5 percent of all teenagers and junior high school students in Indonesia are getting real sex education, and probably only in a few major cities. Another 30 to 40 percent are getting basic reproductive health information through the religious organizations, but rarely any mention of sex.

Yet, teenage sexual intercourse by age sixteen is nearly 60 percent in Indonesian cities, as much as in most Western countries. The government maintains an ostrich policy, all the more remarkable considering that there are an estimated three million unwed mothers in the country, and over twenty-five thousand known cases of AIDS. When I asked officials at the BKKBN about teenage unwed mothers, they said there were none—the same response as in China.

At various private clinics throughout the city, a woman can get an abortion (menstrual regulation, they call it) under conditions of secrecy. It costs about one hundred dollars—on average, nearly two months of a working woman's wages.

"Abortion, if legalized, would have a big impact in Indonesia," Dr. Lubis explained. He cited South Korea and Peru as two countries with TFRs similar to that of Indonesia, at least several years ago. Peru eventually reached a fertility plateau while South Korea's TFR continued to decline because of government-supported abortion in that nation. In addition, the number of couples obtaining sterilization in those Asian countries with much lower birthrates is significant: 48 percent in South Korea, 40 percent in Sri Lanka, and 25 percent in Thailand.

Dr. Lubis definitely sees the day when abortion will be made legal in his country, and sterilization much more widely obtainable. "It might be fifteen or twenty years. But then again, it could also happen in a year if there were a political shift."[31]

Whether Indonesia legalizes abortion anytime soon will not, by itself, solve the country's population dilemma. In an ideal world, abortion should obviously never be deemed a *solution*. What might more realistically inhibit the fourth largest country in the world from becoming the third largest, is a combination of concerted factors: Educational outreach that ties birth control to an awareness of ecological conservation; maternal and child health services enrichment; and more assured access to quality contraceptives, particularly long-term ones. Haryono Suyono, the long-standing charismatic chairman of the BKKBN, was named minister of Population in March 1993, a new post for the country. With partial funding from UNFPA, one of his first steps has been to create an inner-agency task force that will ensure the dissemination and focus of population issues within several government ministries. More specifically, Indonesia has set various quantitative family planning goals for itself. By 2005, for example, the entire country hopes to reach a two-child family norm, or 2.1 TFR. That will require a 30 percent reduction in total fertility rate (based on the lowest AID estimate). To achieve it, the contraceptive prevalence rate (CPR) must rise from the current 49.7 percent (15.6 million female users) to at least 62.7 percent (25.4 million users), based upon the realization that eight hundred thousand newly married women of reproductive age are being added to the general population every year. The 62.7 percent benchmark, if accurate, would be quite unusual considering that most countries have not reached replacement fertility until attaining a CPR of 70 percent for all childbearing women between ages fifteen and forty-nine. But for Indonesia, this means that by 2005, roughly 85 million new contraceptive users—more new users than in all of sub-Saharan Africa—will have to be recruited, based upon the recognized allowance for large numbers of acceptors who discontinue usage every year. That figure represents a huge amount of work—consciousness raising, door-to-door lobbying, mass production of contraceptives, a policy of technical and cultural suasion. What makes this goal particularly ambitious (or just plain unrealistic) is that at present the rate of new acceptor use appears to be *down* by a whopping 75 percent since 1988. And while an estimated 95 percent of all married Indonesian women know of at least one form of contraception, more than 14 million of those women have unmet contraceptive needs.

Malaysia and the Philippines both stalled on their fertility declines once the TFR fell below 3.5 children and the number of women of reproductive age using contraception of some kind reached 50 percent. Indonesia now faces that situation. The good news in one donor-sponsored poll is that some 65 to 70 percent of the women throughout the country expressed a desire to stop having children, or at least delay them.[32]

As 1995 approaches, the density per square kilometer throughout Java is likely to exceed nine hundred, as 3.5 million plus people are added each year to the ranks of needy consumers. Forty percent of the population currently is under the age of fifteen, 25 percent under the age of ten. According to the World Bank, even if all the targets were to be reached by family planners, an NRR (natural reproductive rate) of one would not set in for at least another century. In this respect, population momentum behaves in a fashion analogous to the accretion of ozone-depleting chemicals in the upper atmosphere. We may diminish production of, say, chlorofluorocarbons—which cannibalize the ozone molecule in a ratio of one-to-one hundred thousand—but those CFCs previously produced are still rising into the skies where they will do their damage to the planet for at least another century, no matter what our current strategies. All we can hope to accomplish is a slowing down of our aggression—the beginning perhaps, of a self-discipline that seeks assiduously to reverse the destruction, to act gently, to temper our lifestyles.

Thus, regardless of the BKKBN's best efforts, the intimation of between 300/400 million eventual Indonesians grappling after a Western-style standard of living has cast an enormous pall over the land. All of the hard-won victories of official Indonesian family planning, and of committed individuals like Lubis and Yusef and Wahid, will surely diminish, but cannot hold back, the inevitability of many more people, and what those masses bode for one of the most spectacular biological regions in the world.

## THE HOSTAGE SITUATION

The speed of human appropriation within a single generation will be slowed down by a combination of vigilant enhanced contraceptive prevalence, the aggressive participation of private businesses, foreign capital assistance, ecologically sound technology transfers, and domestic enforcement of all conservation laws. But unless such relatively benign employment as ecotourism, computer biogenetic prospecting, non-petroleum-based sustainable agriculture, and noninvasive cottage industries, can be elevated to the economic equivalent of the lumber trade—and quickly—Indonesia is likely to follow much the same disastrous path as India. The war against nature in Indonesia is being waged strictly for money.

Consider what's at stake here. Occupying 1.3 percent of the planet's surface, Indonesia's forty-seven distinct ecosystems encompass marine, rainforest, and alpine habitat. They host a fabulous plenitude of creatures, all vulnerable, that includes panthers, bird-winged butterflies, black monkeys, Komodo dragons, green pythons, the largest flowers in the world, snake-eating eagles, leopards, a proliferation of orchids, and a few remaining rhinos. The country contains peat, freshwater swamp, heath, montane, and evergreen dipterocarp forests that harbor more than seven thousand tree species. In addition, the nation's lowland rainforests are the most expansive in all of tropical Asia. Indonesia also contains the third largest atoll in the world at Taka Bone Rata in the

Flores Sea. Mangroves and beds of sea grass are to be found throughout the country's coastal waters, whether in Sumatra, Java and Bali, Kalimantan, Sulawesi, Nusa Tenggara, Maluku, Irian Jaya, or any of the thousands of smaller islands. The Sulawesi and Maluku coral reefs are some of the most biologically productive on Earth.[33]

As with all other biomes on the planet, rainforests particularly so, the *known* number of species may be ten to seventy times fewer in number than what is actually out there. This is not so much the case with larger, more obvious mammals, but certainly is with respect to smaller organisms. For example, in the rainforest of Sulawesi, two British taxonomists studying the hemipterans, or true bugs, recorded 1690 species, of which 63 percent were unknown to science.[34] Only four countries in the world can be described as "megadiverse," namely, Indonesia, Brazil, Zaire, and Madagascar.[35] What is known about Indonesia's megadiversity is that the archipelago hosts 10 percent of all flowering plants (an estimated 25,000 species), 12 percent of all mammals (515 species, 36 percent of them endemic), 16 percent of all reptiles (600+ species) and amphibians (270 species), 17 percent of all birds (1519 species, 28 percent of them endemic), and more than 25 percent of all marine and freshwater fish in the world.[36] The natural heritage is so "valuable," in fact, that it provides the direct, daily sustenance in one way or another for some 22 percent, or 40 million, of the country's population, 30 percent of whom live in and around forests.

In Brazil, fewer than two million individuals are directly dependent on the forests, yet the devastation—fueled by multinational corporations and government colonization schemes—has been horrendous. In India, several hundred million people rely on wood for basic needs. Indonesia tragically combines both aspects of exploitation, which makes conflict resolution and the introduction of a nonviolent way of life most difficult.

Three crucial facts about Indonesia's battleground need to be considered. First, Indonesia has 0.0 percent wilderness, according to the World Resources Institute definition of the term.[37] Second, this staggering absence of undefiled territory coincides with the fact that the country has more threatened species than any other nation (with the possible exception of Madagascar).[38] In the 1980s the Javan Tiger went extinct. Eighteen known Indonesian bird species have vanished just recently. The Red Data Books of the World Conservation Union list 126 birds, 63 mammals, 21 reptiles, and 65 other Indonesian species as threatened with extinction.[39] Among these is the Javan rhino—less than sixty are known to exist in four areas of Ujung Kulon on the southwestern tip of Java.[40] Third, the government is committed to protecting 10 percent of its land area, though, according to the WWF, the official agenda actually only covers 7 percent. But even 10 percent would mean that 90 percent will not be protected, and even that 10 percent will require extremely ambitious efforts yet to be undertaken, in most cases. While a goal of 10 percent is far superior to that of most nations, from the broad perspective of "simple ecosystem" biology it means that Indonesia will lose a significant share of its biodiversity, despite an existing network of biological corridors, some on paper, some in reality. As with India and China, all that can now be undertaken is a systematic initiative of economically devised damage control in concert with continued family planning innovations.

Further hampering even basic maintenance of compromised biological integrity are diverse cultural approaches to nature that characterize the expansive range of human psychology, perception, and subsequent behavior. While the occasional natural disaster might stun our sense of hubris, make no mistake: the human psyche is in control. Its vagaries and moodswings dictate the near fate of countries like Indonesia. Embracing the philosophy of compromise as a triumph, that country's government has run the following advertisement in the nation's daily newspapers: "280 million acres PRESERVED forever—By designating 280 million acres as permanent forest, Indonesia is protecting and preserving its forests in their natural state. Over 43 percent of these forests will remain untouched, forever. The other 57 percent will be carefully managed for sustainable growth. . . . " The ad, which ran extensively throughout the winter of 1993, was headlined with the following statement: "Indonesian Forests Forever; Indonesian

Forestry Community and The Jakarta Post present this advertisement as a public service."[41]

While the intention is certainly commendable, inviting many ecologists to declare Indonesia the global leader in Green politics, the numbers, when analyzed, actually conceal a slippery slope that evokes a more queasy response, somewhere between outrage, a stiff upper lip, and metaphysics. After Brazil, Indonesia is the second largest exporter of hardwoods in the world. "Untouched forever" scarcely exists in a human-dominated world, certainly not in a region focused on converting tropical hardwoods into money. The effectiveness of "damage control," even of a holding pattern, hinges upon the quality of criteria applied not just to conservation, but to the very *idea of nature.*

What are the rights of nature? Are they comparative, or absolute? Are they more relevant among the rich than the poor? Supreme Court justice William O. Douglas proposed a "wilderness bill of rights" in 1965 and a few years later argued that people in "intimate relation with the inanimate object about to be injured . . . are its legitimate spokesmen."[42] Back in 1933, conveners met in London to formulate a consensus on the *Protection of African Fauna and Flora* that endorsed the idea of "strict natural reserves," which—in contrast to mere parks— banned all human visitors other than "qualified" scientists whose presence would be carefully regulated.[43] The arts and sciences have long acknowledged the concept of "pure wilderness" and its gradations, or, as Paul Simon and Art Garfunkel sang in 1970, "I'd rather be a forest than a street." Conservationists have classified wild lands and marine habitats according to the extent of strictly defined levels of human encroachment and compromise. But for the general public too often the notion of pure wilderness has been cloaked in nostalgia for what is lost, a phrase more given to longing, reduced to "mere" paradise in pastoral literature and mythopoetic pictorials. Or, by recasting the preexisting wilderness as some *future*, unreachable utopia, we have conveniently disavowed our present responsibilities as powerful stewards, embracing the human condition as the only true reality, an excuse for everything we do. This has served us as a "rational" defense of destruction, an anthropocentric syndrome that is another form of the aforementioned "lie," which we repeat to ourselves under the brazen, or apologetic rubrics of our survival needs, or inquisitive appetites.

If we are to rephrase our standards and practices, to become truthful denizens of an interdependent world, we must first insist upon the healing process, that would mean safeguarding as much wild land as possible, to cease irritating an open wound. Given that some 45 to 50 percent of the planet's net primary production (Earth's natural green growth) has already been co-opted by our species, at the very least, we must ensure that *Homo sapiens* is prohibited from expanding its hegemony any farther. And once that restrained threshold has been stabilized, we must quickly act to take several steps back from our massive, planetary swathe, curbing all future extraction, manipulation, and exploitation of living beings and inorganic deposits. We must recede, the way a destructive tidal wave recedes. But how? If we have not managed to accomplish even a remote semblance of this healing process in a rich nation like the United States, what is possible in a dollar-poor country such as Indonesia?

At no other time in our history have philosophy and spirituality come to play such crucial roles in our politics and our relationships with the natural world. Until we forcefully assert new international laws that uphold the inherent rights of nature, as Justice Douglas not unambiguously called for, we will blindly justify any action and hail all compromise as manifest destiny. In India, the World Resources Institute definition of "wilderness" has been ideologically diminished in scope, scaled back from four thousand square kilometers to five hundred square kilometers with little or no social commentary. That country felt it had no choice because of the sheer pervasiveness of its people.

What is one man's suburbia is another's wilderness. Personal experience and economic circumstances, ethics, aesthetics, sensitivity, and political persuasion all cloud and vary a realistic definition of what constitutes "wilderness" and "biological integrity." Given nature's continuous dynamic, its ceaseless transformations, governments are easily able to justify expedient ecological definitions. Even when those marginal, haphazard, and largely false definitions backfire; when one learns, for example, that all the mammals have vanished from a region, there is other fodder for argument and self-congratulation, a wealth of weed, grass, and bacterial species to sustain the deluded notion of persistent biodiversity. Human beings, having banished most other visible life forms, might even console themselves with the realization of millions of mites in their eye lashes and *E. coli* in their guts.

These are cynical remarks, I realize, but they point to the fundamental truth of this war: the inevitability of compromise, however extreme, given the presence of too many human beings. Living with compromise, and with contradiction, is what the law of averages, and any human collective, imposes. There is simply no known method whereby 5.6 billion large consumers can forge obvious or harmonious points of consensus. Democratic demographics is inherently dangerous, and there are no compassionate alternatives that come readily to the governing mind. The role of science is to constantly remind us of our current compass reading in the vast sea of compromise; while the arts hold the promise of evoking and championing a more pure course. The difficulty with attaining or maintaining a viable course in a world politically, culturally, and linguistically fragmented, is the absence of a true processing, or *feeling* center. An individual's brain processes injury to its body almost immediately, sending out signals of alert and damage control. Even before the brain responds, antibodies have already set to work. But in the case of the planet, the feeling centers are usually not connected to our person. A North American cannot know to respond to injury in an Indonesian wetland. Time, distance, and political boundaries separate us from the causes of crisis, despite our superficial acquaintances through the media. And because destruction of the world occurs under much broader regimes than the course of an individual injury, we are largely disconnected from the Earth's needs and vulnerability, mired as we are in a hierarchy of denials. Emotionally, there are myriad other "priorities" that societies like Indonesia's have placed before some vague balance of nature. Indoctrinated with compromise, ethically imbued with a world biologically overrun by *Homo sapiens*, people have come to view money—blunt, cold, hard cash—as far more valuable than any life forms. In the sixteenth century George Pettie, a writer expressing the first Western frenzy of consumption, wrote, "What disease is so desperate which money may not medicine? what wound so deadly which coin cannot cure?"[44] Can this illusory and disastrous mindset be reversed? There are few places on Earth where such a reversal is more critical to ecological stability than in Indonesia.

An Indonesian reversal would require the political willpower to enact massive restraints. Ten percent conservation is a laudable goal, given the fact of much worse global compromise. But even those 10 percent will require either a renaissance in voluntarism (not likely), or large infusions of capital, not merely of management. So Indonesia is looking for money to increase the power of its infrastructure and improve the lifestyle of its people; and much of that wealth is expected to come from the forests. That, of course, is the problem. Even if Indonesia were to quadruple its GNP, it would find itself in an unenviable position—the economic threshold for ecological balance is simply unknown. For any given area, one trillion dollars might not redress what a no-cost policy of preventative restraint could have brought about.

Nobody knows for certain how much forest is left in Indonesia. Based upon dated surveys by the Dutch, officially (and in keeping with the government's newspaper advertisements), 66 percent of land area is still covered by trees. That means 143 million hectares. One hectare is equal to 2.4 acres so according to the government of Indonesia, that country still has 343 million acres, 10 percent of the planetary total, the third largest timber estate in the world. The government is saying two things: that 280 million of the 343 million acres will be protected and preserved forever (that's 86 percent of existing forests!) and that only 43 percent of that, or 147 million acres, will be untouched forever.

What is the general public to make of the difference between "protected and preserved forever" and "untouched forever"? The government spells it out. Bear with me here because the very uncertainty of the policy-dependent numbers and intentions are part of the problem, and hence the message. Of those 143 million hectares (or 343 million acres), under the four land use categories stipulated in the 1984 *Indonesian Ministry of Forestry's Consensus Forest Land Use Plan* currently in effect, 30.3 million hectares are designated for protection (that means approximately 72 million acres, not the advertised 280 million acres—a rather appalling discrepancy), 18.7 million hectares for conservation (hunting and other activities will be allowed), and 64.3 million for out-and-out production. Another 30 million hectares are earmarked for agricultural conversion (total appropriation). What this dense cluster of apparent contradictions reveals is a dual usage of language, fortified by public relations, and misleading calculations.

Other data by WALHI,[45] the country's leading environmentalist organization, show Indonesia's forests covering between 100 and 109 million hectares as of 1992, not 143. That is nearly a 33 percent difference. Because the estimated rate of loss was 1.2 million hectares per year in the early 1980s, prior to what has become an even more aggressive lumbering industry, it means that there could easily be less than 100 million hectares today. If that is the case, rather than the high government estimate, all total forest cover figures would correspondingly be reduced by nearly a third—what the government had promised as 280 million protected acres (corrected to 72 million) is actually closer to 50 million, or nearly six times less than claimed!

Now the government has said it intends to protect 10 percent of the country, of which supposedly two thirds is forest, the other designated areas covering a broad range of ecosystems, of plant and animal life. With a 6:1 discrepancy in actual "untouched" protection status, that 10 percent is probably closer, in reality, to 2 percent. On an international scale, and forgetting the far greater need for safeguarding a megadiverse region, as opposed to, say, central Greenland's ice shelf, 2 percent is far below the already depressing world average of a mere 4.3 percent of national land area that is protected.[46]

Given the incidental loss of forests, the uncertainty with respect to road-cutting damage (edge effects), watershed disturbance, the ratio of official selective cutting to official and unofficial clear-cutting, and the additional burden of shifting cultivation (nearly 40 million hectares of forest affected), nobody knows with any precision how compromised the official figures actually are. But the feeling is one of growing desperation among many ecologists in the country with whom

I met. Selective cutting laws were introduced in the 1960s, at the beginning of the logging boom. But enforcing the laws has proved nearly impossible. Those laws generally stipulate a seventy-year rotation period, a thirty-five-year cutting cycle, and a minimum limit of a fifty-centimeter trunk diameter before any tree can be felled. Thirty-five years has been the designated regeneration time, part of the stipulated condition that concessionaires replant logged-over areas. As of 1991, only about 20 percent of all industrialists had fulfilled the requirements of the selective cutting rulings, and there was still no adequate regulatory mechanism in place to enforce them. According to one United Nations Food and Agriculture Organization expert I interviewed in Jakarta, policing manpower is at the negligible level of one man per fifty thousand hectares of forest (about the same as for India), so you can forget management of the selective cutting. Uncontrolled forest clearance is escalating. Wild fires are beginning to resemble the "controlled" fires of Brazil.

The speed of planting new land is double the country's population growth rate; in other words, for every new mouth to feed, two additional units of land are stripped for crops. Such clearance for agriculture has been deemed "the greatest threat facing Indonesia's biodiversity." It is impossible to control hungry people, or effectively impart sustainable practices, or find the necessary funding for rehabilitating destroyed lands.[47] Government-sponsored habitat "conversion" for food production will consume at least 17.4 million hectares by the year 2000, in addition to the uncontrolled land use, and the commercial "production" of forest. But these figures are all based upon current demographics, energy capacity, housing starts, the price of petroleum-based fertilizers, the proportion of national budget allocations to affordable housing, and any number of other indicators likely to change. With respect to demographics, if anything the numbers will be worse, and thus, all the other sectors dependent upon the number of mouths to feed, clothe, and house will also escalate.[48] Plantation lands cover 35 million hectares of Indonesia already, 14 percent of them devoted to rubber and coconut. Monoculture is anticipated to eat up more than 5.6 million hectares by the year 2000, with a predictably devastating biological impact on resident life forms. Already, fifteen hundred Indonesian rice species are known to have gone extinct in the race for fast-growing monocultural strains. Similar trends toward depletion can be inventoried for Indonesia's marine environment. Overfishing, dynamiting of coral reefs to maximize fish kill, and the conversion of mangroves to chipwood forests and brackish prawn pools have all taken continuing tolls on the environment. These, too, will only increase.

In attempting to diffuse dense urban populations, or deflect transmigrants looking for jobs, the government has encouraged emigration to the so-called Outer Islands. Irian Jaya, for example, with a mere three people per square kilometer at present, has been viewed as an ecological mecca for overcrowded Java. But the experience so far has been negative. The settlers—like those in the Amazon—have planted inappropriately in places like Irian Jaya, hoping to force short-term yields and clearing forest to plant crops that are mismatched with local soils. Because these millions of settlers are migrants, without a thorough knowledge of regional conditions, they have tended to wreak inordinate damage; many have returned frustrated, and poorer for their efforts, to the big cities on Java, leaving a path of annihilation behind. For every three people who have evacuated the cities, two have returned.

The existing legal framework for conservation in Indonesia is extraction oriented. While the 1990 *Conservation of Natural Resources and Ecosystems Law* recognizes biological integrity, it does so with an eye toward manifesting future exploitation. Over one-fifth, or 21.8 percent, of Indonesia's gross domestic product (GDP) comes from agriculture, forestry, and fishing.[49] The construction industry has enjoyed a large gain in its share of Indonesia's GDP, from 2.4 to 6 percent, over the last twenty years. Construction and housebuilding are rapidly rising industries. In 1988, timber exports (mostly to Japan, Singapore, South Korea, and Taiwan) reaped between $3.8 and $4.5 billion for Indonesia (once again, the data are varied) second only to oil and liquefied natural gas. Domestic consumption of forest products was worth easily another $4 billion. Export of textiles exceeded four hundred thousand tons in 1991, earning $4.1 billion dollars, making it the third largest income-generating domain in the country. But Indonesian textiles are endangered—the domestic market is virtually saturated.

Complying with the call for global free trade, Indonesia signed the GATT *Code on Subsidies and Countervailing Duties* in 1985, as well as a bilateral agreement with the U.S. pledging to end all kinds of export subsidies, knowing that this would wipe out many lowscale industries in the country that could not survive without such domestic assistance. But Indonesia honored its international pledge, hoping to gain a larger world niche for sales of its value-added products. However tariff barriers in Japan and Australia, quota restrictions in the U.S. and Europe, and new competition from throughout Asia have largely negated any benefits of the GATT *Code on Subsidies* and forced Indonesia to aggressively focus on the production and marketing of domestic timber by-products, hence adding to the country's pace of rapid forest encroachment.[50]

Given this financial balance sheet, Indonesia has been saying for years that it needs donor assistance if it is to restrain its lumber industry. In over twenty studies that examined monitoring, manpower, and agroforestry criteria, the government arrived at a minimum figure of $20 billion in grants needed to ensure a sustainable forestry program in the country. But that program itself is no panacea for nature. In fact, as will be shown, it is fraught with problems. Yet one can imagine that without that $20 billion in assistance (which is unlikely to be offered), the protection, and conservation, and production areas will go the same route as the conversion and production areas.

The Indonesian government designates a ridiculously small proportion of its budget for conservation—$5.6 million in 1990–91. And even this pittance is largely spent on administrative activities, rather than upon actual protection or restoration. To equal the preservation investments of, say, China, Indonesia would need to increase its own allocations by at least $125 million per year according to Julian Caldecott.[51] The largest national park in Indonesia (Gunung Leuser in Irian Jaya, a Biosphere Reserve—one of seven in the country, with 10,946 square kilometers), has an annual operating budget of $232,357. The PHPA—Directorate General of Forest Protection and Nature Conservation within the Ministry of Forestry—has few trained field staff and virtually no marine biologists. For a country with seventeen thousand islands, that is not an ideal situation.

In spite of these limitations and population and commercial pressures, Indonesia is trying hard to implement strong global alliances for conservation, having joined the Global Biodiversity Conservation Strategy of UNEP, IUCN, and the World Resources Institute. It participates in the UNESCO Man and the Biosphere program, as well as working closely with the World Wide Fund for Nature (WWF). In all, 366 terrestrial conservation areas have been created in Indonesia. A total of 16.2 million hectares, or 8.2 percent of the country's land area, has been designated as preserved. Most endemic species and all known Indonesian birds appear to be represented in at least one or another of the preserves. Another 22.7 million hectares are being considered for preservation. Together, these existing and proposed preservation sites would add up to the official 10 percent figure. So far, however, many of the parks and reserves "exist on paper only." No new reserves have actually been created since 1984. In addition, many of the more remote parks and preserves have virtually no staff in place. There is no protection, no way to ensure that such lands are not usurped by other over-zealous conversion or production schemes. Poaching and illegal trade are widespread. Backtrack further through the aforementioned numeric "discrepancies" and a kind of fuzzy illogic begins to surface, a house of cards that betrays any solidity or conservation comfort zone.[52]

As of early 1993, 575 concessionaires had legal claim to over 60 million hectares of Indonesian rainforest. As of 1990, a known 25 million hectares of trees had already been mowed down, fed to some 2,724 sawmills in the country. The World Bank and the U.N. Food and Agriculture Organization have estimated a loss of 1 million hectares per year. Official Indonesian figures show timber concessions as comprising more than half of all existing forests, which would mean not 60 million, but at least 71.6 million hectares. Since every last concessionary acre is destined to be clear cut, probably within thirty years, that means the overall protection figures are even further compromised.

Indonesia is now the world's largest plywood producer, having secured 50 percent of the market. Because this quasi-cartel can determine its own pricing strategies, thus exercising enormous power over the country, it has managed to eschew environmental concerns. As of 1994, according to WALHI, there has been "little replanting" of demolished regions, while 106 primary dipterocarp tree species are now threatened with extinction.[53] According to the Ministry of Forestry, however, the country has been allocating roughly $300 million a year (or nearly sixty times the whole Ministry of Environment budget) on reforestation efforts, the equivalent of five hundred thousand annual incomes. The forestry minister says the country is taking steps to convert two decades of reckless exploitation into sustainable management. And yet, the total *Reboisasi* (reforestation) and *Penghijauan* (afforestation) figures in the last few years have been lower than ever before in the history of the Indonesian forestry program. For example, the *jumlah* (total) for 1979 was 990,633 hectares. For 1989, it was a paltry 93,654 hectares.[54] Of the $3.8 to $4.5 billion in foreign exchange, which value-added timber exports now annually bring into the country, only $416 million is actually collected by the government, though that is more than enough to cover the alleged $300 million reforestation expenditures. The rest is profit for an entrepreneurial elite. As for any perceived benefits to Indonesian labor, slave wages are paid to a work force that numbers less than two hundred thousand people (one-third-of-one percent of the country's total employee sector). This hardly begins to justify the irreversible ecological devastation taking place. Nor have the reforestation efforts as yet shown much progress.

    As in India, the Indonesian government is attempting to convert its rapidly expanding burden of wasteland into commercial plantations for fast-growing wood fibers to support paper, pulp, and rayon products (HTIs, or *Hutan Tanaman Industries*). So far there are about 4.8 million acres of existing plantations, mostly devoted to teak, spread across the island of Java. But these are expected to expand rapidly (by three-fold) by the end of the twentieth century, and that's just the beginning. The country is developing the largest seedling nursery in the world—it will produce 72 million seedlings a year. The HTI are estimated to yield 24.5 million cubic meters of wood per year, compared with the current 40 million cubic meter yield from selective cutting estates. Given Indonesia's nearly $80 billion debt, and an annual demand for wood of nearly 44 million cubic meters, there is little doubt that primary forests will continue to be exploited. Indeed, the government has already calculated that 10 percent of the entire country will have to be turned into such plantations, in addition to the 65 percent of the country allocated to production and conversion. Even presuming that the 65 percent includes existing wasteland, the bog of "fuzzy illogic" deepens. Now, in addition to the slew of other compromises, we can project 10 percent of the country, or 18,115,700 hectares devoted to these plantations that totally strip the land of its previous biodiversity. WAHLI and YLBHI have exhaustively analyzed the plantation scenario in terms of its ecological and economic impacts. Pulp and paper will simply go for export, in view of the fact that oil revenues have been steadily declining in Indonesia since 1982. The plantation trees grow to maturity in eight years in Indonesia, whereas the same trees would take three times that long to reach maturity in, say, Finland, where one metric ton of paper costs $485, versus $285 in Indonesia.[55] Because Indonesian firms can easily outcompete foreign markets by discounting and devaluating their own domestic stocks while still making a large profit, at least one country—Australia—has threatened to boycott Indonesian stationery. If every country stopped buying tropical wood products, that would provide an answer to the loss of biodiversity. But that is an unlikely response from the developed world.

In principle reclaiming wastelands is an admirable goal, a form of poetic redemption. However, most investors to date have shown little enthusiasm for redemption. Many of the fast-growing tree varieties such as eucalyptus and acacia are subject to pest infestations. The *Acacia magnum*, for example, is attacked by as many as nineteen species of insects while still a seedling. In fact, no pest-resistant fast-growing varieties have yet been found.[56] The HTI fast-growing wood fiber programs, rather than deflecting destruction from natural hardwood forests, are encouraging it. Industries are clear-cutting existing forests, selling the lumber, the captured pigs, deer, monkeys, and other creatures they can easily get their hands on, and only then planting the fast-growing monocultural varieties that spell the total doom of the rainforest eco-system. In other words, plantations are not precluding primary forest encroachment. Furthermore, it is in the timber estates that most of the tropical forest fires start, like the disastrous fire in 1983 that wiped out an area the size of Belgium in east Kalimantan. This biodiversity disaster received scant attention in the world media at the time.

There are other, secondary, mal-effects of the plantations and primary forest estates. While some concern has been expressed by the government of Indonesia, to date little has been done to curb the ill effects of chemical pulping with its release into the soil and rivers of numerous toxins, including caustic soda, toxic organochlorides, and dioxins. For new pulp mills, pollution control equipment is either not installed or not employed.

Indigenous rights are also being perverted. The government has avowed its right to regulate the legal relationships between people and the forests. Inasmuch as most indigenous groups have neither maps of their community tribal lands nor effective legal representation, their claims to traditional sovereignty are at the mercy of the government's own agenda. That agenda is dominated by entrepreneurs and foreign investors. Millions of animals will go extinct so that we can all avail ourselves of Indonesian toilet paper, unless there are international boycotts, trade barriers, or other kinds of pressure exerted. Writes WALHI, "HTIs, promoted as the solution to Indonesia's tropical deforestation crisis, are no solution at all. Instead, they are likely to escalate our environmental and sociocultural crisis. Pulp and paper development, hailed as a dynamic mechanism for economic develop-ment, maintains the status quo by further empowering the wealthy and further threatening the status of the poor."[57]

In one Indonesian village, a paper company clear-cut and burned 3,379 hectares, destroying every last remnant of the local Sakai peo-ple's livelihood and legacy—its farms, even its burial grounds. In the Lake Toba region of north Sumatra, a once gorgeous watershed, a pulp company completely ruined the area near the village of Sugapa and displaced the indigenous Batak people.

Foreign investors bankrolled an Indonesian conglomerate that went into Sumatra in order to produce pulp cheaper per ton than any other mill in the world ($226 per ton). It did this by securing Swiss convertible bonds, Finnish thirteen-year interest-free loans, and Canadian consultants. The company avoided any environmental controls, according to WALHI, and manipulated Indonesia's relatively "innocent" bureaucratic system, in spite of the staunch objections of Indonesia's own Minister of Population and Environment. The company established a 520-acre factory complex, promptly clear-cut whole ridges—inducing major landslides—turned a potable river from blue to dark brown, contaminated all the fish, and bought off sixteen locals who were threatening to sue the company. Before that WALHI believes, they manipulated the local system of land transfer, excluding many aboriginal residents and bypassing the local custom in which the village head-person must witness such transactions. They did this by engaging two allegedly corrupt local administrators. The company took ownership of the land for thirty years. The Batak community protested to the government and—as in several similar cases in India—were harassed and beaten. Then a village girl was allegedly raped by company employees. The villagers retaliated by ripping out thousands of company eucalyptus seedlings. The police responded this time, arresting ten women who were sentenced to six months in jail. They appealed to the National Supreme Court and got the decision reversed, but still had to agree to replant the company's seedlings. The women have refused. Two years later, the case is still active, the community of Sugapa has been utterly traumatized and exhausted by a struggle for survival that is ineluctably tied to the immediate and long-term welfare of the forests and wildlife. Meanwhile, the company in question is poised to become the largest pulp and paper producer in all of Indonesia.[58]

Throughout much of the country, economic imperatives have totally replaced sound ecological strategies. Whether in Sugapa, or at the immense Freeport copper mine in Irian Jaya, Indonesia is struggling for revenues. Overpopulation has saturated many of the domestic markets, and the global trading blocs have reduced the country's revenue stream from textiles, shoes, and electronics. Whether passage of the GATT will change this is yet to be seen. As it stands, forests are Indonesia's most profitable commodity, its green gold. Indonesians still see enough rainforest in place to imagine that the alleged crisis of deforestation will be taken care of by the implementation of a few parks, pious proclamations, and largely symbolic legislation.

According to Dr. J. Harger, an outspoken program specialist in the UNESCO Jakarta Office for Science and Technology for Southeast Asia, there is a new generation of ecologists in Indonesian government who were trained outside the country and understand quite thoroughly what needs to be done. But they are too few, says Harger, to engender anything like a national consciousness. In the late 1960s, while at the University of British Columbia, Dr. Harger, a New Zealander, was one of the initiators of the Greenpeace concept. For him, the saving of Indonesia is the most important ecological battle to be fought anywhere on the planet.

"Most ecological monies in the country come with donor assistance," he told me impassionedly one morning. "The Indonesians need help, they are begging for help, they are prepared to make adjustments, but they need help. The onus of arriving at a solution must cut just as deeply in developed, as developing countries. Until that happens, nothing will change. For years now the U.S. policy has stated, 'Not open to negotiation.' OK, says the South, we'll all go down together if that's how you want to play it. Vice President Gore's statements are good," Harger avowed, "but it's easy to write this stuff. Has he paid his dues? It's a hell of a lot harder to take heat, harder than they ever thought. You take Mrs. Gro Harlem Bruntland. She's zipped out. She says catching whales is great. Any ecologist can tell you catching whales isn't great. If those guys can't find some other way to operate, how can the Indonesians? This is insane. Loosen up on some of the strategies that have brought you to dominance and give us some of the difference so that we can move forward and we can make some kind of a joint plan. You keep holding us by the throat and we can't move. North Americans have been brainwashed. They DO NOT understand the world rightly. Indonesians who have grown up with nature know what is the problem. The senior people in the developing countries understand—they live it, they see what's happening. They know that we are trapped in the ideological crisis of all times, a serious intellectual conflict of what should be done here in Indonesia."

One of those individuals very much immersed in the conflict is the Minister of Environment, Dr. Emil Salim, who met with me for an afternoon at his office in Jakarta. He calculated in one hand that by the year 2020 the country's population would exceed 266 million. With his other hand, he made a gesture to indicate his country's finite resources. "In spite of active family planning we are going to see 80 million new babies," he said. "How do you accommodate these people? Either the developed nations open for migration, or for non-forest exports—like textiles, foods, and electronics. We are a country of $600 per capita. Our infant mortality rate is 70 per thousand. It will come down by half again but the population will go up. How do you raise income to cope with the increased population? Either you go outside Indonesia—but be realistic, that's not going to happen—or you open the door for industrialization outside the forests—shoes, electronics, textiles, manual labor. But what are the facts: We are now facing an embargo in the U.S., while the U.S. talks about a free market economy. It's hypocritical. Everyone protects their own economy so we cannot move fast in nonforest products. Our own economy is not enough to sustain the nonforest products market. So what happens now? Clinton increases protection for his domestic economy. A tariff wall. Indonesia is competitive, despite the fact that China spends thirty-four cents per hour on its labor. Import duties in developed countries have penalized higher value exports by as much as 15 to 20 percent, whereas logs have zero duties. We banned exports of logs in 1986, we shifted to higher value-added products, we boosted export of furniture—but we hit a tariff wall."

Because Indonesia had built up a number of small-scale industries around that of value-added furniture, the tariffs have hurt the country badly, while rewarding the mowing down of tropical hardwoods so that Western manufacturers could derive the economic benefits of raw lumber. There are two profound paradoxes in all of this. While the developed countries are crying for protection of tropical rainforests, they are, as Salim says, also raising import duties for higher value-added products, thus depriving the developing countries like Indonesia of their due revenues. This is a form of imperialistic schizophrenia. But there is an even deeper ecological malaise that can be read into it, and which even passage of the GATT in December 1993, with its associated 33 percent—on average—tariff reduction, is unlikely to mollify: whether the U.S., or Indonesia fashions the furniture, the trees still get chopped down. There are no guarantees that higher incomes in Indonesia derived from value-added furniture products would actually favor less exploitation of the forests! Now that those import duties have actually come down, the likely incentive for manufacturing more and more furniture in a competitive arena is likely to result in increasingly aggressive conversion of primary forests to furniture for export.[59]

Will the GATT safeguard those forests? "No," says Salim. "It's all politics. What did Austria do? They passed a law requiring labels on tropical hardwoods. My government asked me to go to Austria. We all agreed in Rio on no unilateral action. GATT says you cannot build nontariff barriers. If forests are important, why not temperate forests? I do not get satisfactory answers."

I reminded Salim that a rainforest possesses considerably more biodiversity than a temperate forest. And that just because the West has made colossal mistakes is no reason for Indonesia to repeat them.

His response was direct. "We are told, 'Look, Indonesia, you are a treasure for the world, maintain your forests. . . . ' If the tropical forest plays an important role, what do we get for not exploiting it? Everybody shouts, 'Save the forest.' But when you ask how do we meet the needs of the poor people, there is silence. 'I won't give you technology, I won't give you gene patents, you remain poor, you go to hell.' That is how your people (the Americans) look at us."

How does Salim rate Indonesian poverty? One U.S. baby equals twenty Indonesian babies, he says, according to the IPAT equation (population impact is equal to the population multiplied by its affluence, times the level of its technology). In the U.S. sense of numbers, then, Indonesia's *effective population* would be less than 10 million.

What are Salim's solutions to Indonesia's complex dilemmas? In general he emphasizes resource management and resourceful utilization. Exploit the resource below the level of its sustained yield, he advocates. Maintain ecological carrying capacity. But, as I pointed out to him, there are no obvious benchmarks for carrying capacity, at least not until a species goes extinct or a habitat is clearly transformed. Economic indicators rarely, if ever, account for carrying capacity. The best example was that of the Exxon *Valdez* disaster, which actually yielded a picture—on paper—of a surge in economic development, progress, jobs, and the marked-up dollar value of oil. One of the great environmental resource economists, Robert Repetto of the World Resources Institute in Washington, analyzed Indonesia's economic indicators and found that depletion of natural resources was lacking from the calculations. In Java, the GDP (gross domestic product, the measure of the region's wealth excluding income generated by sales abroad) was said to have risen more than 7 percent each year between 1971 and 1984. Yet, study of the depreciation of just three environmental sectors during that time, namely, petroleum, timber, and soil, showed the GDP actually only increased by 4 percent.[60]

There are dozens of other sectors that need to have an economic tabulation to account for their depreciation—air, water, and vegetative cover, biodiversity, ecosystem-specific integrity, pollution-related epidemiology, and lost work days—even biological oxygen demand in bodies of previously pure water. Add all these varied depreciations together and the GDP plunges into a deficit realm, reflecting an even more mammoth loss of nature.[61]

Twelve years before Indonesia ever acknowledged a carrying capacity or population problem, the country was losing a known 50 million tons of soil each year, according to a 1959 U.N. Food and Agriculture Organization study. Ironically, says Repetto, the expenses associated with soil erosion, such as the dredging of clogged irrigation channels, now drain the country of as much as $50 million annually, yet these same expenditures show up on the national income accounts as *positive* additions to Indonesia's GNP.[62]

According to Salim, Repetto has taken the depletion figures but ignored the contributory ones, such as the inherent importance to global oxygen balance of the rainforests themselves, an Indonesian carbon sink value not included in the GNP. "Free services should be part of the equation," says Salim. While he agrees that so-called "external diseconomies" (resource depletion) need to be carefully estimated and factored in, you need to incorporate external economies as well, he argued. But this is a logical trap. It would be like insisting that Norway could kill all the whales on Earth because the value of Norway's glaciers to the world's meteorological cycles somehow more than compensated for the loss. There are countless twists and turns to the "survival" argument. Some are absurd.

Salim himself has recognized the fallibility of balance sheets when examining the natural world. Already, Indonesian economists have tried to pin him down for the benefit-cost ratio of certain species that have gotten in the way of development, whether bird, tree, or monkey. "Benefit-cost ratio nobody can say," Salim told me. He explained how the Ministry of Mining was looking for new opportunities and demanded to exploit a certain region. Salim said there were alternative regions for coal but no alternative regions for orangutans. Open pit mining was determined to be the most cost effective means of getting at the coal. No rehabilitation was planned. Said Salim, "The fact we don't know the value of a monkey should not mean that we can thus exploit it." And quoting the Qur'an, he reiterated, "God does not create something without a purpose. Man with limited knowledge does not know the purpose. God knows. There must be a reason God created a long-nosed monkey." Based on that reading of his religion, Salim is the one who has tenaciously pressed for exploitation-free forest preserves. He told me he wants Indonesia to be the "best in the world" when it comes to conservation. But he was not able to stop developers from eliminating Jakarta's last mangrove right in his own backyard. The battle is being waged, then, between those who have recognized and appreciated God's creation (whatever, or whomever God is—YHWH, Gaia, the life force, the cosmic dance, Allah) and those whose blinded, unessential, financial obsessions have bathed the world in a destructive, self-gratifying religion of expediency.

Other forces at work are bound to increase this conflict, certainly in Indonesia. Salim believes that the country will run out of oil as

early as 2010; that natural gas from the South China Sea is too expensive to pipe in; that solar energy may work for hot water, but not for
commercial industries; that they have already utilized their full
hydropower capacity. He argues for nuclear energy as the only possibility
of supporting economic growth painlessly. Despite opposition from
WAHLI, the National Atomic Energy Board (BATAN) is pressing for
the country's first power plant on the northern coast of Central Java
in the Mount Muria region. A scientist with Greenpeace pointed out
that Mount Muria is an active volcano, but the Mines and Energy
Minister evidently sideskipped this piece of information and cited instead his belief that only two out of 420 existing nuclear power plants
in the world have ever leaked (a sadly erroneous assumption). The bottom line, according to the government, is the soaring electrical needs
of Java and the inability of PLN—the state electrical company—to meet
that demand with coal, oil, hydro, or geothermal.[63] But WALHI has
argued that the government has grossly overestimated energy needs for
the year 2000, even under the most ambitious GNP projections.
"Indonesia's status as a developing country is often presented as an
excuse for rapid energy growth," says WALHI. "But the fact that Indonesia is at an early stage of industrialization should instead be viewed
as an opportunity to avoid the energy-inefficient history of the more
industrialized nations." By applying more efficient technologies across
the spectrum of Indonesia's nearly thirteen thousand large and mid-
size industrial ventures (which in turn use over 50 percent of all commercial energy) WALHI found numerous conservation opportunities—
from the use of waste heat for electrical generation, to the simple utilization, where practical, of natural sunlight, the introduction of energy-
saving light bulbs, and the replacement of clear glass with dark glass
to cut the air conditioning demand by 4 percent, and so on. Some 40
percent of total energy demand could be saved in Indonesia at a
reasonable price, providing for sufficient megawatt capacity to accommodate the country's fast-growing population while obviating the need
for embarking on the nuclear alternative. In the rural sector, small-scale,
locally managed, renewable energy sources have been identified

throughout the country, including microhydro, solar, wave power, the use of agricultural wastes, wind power, and geothermal.

The debate rages on. Seventy-six percent of cooking fuels come from wood, not electricity. All the more reason, says Salim, to promote fast growing trees species, particularly bamboo in conjunction with limited nuclear. He advocates a Chinese-style model of Green Walls in every province, and would like to see more Indonesians using kerosene for cooking. In deflecting demographic pressure away from the forests, Salim acknowledges that the urban population will swell from a current 17 percent to 30 percent by the end of the 1990s. His goal, and that of the BKKBN, is to reduce pressure on the land by guiding urbanization into suburban buffer zones surrounding the major cities.

At the same time, says Salim, prudent resource management will be bolstered by new environmental laws. Integrated pest management, environmental impact statements, the first automobile emission controls in the country as of 1993, and unleaded auto fuels beginning in 1996, are all aspects of this quickly evolving environmental sensibility. He explained that public officials can now be sued for alleged environmental transgressions. "I must be very careful," he said. He has already been sued twice as Minister of Environment. In addition, the Indonesian public now has total access to environmental impact documentation. Salim himself is in charge of the country's environmental impact statements. There is stricter liability, though public hearings have not yet been introduced. Those who are accused must prove their innocence. A poor farmer, theoretically, can sue a multinational and send a CEO to prison for up to ten years. But the judge always looks for strict causation. So far very few cases have been prosecuted. If nobody dies, it is difficult to find a victim. The judges normally side with the accused, not the accuser. There is no third party arbiter. In twelve years since the country's major environmental legislation was passed, all of three people have actually gone to jail.[64]

Part of the problem is the referral laboratories where evidence of an environmental victim is judged. While there are six laboratories in the country, there is only one that—according to Dr. Harger of UNESCO—is even 90 percent capable. The rest have equipment that came from various donors. If it's Japanese, it's junk and the Indonesians will spend all their time trying to figure out how to use it and begging for spare parts, though the Japanese have stopped manufacturing that equipment, which is why they've given it to the Indonesians in the first place.[65]

Whether an urban factory town or the distant tropics, because the economic stakes are so large and the players so powerful, "evidence" or "culpability" is difficult to prove and the prospects of prosecution very slim. Ending the war against the rainforests of Indonesia transcends the country's lawmakers because the demographic crisis is so pervasive. Moreover, those demographics are tied to a global economy, a global population bomb, a world war, which means that trading nations and consumers everywhere are "culpable." Indonesia is merely a front for others who would try to remain blameless and whose consciences, evidently, have become numb to their involvement.

Such psychic disassociation on the part of corporations, governments, and consumers allows for both knowing and unwitting conspiracy. Collaborators in the destruction of the rain forests are those who somehow encourage the market for tropical timber. But it is the governments who are most to blame for they are the only concentrated force that can intervene to halt the cycle. So far, the cry of global recession and unemployment has stifled such mediation.

The problem can be summarized: even selective cutting that supports at least a partial biodiversity has long proved to be erratically implemented, at best. Corporate greed, laziness, deception, and ignorance, coupled with most developing countries' lack of enforcement capabilities, has perpetuated the clear-cutting mentality. Energy consumption in Indonesia grew at 15.6 percent annually throughout much of the 1980s.[66] While the export of raw logs, as well as clear-cutting, has been banned, and a tax levied on sawn timber (except in Irian Jaya, which happens to contain some of the largest tracts of forest in the whole country) settlers are nevertheless building roads and burning down large sections of forest for agriculture. "Wasteful logging" operations—in addition to the official quasi-selective and clear-cutting figures—are resulting in some four hundred thousand hectares of lost forest each year in Indonesia, illegal encroachment continues unabated for lack of policing staff, and it is also well understood that the remedial tree-planting schemes, in which artificial ecosystems supplant real ones, are biological tragedies in terms of biodiversity loss.[67]

An enforcable moratorium on all new exploitation of primary forest —the goal of India—would mean that Indonesia's entrepreneurial impulses and huge debt, would have to be: 1) repaid via other non-forest means, 2) repaid through fast-growing timber estates created on existing wasteland, with an eye toward regenerating the forest floor through a decisive diversity of planting, or 3) hope that the developed nations might forgive the debt in perpetuity.

But consider that together, tropical countries are besieged by a nearly one trillion dollar debt, half of which is owed by those nations responsible for two-thirds of all deforestation.[68] Any sizable level of debt forgiveness is not likely to occur in today's economically depressed climate. Conversely, no nation has thus far indicated its willingness to compensate tropical countries for income voluntarily resisted, despite the fact that forest preservation benefits everyone.[69] India's recent resistance to the commercialization of its forests has made that country available for high-interest loans from the International Monetary Fund and the World Bank, but not outright aid. Countries are selfish, just as genes are selfish. If the developed world suddenly "got religion" and agreed to abolish all outstanding debts, another difficult problem would confront them: the world has not yet technically, morally, or aesthetically figured out how to properly evaluate the true worth, in terms of dollars spent or dollars compensated, of the rainforests, or of any "line item" in nature. The United Nations SNA (system of national accounts) has been in place since 1968 but it does not acknowledge natural resources in the public domain. Wilderness, in other words, is not considered an "asset." At the same time, nearly all Western nations have been reducing their aid by as much as 10 percent each year. Only Japan has increased its promises of Official Development Assistance (ODA), but much of the money is earmarked for Russia. In the long-term, considering the inherent nuclear risks associated with the breakup of the Warsaw Pact, that may be money well-spent. In the short-term, stabilizing the Commonwealth of Independent States with foreign aid will only diminish the available dollars that might have helped a country like Indonesia resist the impulse to exploit its forests. Several international organizations have been pumping money into conservation programs, largely targeted at biodiversity hotspots, like Indonesia. But, collectively, such efforts do not come remotely close to the estimated $52 billion a year needed just to get consistent and effective biodiversity management initiated in the developing world. And to repeat, Indonesia, by itself, is calling for a minimum of $20 billion a year in no-strings-attached assistance to get conservation moving.[70]

The much talked about debt-for-nature swaps, an alternative form of discounting debt in exchange for the preservation of endangered lands, has thus far elicited a meager success rate. The total value of all such swaps was about $100 million as of 1992.

As far as ecotourism is concerned, the total world proceeds are valued at approximately $12 billion a year currently. Nearly 17 percent of that total occurs on Bali, making it Indonesia's fourth largest revenue earner, with enormous countrywide opportunities for increased revenues. However, very little of that sum actually goes back into environmental preservation. An equitable system for linking tourist revenues to ecosystem protection and social justice is only in its infancy at present. In a study of eighteen efforts in the tropics to merge conservation with local sustainable development, widespread infringements of the goal were observed, typically by outsiders.[71] The one prominent test case in Indonesia, which specialists are watching, concerns the Arfak Mountains Nature Reserve in the rainforests of Irian Jaya. It is the brainchild of the indigenous Hatam tribe in concert with the World Wide Fund for Nature and the Indonesian Directorate for Forest Protection and Nature Conservation. The Hatam people have all signed an agreement to guard their forest from excess (illegal) cutting and to protect many of the species therein. Whether, over the long haul, the working understanding of "excess" will in fact comply with the many ingredients of "carrying capacity" is yet to be seen.[72]

Otto Soemarwoto, professor of biomanagement at the Padjadjaran University in Bandung has argued that because temperate forests comprise about 50 percent of the world's timber, they require the same sustainable practices of use if a total world conservation system is to be reliably implemented. Under the U.N. Conference on Trade and Development, negotiations have been underway for an international timber accord that would cover all forests. The Rio Summit of June 1992, failed to achieve that, managing only to reassert each country's inalienable rights to determine the fate of its own territory. Instead of serving to ensure the sustainable management of smaller ecological regions, political self-determination have fragmented the global commons. In the guise of local know-how, heritage, and self-defense, "independence" and democratic self-expression have had the tragic effect of permitting environmental destruction under international law. Much of that fragmentation has resulted from Western domination, of course. Postcolonial Indonesia, India, and China have each had to make arduous comebacks, in the wake of Dutch and British rule and ruin.

Yet, given the current escalation of "World War III", "sustainability" is a tragic misnomer that has been thoughtlessly adopted, to varying degrees, by ecologists and governments. The fact is, in a state of war, sustainability means the continuation of that war, a profit-driven politic whose compulsions and business-as-usual ethic is simply disguised, in most cases, by tact and respectable lip service. Given what we now know is actually taking place in the forests—the immense losses, the permanent transmogrification—the goal must be to prohibit the cutting of any and all trees, whether temperate or tropical. Selective cutting will not do. World leaders would do well to impose a global ban on all extraction, export, and import of primary forests, while compensating concessionaires with alternative income-generating land giveaways, interest-free loans, and grants. Antiforest poaching patrols must be trained and dispatched in the same manner as antianimal poaching units. Up until now, such interventionism has been the province of radical fringe groups (i.e., Earth First!). But, I can well envision international antipoaching brigades, much like the U.N. Blue Helmets, financed by a cooperative world body. Skeptics will argue that the existing eighty thousand U.N. peacekeeping forces are insufficiently financed. The counterargument rests upon the fact that the prevention of a few dozen multinational corporations from building roads into forests would be considerably easier to exact than, say, a human war in Bosnia or Somalia. The number of needed troops—in cooperation with the countries involved—would be relatively small.

A globally enforced ban would invite a more concentrated timber estate industry (such as exists in the southeastern U.S.) in conjunction with the badly needed restoration of existing ruined lands. In addition, biogenetic engineering of hardwood surrogates would become more bankable.

In this age of ecological doublespeak, the alleged practice of "sustainability" has shielded countries from criticism. Going much farther by implementing a total ban, as I am proposing, will require tenacious diplomatic finesse and economic counterincentives by the world's lending organizations and the major banks (many of which are in Japan). The argument *against* such a ban rests upon several emphatic assumptions. Mohammad Bob Hasan, chairman of the Indonesian Forestry Community, points out that most northern countries have destroyed more than 95 percent of their forests whereas some tropical countries have as much as 40 percent left. In addition, the North has safeguarded a total of less than 60 million hectares, versus 87 million hectares in the South (discrepancies and vague sources of data—the fuzzy illogic syndrome—lend little confidence to these figures, however). Indonesian officials say that ecolabeling, punitive measures by the North against the South (such as trade boycotts, let alone a total ban), would further stress an already impoverished system, diminishing the perceived value of timber and thus the incentives to protect it. "If we don't take care of the poor, they will take care of the forests in their own way," says Hasan. Unless the landless farmers and increasingly unemployed poor can realize immediate financial gain from the forests, conservation is not going to happen. The government says that if the tropical rainforests are to be truly understood as a global commons (a notion India rejects), then the rest of the world needs to help finance their survival.[73]

Environmentalists insist that rainforest destruction is the result not of poverty, but of greed, corruption, and the lack of willpower to police logging by governments in tropical countries. As for the developed countries, those same environmentalists have argued that governments, pressured by corporate lobbies and the much bloated emphasis on unemployment, have willfully encouraged the destruction of old growth forests while ignoring selective cutting abuses or higher market incentives for timber estates. The United States, for example, has been incapable of ensuring the lives of the old growth coniferous stands with their thousands of precious species, including the *Strix occidentalis*, or spotted owl, the rare tailed frog, the Del Norte and Olympic salamanders, and the *Taxus brevifolia*, or Western yew tree, noted for its anticancer taxol substance.[74]

By now it should be obvious that such arguments hinge upon several glaring contradictions. For example, all the wealth in the world has not protected temperate forests. Alleviating poverty is thus no guarantee of anything. And while India has attempted to limit forest commercialization, those efforts have been overwhelmed by the personal foraging of nearly half-a-billion persons dependent upon wood. The argument smacks of lunacy: one side is asking to extract several liters of blood, the other just a few liters. Both sides are ignoring the fact that the patient is already hemorrhaging to death. In sixty years, Indonesia's population will have doubled, encouraged by Indonesia's legal system, which has unambiguously stated that "The large number of people is an asset in the implementation of and a potential for the improvement of development in all fields."[75] That is the *large* picture, embroiled in grand conflict resolution schemes. But, during a realistic day and night of an average village, in the private quarters of a real human life, where troubles are strictly personal, these statistical data and sophisticated *arguments* are somebody else's problem, a kind of morbid poetry.

## INDONESIA'S PAST AND FUTURE

The train trip past partially denuded hills and intensively farmed paddies to Bandung two hours southeast of Jakarta yields a composite picture of Indonesia's problems and possibilities, cloaked in haze and human high-density. Bandung used to be a mountain-cooled Dutch resort, tranquil and uncrowded, nestled beneath a still-steaming caldera. The city maintains its village feel, though it is the second largest in the country now, with some 2.5 million residents. In the last few years, ten thousand acres of volcanic rainforest were mowed down to make way for migrants who put up bungalows and planted crops. Another forty-two thousand acres will soon be similarly converted. Traffic is already gnarled, babies are everywhere, and the population increase is on a par with Los Angeles, namely, 4 percent annually. The "village" already has its share of ghettos where the BKKBN has instituted income-generating schemes for impoverished women. The pattern of forest encroachment is chaotic—imperceptible today, a gigantic bald spot tomorrow—as it is in India. But what distinguishes Indonesia's plight from that of the Indian subcontinent, is the amount of forest still remaining. Encroachment thus enjoys, for now, a luxury of proportion that masks its seriousness, and can only be fully diagnosed from a distant vantage point, or a satellite.

Descending toward Bandung one day from twenty miles away along one of the surrounding volcanos, Tangkuban Perahu, I ventured off the road by foot into the jungle. I walked several hundred meters into the thickest tropic I could discover, kneeling, crawling, rubbing my face in it, taking in every penetrant odor, sight, and sound; sat down amid a teeming puzzle of life forms, took off my clothes in a faint drizzle, folded them neatly in a bundle beneath what appeared to be a rather rare Anggrek Bulan orchid, and waited.

Ants investigated my arms and neck, though none were aggressive. Butterflies of pale green and Antwerp blue dallied all around me. A dark snake moved through the thicket, just in sight, while a medley of determined and unrecognizable birds went about their business overhead. I dug through the grass and weeds and vines. Layers of dazzling life egressed with each additional millimeter of exposure. In the rotting bark of an immense softwood, beetles multiplied in a rousing chorus of motion, shape, and color. Here, all around me, were universes within universes. I closed my eyes, grateful and spellbound.

It was in this very archipelago that Alfred Russell Wallace, sick with fever, contemplating Malthus's *Essay on Population*, and noting a more remarkable divergence of species than anywhere else on the known Earth, first postulated his theory of "natural selection" in the mid-1850s. In his writings, he recognized that the existence of any organism depends upon "pre-existing closely-allied species" forming an interdependent web of life. "Closely-allied" pertained to a kind of coexistence, or harmony. But what harmoniously admixed species immediately preceded man, let alone a few hundred million of them? The mid-Pliocene *Pithecantropus erectus* (Java man) of seven hundred thousand years ago? Dr. Eugene Dubois who made the discovery during the early 1890s, found hundreds of other skeletal remains in the lapili bed along the now-polluted Solo River in east Java—rhinos, cats, hippos, crocodiles, hyenas, elephants, a gigantic pangolin, and countless smaller mammals that have all since gone extinct. Did Java man kill them? Was he as ruthless a hunter then as now? *Homo soloensis*, or Ngandong man, a later Javanese–Neanderthal type, the possible precursor of both Indonesian and Australian aborigines, slaughtered mammals, fashioned hunting tools out of deer antler, and—based upon skull remains—was probably a cannibal. Evidence from both of these paleontological layers suggests anything but prehuman harmony.[76]

Wallace, who was not yet aware of Java man, or the Ngandong cannibals, meditated for three years on the interspecies relationships throughout these rain forests and hit upon the notion of the "survival of the fittest." It was a concept that seemed appropriate to a balanced ecosystem. The population on Java by the time Wallace sojourned among the islands was probably around 9 million. What would he have said to an Indonesian human population of more than 60 million, the estimate following the country's first census in 1930? Would it have confirmed his and Darwin's analysis of evolution? Or completely negated it? If 60 million humans have a fitting place in a balanced ecosystem, what is it, where is it? Is there some cosmic purpose to human folly and destruction that we're not appreciating?

More likely, the theory itself is inappropriate for describing or predicting the human maelstrom, which has drifted apart from evolution like a continent coming unglued.

The forces of natural selection that had lavished such spectacular biological largesse on the rain forests were already being grotesquely skewed by one type of species that stood out from all the others without precedent, even thousands, possibly hundreds of thousands of years ago. To sit alone in that sylvan cornucopia on Mount Tangkuban Perahu was to feel the brunt of this conclusion, to hear the distant sounds of human industry, and to know that it was only a matter of time before Bandung's quickly growing population would overwhelm the very garden spot of creation whereupon I lay, a paradise in which the Earth had invested several billion years of loving patience and creative impulses.

Now, all is impatience and harm. As in the Amazon, Indonesian transmigration and colonization have meant the massive construction of new highways through primeval rainforest, like the trans-Sumatra and trans-Sulawesi. Fifteen cities over half-a-million in population are rapidly absorbing Indonesians who harbor a dream of mercantilist security and gain. These are not hunters and gatherers content to wield tools of stone or antler. There are two important ways to view the growing city populations. Certainly it can be held that urbanization lessens the immediate burden of the human presence on the forests and lowlands. In that respect, the more *metropolitan* a country, the better. But from the standpoint of the concentration of power, which is then employed in prospecting for wealth out beyond the urban peripheries, cities pose an inherent liability of monumental proportions.

Some policies have the weight of overnight reemphasis. Others are mere goals toward which the decade, or the generation, is slowly turned. Wallace's rather pleasant "natural selection" is no longer an option, even under the most rigorous one-child policy, which in any event, nobody at the BKKBN has advocated. Like Antarctica, Indonesia is a proving ground for both types of policies: for alternative, ecologically prudent scenarios that can, and will, reverse the immediate injuries to nature; and policies for the long-term which intuitively recognize that "more is less," more development, more people, more progress, less nature. "More" actually means "pain."

Much of Alfred Wallace's splendid wilderness has been lost; yet much remains to be saved. In 1980 there wasn't even a national park in this country. Not until 1990 were parks incorporated under any legal framework for protection. Now there are twenty-eight national parks. At the Caracas parks conference in the spring of 1992, ecologists everywhere were amazed by the commitments of Indonesia, said to be among the best and brightest in the world. There are brilliant, well-meaning, and motivated people at the highest levels of Indonesian governance. Many are working in the country's park system, yet must habitually come to the World Wide Fund for Nature offices to borrow paper clips, as one official described the situation. Because the Republic of Indonesia is politically accustomed to forceful executive orders, this nation has the necessary hierarchy and willpower to enact far-reaching legislation that would truly set a global example, both in terms of fervent population control and conservation.

For many years President Suharto has vigorously pushed family planning, motivating lower ranking officials from provincial governors to village headmen.[77] At the same time, relative to other developing countries, Indonesia's national budget is weighted in favor of education, health, and social welfare, which account for 10 percent of annual expenditures versus 8 percent for defense. In the same spirit, this is the crucial decade for Indonesia to greatly enhance its national budget expenditures for conservation, tying those costs to the very preconditions propitious to population stabilization early in the twenty-first century. The country needs more medical clinics as it prepares to receive the next baby boom. But, according to Pathfinder's Does Sampoerno, the key is *education*. Better teachers are needed, and a government commitment to compulsory secondary school education.

UNFPA says that the majority of Indonesians are in fact already aware of the fertility problem. It's the access to medical services that is the key, whether in Indonesia, the Philippines, or Bangladesh.

According to the BKKBN, the best way to meet the unmet demands for better education and access to medical services, in an effort to lower the country's total fertility rate and thus its negative ecological impact, will be through that agency's program to dispatch trained midwives to all of Indonesia's sixty-seven thousand villages.

For their part, the World Bank and AID recognize that without a greater commitment to long-term contraceptives, Indonesia will not be able to further lower its birthrate. That commitment will necessarily involve political inroads to the conservative branches of Islam within the country.

All of these components of the fertility crisis are half the story, of course. An official recognition of the fact that under most Islamic tradition, "ensoulment" does not occur before the beginning of the fourth month of pregnancy, would pave the way for the legalization of abortion. That would significantly reduce the TFR as well, as it has in places like Italy, the Commonwealth of Independent States, South Korea, Germany, and Spain.

Yet, in spite of such gains, the country is going to witness a doubling of its numbers. Sooner or later there will be between 350 and 400 million Indonesians, or a small village for every *individual* person Alfred Wallace encountered. In the planetary sphere of things, such myriad human lives and wants challenge the status quo of tropical Asia's 600-million-year-old wilderness, the Earth's heartland.

In my meetings with Mr. Zulkarnaen, the head of WALHI, he reiterated the urgent need for ten very compelling changes to existing ecological policies in the country, WALHI's published prescriptions, applicable, in fact, to every country with remaining forest. They are:

1) Base all economic decisions on ecologically sustainable development, determined by local and regional environmental carrying capacity parameters.

2) Reduce the demand for raw materials by recycling and through the maximal use of viable fiber residues, normally left by loggers, as well as by increasing the quality and efficiency of the harvesting technology.

3) Redirect a substantial proportion of all in-country paper production toward educational materials, particularly for those in the Outer Islands where school books, dictionaries, and writing pads are essentially non-existent.

4) Integrate local cultures and economies into plantation management.

5) Substitute the country's excess of agricultural residues for wood fiber. Bagasse sugar cane residues, rice straw, and bamboo are all viable alternatives.

6) Create community cooperatives owned by the local shareholders.

7) Involve those communities in all aspects of the decision-making process, such as whether to establish a factory and plantation or not, with a thorough presentation of the ecological pros and cons.

8) The national government should enforce its existing laws mandating environmental impact assessments.

9) Phase out all toxins, such as those associated with chlorine bleaching, and implement a law of best-available-technologies for mitigating all pollution.

10) And finally, establish a system whereby the investor and the beneficiaries (i.e., consumers in the North) must pay for natural resource depreciation in the South, much in the "polluter pays" and "energy tax" manner. At the same time, northern countries must not be allowed to export any technology that is not at the standard of its own domestic environmental control levels.[78]

As for the World Wide Fund for Nature, which wrote the important *Biodiversity Action Plan for Indonesia*, it too recommends a wide spectrum of viable political solutions. These include developing wood, charcoal, and rattan plantations in nonforest areas; community involvement to establish buffer zones around parks (as in India); increasing the size of preserves; and involving youth in tree planting campaigns. The *Action Plan* would encourage more community seed and data banks, and the local planting of wild-collected fruit trees and mixed agro-forest gardens. It would redivert all zoo resources toward captive breeding for eventual reintroduction into the wild of those few species that are in danger of imminent extinction, like the Sumatran rhino, the Bali *starling, the Komodo dragon, and the Indonesian birds of paradise. The Action Plan* recognizes that an opportunity exists to experiment with new breeds of mango, durian, various grains, legumes, and local vegetables that could enrich Indonesian agriculture, while serving at the same time as cash crops along with rubber and spices. These crops might be planted in the buffer zones themselves, which would thus be perceived as extractive reserves of nontimber products for local consumption. The Herbarium Bogoriense and Zoological Museum in Bogor has the largest plant collection in Southeast Asia, one of the best in the world. With technical computer support from donor nations, Indonesia could greatly enrich the quality of its wild and cash crop strains while engendering valuable "intellectual properties," the software for which could be leased to multinationals for noninvasive biodiversity prospecting purposes.

To this end, the WWF stresses the country's need to reassess the economic potential of native wild strains heretofore unexploited—such as various orchids and dipterocarps—by providing financial incentives to farmers and eliminating those disincentives to try new strains and natural fertilizers and integrated pest controls, even if it should mean lower yields.

The *Action Plan* points out that in each of the six major regions of Indonesia there are hot spots, or preserves in especially dire need of financial help. They include Kerinci-Seblat in Sumatra, Gunung Palung in Kalimantan, Gunung Halimun in Java, Gunung Rinjani in Nusa Tenggara, Rawa Aopa in Sulawesi, Manusela in Maluku and Gunung Lorentz in Irian Jaya. Each of these preserves desperately requires immediate assistance, and the adjacent buffer zones and habitat corridors are as well. One way to help ensure steady financial support for such conservation efforts would be to effect a system whereby those growing ecotourism revenues would be put back into protected areas themselves. The country needs to expand its marine protection areas so as to make all Indonesian waters a whale sanctuary, while providing legal relief for dugongs, turtle nesting sites, and sea grass, coral, mangrove and coastal habitats under an integrated policing and pollution abatement regime. This will surely entail developing a method for cost-benefit analysis of conserving biodiversity in both existing protected areas and nonprotected areas. And it will require a much greater commitment to funding the many more police, monitoring, and anti-poaching units referred to earlier in this chapter.[79]

Finally, the WWF's *Action Plan* has advised the government to introduce biodiversity studies in all schools and universities throughout the country, to initiate a national conservation media campaign aimed at community leaders, politicians, parents and their children, and to enact a host of administrative and infrastructural changes that would facilitate increased data linkage within national and provincial governments in order to better enforce existing laws. Such linkage would ideally avoid situations where one department (i.e., agriculture) subsidized activities that were forbidden under another department (i.e., national parks). In essence, this is a call for "a holistic and consistent management approach to conserve biodiversity," which needs to be approved by the DPR (House of Representatives) and signed by President Suharto on behalf of the Indonesian people.[80]

I lingered there, above Bandung, sprawled reflectively amidst the heartbreaking splendor of a fertile planet, trying to put myself in the place of those at the highest levels of government who must work with, and transcend, the shortcomings of human nature to effect curative policies. However one might visualize and practice the *Zen of conservation* and restraint, out beyond, down in the valleys, other human beings were making more noise, chopping down trees, chewing up the land, swarming far more furiously than those few ants that quietly meandered through the hair on my arms. If the fragmentary skulls of prehistoric Java man teach us anything, the lesson is that while our personal spans are remarkably brief, our footprints—the damage we are capable of—may last forever. That message is a challenge each of us must confront.

# 5

# The Forgotten Ones: Africa

## EVOLUTION AND ADVERSITY

Beneath 19,340-foot Mount Kilimanjaro, the warm, dry dawns are accompanied by a chorus of life forms that have found their voice. Their gait and tempo are as well-worked as a pearl, as fixed as amber, even if their population sizes are totally vulnerable to a host of vagaries, normal fluctuations that only recently have been skewed and subjected to shock. Tateva golden weavers busily flit about the dusty, sun-drenched acacia thickets, their chattering as involved as any stock exchange. Elephants, who as adults have no natural enemies, sojourn among the yellow-barked fever trees along seasonal marshes and permanent swamps. The water, amid expanses of desert and salt flats, is fed by underground springs from the perpetual snowmelt atop Kili. The elephants sink neck-deep into the rush- and weed-steeped pools, satisfying their vegetarian diets, as do hippos, which rummage and snorkel blissfully. Above ground, churning up the volcanic soils, are thousands of brindled gnu (or blue wildebeest, as they are also called) heading out in massive phalanxes or in single file, toward their favored watering holes that day, along with Grant's and Thomson gazelles, impala, cape buffalo, zebra, and more discrete medleys of hogs, bushbucks, waterbucks, reedbucks, elands, gerenuks, dikdik, addax, oryx, roan antelope. In all of Kenya there are known to be 314 species of mammals, 1067 species of birds, 191 of reptiles, 88 of amphibians, and 180 of freshwater fish. In Africa, only Zaire has a slightly greater variety of animals, but nowhere near Kenya's migratory density. Newborn Grant's zebras prance delicately, testing their still unpredictable legs and the endurance of their mothers' patience. Giraffes, in small groups, usually less than six, forage on tall *Leguminosae* trees, browsing for twigs and leaves twenty feet in the air. The males are constantly tasting the female's urine, as well; this is presumed to be "sexual behavior."[1] Overhead, Madagascar squacco herons swoop in formation, croaking raucously throughout the day. Their plaint is often perceived (by humans, at least) to be a kind of primeval laughter.

In a matter of a few hours I witnessed well over forty exotic species—and probably twenty thousand individuals—at Kenya's Amboseli National Park, each of whom, upon even the most cursory observation, can be said confidently to be endowed with often memorable personalities. A family of spotted hyena conferred at dusk around their sandy burrow, father and pups playfully exploring the fringes of their family center. He was extraordinarily indulgent, his pups great charmers aware of their effect on their parents. To see the elephant young nursing, or the Kori bustards, Maribou storks, and cape buffalo commingling, is to recall—in its most basic configuration—the conscience and all that it is capable of dreaming. A similar crescendo of feeling, however vaguely defined its perimeters, seems to be evoked by, and in, every species, which must share these precious oases and desert blooms; and which, together in seemingly ideal choreographies, meander at intermittent ease. An innate interdependency conjoins this glimmering ballet that suggests a widespread tolerance, even affability, as in the case of a white rhino that had joined a zebra herd, or the various teeth-cleaning birds that hang around the grinning crocodiles. There are scientific reasons, of course, for such relationships in the wild. But no precise rationale can altogether explain the obvious pleasure egrets exhibit whilst riding through the marshes at an even keel atop the hippos and elephants, feathery white Washingtons crossing their Delaware; the seeming relish the wildebeest and gazelle, or ostrich and wart hog, take in one another's company. The biophilia, or natural affiliation among different species, may be commensal, biologically mutualistic, a shrewd, even necessary compliance with the fact of a broad, ecological community. Evolution has exploited the wisdom of biophilia. On the plains of East Africa, where the greatest terrestrial mammalian biomass on Earth is concentrated, it has come down to being poignant, emotional, and heartfelt.

One night while I was lost amid thought, gazing at a circus of platinum-white hanks, that loom of the Milky Way, the warm winds across Amboseli accosted my senses with the penetrating stench of a kill somewhere nearby. In the morning, I found my prancing little zebra, or one identical to it, lying near a dirt road, its body stiff as a museum piece, eyes mute and dark with wonder. No predators had come, leading me to believe this poor foal had inadvertently roused the ire of a green momba, one of the most poisonous snakes of the desert grasslands.

Later on I passed a lone wildebeest, abandoned by its fellows, legs positioned defensively, awaiting its fate. It was clearly ill, trembling, half-starved, no more than a year old, but intent upon standing up. It knew that soon—perhaps that night—the end would come, probably in the form of hyenas, about the only predators left in this part of Kenya. Hyenas, and man.

Amboseli is the most frequented national park in Africa, with over 208,000 human/day visits per year. A short flight, or few hours drive southwest of Nairobi, Amboseli elicits that part longing for, part horror of, Nature, and complicates the picture with the advance of *Homo sapiens*. In their wake, what was once pandemic to the African continent has become a mere island of ecological memory, increasingly beleaguered. Like Conan Doyle's "Lost World," the once vast canvas of life has been framed edgewise, amounting now to no more than a few primitive vignettes, an anachronism like Ishi, the last Yahi Indian of California, or the Apache, Geronimo. A glorious, but dangling modifier of another time, lodged pathetically in the twentieth century.

Amboseli was reduced from 3260 square kilometers to its present 392 square kilometers following protests to the Kenyan government by the dispossessed local Masai tribe. Africa's much-romanticized "warriors" had been forcibly relocated by the English and Germans in the late nineteenth century to the area of Amboseli, only to be relocated again by the newly independent Kenyan government. Outraged by what they deemed to be a gross injustice, their anger augmented by the government's construction of a faulty pipeline intended to compensate the Masai for lost water rights, the tribe slaughtered many endangered rhinos in frustration. The government caved in—the Masai now water their domestic livestock inside the park, which means added pressure on wildlife. Until their displacement, the Masai ranged across the more northerly Rift Valley, hunting lions, giraffes, birds, and the many species of ungulates. With their swords and spears, the young *moran*, or bachelor-soldiers, were legendary for their ferocity. Today they subsist on blood and milk from domestic animals. Masai hunting inside the park is illegal. More recently, the tribe has begun consuming purchased cornmeal. Goats and sheep are killed far more often than the cattle, which are singled out for special occasions.

Humans and our ancestral hominids have been cohabiting with other wildlife in East Africa for at least three million years. Some *Homos* were big meat-eaters, others probably were not, such as *Homo habilis*, a contemporary of *Australopithecus boisei*, *Australopithecus africanus*, and *Homo erectus*. At Koobi Fora, east of Kenya's Lake Turkana, and possibly at Amboseli, between 2 and 1.6 million years ago these precursors of our species shared the same turf. *Homo habilis* and *erectus* applied different tools to their environment (the so-called KBS and Karari industries), while a few hundred miles away at Tanzanian Olduvai, yet a third industry, that of the Acheulian, was in play. These nascent technocracies perfected the use of fire and advanced the course of toolmaking by deft chiseling and scraping, pulverizing and granulating and grinding, pounding and rasping, grating and crumbling.

Meat-eating among early *Homos* is revealed—whether in Africa, China, or Java—by the dietary remains, including at least ninety known species of mammals that they preyed upon. Mutilation of the cranial region surrounding the foramen magnum suggests that *Homo erectus*—our most pronounced forebear—also ate other *Homo erectuses*. In fact, it is believed that a well-cooked *Homo* could serve sixty participants at a feast, all dining from bowls made of skull bones. But cooked ungulates were the most widely consumed animal.

Abetted by the first manipulation and maintenance of fire (whose earliest African evidence, 1.5 million years ago, comes from an ash-pit at Chesowanja, near Mount Kenya) this orgy of meat-eating ultimately impacted the cranial shape, and population size and density, or neurons in *Homo erectus'* brain. Forty million neurons per cubic centimeter, and seven hundred cubic centimeters—a pound of gray matter.[2] In early *Homo,* natural selection chose the act of meditation, countless dreams and metaphors, youthful energy looking ahead, a mind capable of determining causes and consequences. Yet, even in the presence of such options, *Homo's* future could have gone in many directions.

The Broca region, that area in the brain for coordinating the muscles of the mouth, tongue, and throat during speech, was (relatively speaking) rapidly re-formed during the reign of *Homo erectus.* And not surprisingly a large portion of the cortex was reserved for controlling the thumb, whose most important muscle, the *flexor pollicis longus,* ensured *Homo's* broadly precise grip, grasping, and tool-fashioning behavior.

*Australopithecines* used at least eleven tool types, later *Homo erectus* even more. Much of our backbone, jaw, cranium, skin, sex, vocal tract and birth process, pelvic structure, gluteal muscles, foot skeleton, and movement was rearranged according to the opportunism inherent to aggressive tool use, and with it, the sudden emergence of substantive, cultural designs.[3] In spite of the advent of agriculture, this meat-eating quality of mind is more with us than ever before. But only in the last few hundred years, as populations like those of Africa rapidly increased, has this bloodletting demonstrably altered a long pre-existing homeostasis of numbers.[4] It is the accelerated, graceless rate of change that is particularly vexatious—that shift, driven by myriad local human population explosions, far exceeds any historical or biological precedent, and has today imposed such cumulative adversity upon the ongoing evolution of the diverse African environment as to intimate a tragic simplification. The suffering inflicted by a large number of carnivorous consumers on surrounding life forms is unimaginable. It is as if *Homo sapiens,* through cultural evolution, had become an alien species, of which many such creatures—non-indigenous organisms that relocate and cause havoc—are all too familiar to modern biology: the Asian tiger mosquito that has caused outbreaks of malaria throughout the southeast United States; the kudzu vine and gypsy moth, to name a few well known pariahs; the Australian melaleuca tree in Florida's wetlands that pushes out native species at the rate of fifty acres a day; and the brown tree snake from New Guinea that has exterminated nine bird species in Guam since 1945.[5] By analogy, on the sub-Saharan island of Madagascar, where humans have been settled since the eleventh century, 90 percent of all native vegetation has subsequently been destroyed, as well as sixteen known primates now extinct because of human aggression.

What is unprecedented, and hence requires unprecedented solutions-oriented nerve and thoughtfulness, is the fact of our sole responsibility for a disaster that greatly transcends mere human destiny. In Africa we are viscerally haunted by our own paleontological record. Creatures of dual nature and behavioral ambivalence, fond of the gazelle's flesh and the elephant's ivory, we are also moved by the elephant's weary, forgiving eyes, and tempted to believe in some "better" past. We are nearly sabotaged by who we are, or what we have become, and by what we are so rapidly losing. Over the course of biologic eons in the Samburu Hills near Mount Kenya, at Sahabi in the Libyan Sahara, throughout the Omo Valley, at a place known as Hadar in the Afar Triangle of northern Ethiopia, where 3 million years ago little Lucy once walked with her mother, and more recently, in every politically defined African nation—the evidence continues to mount: humans are violent, and this violence is ultimately self-destructive. Like the earlier forms of beetle-browed *Homo*, 40 percent of whom died before they were fourteen years of age, we too will no doubt become something else, regardless of our diet, our tools, our mean-spiritedness, or ultimate goodness.[6] But, in the meantime, we have a choice to make on behalf of all other species; the opportunity to re-invent a biological equipoise.

Nowhere is diagnosis of ecological agitation, human discontent, and strategies for coping with contradiction and subversion more urgently needed than in Africa, particularly East Africa, where the collision between rapidly expanding human populations and most other life forms is at a fever pitch. Human fertility in Africa is like a train screaming through the dark night, its headlights turned inward, to paraphrase Boris Pasternak. While the extent of population-induced devastation has been worse in West Africa than probably anywhere else in the world, there is an enormous amount of wildlife yet to lose—and thus, potentially, to save—in East Africa and much of the Central African rainforest countries. Forceful but empathetic African family planning, and a comprehensive nonviolence toward all remaining biodiversity on that continent, are, without doubt, among the most serious tasks confronting twentieth-century humanity. It is a summons to the whole world. The plenitude of wildlife is not the only locus of pain and obliteration: human children are the other innocent victims. Children who are "led on" by all that being human means, to believe and trust in the aptitude and stewardship of their parents and the adult community, only to discover, too frequently, that such faith is premature.

Twelve percent of the world's population lives in Africa. That proportion will expand to beyond 16 percent of the world total by the year 2025.[7] Africa's mean Total Fertility Rate (TFR) is over 6, currently resulting in 21 million innocent newborns each year. That annual increment will continue to rise. Africa's population density of 21 per square kilometer equals that of Latin America, but compares favorably with the Middle East (29 per square kilometer) and Asia and the Pacific Rim (each 113 per square kilometer). But that density masks the extreme competition for scarce resources around lush mountain areas, forests, lakes, water holes, and coastal regions, which serve as magnets not only for human consumers, but other wildlife as well, a situation not unlike that in Antarctica. Eighty percent of Kenyans, and 50 percent of their livestock, dwell on some 20 percent of available agricultural lands, a ratio about equal to the whole planet. Outside these regions, other arid land is being "marginalized" as Kenya adds more than one million residents each year and the fast-growing population seeks out new territory to plow, deforest, and graze upon. The real problem is that some 95 percent of all Kenya Wildlife Service protected reserves, such as Amboseli, lie in those very outer areas (keeping in mind that the animals have *already* been repeatedly evicted from one location after another for well over a century), thus posing inevitable daily pressures on the already declining genetic stocks of wildlife. There are well over one million pastoralists in Kenya, each household, on average, possessing about 100 animals, 86 of them sheep and goats, the rest camels and cattle. These people, who include such tribes as the Rendille, the Samburu, Gabbra, Boran, and Turkana, are competing for space with a tidal wave of farmers. As arable land vanishes, the competition gets worse, and wildlife is the ultimate loser.[8] The patchwork nature of human survival almost guarantees that whatever remaining crucial migration corridors for animals exist will be rapidly infiltrated and destroyed, leaving the official reserves as isolated islands cut off from the rest of Africa, from migratory genes that might otherwise revivify dwindling populations.[9]

Average densities also fail to account for the rapid urbanization of Africa—24 percent to 31 percent in just fifteen years—with countless expanding slum populations, and the increased spread of diseases. By 2025, 56 percent, or 763 million people, of Africa is projected to be urban.[10] In sub-Saharan Africa, six women are becoming infected with AIDS for every five men. In Nairobi, one bed in five is now occupied by an AIDS patient. In fact 8 million of the WHO-estimated 14 million global AIDS infections are in sub-Saharan Africa. And that, of course, does not include the "unofficial" infections, those who have not been recorded in any hospital census. The Kenyan Minister of Health has predicted that every hospital bed in Kenya will be needed for the forty thousand AIDS victims in the year 2000. And yet, in spite of AIDS, malaria, TB, hunger, and wars, the population will triple across Africa by the year 2025. Everything about such calamities will have a *negative* impact on family planning, and on twenty-five years of efforts by the governments and NGOs to bring down child, infant, and maternal mortality.

At present, life expectancy on the continent has been elevated to an average of 51.4 years for males, 54.6 for females (much less in many African countries), but that figure will fall as AIDS takes its toll. And as it drops, more and more women will have more and more children to compensate. As it is, Africa's Contraceptive Prevalence Rate (CPR) as of mid-1994 was a mere 4 percent, the lowest, by far, in the world.[11]

It has been estimated that for every 10 percent increase in the CPR, the fertility rate declines by 0.62, meaning that fertility replacement—

or approximately 2.3 children per couple—requires about a 75 percent CPR, or 71 percent yet to be achieved across Africa.[12] Unfortunately, as Thomas Goliber has pointed out, a slightly increased CPR in a few African countries, like Zaire, has not pushed down the TFR (total fertility rate) but merely heightened "the decline in postpartum abstinence," thus increasing "the time a woman is at risk of pregnancy and inevitably the fertility rate as well." In essence, then, the traditional tribal and Islamic two-year abstinence between births has been diminished with the advent of contraceptives, but not the number of children or desired family size. And thus, concludes Goliber, "at the outset of Africa's transition from high to lower fertility, there is no direct connection between contraceptive prevalence and fertility. As family planning programs expand in Africa, they are not necessarily going to result in immediate fertility decline; indeed, fertility may actually rise in the short term."[13] There is no other situation quite like that anywhere in the world.

The paradox is a gripping one: if the African TFR remains at 6 for just six more years (until 2000), fueling intense population kindling, it is almost guaranteed that by the year 2045, there will be 2.1 billion sub-Saharan Africans (many of them unhealthy, uncelebrated, unfed, uneducated young people), and an additional 250 million restless North Africans, many undernourished. The stress and civil unrest of population-induced problems is *already* being felt across North Africa, from the Sudan to Algeria. If by some miracle, a TFR of 5.2 were achieved by the year 2000, there would be 500 million fewer people throughout Africa by the year 2045. A TFR of 4—the short-term goal of Kenya—would mean 900 million less persons.[14]

In all Africa, only Zimbabwe, Botswana, Mauritius, and Kenya have achieved high CPRs, though the figures are taken from married women, not those unmarried, with the exception of Kenya: 43 percent in Zimbabwe, 33 percent in Botswana and Kenya. Only the remote island of Mauritius has managed to achieve a high CPR (by international standards) of 75 percent. But Mauritius is anomalous, statistically minute. And even its population of just over a million has exceeded the carrying capacity of the island, in a manner less discussed, less dramatic, but no less unfortunate than the crisis in Madagascar to the west.[15]

As for Zimbabwe, Botswana, and Kenya—many of those who study these trends have already perceived that a plateau may have been reached, beyond which CPR is unlikely to increase, or TFR to decrease. Modern methods of birth control are frowned upon throughout Africa, and many are unaffordable. Condoms are used to fight AIDS in some cities, but not for birth control; social marketing is new and, as yet, unproven. And while more and more women are indicating their uneasiness with large families, in most African nations, with the exception of Kenya, the global recession and increased poverty is enhancing the perceived economic value of more young ones. Indeed, over a third of all governments in Africa believe their respective national population growth rates to be "satisfactory." Seven percent of them even feel their growth to be "too low." At least half of all African nations oppose any family planning intervention.[16] This policy inertia is economic in nature, motivated by the male-dominated laws of inheritance, the lack of social welfare, and the self-perpetuating tyrannies of poverty, whose primary victims are women, and then their many children. The syndrome has condemned much of the continent, transforming women into fertility slaves, their bodies frequently denied even the pleasure of orgasm by painful and forced excision of their clitoris and labia at an early age. Brainwashed and brutalized, African women— like so many women throughout the world—have been sexually indentured, rounded up and made agents of their cultures' de facto warfare against nature. Wildlife and human children are the ultimate victims of this complex web of egotism and cruelty that has ensnared the Mother.

## THE EAST AFRICAN CONUNDRUM: KENYA

East Africa's population—in fourteen nations—currently encompasses some 312 million people; its mean TFR is above 6.55. United Nations projections for the region in the year 2025 range from a low variant of 476 million, to a constant fertility rate of 674 million.[17] The World Bank has published an even higher year-2025 projection of 701 million.[18] More than 80 percent of all those people are rural, usually in direct competition with wildlife for fast dwindling resources, a situation that is most pronounced in Kenya, where 85 percent of rural households fall below the poverty line (3000 Kenyan shillings, or $42.75 per year). The total income generated domestically by 450 million sub-Saharan Africans is less than that of Belgium, with its 11 million residents.[19] At least twenty sub-Saharan countries now face food emergencies, yet the annual food deficit is predicted to spiral downward fifteen times below even its current debacle. In war-torn Rwanda, for example, (where the few remaining mountain gorillas are being slaughtered out of spite by the combatants in that country's ongoing civil war—a means of crippling the nation's source of ecotourism revenues), a few hundred thousand Burundian refugees added to the over 7.5 million people in the country. Rwanda has virtually the highest TFR in the world, somewhere between 7 and 8.3, yet it cannot even produce enough food for 5 million people. And there is no more arable land to cultivate.

It used to cost a Kenyan one shilling for salt each month. Now the cost is fourteen shillings. People are doing without sugar. In 1989, one U.S. dollar got you eleven Kenyan shillings; in 1994 the exchange rate is 1:70. The minimum wage in urban Kenya is 1500 Kenyan shillings (KS) per month, or about $22, but perhaps only 30 to 40 percent of adult Kenyans in cities are employed. Urban rent, on average, is 500 KS a month, bread for a family of six, 600 KS a month, milk 300 KS. The parents walk to work, take the bus home—another 300 KS a month. Already a man's salary is exceeded by his expenditures. If both parents are working, then there is no breast-feeding. Eventually, the children will drift out onto the street because the parents cannot remotely afford the price of sending children even to primary school, not with fees for materials starting at well over 1000 KS per child per year. The government used to pay all school costs, but now it can afford very little.

This economic downturn will thus confound two of the key solutions that some family planners and sociologists had advocated for Africa: breast-feeding and rural electrification. Get Africans watching television at night and they'll be too busy to make love—or so rang the aphorisms ("the rich light candles, the poor make babies"), though the more universally acquired radio has not proved the case. Furthermore, Africa's economic stagnation does not lend itself to electronic consumerism. No one will be able to afford to buy a TV.

The high literacy rate for which Kenya was famed all over Africa is now slipping steadily just at the time when the country needs it most. As of 1989, literacy rates varied from 29 percent to 74 percent; in some areas, an overall estimated average of 50 percent. But now, 60 percent of females are dropping out of school. The importance of this slippage to family planners is this: in Kenya married women with at least some secondary education are three times more likely to use a modern method of birth control than those with no formal education. Moreover, less than 25 percent of all babies born to women with no schooling end up receiving any kind of medical assistance at delivery, versus 75 percent of babies born to mothers who have had secondary education.[20]

As of 1989, life expectancy in Kenya had attained virtually the highest level in all of Africa, fifty-nine—up from a mere thirty-five in 1948—and the Infant Mortality Rate (IMR) was less than 70 per thousand, as compared with 262 in 1962 and 105 in 1989. But these figures are also bound to change as the economic crisis continues to worsen. Already, the IMR appears to be going up in the slums and the poorest provinces.

The law says you cannot uproot coffee trees for indigenous tree nurseries. Coffee and tea are two of Kenya's major cash crops. Yet the cash economy is volatile. Just as copper prices crashed in Zambia, and the declining market for oil has wreaked havoc in Nigeria and Gabon, so too the global coffee crash has devastated Kenya. The country's exports are doing miserably.[21] Civil servants are being retired at age forty. Fifty thousand such workers were let go in 1992; the government simply cannot afford to pay them.

Debt continues to increase across sub-Saharan Africa, where debt ratios are the highest in the world. Long-term debt exceeds $150 billion and annual repayment continues to degenerate at less than half of its stipulated conditions. Only Botswana and Mauritius have maintained liquidity; but most African countries are mired in negative ratios. Mozambique, for example, would require eleven years of its full export earnings to pay back its lingering debt. Capital outflow is now 95 percent from the sub-Sahara. Those who can are fleeing the continent. Conversely, few outside investors or banks look to Africa with any interest or confidence. The so-called Paris Club of creditor governments to Africa (of which the U.S. accounts for 15 percent of all Paris Club loans) is looking to reduce debt payments by encouraging "transparency, accountability, rule of law, and public participation," according to Jeffrey Shafer, U.S. Assistant Secretary for International Affairs in the Treasury.[22] In Kenya the situation is particularly tenuous.

Every year Kenya's auditor-general reports on hundreds of millions of missing or misappropriated funds—yet nothing is done because many of those in Parliament are alleged to be on the dole. This explains the 1991 multilateral ban on Official Development Assistance to Kenya by the Paris Club, pending democratic reforms by Daniel arap Moi's administration. According to Sheila Macrae, head of UNFPA in Kenya, these measures were taken to punish the government and have not affected humanitarian assistance, such as family planning. But inevitably, the punishment trickles down through a corrupt infrastructure. Forty percent of all Kenyan government spending used to go to education. Then the International Monetary Fund forced that line item down to 27 percent in order to pressure the country into quicker debt repayment. It is expected that the educational budget will be down to 20 percent by 1994, while Family Planning has all but vanished from the government's specific annual budget. Contrary to its policy against purchasing "expendables," UNFPA was forced to provide the Kenyan government with surgical gloves in order to facilitate medical exams, which the law insists upon prior to any family planning or birth control counseling or procedures.

In the village of Gaatia on the luscious slopes of Mt. Kenya, midwives complained to me that they must cut umbilical cords with sharp pieces of wood or stone, because the government does not even have the money to outfit them with razors or knives. In the capital, Nairobi, which is growing at 4.86 percent per year and where 40 percent of the population is below the absolute poverty line, many of these denizens crowd half a dozen of the largest slums, like Kariobangi, Mathare, and Keebira where only one in five households has a latrine.[23] Everyone speaks only of the economic crisis—and of *chai*, the word for bribe in Swahili.

When I asked one still young-looking Masai father at Ekongu Narok Manyatta village how he expected to feed and care for his two wives and eight children, he smiled and dismissed my alarm by referring to a much-lauded government minister from his tribe who has eighty-four children by four wives. There is no limit to the number of wives a man can have. Male TFRs are never discussed in family planning literature. Like most of Africa, Kenya is polygynous. Meanwhile, the growth rate in Kenya overall had risen slightly from 3.6 percent in the 1980s to 3.7 percent in 1990, one of the fastest growth rates in the world.

This same graceful and prolific Masai husband complained to me that his wives and eight children were not the problem, but rather the prohibitive costs for their health and education. How then, given such decomposing economics (and local interpretations of bad times), do family planners reasonably expect to meet targets in Kenya that include reducing population growth rate to 2.5 percent per year and increasing the CPR to 40 percent by the year 2000?[24] To increase the number of Norplant users from 5400 to 24,000, and birth control pill acceptors from 217,000 to 320,000. Where will the money come from?[25]

The Masai father's eight children by two wives is actually a restrained family size for Kenya, where the average TFR in 1993 is closer to 5.4 children per wife, 10.8 for two, down from 6.7 (or 13.4 per two wives) just four years ago, if one accepts the most recent *1993 Kenya Demographic and Health Survey* (KDHS).[26] That survey has declared the new 5.4 TFR to be the continuing result of "one of the most precipitous declines in fertility ever recorded."[27]

In addition to the 20 percent decline in fertility (5 percent per year since 1989), eight other encouraging facts emerged in the new *Kenya Demographic and Health Survey:*

1) The pill is recognized now by 95 percent of all married women in Kenya and accounts for 9.6 percent of the CPR. Most of

the pills, at least until recently, have come free from govern-ment of Kenya sources.

2) Fifty-six percent of all married women in Kenya have used one form or another of family planning at some point in their lives; 99 percent of all men in Kenya are aware of family planning; and modern methods of contraception use are up 10 percent in the past four years.

3) In the most densely populated parts of the country—Nairobi Province, and the Central Province—the CPR is at 45.4 per-cent and 56.0 percent respectively.

4) Forty-six percent of married women want no more children; another 26 percent say they want spacing of at least two years. Mrs. Kalini Miworia of the International Planned Parenthood Federation in London and former head of the Family Planning Association of Kenya, explained to me how women in Kenya were crying out for contraceptives. At the Sulmac flower plan-tation south of Nairobi, a private sector family planning proj-ect, 90 percent of the women workers accepted family planning almost immediately. Taken together, such facts imply enormous potential in the very near future for further increasing CPR rates and a more rapid decrease in the fertility rate. In fact, if the CPR continues to increase at its current rate, three times more married Kenyan women will be using modern methods of contraception in the year 2000 as in 1989, or 1.7 million compared with 575,000.

5) There is antenatal care in at least 95 percent of Kenya. While the quality and type of assistance varies greatly, from say the economically disadvantaged Nyanza and the western provinces to the wealthier Central Province, the fact of such near-universal care is unique with respect to the rest of Africa, or most Lesser Developed Countries (LDCs) for that matter.

6) The preferred family norm is alleged to have gone from 4.2 to 3.0, a very important decline.

7) In 1989, 90 percent of men said they approved of family plan-ning; 60 percent of the women said their husbands did not ap-prove. So now, seeking to optimize this apparent window of

opportunity, the Family Planning Assocation of Kenya (FPAK) is encouraging couples to talk with one another in counselling sessions set up to break through the discrepancy, the many myths and misconceptions regarding birth control.

8) The commitment of President Moi to a working TFR of 4, and the structuring of the National Council for Population and Development in 1984, suggest that Kenya may indeed be able to do what only Mauritius in the developing world has done: turn around what was one of the fastest growth rates in the world.

There are some within family planning circles who believe that these new trends in Kenya are actually the result not of a particular slogan or outreach effort, but of diminishing per capita income.[28] Once the economy begins to flourish again, nobody knows whether the TFR will also rise, as it has in China, Costa Rica, Singapore, and elsewhere. Does international largesse toward Kenya, and the rest of Africa, engender long-term stability or simply induce a dependency that perpetuates the fertility crisis? Are there scenarios according to which poverty is best suited toward accomplishing population control, or is economic prosperity essential? Third World governments have argued for decades that the best form of birth control was wealth. But in Kenya, *poverty* is pushing down the TFR. At the same time, NGOs and multilateral organizations are trying desperately to conceive of ways to help wean governments like Kenya off donor assistance. Unfortunately these efforts are occurring at the very moment that the country is economically on its knees. To move Kenya from mere awareness to action, and to discover ways in which to link population programs to environmental restoration in recessionary times, is indeed a challenge.

But from the family planning perspective, Kenya has enormous promise for holding back total tragedy. The signs are extremely positive. It is important to realize that family planning began in Kenya when there were 5 million people, not 300 million as in India. While still characterized by a high TFR and percentage of growth, the country seems to be making unprecedented strides to curb a runaway population boom. That's the positive side.

But there is discouraging news as well: in spite of the fact that there are currently some 154 separate projects in Kenya addressing population concerns, the population is expected to double (again) in eighteen years or less.[29] Part of the problem is Kenya's enormous population momentum, with 60 percent of the country under the age of fifteen. By the year 2025, there will be nearly 61 million Kenyans, by 2050, 86.5 million. And, according to the World Bank assessment of recent trends, Kenya's population in the year 2150 could be over 112 million, or more than four times its current size.[30] Moreover, if the KDHS is overly optimistic (as some officials have suggested) and the TFR is actually higher than stated, or it should begin to rise, the U.N. "constant fertility variant" for the year 2025 shows 94,409,000 Kenyans.[31]

Now add to these ill-boding premonitions the debilitating influence of Catholicism. Recently a bishop in Nairobi found a carton of condoms in a school locker and was outraged. Within a week the Minister of Education vowed to permanently ban all formal discussion of sex in any classroom in Kenya, thus undoing ten years of curriculum development work by UNFPA and UNDP. Only UNICEF has managed to incorporate condoms into a kit that is handed out to Kenya's armed forces. More than 25 percent of the country is Roman Catholic. On August 25, 1993, in commemorating the twenty-fifth anniversary of the Papal Encyclical—Humanae Vitae—the Kenyan Catholic Church and its Archbishop and Cardinal attacked abortion, population control, condoms, and artificial contraceptives. They argued that what was necessary to control population was merely "the proper use of economic resources for development of the people in the spirit of transparency and accountability." Needless to say, it is doubtful that such policy recommendations will be more useful than, say, the Catholic Church simply owning up to its responsibilities to the planet and coming out in unilateral support of contraception. But that is not likely, or not under the present administration in the Vatican.

While Kenyan NGOs attempt to circumvent Catholic-dominated policies regarding sex, the country lacks any official policy concerning high-risk adolescents, despite the fact it is understood that the

secondary school ratio of 144 boys to 100 girls is, in part, directly at-
tributable to the high dropout rate for girls due to early pregnancies.[32]
Though the YMCA, the Boy Scouts[33], the Salvation Army, and
Seventh Day Adventists all disseminate information regarding reproduction to young people, in some provinces there are at least fourteen
known teenage pregnancies per 1000 girls. In fact 20 percent of the
birthrate is among teenagers, who constitute approximately 3.5 million
hormonally-excited individuals, not including the larger population
of incoming sexually active ten-to-fifteen year olds, currently in the
country.

Family planning has failed the youth in Kenya. While the legal age
of marriage is sixteen for girls—eighteen for boys—in places like Kilifi
on the coast, girls are getting married at the age of nine. The country
meanwhile, is expanding at a rate of a million a year. The data is so
disturbing that the government has still not released the 1989 census,
which at the time was considered by insiders to be an economic, and
thus political, bombshell. Nevertheless abortion is encoded in the legal
system as a criminal offence in Kenya; a woman and her doctor go to
prison if they are caught. Still, there are at least twelve abortions performed every day of the year just at one hospital in downtown Nairobi.
Those women who don't make it to a hospital (most do not) often die
during a botched procedure.

Already, with some 28 million people in the country, 30 percent
of the population is undernourished. The demand for fuelwood and
charcoal, 73 percent of all energy supplies in Kenya, is now three times
that which is available. There is severe desertification over 19 percent
of the country; forests cover only 3 percent of the nation, arable land
only 17 percent. The rest is arid. Agricultural production has been dropping more than 1 percent per year. In some of the northern districts
with the most marginal land (Mandera, Wajir, Turkana, and Marsabit)
child malnutrition is as high as 80 percent, unprecedented in the
history of Kenya. Accelerated poverty is already beginning to nullify
the very triumphs recorded in the 1993 KDHS. Out beyond the city
of Nairobi, in the Dagoretti Constituency, the TFR is 8 in a population base of 200,000. Whatever inroads had been accomplished to

help control maternal and infant mortality in an effort to bring down the fertility are fast disappearing.

"We don't know what is going to happen," says Ms. Leonola, who trains volunteers out of the Chandaria Health Centre in Dagoretti. "We regard porridge for lunch as the single meal every twenty-four hours. It's at least enough to get the children through the night," she says. "So they can sleep a little. There is simply no food, no money, no jobs. We're defeated." As of 1994, Ms. Leonola doesn't know if U.S. AID, and the Minnesota International Health Volunteers organization will be able to refinance them. Yet, without the Chandaria Centre, one might as well write off this suburban district, where 70 percent of all births are served not by doctors, nurses, or any kind of hospitalization, but by traditional birth attendants, many untrained and superstitious. The program at Chandaria hopes to teach those attendants, to impart modern skills and make available a modicum of tools; to encourage antenatal visits and postnatal follow-ups. Thirty-six thousand women and children are served each year by the volunteers and paid staff at Chandaria, who now must spend most of their time writing grant proposals, or referring patients to other centers, and giving out condoms, as available. Money in Africa, it would seem, is everything: fertility control, sustainable development, and the ability to sleep at night. Neither cultural norms nor indigenous wisdom seems capable of surmounting the daunting obstacle of persistent poverty here.

But perhaps the most disturbing report of the impact of recent economic hard times on family planning services comes from the region of Mount Kenya, one of the country's premier farming belts, a lush tropical highland fed by glacier streams. On the mountain's eastern slopes, 120 miles north of Nairobi in the Tharaka-Nithi District, Chogoria hospital is still considered the busiest and best family planning clinic in all of Africa, serving a population base of 400,000 people.[34] As of 1985, the CPR in the region was 42.5 percent, the TFR 5.2, and IMR about 50—at least a decade ahead of the rest of the country. Moreover 34 percent of those in the Chogoria catchment area were using modern contraceptive methods. (At that time there was only a 1 percent usage throughout the rest of Kenya.)[35] Since 1985 statistics at Chogoria have gotten even better. The key explanation seems to hinge upon four critical strategies: community involvement, including community leaders who themselves pass out supplies; dependable service and provisions; integration of primary health care and family planning; and a thorough system of monitoring.[36] Today, CPR in and around Chogoria are the highest for all of rural sub-Saharan Africa—over 54 percent. But there are terrible changes in the wind.

The collapse of tea and coffee prices has shocked the system. The government gives Chogoria hospital no subsidies. Clients pay for services, though at a minimal rate. But worsening inflation, the escalating cost of education, the loss of government jobs, and the continued marginalization of lands, have had a chilling effect on the Chogoria hospital program, and on demographics in general. Landless women are particularly hard hit. Many have wandered into cities like Embu and Nairobi, which has driven up the HIV seropositivity among such migrants.[37] Five-to-ten percent of all adults in Kenya are now estimated to be HIV positive.

"We are broke," was the constant refrain I heard from hospital staffers. Though over 218,000 patient contacts were made in 1992, the number was less in 1993 and is continuing to drop. Locals simply cannot continue to afford to pay for services, and the hospital is finding it difficult to pay for its staff.

Thirty minutes away from the hospital, in the village of Gaatia, one of Chogoria's sublocations, I sat with thirty-three traditional birth attendants who described their current difficulties. They have no kits—neither sterile gauze nor razors, neither scissors nor medicine, no pain killers, nothing. They have virtually no money. The average per capita income for a large family is 4000 KS a year, or $52. The Kenya Planters Union has not paid the Gaatia coffee growers for two years. Those growers are mostly women. "Men plan, women plant," goes the saying in the village. "Men are for looking at cows, not work," the women add.

The effects of poverty on fertility in Gaatia may be very different than in Nairobi. Out of a total village population of four thousand, three thousand are children. With few jobs, little income, and no family planning services, those three thousand children are more likely to have sex early on. Their numbers will grow. Family planning successes will vanish. At the same time unemployed women, spending more and more time with unemployed husbands, will also find themselves more frequently pregnant.

"Why have more children?" I asked. Those males among the company looked away, the women winced, then launched into an attack on what at first seemed to me to be a side issue, though in fact it is not, namely, the male-dominated policy of inheritance. Kenyan inheritance customs favor each son with part of his father's property. Women do not yet have the right to inherit family property. Normally, divorced and widowed women also lose title to their husband's land unless they have sons. In addition, the men believe that the benefits of educating a son far outweigh those of educating a daughter. After all, would it not be sheer lunacy to invest in educating a girl who will marry and benefit another family?[38] The same logic prevails in India and China. It seemed likely that the women who were speaking, in their colorful homespun dresses and sneakers, were publicly voicing pent-up grievances for the first time. For hours I listened, trying to understand how poverty worked on the village level, where the sexual urge, hunger, and restlessness were involved. Gaatia seems, through a camera lens, to be idyllic; a Renaissance pastoral vision. Families lie out together under dense palm trees at noon, resting from their labors. The breezes off the mountain tone down, or filter, the light through those groves, a photographer's dream. By comparison with the deserts fifty miles to the north, where the Samburu people struggle to find any wood or water at all, it is a paradise—protected, forested, the ferruginous soil saturated with biological bounty. Mount Kenya's altitude ensures a temperate, Southern California–type climate. But, for all of its apparent agricultural industriousness and aesthetic tranquility, the women are tired long before their time, the children are not well-fed, and happiness languishes in the anthropological realm of the subjective.

One might be tempted to wish greater prosperity on these villagers, to "lift" them from their present circumstances into a faster lane. But, of course, that is precisely the tragedy of every village, and every confrontation with greater wealth, throughout the world. Gaatia, as I saw it, had four salient problems: first, the women were not equal partners in life and did not seem to share to the same extent in the proceeds obtainable from nature, as did the men;[39] second, coffee and tea, to which much of the land had been devoted, *are not edible*, and with the market having collapsed, Gaatians are in an economic cul-de-sac characteristic of all postcolonial cash crop dependencies in Africa; third, there are too many children, and too high a continuing fertility rate, given the ecological boundaries of the village and of surrounding villages; fourth, the traditional birth assistants and the nearby Chogoria hospital are lacking the necessary resources to slow down this micropopulation boom, and hence are unable to counteract growing malnutrition, landlessness, and unemployment. A person caught out in these high seas of powerlessness and frustration is likely to become paralyzed by depression and inertia. Without money, or any kind of communal self-help, or government assistance, Chogoria hospital can do less and less to support the very midwives it has trained; neither can it protect mothers or children. Hence—by the subliminal forces at work in families, between men and women, and in the hearts and minds of breeding individuals—more and more children are brought into this world as a kind of blind universal buffer against despair. Money well spent toward inhibiting more children in a place like Gaatia would mean the difference between contraception and desperation. That difference would directly connect to the fact that Gaatians, and others like them throughout Africa, throughout the world, are compelled to kill off other life forms in search of food, shelter, and economic breathing space.

As the economic pressures escalate, those politically in power throughout Africa have typically skimmed more and more money off the top of the infrastructure, and manipulated tribal differences and coalitions to remain in control. The breakdown of the neighboring Somalian government occurred after thirty years of independence, largely as a result of former leader Mohammed Barre's exploitation of tribal differences. Recently, Kenyan politics have witnessed a clamor to rewrite the country's thirty-year-old constitution, and a call for new pluralism, for tribalism and multiparty democracy, beginning with a new constitutional convention in Nairobi. The attempted revival of GEMA—the powerful Gikuyu-Embu-Meru Association—suggests that the govern-ment has indeed become vulnerable. But that vulnerability, that empty pit in the political stomach, is merely a reflection of the ecological and population perils assailing the country. The superficial Eden that is Gaatia haunts the nation precisely because there are already too many people, and insufficient white maize, wheat, millet, sorghum, rice, pulses, starchy roots, tubers, and milk (the main diet in Kenya, along with a heavy preponderance of meat) to feed them. Micro- and macroeconomic analysis of the nutritional carrying capacity, energy intensity, and future productivities, however interpreted, fails to recon-cile nearly 30 million meat-eating Kenyans.[40] While *nature* in East Africa has not changed in millions of years—evolution adhering to its biological code of ethics—suddenly, in the course of one century, a spark of time, the populations of heavily consumptive *Homo sapiens* in East Africa have exploded.

I have endeavored in the foregoing to highlight some of the human difficulties brought on by human multitudes. But the injury to that pre-existing nature—evolution that was timed and honed by graceful change, now suddenly convulsed, the rate of change perniciously accelerated—is equally, if not more disconcerting. There are many voices speaking up for orphans, AIDS victims, disheartened men, in-nocent, hungry children and their oppressed, desperate mothers. But who speaks for those with no voices—be they gazelle or rhino? The dynamics of boom and bust, as described earlier on, whether among lemmings or deer, hares or locusts, dinosaurs or butterflies, always leave—as in the ecology of fire—an aftermath of splendid resurrection and realignment. Only one species, *Homo sapiens*, has flatly disrupted this biology of reincarnation, the ratios of predator to prey, the recy-cling of nutrients, the balancing of populations. Humankind's own per-sonal willfulness has violated the karma of whole ecosystems. If some greater law, or destiny, has ordained such wholesale carnage, then more than likely, what I call the "small claims courts" of human ethics—our feeble laws and meritorious treaties—are unlikely to alter cosmic necessi-ty. But if—as many today believe—there are solid grounds for an alter-native future, then what is occurring in Africa must be viewed as but one more horrendous battleground in this criminal and useless global war. The combatants are desperate, the stakes as high as anywhere on the planet.

One way to gauge the dimensions of this war and of our species' population overshoot in Africa is by considering the needs of other carnivores (which, with the exception of leopards and lions and crocodiles, are smaller than man). Five hundred cheetahs (the suggested absolute minimum number needed to sustain the genetic population of any local group of carnivores, and a tenuous number at best) require 30,000 square kilometers ($km^2$) of unencumbered habitat; 500 lions 6250 $km^2$, 500 leopards, 12,500 $km^2$, 500 spotted hyaenas, 3333 $km^2$, and 500 wild dogs, 100,000 $km^2$![41] Now, consider that the total amount of preserved land in all of Kenya is a little more than 31,000 square kilometers! Tanzania has about 100,000 $km^2$, Uganda a mere 13,000 $km^2$. Tanzania's Serengeti, which is connected to Kenya's Masai Mara (for a total of 30,000 $km^2$) is the only area left in all of Africa that can be said to contain a sizable constituency of mammals— an estimated 1.5 million wildebeest, 250,000 zebra, 500,000 gazelles, and several thousand predators, for a total population on a par with human Nairobi.[42] Those proportions in the wild are indicative of the carrying (and caring) capacity of the natural world. Only a very small number (relatively speaking) of predators can be accommodated. If there are, say, 2.5 million mammalian vegetarian inhabitants of the Serengeti, and at most 5000 large carnivores, that figures out to 0.20 percent meat-eaters. Forgetting, for a moment, the technology of human carnage, which has completely eclipsed the evolutionary rules of behavior on the killing fields of Africa, there is simply no way to incorporate the nearly 70 million human carnivores of Kenya, Uganda, and Tanzania in just those three biodiverse, politically defined regions. Mass starvation, habitat destruction and extinctions of other life forms are inevitable. Now add the fires of deforestation, the greatly shortened fallow periods, the poacher's machine gun, and the global market for rhino horns (one horn is worth $100,000 in Asia), elephant tusks, furs, and skins to the mix, and the full scope of the ecological holocaust emerges.

The end result is a dark tunnel of human appropriation, a meteorite shower of consumption, unchoreographed, propelled by short-term exigencies. Caught up haplessly in this peppering of many conflicts is all of nonhuman Africa.

Poaching is escalating in Angola, Botswana, Cameroon, the Central African Republic, Ethiopia, Malawi, Mozambique, Rwanda, Somalia, Sudan, Swaziland, Tanzania, Uganda, Zimbabwe, and Zambia. In 1970 the black rhino population stood at sixty-five thousand throughout Africa; today, there are less than four thousand. In just one survey along the Zambezi in 1991, indications suggest extensive poaching still going on (a carcass-to-live-elephant ratio of some 13 percent).[43] And only since 1990 has poaching been controlled in Kenya, Namibia, South Africa, and to a lesser extent, Zaire. Controlled, but not halted. The Kenya Wildlife Service is taking no chances: it has built up an antipoaching air force of eleven planes.

In 1988, at the Mount Meru National Park in Kenya, poachers killed every white rhino, as well as one of their human bodyguards. These rhinos had become tame, following their transport from South Africa in an effort to breed them against the likely prospect of extinction. From 1973 to 1989 ivory poaching reduced Kenya's elephant population from 170,000 to 24,000. At Tsavo National Park some 36,000 elephants were reduced to 5000 (6000 died of famine, then 25,000 more were poached). The 1989 U.N. ban on the ivory trade has meant that most of the elephant poaching—at least in Kenya— has stopped.

But recently the Tanzanian government issued a license to the United Arab Emirates deputy minister of defense permitting the slaughter of endangered species. Under the guise of "Wildlife Conservation, Management, and Rural Development," the brigadier in question, as reported in the New York Times[44] now has the purchased right for a decade to kill lions, leopards, cheetahs, rare gerenuks—in short, anything in the renowned fourteen-hundred-square-mile Loliondo Game Controlled Area bordering the Serengeti National Park. In 1984 there were an estimated five thousand black rhino in the Selous game reserve of Tanzania, the largest in East Africa. Now there are fewer than three hundred. Again, the work of poachers. According to Dr. Michael

Fox, a vice president of the Humane Society of the United States, behind "a facade of concern for wildlife" that has "lured funds from non-African nations, for supposed conservation programs" there is abundant "documented evidence of the mass slaughter of wildlife—including animals belonging to threatened and endangered species—by wealthy safari hunters and even local government officials."[45] Fox saw "stockpiles of ivory and skins in some government and private vaults" throughout the country.

In Zimbabwe, where the rhino population plummeted from fifty thousand in the mid-1970s to less than four hundred in 1990, officials are now surgically dehorning the tranquilized animals with chainsaws in an attempt to de-motivate poachers of that one species. But the avid, official issuing of licenses for trophy hunting is routine, countless species are "culled," and the widespread killing of wild animals for meat has become something of an industry there. In fact, the government wants to cut its elephant population from around sixty thousand to half that; it is also threatening to breach its pledge under the 1989 U.N. (CITES) ban on ivory trade and sell off its existing stockpiles on the world market. With that money, say Zimbabwe officials, millions of dollars could be earmarked for conservation, notwithstanding the inducement to the market for additional ivory that such a strategy would inflict. In South Africa's Kruger Park, more than four thousand "surplus" elephants, hippos, Cape buffalo and other species are butchered by authorities in a park abbatoir every year under a similar economic management philosophy.[46] The one hundred twenty member nations of the United Nations Convention on International Trade in Endangered Species (CITES) have not yet figured out how to contend realistically with poor nations.

As for Kenya, which has rejected killing-as-management, Richard Leakey, head of the Kenya Wildlife Service, insists that ecotourism is the only ethical and ecological way to finance conservation. I totally agree, assuming ecotourism can be controlled, as it has been in, say, Bhutan, so that the animals are not agitated, the natural vegetative cover is left undisturbed, and the indigenous people are justly compensated and involved in maintenance of the park. Unfortunately,

Kenyan ecotourism has for several years been placing an unsustainable and senseless burden on many of the country's preserves, particularly at Amboseli, Samburu, Mt. Kenya, Masai Mara, and Nairobi National Park. Considering how relatively little land has been set aside for such sanctuaries, ANY disturbance by tourists is outrageous. But, in fact, what is called ecotourism in Kenya is too frequently nothing short of an all-out assault by photo-hungry off-road caravans. Furthermore, at least half of all estimated wildlife in Kenya exists outside the parks and is thus prey to the same vast assemblage of forces known collectively as development. Leakey and staff summarize "Kenya's single greatest problem as its increasing population." And, they add, "The future prospects for wildlife are undoubtedly partly dependent on the success the government may have in bringing population growth under control."[47]

The Kenya Wildlife Service (KWS) currently projects its annual budget at $34 million per year, a small price considered in relation to the rewards of tourism. Eighty-eight percent of all foreign visits to Kenya are motivated by a desire to view wildlife; they generate about $200 million a year for the country. As of 1989 that was approximately 2.5 percent of the nation's GNP—not a lot. With the crash of coffee, that percentage perhaps has risen, though it has been correspondingly dampened by a 20 percent decrease in tourist revenues since the Gulf War. In addition, a wildlife "existence value" can be extrapolated— that amount of charitable, conservation donations from throughout the world by those who have not visited, and perhaps will never visit Kenya. One study estimated a $30 million existence value a year for elephants alone.[48] The economics of so-called ecotourism are no end-all panacea. Furthermore, little of those revenues is rechanneled into local indigenous communities who might be charged with helping to manage the preserves. Quite to the contrary—hotels, foreign investors, and government officials are the beneficiaries of most of the money.

Amboseli, home to the Masai, is the country's most popular tourist haven. Yet, in 1993 a mere half-million dollars was put back into Amboseli communities.[49] At the same time, as described by the Kenya Wildlife Service itself, deteriorating conditions in the park are such that tourism is predicted to fall off 10 percent by 1995, that is, to 184,000 human per day visits.[50] To reverse this trend, the Kenya Wildlife Service intends to invest an additional half-million dollars a year over a four-year period in Amboseli, based upon a 29 percent projected return, and the presumption that Amboseli can withstand 368,157 visitor days a year.[51] If the KWS does reverse the post-Gulf War ecotourist slump, it does so with an acutely conscious dollar sign attached to each visitor. For example, if the increase were 5 percent rather than 10 percent a year, the financial return would only be 14 percent, not 29 percent. And since there is increasing competition from other East African countries for tourists, parks like Amboseli are determined to do their part to attract more people. That will mean increasing the opportunities for visitors to get near the animals, off the road, day and night. While it is noteworthy that off-road travel, travel by foot, and night travel are illegal in all of Kenya's parks, more and more instances of it are occuring, again because that's what tourists demand.

There are those who will argue that tourism has had no measurable impact on the wildlife. But, even to an unschooled eye, such theories are patently fallacious. I saw lions deterred from mating during their two-week rutting season by the frenzy of loud tourists a few yards away, laughing from their secure semicircle of elevated vans; and convoys of dust-spewing vehicles digging up plants, disrupting quiet twilight gatherings of baboons, frightening elephants and Cape buffalo, pursuing feeding giraffes across fragile terrain, the tourists coaxing drivers with the veiled promise of larger tips to get that last photograph, whatever it takes. The sheer presence of hundreds of thousands of people that close to wild animals has engendered a time bomb within the ecosystem. The checkerboard of fast-moving vehicles across the preserves, all fanning out like clockwork from lavish hotel epicenters where animals come begging Yosemite raccoon–style, is perverse. Animal behavior is being thwarted, crowded out, incalculable hormones, pheromones, and other crucial emotional and psychological signals, desires, and needs, inhibited, while the land itself is being mauled. Lions, cheetahs, and rhinos are no longer to be found at overworked Amboseli, though many of the glossy international tour brochures still promise whole cheetah families napping right atop the hood of one's Land Rover.

The parks constitute approximately 6 percent of the total land area of the country (where, as stated, an estimated 50 percent of all remaining wildlife reside). Of course, the percentages are inexact, considering the nature of animal movement and migration. What is more precise is the fact that Kenyan population growth, with all that it implies, is rapidly shrinking the available territory for all species. While most "official" poaching has ceased for the time being, the increasingly severe economic times will force more and more of those Kenyans in rural areas to hunt for their dinner. At the same time, commercial export of live primates is the highest in all of Africa—over thirty-three hundred a year—with Tanzania in second place at over twenty-one hundred annually. Around rapidly expanding Nairobi, particularly in the Kitengela area, animal corridors have not been provided. A ten-kilometer-wide greenbelt—free of farmers, houses, roads, and fires—leading from the plains to the mountains, could have limited the devastation. But it's too late. The Kenya Wildlife Service only came into being in 1989. What animals remain in the Nairobi area have mostly taken cover, or been hand-placed in Nairobi National Park, where their gene pools are condemned to something like a melting iceberg.

North of Nairobi, the Rift Valley lakes of Naivasha, Victoria, and Nakuru are also under siege. In Lake Victoria, algal blooms from untreated sewage, nitrates, phosphates, heavy metals, and undegraded organic materials flowing out of Kisumu, Kenya's third largest city, are suffocating large quantities of fish. At Lake Naivasha, water levels are being reduced because of intense horticultural activity, while water hyacinth are proliferating because of the increase in nutrient concentrations. The famed flamingos are dying out at Lake Nakuru, allegedly from sewage contamination.

Southwest of Nairobi, along the coast, some thirty remaining groves of forest are now embattled. Known as the Kayas enclaves—considered sacred by the nine coastal tribes, the Mijikenda of the Kwale and Kilifi Districts—much of the forest was recently designated as reserves, or—better still—national monuments. What used to spread for hundreds of miles, from Somalia to Tanzania, is now reduced to a desperate total of no more than five thousand acres. Yet, the biodiversity is rich, with many endemic species of plant and animal and insect. A special Coastal Forest Conservation Unit funded by the WWF (World Wide Fund For Nature) is looking after them, but developers are still trying to get rid of the forests and put up high-rise hotels overlooking the beaches. One Kaya on Chale Island, was sold to a foreign land speculator who cleared the "bush" and instantly put up a tourist mecca, despite an outcry from the press, the elders of the community, and ecologists.[52]

Elsewhere in the country, the spread of wheat cultivation and concentration of pastoralists and their herds of cattle into group ranches adjacent to national reserves has diminished wildlife dispersal options. In addition, rapid subdivision of ranches and trust lands—a form of economic fallout—has been especially pronounced near Amboseli, Nairobi National Park, the Narok district adjoining the Mara National Park dispersal area, the Laikipia district, and Samburu—all crucial biodiversity oases.

Most subdivision occurs without environmental impact assessment or appropriate land use consideration, purely haphazardly, for profit, or for the perceived survival needs of locals. Beginning in the 1960s, one of the richest farming regions in the country, the Kisii area of western Kenya, was divided into thousands of small plots (as the TFR of the the Gusii people exceeded 8). Landholdings diminished in size by more than 100 percent, but the local response was not (as family planners might expect) to seek ways to check human fertility, but rather to increase the fertility of the land to meet expanding human numbers. That proved to be impossible after two generations.[53]

And yet there is still time for Kenya to consider legal means that would nationally mandate that the assets of land, rather than the land itself, be subdivided; and that any exchange of land title be dependent upon appropriate wildlife management use. Both of these alternatives have been proposed by the Kenya Wildlife Service.[54]

Many elephants have knowingly flocked to the parks to escape poachers, but there are thousands of others who remain outside those preserves. It is largely on their account—as well as that of the effectively disinherited, hungry scions of small landholders—that more enlightened land division principles need to be enacted. But such proposals invoke a *long-term* tapestry of usefulness, whereas the human-elephant conflict requires *immediate* resolution as the more than one million Kenyan pastoralists begin, of necessity, taking to agriculture, and the government increasingly "degazettes" segments of previously preserved forest in order to provide land for the landless. This syndrome is engendering veritable island farms in the middle of elephant habitats, or forest peninsulas surrounded by farms. And since the best arable land in Kenya exists around the forests, the competition between the several thousand nonpark elephants and people is critical, particularly in several northern districts.[55]

Hence, the Kenya Wildlife Service has had to plot out an appropriate future for elephant populations. It has rejected the idea of culling herds (employed in other African countries) for three reasons: on ethical grounds; because of the adverse effect it has on tourism; and because they believe that periodically killing the animals can have a very "destabilizing effect . . . on population dynamics."[56] Instead, the KWS has decided, in some cases, to literally fence the elephants into certain parks and forest reserves (at Mt. Kenya, among the Shimba Hills, the Aberdares forest, and Mwea Reserve). In the Laikipia–Samburu area, where there are many farms, some three thousand elephants move twice a year between the north and south, during the short and long rainy seasons. So far, farmers have been tolerant of the elephants in Laikipia but the KWS is concerned that should the elephant population increase, it is likely to tip the scales of human tolerance against them. Already, demands for compensation (many probably fraudulent) exceeding $3 million have been made by landholders against the Ministry of Tourism and Wildlife for alleged damages by wild elephants. But it is especially important to remember that while humans have cold-heartedly massacred over 146,000 elephants in Kenya in the last twenty years, elephants have only killed between 60 and 80 humans, in nearly all cases by accident, or having first been provoked.

Pressured by rising conflicts, officials are considering methods of elephant "family planning" so as to prevent any harsher "solutions" being proposed in the future, along lines taken in countries like Zimbabwe. Such fertility regulation has already been undertaken elsewhere in the world with many other species, such as white-tailed deer, elephant seals, and, of course, cats and dogs. In the case of the elephant, the KWS is examining the use of RU-486 embedded in fruit, of exogenous progestin, steroidal hormones, and immunocontraception. Delivery of the last three would be by darts shot into the elephants' muscle tissue. Manipulating the duration between births, and somehow increasing the age of first reproduction, have also been studied. At Amboseli it was determined by scientists that "increasing the average age of reproduction by two years (from thirteen to fifteen years old), and increasing the birth interval from four to six years, would be sufficient to hold most populations at a constant size."[57] Techniques that induce abortion, or sterility, in a percentage of females, are also being considered.

The call for fertility control seems to contradict the alleged "existence value" of wildlife, but in the real world international largesse and the spoils of ecotourism are not enough to reengineer the social, moral, and pragmatic norms of East Africa. More than anything, the suggestion that family planning be implemented among the pachyderms (albeit, mercifully), a species that has already been devastated by Homo sapiens, points out the degree to which the human war against nature in East Africa has escalated. It would be like saying, give Bosnians birth control pills so that the Serbians will have fewer to kill. But this is precisely the challenge confronting every human being, burning activist and bewildered armchair conservationists alike.

## VICIOUS SPIRALS

However problematic the conflicts between overpopulation and the environment in Kenya, elsewhere in Africa they are even worse. Africa replaced Europe as the second most populated region in the world as of 1992, with over 680 million residents. But in the year 2025, the continent's projected population ranges from a low variant of 1 billion, to the unlikely possibility of a constant fertility variant of 2.1 billion. The median projection by the U.N. is 1.58 billion for the year 2025.

Throughout Africa populations are growing at about 3 percent per year, despite the highest death rates in the world—15-to-18 per thousand. Life expectancy in nearly half of all African nations is under fifty, a benchmark, according to demographers, for projecting any kind of positive inroads that might be effected by family planning. Keep in mind that death rate data does not include infant mortality rates. An IMR below seventy has been discerned to be key to resolving the fear of family extinction. However, throughout most of sub-Saharan Africa, the IMR is 120 per thousand (versus 10 in the U.S. and 6 in Sweden). In most countries of the world, the IMR (which pertains to the first twelve months of life) is deemed the critical hurdle—surviving the first year, a child is likely to survive, period. This is not true in Africa, where a child has the same likelihood of dying within four years as during those first twelve months. Even in cases where infant or child mortality rates have declined (as in Kenya), the persistence of low life expectancy will inevitably supersede any "awareness" of contraception. The innate compulsion to perpetuate a family will not obey slogans handed out by bureaucrats recommending small families. In several sub-Saharan countries, the preferred family norm (PFN) is nearly 9 children, more than the existing Total Fertility Rates. In Mali, PFN is 6.9; in Mauritania, 8.8. The percentage of use of modern methods of birth control in Mauritania is 0.0. In addition, while the number of students enrolled in both primary and secondary school has increased markedly in many African countries over the past twenty-five years, in some nations— the Central African Republic, Malawi, Lesotho, Tanzania, Somalia, Kenya, and Mali—it has actually declined.

At least 75 percent of all chronic and acute hunger on the planet (more severe conditions than mere malnutrition or iron-deficient anemia) is to be found in sub-Saharan Africa and South Asia. But only in Africa are tens of millions of people faced with actual *imminent* death. Those in the "hunger business" claim that 36 percent of the developing world was deemed chronically undernourished in 1970, compared with what is alleged to be 20 percent today. However, computing approximate population growth since 1970, the numbers of afflicted turn out to be about the same, and the majority are Africans.

Africa comprises 20 percent of the land surface of the planet. Most of the continent is plagued with drought and desertification. North of the Sahara, more than half of all grain has to be imported with oil money, which is running dry. The U.N. Food and Agriculture Organization in 1982 stated that there was sufficient cultivable land across the continent to feed 940 million people, without any fertilizers or conservation effort. Yet, by 1975 half of all sub-Saharan populations, inhabiting a third of the land area, were losing the battle to maintain food self-sufficiency, and today they lack the cash liquidity or even credit to import food to stave off chronic malnourishment. "There were some exceptions," says the World Bank, "but sub-Saharan Africa has now witnessed a decade of falling per capita incomes, increasing hunger, and accelerating ecological degradation. The earlier progress in social development is now being eroded. Overall, Africans are almost as poor today as they were thirty years ago." According to the World Bank, if Africa is to avert total starvation its economies need to grow by at least 4 to 5 percent a year as a minimum target; and the target growth for industry should actually go to 7 to 8 percent a year. Nothing like that is taking place; in fact, the economic growth is declining, not increasing.[58]

In Nigeria, an early oil bonanza utterly undermined population control programs in the country by providing a false sense of economic security. When oil prices fell away, however, the government panicked and vowed to achieve a TFR of 4.0 by the year 2000, as in Kenya. In actual fact, Nigeria is likely to become the third largest country in the world by the middle of twenty-first century, with a population greater than in all of Africa in 1984. There are 3.6 million Nigerians born every year. Currently, the population exceeds 125 million people, with a TFR of more than 6. There are two hundred fifty tribal groups and four hundred languages, making any consistent population policy next to impossible. The country is more than likely to exceed 400 million residents long before it ever achieves fertility replacement. That can only mean economic disaster. And the population is expected to approach 620 million by the middle of the twenty-second century. Already there is virtually no wildlife left in the country, and crop yields are lower than they have been in thirty-five years. Consistent with this colossal conflict between human numbers and the environment is the fact that Nigeria's crime rate is already the highest in Africa, and possibly in the world.[59]

With the exception of Zimbabwe, Botswana, Kenya, and a small region of southern Nigeria—consisting of 30 million people[60]—as well as in a few local areas of Togo, Cameroon, and Senegal, such blatant conflicts appear inherent to sub-Saharan Africa, where the contraceptive prevalence rate is below 5 percent, and where a woman has a 1-in-21 chance of dying from pregnancy, versus a 1-in-6,366 chance, in the U.S. If fertility is such a roulette game for the African mother, why does she persist? The answer has as much to do with the psychology of poverty as it does with male oppression. Ancestor worship, the economics of fertility descent, polygyny—which affects nearly half of all African marriages—separation in the man's mind of "reproductive decisions and the cost of child raising," weak political initiatives in the family planning arena, the absence of sex education in schools, the inaccessibility of contraceptives for young people, the woman's inability, or fear, to take the lead in contraceptive use, and the African bias against modern contraceptives, have all contributed to this low CPR.[61]

As in every country, decisions pertaining to reproduction are politically charged. In Africa, political willpower is frequently moderated, or stymied, by the very diversity of its people: more than eight hundred ethnic groups; one thousand languages and dialects; intensely separatist tribal affiliations and customs, the duly paranoid aftermath of centuries of colonialist-induced internecine slave trading. In a survey of demographic pressure and social unrest in one hundred twenty countries by the Population Crisis Committee in Washington (now called Population Action International—a sign, hopefully, of the times), it was found that the combination of dissatisfied youth (frustrated expectations), communal violence, the breakdown of civil liberties and the frequency of governmental changes all added up to the dangerous potential of destabilization and war.The countries most vulnerable were African—Mauritania, Ethiopia, Zaire, Burundi, and the Sudan.

At the same time, the most troubling demographic pressures upon governments are also in Africa (with the added exception of Saudi Arabia). Those countries most afflicted are Kenya, Cote d'Ivoire, Tanzania, Nigeria, Libya, Uganda, Rwanda, Liberia, Zaire, Sierra Leone, and South Africa.

But in stricter ecological terms, nearly every nation in Africa is already at war, engulfed in a somber haze of self-destruction; and not merely among those tribes with a history of unfortunate conflict, such as the Igbo, Hausa, and Fulani of Nigeria, the Shona and Ntebele in Zimbabwe, or the Tutsi and Hutu in Burundi.

Consider the decimation of African forests, where so much biodiversity still remains. In countries like Brazil and Indonesia, approximately ten tropical trees are burned or cut down for each one replanted, a disastrous ratio. But in Africa the ratio is twenty-nine to one! Add to that the damage wrought by the browsing of the largest number of free-ranging livestock in the world. In every country of Africa where there are forests, the charred remains, or actual flames of the conflict are easily observed.

In Tanzania, forest cover is declining at a rate of 300-to-400,000 hectares (ha) per year. Half of Tanzania is expected to be desert by the year 2000. All its natural forests will vanish by 2015.[62]

Across the seven tropical forest countries of Central and West Africa—Cameroon, Central African Republic, Congo, Equatorial Guinea, Gabon, Zaire, and Cote d'Ivoire which make up the second largest contiguous moist tropical forest region in the world—inappropriate microeconomic policies, and rapid population growth, economic downturns, total forest-based economies, weak and corrupt management, and administration incapacity are causing deforestation rates to increase in a violent crescendo.[63] In Zaire, 1 million square kilometers of closed forest are often cited as the current remaining amount; however, as in Indonesia, there are vast discrepancies in data. The IUCN suggests that half that much remains. In Equatorial Guinea, forest cover has plummeted from twenty-two thousand square kilometers during the colonial era to thirteen thousand today. In the Congo, having seen the price of its oil plunge to below $15 a barrel (the lowest in many years) the government is looking to its 220,000 square kilometers of forest for income, of which at least 23 percent has already been subjected to intensive logging. (With a large percentage of forest still remaining, the Congo deserves particular attention as a noted tropical hotspot.) The same sorry syndrome can be described in Gabon—Africa's richest nation, but one dependent on oil until recently. In Cameroon, between 200,000 and 260,000 square kilometers of forest are under rapid conversion to logging areas for timber products and clearance for agriculture. In the Cote d'Ivoire, with a population growth rate of an astronomical 4.6 percent between 1965 and 1985, 95 percent of all tropical rainforest (140,000 square kilometers) has been burned or cleared, the result of shifting cultivators. Less than ten thousand square kilometers of forest are still standing.

Beyond the rainforests to the west, Africa's woes multiply even faster. In Nigeria the last black rhinos were exterminated at least fifty years ago, and countless other large mammals are also gone. Even nonhuman scavengers, such as jackals, are now threatened with extinction. In the local dialect of Hausa, *nama* means both animal and meat, a perceptual bias ingrained in the nomenclature of biodiversity loss. In March of 1978, when author Peter Matthiessen traveled through Senegal and Cote d'Ivoire, he reported that there were an estimated 4 million muzzle loader rifles used for hunting meat in the back country of those regions.[64] Increasingly, rural Africans are killing whatever they can. The situation is grimmest in the seventeen countries of West Africa, which currently host some 275 million human residents. And that number is expected to reach 925 million by the year 2050, and 1.36 billion in the year 2150.

Particularly hard hit—however self-inflicted—is Mali, a country too overwhelmed by poverty to even afford the pretense of any kind of native ecological movement.[65] Climate, the global greenhouse effect, unrelieved demographic pressures, a cumulative pattern throughout history of environmental degradation and simplistic solutions (outmigration) no longer applicable anywhere in postcolonial Africa, have left this particular country virtually bankrupt.

While the varied Sudanese cultures of Mali and most of West Africa have championed glorious cultural traditions for many centuries, the human-induced changes in biodiversity have been appalling. Not surprisingly, negative ecological feedback in West Africa has made for considerable human suffering.[66]

Under section 118/119 of the U.S. Foreign Assistance Act, the assessment of wild flora and fauna as part of any integrated development strategy is mandated for those countries receiving aid. Peter Warshall, a scientist with the University of Arizona Office of Arid Lands Studies, under contract to the Natural Resources Management Support Project of the U.S. Agency for International Development, surveyed Mali's biodiversity in 1989. His report is the first such comprehensive investigation in West Africa and stands out as a unique and critical contribution to biodiversity research and policy analysis for that region.[67] Warshall's depiction of the impact of the twenty-year drought throughout Mali and the Sahel, with its paralyzing ramifications and ecologically "vicious spiral," reads like the end of the world. In traveling through Mali, a country populated by 9.3 million people, with a TFR exceeding 7, and very little wildlife left, one fears that this land, this tragedy of human impact, is premonitory, a glimpse of the very fate of the Earth. Even before the continuing drought that began in the early 1970s, Mali's problems were salient.

There had been other droughts, other environmental crises. Archaeological evidence suggests that historical cycles of population boom and catastrophe have recurred. Two thousand years ago there were iron smelters in Mali, whole toolmaking assembly lines. One can still find at least twenty slag heaps along the Samonco River just outside the capital city, Bamako. There is evidence that iron residues disrupted the already poor laterite soil, and that extensive deforestation resulted from a widespread charcoal industry, the wood necessary to heat furnaces for the smelting. Inside the cool caves of Kourounkorokotou fanfan, a quasi-preserved site in the Montes Mandingues *Foret Classee* in southwestern Mali, one can recall a time five thousand years ago when stone, not iron, tools were used to skin the plentiful elephants and crocodiles. A still-standing relict tree species from a much wetter era indicates radical climate change over the past centuries. At one time, the nomads who dwelt here were seasonal visitors, moving from camp to camp across West Africa. Today the Bambara tribespeople are permanent and they still use the caves. Now movement is politically and economically restricted. Malians can neither escape nor ignore their own ecological presence.

Today's circumstances are indeed unprecedented. Even its 9 million people are way beyond the apparent carrying capacity of the country. Yet by the year 2050 Mali's population is projected to be nearly 60 million, unless drastic family planning measures are culturally accepted.

Normally the West African sub-Saharan rains come from June to the end of September—about thirty-two inches per year, on average. The country's principal crops—millet, sorghum, rice, peanut, and cotton—are totally dependent on that rainfall, which has come only in meagre bursts during the past twenty years. As Peter Warshall dramatically describes it, the present crisis began as the diminishing rainfall resulted in the incremental disappearance of vegetation cover. Keep in mind that the country is Saharan in the far north but semiarid throughout its central and southern portions; ribboned by vegetated rocky plateaus and occasional small forests; and bisected by the enormous, luxuriant Niger River delta. As the rains failed to materialize, semiarid turned to arid, and herds of mammals concentrated on the remaining grasslands, their hooves harming the fragile soil, impeding further regrowth, increasing surface heat, deterring the viability of any odd seedlings, which were hence prey to winds and erosion and any final spurts of rain.

The combined total of cattle, sheep, and goats, and of people, are approximately the same throughout Africa from region to region. But in Mali specifically, not only is the cattle population as large as the human one, but also there are three times that many sheep and goats. This combined total approaching 40 million animals forage on whatever they can get. Human pastoralists, agitated and wary as a result of the absence of the rains, grazed their animals on the last remaining perennials with a concentrated frenzy, setting in motion genetic changes that engendered annuals, with short maturation times, over those plants requiring years to establish themselves. Thus a new generation of seedlings were born, lived, and died within weeks, adding little biomass to the soil. Trees began to die of thirst, while algal blooms encrusted the surface of the Earth, tending to prevent any remaining moisture from getting in. It is believed that in Mali, groundwater levels sank as much as seventy-five feet, and are continuing to sink.

As tufts of grass, islands of nutritionally critical browse, and weakened groves of trees increasingly took on the appearance of beleaguered islands in the desert, pollination became next to impossible. Three primary agents of seed dispersal were nullified: insect populations crashed, particularly amid the desiccated wetlands; birds out beyond the delta disappeared; and the herds of ruminants (whose alimentation of some shrub seeds is necessary for germination) began starving to death. This massive demolition occurred at the same time that the Niger inner delta—the largest wetland in West Africa—began to shrink, and with it every associated food and habitation resource. Domestic milk production from livestock plummeted. Many of the cattle, as well as the zebu and other wildlife, were already suffering additionally from "nagana," the equivalent, in humans, of sleeping sickness caused by trypanosomes and carried by tsetse flies. Stress from drought and hunger only exacerbates the symptoms of nagana, furthering crippling the animals.

Brushfires were started by farmers in a desperate attempt to force new biomass production. But the results were short-term, their overall impact merely increasing the loss of organic nutrients. Of the two major local herbaceous families, *Capparidaceae* and *Mimosoideae*, the Sahel supports, at best, between one hundred and four hundred "trubs" (plants halfway between trees and shrubs) per hectare. These are the plants whose green leaves provide livestock and other vegetarian ungulates their primary source of protein, while goats and sheep feed primarily on *Pterocarpus lucens* and *Acacia albida*.[68] All of these plants, indeed every Saharan/Sahelian/Sudanian biome were affected, the various ecosystems wilting and dying. And as this house of cards collapsed, the interdunal marshes, the deciduous bushlands and woodlands, the grass, trub, and tree steppes, the aquatic meadows with their all-important "bourgu" grazing grass, the shrub and wooded savanna, and the rarest dry tropical forests in the world, containing several relict species,[69] were all subjected to unprecedented levels of exploitation by the rural denizens. In addition, "drought fall-back plants" were also consumed. At the very moment that nature was most vulnerable, millions of human omnivores exerted their most lethal blow, in the form of uncurtailed predation on anything that moved or grew. Even the addax (Addax nasomaculatus), the most perfectly adapted desert antelope in the world, able to withstand almost any shock to its system, said to have a "special sense" enabling it to find the rarest grasslands, and needing no water, is today close to extinction. Similarly, populations of the most important native Sahelian tree (Balanites aegyptica), likened to the milk cow in usefulness due to the quality of its charcoal, its candy-tasting fruit, nuts, oil, browse, and insect-resistant building material, have also suffered extensively.[70] All of this is due, in large measure, to the traffic congestion on the freeways of Los Angeles, Athens, Bangkok, Mexico City, and Jakarta, to coal-powered electricity in Calcutta, Beijing, and Poland, and to the livestock industries of Australia, Argentina, and Canada, to name but a few pronounced greenhouse gas culprits. And not merely the traffic, but the millions of miles of tarmac—road surfaces in the sun contribute to methane in the atmosphere, which traps heat twenty-five times more effectively than a molecule of carbon dioxide.

Two centuries ago, prior to the industrialization of Europe and North America, the so-called Saharan–Sindian Region, including much of Mali, contained as many as one thousand animal species and sixteen hundred plant species for every 10,000 square kilometers area. Furthermore, the Sudanian Domain, which cuts across southern Mali, contained nearly nine hundred endemic plant species.[71] Over time, some 640 bird species—more than in all of North America—have been spotted in Mali, though 96 of those species have been seen fewer than six times.[72] In addition there used to be at least seventy sizable mammals in the country, but since the drought, such populations have been devastated. Yet, there is not a single protected Saharan area in Mali, and there is only one in the entire Sahara—Tenere, in Niger. Nor has there been any inventory of rare or threatened species of plants in Mali.

Few mammals remain in the delta region—the odd hippo and manatee aside—though over two hundred species of birds and fish still inhabit, or migrate to, the delta's rich wetlands (covering a region of 30,000 km² at its peak). This region is also inhabited by over five hundred thousand people.

Three of the four small herds of West African elephant (*Loxodonta africana*) have gone extinct. Only the Gourma group is left, numbering, as of 1989, as few as two hundred (not ten thousand as stated by Mali's Ministry of Environment). The only elephants in all of the Sahel, they move more than any other elephants because of their constant search for water. It is estimated that they require 33,000 square kilometers of habitat. Like Mali's chimpanzee troop (*Pan Troglodytes versa*, the northernmost chimps in the world, and possibly the only ones who have moved out onto the arid savanna, thus making them extremely important to students of human evolution), the Gourma elephants appear headed for extinction unless emergency measures can be taken. Neither the elephants nor the chimps enjoy any protection whatsoever. Anyone can hunt them. Many of the primates are known to be live-trapped for export, possibly through Guinea to Spain and elsewhere. Proposed parks to safeguard the chimps and pachyderms have been thus far ignored by the government.

In the meantime, at least thirty thousand rare tropical birds are shipped out each year. In Bamako's Dibida and Marche de Medin markets, among others, I photographed animal skins and body parts for sale, along with countless animal-derived medicines.

These by-products of superstition and odious fetishes in the form of hyena skins, python meat, chimpanzee hands, horns of rare ungulates, thousands of bird skulls, pangolin paws, and a proliferation of other distorted animal fragments fill the market stalls. In one negotiation with young poachers across the street from the l'Amitie Hotel (the largest high-rise in Bamako), I acquired forty-eight exotic birds that had been condemned to a waterless, corroding cage, placed on the edge of the diesel-strewn motorway in direct sunlight. When I purchased the birds, which included African greys and Senegalese long-tailed parakeets, for $135 cash, the temperature—though it was only 8 in the morning—

was already over 100 degrees. With the help of Robert Radin, an American photographer, and John Anderson, the chief supervisor of the Montes Mandingues *Foret Classee*, a forester originally from New York, I drove out of town and released the birds along a tranquil, protected stream bank in the Mandingues. All of them survived the bumpy two-hour drive.

While the country, many years ago, passed a Code de Chasse curtailing hunting, it was altered in 1986 so as to prohibit the killing of animals during breeding periods, six months of the year. But, newly democratic Mali has eschewed most enforcement of any kind, at any time. Traditional Islamic culture does not support sanctuaries, but rather encourages hunting preserves. Even the wart hogs—usually exempt from slaughter for the Muslim diet—are fair game in Mali. Long before the onslaught of this most recent drought, a U.N. Food and Agriculture report stated that 65 percent of all protein in rural Mali came from wild game, and that in some areas 90 percent of all men were hunters.[73] Today subsistence hunting is endemic, as are poaching and illegal trade. According to Warshall, "government agents, diplomats, the army and DNEF (Water and Forestry Department) have all been implicated in illegal game and trophy hunting."[74]

It is worth quoting extensively from Warshall's introduction for it highlights the devastation, as of 1989:

"Without significant financial aid by donors and political change by the Government of the Republic of Mali, there will be no significant large mammal populations left in Mali within five to ten years. The Saharan and northern Sahel regions have no protection. Addax, Dorcas gazelle and aoudad (barbary sheep) are threatened or endangered. The oryx is extinct. The slender-horned gazelle is probably extinct. In the mid/south Sahel and Sudanian woodlands, protection has been very weak. Most elephant herds, lion, Derby's eland, giraffe and ostrich populations are locally extinct and declining. Dorcas and Damas gazelles are rare. The cheetah and hunting dog have disappeared. All wetland species have declined from intense use of the floodplains, drought, hunting, fishing, and (the) Manantali dam. These include the African buffalo, manatee, waterbuck, Buffon's kob, hippo, various fish species, and crocodiles."[75]

Warshall's "bleak picture" of 1989 has only gotten worse in the last five years.

As of 1989 at least 80 percent of the population, or 7.2 million people, depended for their livelihood on wildlife in one form or another. The number of humans dependent on the country's natural inheritance has surely grown since then. In 1989 there were said to be two giraffes remaining in the country, a pair at the Asango–Menaka faunal reserve. Two years before, a third one had been killed by a sheikh who believed the fat from the giraffe would give him longevity. (In fact, the sheikh— who had purchased the permit from a local government official—died the following year.) Two round-the-clock wardens were supposed to be guarding those last two giraffes, but even they have since been poached. Another wealthy local, implicated in the capture of chimpanzees, was "single-handedly responsible for the extinction of the Bougouni herd of elephants—the next to the last in Mali"—while yet another, an Arab "prince"—shot forty-five Dama gazelles, which are a seriously endangered species. And all of this was done with official permits. The country's one national park, the Boucle de Baoule complex, has also been overrun by poachers, pastoralists, farmers, locals gathering wood, foreigners seeking big game kills. Those with influence and money can override any of the existing laws. Those who are impoverished find that they *must* do so. Unlike some countries in East Africa, Mali does not pursue poachers.[76]

At the same time, the flood-pulse of the Niger has been severely compromised by three large hydroprojects (the Manantali and Selingue dams and the Canal du Sahel). In the case of the Manantali, the whole project has been nothing short of an ill-conceived disaster, with no funding for restoration of the multiply-damaged ecosystems in its wake. The lack of integrated pest management has resulted in a build-up of toxic pesticides, fungicides, and herbicides in the water. "Risk-avoidance" and conflict resolution over biological resource issues have been low priorities for the GRM (Government of the Republic of Mali). The tragedy of Mali's poverty is not merely its inability to protect its children, but also to do anything on behalf of its wildlife. The choices rest solely with the international donors who are economically forced to select the most urgent candidates for remediation.

The Niokolo Koba National Park of Senegal, and the Tenere protection area of Niger have seen far more interest from the international community than has the Boucle de Baoule, despite the park being made into a "Man and the Biosphere Reserve" by UNESCO in 1982. The reason for its abandonment is that in the years following its designation as Mali's first and only national park, it has been invaded by heavily armed Mauritanians in Land Cruisers who—having killed all the wildlife in their own country—come into Mali to kill whatever animals they can find. "You don't see anything in the park," says John Anderson.[77] In addition to the Mauritanians, local Malians have deforested, planted, and assaulted it with equal vigor. Elsewhere in West Africa—in Benin, Burkina Faso, Nigeria, Chad, and Cameroon—other would-be protected regions compete for scarce funding from outside. Donors who must choose, exercising a financially constrained policy of ecological triage, cannot help but ask themselves whether, in comparison with other needy sites, the Boucle de Baoule may not be beyond saving.

As with the Boucle de Baoule, the laws protecting certain plant and tree species are in place, but Malian management and policing has proven nearly impossible. In the reserves, there is one patrolman, on average, per 85,000 hectares, and he is unarmed. As of 1981, the World Bank calculated that there were 12.9 million hectares of natural forest remaining in the country. Of this amount, some 4.5 percent was protected. But the 12.9 figure was a mere estimate, lacking in precision. During the 1980s, there was zero reforestation going on throughout the country, while at least 36,000 hectares were being deforested, more than 0.5 percent of existing forests and woodlands. As of 1985, 49 percent of Mali was considered untouched, a wilderness area (mostly desert); but such figures are completely misleading. According to foresters in the country, and contrary to World Bank figures, in the early 1980s there were not 19.2 million hectares, but 7.2 million.

In 1994 there is no sense of what is left. There is not even consensus on how many actual forest preserves are in the country. For exam-

ple, when I met with the pleasant Minister of Environment, Moham-
ed Aq Erlaf, he told me there were 250 preserves. John Anderson, who
manages the three largest of them, says there are only thirty-one. Others
who work for Anderson claim there are about one hundred. The discrepancies are all on paper. In truth, only three small forest reserves
in the country have any solid protection or management, and they are
Anderson's. Well-managed oases of hope, they are sterling examples
of what NGOs, working with the government and villagers, can do
to effect balance. But these few hundred thousand hectares (out of a
total Malian land area of 45.9 million hectares) cannot undo the
avalanche of irreparable damage throughout most of the country. And
even these reserves, according to Anderson, are not working out perfectly. The system of village self-monitoring that has been put in place has
its shortcomings in the guise of uncontrolled fires, increasing short-
term agricultural conversion of woodlands, and occasional poaching.
But the biggest problem is that 90 percent of all energy comes from
wood sources.

"If you unclassify *(degazette)* the forest," says Anderson, "you will
not find a toothpick in two years."

The land is unproductive, low in phosphorous and lime, prone to
bushfires, erosion, and leaching, and low in fertility.[78] Impelled by
poverty and hunger, farmers have lessened their fields' fallow periods
from twelve years, to four years. And the population is going to double
by the year 2000.

What is known is that wood consumption throughout Mali as of
1989 was already occurring at twice the rate of annual production, in
spite of impressive strides within the country to introduce more fuel-
efficient woodburning stoves and fast-growing species. Adding to the
toll on the forests is the widespread depletion of presumed deadwood,
which is plucked up daily by villagers, whether in the *foret classee* or
outside. That deadwood is crucial to holding the soil together and maintaining insect populations. Furthermore, in 1988 brief rains revitalized
much presumed deadwood, particularly among the acacias, which
started to show leaves for the first time in a decade, highlighting the
tragedy of its premature consumption.

But deadwood gathering, and deforestation are not likely to be great-ly slowed down, even given the few well-managed fast-growing tree plan-tations. In fact, marginalization of lands and the persistent depletion of forests has been the human story in the Sahel for thousands of years.

But the firewood crisis especially dogs the country, and so, for over a decade John Anderson and his colleagues have been developing in-dustrial plantations, as well as a diversity approach: trees that can be harvested for food, or used as windbreaks, living fences, firewood, or secondary products. The trees are low density white wood that grow eight to ten times faster than indigenous species. Such trees unfortunate-ly, do not encourage much biodiversity, but they do satisfy human needs for fuel and shelter. The Indian broadleaf *Gmelina arborea* has been in-troduced as a firewood species. With its thick growth, it closes the canopy, cooling the soil, and inviting at least some wildlife like the bushbuck to take refuge among the plantations. Eucalyptus, cashew, cassias, acacia, mesquite, and kapok have also been planted. The local Bambara tribes that used to inhabit the forests but were moved out by the French some sixty years ago, are now allowed to come in and utilize the by-products in the form of wood, honey, minimal agriculture along the fireroads, and grazing and secondary browsing of limited livestock. Of the one hundred twenty tree species in the Montes Mandingues— one of three forests Anderson manages—the locals know of over one hundred varieties.

Tragically, that knowledge, that sensitivity, has not ensured sus-tainability or prevented the ongoing ecological holocaust. Three small *foret classee* are not enough to reverse a national paroxysm. I spent two days with the Bambara people of the village of Farana inside the Montes Mandingues. We were all seated under a shade-giving mango, discuss-ing the old days. The chief, Mr. Sadiki Troare, his scores of children (someone said forty-six, though I was unable to verify this), several wives, and compatriots, all sat with us. A little goat cried for her mother, chickens wandered under our feet, pecking at scraps of bread I tossed. I felt slightly guilty feeding the birds, inasmuch as the children in the village looked rather malnourished, many bellies distended, some eyes evidencing sores.

The community evidently contained three hundred children under the age of fourteen and three hundred adults who pay taxes. They were out of agricultural land. They hoped to sell wood from the industrial plantation to turn their lifestyle into something of a cash economy. I asked the chief if he remembered a time when there was wildlife, for other than several bird and butterfly species, the occasional bushbuck, and bats, the forests appeared bereft of life.

"When I was young, lions ate our cattle in the forests. Now, no more lions," he reminisced. "After World War II the military would come in from Bamako and kill everything—gazelles, lions, antelopes, bushbuck, wart hogs. Poaching has killed them all. The law forbids hunting in Mali, but everyone has a gun and everyone hunts."

"Do you tell your children how it used to be?" I asked him.

"No," he replied.

The litany of woes in this much beleaguered land does not equate with the dream of reason supposedly characterizing human existence. Add up all the afflictions: poaching, deforestation, the impact of drought, poverty, malnutrition, poor laterite soils, temperatures exceeding 110, even 120 degrees, throughout much of the year, many of the world's most virulent viruses and bacterial infections, toxic chemicals coming from Europe, Guinea, Senegal, and Nigeria for tie-dying textiles, the improper disposal of alkaline batteries, soap, and pharmaceutical wastes, brewery effluents, rotting corpses from countless slaughterhouses (virtually on every street corner), all of the outdoor latrines and huge open garbage dumps through which countless individuals forage, the dumping or leaching of those toxins and refuse into the river Niger as well as city and village wells, the rapid-fire extinction of species, and the victimization of other innocent humans— add up all this turmoil in broad, dusty daylight, and you will be struck by the contradictions. For there, amid gaiety, and Malian beauty, intoxicating music and Antwerp blue skies, are bountiful vegetable markets, the most colorfully-clad women in any country, and the swaying multitudes of jeans-clad teenagers who dance the delightful *dongee* by moonlight on weekends out on the edges of every town. In these respects, I found Mali culture captivating.

But the fact remains, the country is a human enclave of desperate proportions.

"If no measures are taken," the Minister of Environment told me in a calm voice, "it will be a catastrophe."

## HUMAN ANTIDOTES?

Questions do not necessarily lead to answers. In February 1991, the World Bank estimated that Mali's population was 8,461,000; its TFR 7.0; its crude birthrate 50.0; the life expectancy for men 45.6, for women 48.9; its IMR 169; its annual growth rate 2.7 percent; and its CPR for married women 15 to 49, 5 percent, all disastrous statistics if one is concerned about lowering the birthrate.[79] The report also projected a population in 2020 of 20,785,000. Exactly twenty-four months later, the U.N. published revised data for Mali, and in every category—save CPR—the figures were even worse, boding a higher population still. For example, the population was up to 9,214,000; its TFR was seen to be 7.1; the crude birthrate 50.7; the IMR 159; life expectancy for men 44.4, for women 47.6; and the annual population growth not 2.7, but 3.17 percent. Moreover, the population projections for the year 2020 were 23 million, a demographic discrepancy in just two years—between the two leading statistical organizations in the world—of 2,215,000 persons, or, nearly 100,000 people per month somehow added to the otherwise credible projections, and for a country whose total population is less than that of Los Angeles. What's going on?

Unlike Los Angeles, with its huge immigration gains, Mali shows a cumulative exodus of about 28,000 persons per year leaving the country, or 2300 per month, further calling into question the stated net gain of more than 2 million people from one projection to the other. This all no doubt seems overly fussy, and slightly unforgiving of what is, admittedly, an imprecise social science. But it points to the mathematical inconsistencies inherent to most demographic data and methodologies, thus hampering population policies within any given country. What is additionally striking about the numbers game in Mali is that, on the surface, the country should be doing well. After all, population density is a mere 7 per square kilometer (versus 41 in Kenya), and, with 7 million rural Malians, there are approximately 3.4 hectares of arable land per person, 17 times that of China. With the Niger delta, there should be—theoretically—sufficient volumes of irrigation water to counteract drought without compromising the riverine ecosystem. The country manages to export cotton, ground nut oil, and oilseed cake at a profit.

The problem is that each year the country must absorb another 317,000 Malian newborns for whom there are insufficient means for existence. The official IMR is now 102, way down from the earlier figure (though Isaiah Ebo, the cheerful, seasoned head of UNFPA for Mali, told me that UNFPA still believes it to be 159).[80]

Both standards of nutrition and literacy are on the decline in Mali, where 20 percent of the population is under the age of five, and more than 20 percent of those are affected by stunting and wasting. In 1987 less than 10 percent of all children under the age of one year had been immunized against any of the major child killers, such as measles or diptheria or tetanus.

Only about one in four Malians has access to a health facility and only 56 of the 769 fixed facilities provide any family planning. Ninety-nine out of one hundred married women in Mali have never used a modern contraceptive, though 42 percent are aware of such technologies to some degree. However, such data does not take into account unmarried sexual encounters, where many teen pregnancies occur.[81] Yet, following a visit by Ebo to Gao not long before he and I met, he told me how he'd been surprised by "the number of women wanting contraception. They say, 'I have seven children. I can't walk. That man—my husband—he wants even more. Give me something (contraception). I want to block him!'" It is not the women but their husbands who want still more children, though the men will manage to persuade, commandeer, or force women to accept the "logic" of large-family security. In Mali, a woman spends hundreds of dollars—many years' savings—on her wedding and hair, but next to nothing on contraceptives.

In a survey of the villages Farabana, Faraba, and Mamariboutou, partially funded by the U.N. Food and Agriculture Organization in 1991, a traditional plant classification system was discovered. Malian healers have detailed knowledge of sixty-seven pharmaceutical species, but none would share contraceptive recipes for fear of competition (much like the great chefs of Tokyo).[82] This is not a climate conducive to birth-control policies. Whatever family planning's goals may once have been, in reality, they have only served to further the status quo, trying somehow to piggyback on the medical technologies that have increased average life expectancy from a mere thirty-five in 1960 to forty-six in 1994.[83] That has meant, of course, a surge in population.

But the fact remains that 50 percent of all women who die between the ages of fifteen and forty-nine, die during or just after childbirth. Abortion is illegal. Teenage pregnancy is at epidemic proportions. Only 11 percent of all adult women, and 27 percent of the men, are literate, yet school enrollment of any kind in 1994 remains sadly stagnant: 19 percent for girls, 38 percent for boys. Girls are routinely, forceably circumcised so as to prevent their enjoyment of sex, prior to being married off—often into polygynous unions—by the time they are fifteen, if not younger. Despite clear proclamations by the Companions of the Prophet, the *Hadith*, and by Mohammed himself, the interpretation of Islam in Mali and surrounding countries is such that women must obtain their husbands' consent to use contraception, and are discriminated against by inheritance laws.[84] Estimates for mean per capita income vary from $210 to $270, placing it among the nine poorest countries in Africa, along with Burkina Faso, Tanzania, Mozambique, Malawi, Guinea-Bissau, Zaire, Chad, and Ethiopia.

Incredibly, a French colonial law prohibiting family planning is still in effect in the country. However, the Government of Mali insists that it intends to increase the CPR from a current 1.2 percent to an astonishingly hopeful 60 percent in the year 2020, but UNFPA finds such optimism less than plausible. In fact, UNFPA in Mali acknowledges six different sets of conflicting demographic data. CPR, at present, according to Isaiah Ebo, is probably not 5.0 percent, as both the U.N. in New York and the World Bank in Washington have estimated, but closer to 4.2 percent. Are such minutia and discrepancies that important? Yes. They determine the tenor, zeal, and emphasis of policies. It was UNFPA that first established the country's only population censuses of 1976 and 1987 and helped the government formulate its first population agenda, but which now finds itself begging the world community to obtain a mere $5.2 million just to keep going. As in the case of donor decisions pertaining to the refurbishing of the Boucle de Baoule National Park, so too, human health and human ecology are the victims of international triage in a country like Mali.

There is only so much donor money to go around. Many find it easier to simply "write off" sub-Saharan Africa. Indeed, Mali's own economics do not easily recommend heavy cash infusions from outside, though there is some Chinese and Middle Eastern investment capital in the country, and much could change in the coming years.[85] Without such monies—donor, or investment-guided—nothing in the ecological or family planning arenas is likely to improve in Mali. This is a crucial fact, at an unfortunate time when more and more triage-oriented thinking and investing is focused on instant gratification of one form or another. Even traditional "planned giving" now views charity as the perpetuation of dependency; a perspective which holds, in addition, that dependents—trusting to the outside world for their *allowance*—will feel comfortable maintaining a high fertility rate.[86]

But for those donors who are actually targeting their money, their labors, and their professional idealism, in countries like Mali there is no ambivalence, no confusion: these are not donations, but lifesaving transfusions. As one foreign official told me, "We're doing this for individuals, not for principles."

Along with UNFPA, the other major multilateral donors to the country are UNDP, UNICEF, and the World Bank. In addition to the World Health Organization, the Canadian International Development Agency, the government of Germany, and U.S. AID, there are several NGOs in Mali assisting with forty-five different projects.[87] These include integrating women into development programs, sensitizing the populations of hundreds of villages to health and home economics, strengthening cooperatives, analyzing data from the 1987 population census, supporting self-help organizations for rural women, strengthening MCH/FP clinics and improving service delivery, the introduction of postpartum IUDs, counseling to prevent unplanned pregnancies, and the social marketing of contraceptives.

One of the most exciting of these initiatives—a true reason for hope in West Africa (much in keeping with the cries for help Isaiah Ebo heard in the city of Gao)—is the Katibougou Family Health Project, funded by the Centre for Development and Population Activities. Since 1986 this program has provided community-based distribution of contraceptives in rural areas of Mali. The Contraceptive Prevalence Rate in those regions covered (nineteen villages, fourteen thousand people) went from 0 to 57 percent! This suggests a remarkable unmet need that one presumes could easily span the whole sub-Sahara, if there were sensitive financial targeting.

UNFPA maintains the largest platter of population projects in the country (twelve as of 1993). Its overall goal is to bring down the TFR to about 3.5 by the year 2020, which would mean a million less people in the country at that time. But since most government ministers themselves have three or four wives, and five children by each, UNFPA—hoping the government will set an example—is going to have its work cut out for it.[88]

Economically the country needs urgently to expand its production of paddy rice and sugar cane, of textile fibers from seed cotton, of oil crops and sesame, tobacco, and tea. In addition, butter extraction, flower milling, spinning and weaving, hosiery, and clothing, cashew nut shelling, fruit juices and concentrates, and a variety of cottage industries present themselves. These aspirations will be tested by the current fertility pressure. Because 89 percent of all farming in Africa is by hand (there being no money for gas-powered machinery, and no spare parts) human health is essential for productivity.[89] Mali's defense budget uses up 11 percent of the country's income each year, whereas the total health package, including housing, social security and welfare, accounts for only half that amount. The reason is that Mali's adversarial, impoverished, hungry neighbor, Mauritania, is pumping 29.4 percent of its budget into defense. The same demographic explosion in Mali has ripped through all of sub-Saharan Africa, a vast region that can least afford it but that is caught by the global arms-profiteering mentality, an expensive trigger-finger treadmill that steals from other, crucial sectors like family planning and ecological remediation.

Whether in East or West Africa, antidotes to the population crisis have been set forth by many experts. There is little mystery about the methods, only frustration enshrouding their execution.

In Kenya, for example, the thrust of UNFPA's program hinges upon "achieving future increases in the contraceptive prevalence rate," which it says "will depend upon creating specific messages to fill identified information gaps, in addition to improving upon the quality and coverage of family planning services."[90] In achieving these messages, UNFPA would conduct research into the "wide gap between knowledge and use of family planning," the awareness threshold that is proving so difficult to define, worldwide. In Kenya, 96 percent of all women (married and unmarried, from age 15 to 49) have evidently "heard of" at least some method of contraception, be it traditional or modern. And while 89.8 percent of all women know where to get pills, only 9.6 percent of them actually do so. Awareness is slowly closing the gap, at the rate of about 2.25 percent per year of increased usage, but UNFPA wants to speed it up. Among men, the awareness gap concerning condoms is smaller (99 percent awareness, 54.8 percent current practice), yet it is unknown to what extent these same men were actually using condoms with women other than their wives, to avoid contacting an STD rather than to prevent pregnancies. Furthermore, only 38 percent of all married men indicated they wanted no more children, compared with 46 percent of the women, further suggesting a large discrepancy in the awareness factor among men.[91] Closing the discrepancies between awareness and action presents the greatest challenge facing family planners.

In the Kariobangi slum of Nairobi, I walked with Alice Githaes, a nurse who has started her own family planning clinic for the poor. She had a vision, borrowed 27,000 Kenyan shillings (about $1000 at the time) from her former cooperative, leased a small office, installed four beds, worked nights, and raised funds for drugs. Seventy to eighty patients can now be accommodated. Equipped with a surgical theater, her clinic has literally provided a medical mecca in Kariobangi. Over twenty thousand patients visited her in 1992. She offers free steriliza-tions (though in 1992 only one male accepted, while one hundred thirty-one females did so, typical of all of Africa). Birth control pills cost thirty cents each, a Norplant insertion, roughly $4.25, if the pa-tients can afford it. If not, she provides the service anyway. When she came to this slum in 1969, everyone was having eight to ten children. You still find up to seven. But not ten. Outside her door hangs a con-dom dispenser. They are free for the taking. Her annual budget is $30,000. Alice has a big heart, and a generous hug for everyone. She is precisely the sort of person to break the crisis of confidence throughout Africa, that pessimism, which so easily breeds self-fulfilling defeat.[92]

Family planners all know many such "Alices," women who are, in fact, closing that awareness/action gap. They lighten an otherwise over-whelming horizon, in Africa and elsewhere, and remind family plan-ners and the whole community of demographers and health workers what it is their profession is all about.

Indeed, to repeat, family planners know what needs to be done. They have urged host governments to maintain vigilance regarding the type and quality of contraceptives; to increase sensitive counseling, examinations, follow-ups, and the implementation of national programs for comunity-based distribution of the contraceptives; they have em-phasized working with NGOs and establishing clinics like Alice's that are solely devoted to fertility control; and they have experimented with countless methods of popularizing family planning. But all of these essential components evaporate, ultimately, if women themselves— who consistently bear the brunt of responsiblity for population con-trol—are not free: free to be educated beyond primary school, free to work at the job they choose, free to find happiness, dignity, and love. That freedom is not solely dependent upon money, of course. But money helps. It surely does. Alice could not, and would not, have started her clinic in Nairobi's slums, I suspect, without her initial $1000 loan. And, in Zimbabwe, the government could not have increased the distribu-tion of condoms from half-a-million in 1986 to 65 million in 1992—

which in turn tripled CPR from 14 to 43 percent—without foreign assistance.

And yet, by itself—as I have been arguing throughout this book—family planning is not enough. The rise of contraceptive prevalence in Zimbabwe has not deterred poaching and culling of wildlife in that country and there is no reason to believe that it would have done so in Kenya without Richard Leakey's inspired efforts at the helm of the Kenya Wildlife Service. Not even the CITES ban on ivory trade can control the Asian obsession with amulets and aphrodisiacs.

In southwestern Madagascar, it was a group of biologists concerned with the fate of three rare primate species—the Golden Bamboo, the Gentle, and the Greater Bamboo lemur—which resulted in Ramoma-fana, one of the very few "sustainable parks" in all of Africa. But again, money was needed from the outside world. Inaugurated in 1991, Madagascar's fourth national park now supports a village-based economy ensuring dignified employment and nonimpactful labor—beekeeping, local gardens, rice paddies, the cultivation of endemic trees and fruit-bearing trees rather than eucalyptus and pine, the offering of park-related careers to the locals, the building of kilns and making of locally stamped ceramics, and the founding of women's cooperatives. In addition, the park consortium is offering training to a new genera-tion of local biodiversity systematists. And finally, strong health and family planning elements have been woven into the overall delivery system.[93]

In Mali there is enormous potential for such parks. For example, the last remaining northern chimpanzees and elephants of that country—if they can be rescued soon enough—would be of enormous interest to tourists.[94] In addition, the birds and butterflies of Mali—even right outside Bamako at the Montes Mandingues *Foret Classee*—have much to offer visitors, as do the shores of Lake Selingue, the islets of Gao, and the region of Lake Faguibine.[95]

Peter Warshall recommends four priority projects for Mali: the setting up of a Gourma National Park (for the elephants); a Bafing National Park (that would provide some sanctuary for the chimpan-zees and Derby's eland); the redevelopment of Boucle de Baoule

National Park; and the breeding of water bird, hippo, and manatee sites, as well as fisheries in the inner delta.[96]

Total foreign assistance to Mali is in the $200 to $350 million range. Between 6 and 7 percent of this comes from the U.S. In the case of U.S. AID, only about 5.3 percent has been devoted to natural resource management, according to Warshall, and virtually none to biodiversity remediation or ecological mitigation. In truth, Africa, Africans, and most of Africa's wildlife, have been forgotten by the rest of the world, regardless of the jubilant tour packages and glossy nature documentaries on television.

Unless biodiversity concerns become the focus of all long-term economic strategies (whether the funding comes from other missions or multilateral banks such as the African Development Bank, the IMF or the World Bank), Mali's future is grim. Again, money and vision are the keys. Local conflicts over natural resources require enlightened resolution that will ensure social justice for both the transhuman grazers and the permanently based communities with whom they are increasingly at odds. Finally, the GRM (Government of the Republic of Mali) continues to spurn the CITES III international treaty, allowing influential locals to hunt elephants and all other species. Even Eaux et Forets agents are involved in the killing. "There is need to make a decision quickly," says Warshall, or "there will be no large mammal reserves in Mali."[97]

Which brings me back to endangered Amboseli, beneath eternal Mt. Kilimanjaro, and that ancient sunrise. The record of human biology, both in its origins and current plight, can be interpreted in support of an evolution favoring violence, or nonviolence. However accurate the former notion—steeped in predation, meat eating, and self-aggrandizement—it has ecologically speaking, become an untenable anachronism, to be replaced by a new species of individuals driven not by destruction but by creation myths; souls that must, in turn, embrace a new credo of diversity and compassion.

Accepting the premise that human evolution is now about sustainable ideas, not physical force or digital thumbs, implies that we can alter our mindsets, choosing love over hate, sharing over greed. Evolution, I have stated repeatedly, neither condemns nor liberates us: only our choices can do that. All people harbor an ideal of tranquillity toward which we have been moving—in fits and starts—for countless millennia. That ideal today marks the fragile birth of a new human nature, of a noble creature of aspiring grace, who is fraught with ecological anxiety but little time left for conceptual, let alone practical, reconciliation. But we can, and we must, effect this grand reconciling gesture. The clear high snows of Kilimanjaro, and the blessed animal plains below, challenge the human heart to get it right.

# The Price of Development

*"Know ye that on the right hand of the Indies there is an island called California, very near the Terrestrial Paradise . . . "*
*Garci Rodriguez Ordonez de Montalvo, 1510[1]*

## THE GLOBAL CRASH

Since the early Neolithic period ten to fifteen thousand years ago, all domesticated societies have meticulously organized their powers of aggression—be it with spears, hatchets, arrows, or ammunition, or with fishnets—in order to exert greater power over nature, and each other. In humans, the territorial imperative has become a quest for dominion. Why is that? Some social scientists have claimed that such aggression is a "drive," subject to premeditation and choice, as opposed to an instinct, which more closely obeys the implacable, but, by comparison restraining, edicts of biology and genetic inheritance. "Drives," writes New Zealand anthropologist Peter J. Wilson, "unlike instincts, can be sublimated, rechanneled, repressed, and metamorphosed . . . drives can be directed toward a larger and more complex variety of final aims . . . "[2] That insight implies, of course, that humanity can work together for the common good, or conversely, chose to engineer more destructive weapons and systems of environmental and human exploitation. Even the "common good" fails to account for the larger good, namely the natural world we have insistently tended to want to subsume within our own order—biblically, economically, egotistically.

Most of the earliest civilizations seem to have piggybacked the energies of "drive" atop "instinct" by shrewdly calculating the potential for economic gain, and devising the means, the weaponry, the military posture, or some other attack plan, for obtaining it. At the same time, driven by political megalomania, or a culture's very perception of its role in the cosmos, one Tower of Babel—and the surrounding city that gave birth to it—after another rose up across the land to meet the heavens and secure for human hegemony the promise of an imagined divine ordination. Great walls, fortresses, castles, shrines. In Mesopotamia, Mexico, Java, Egypt, and Angkor, countless generations of indentured slaves worked hundreds of millions of cubic feet of brick into colossal gestures of self-edification, the precursors of today's mammoth urban landscapes. In terms of the history of economics, form—the look and purpose of the megacity—clearly follows the functions that support arrogation and control. From the ecological perspective, modern architecture has served humanity's greed and hubris far more than its artistic dreams.

What began as temporary entrepots, international trading centers, tents erected for their season, soon became permanent places of business, as the surpluses of agriculture, sailing expeditions, a metallurgical revolution, and a growing slave trade invited the densification of wealth, central to the fact of human crowding. Finally, as population densities increased the unpredictability of human behavior, the politics of control and manipulation evolved as a means of protecting that wealth. Not surprisingly, more and more evidence indicates that the downfall of early civilizations and cities may have had less to do with political unrest than the environmental fallout of overpopulation. Not far from the ancient Mesopotamian capital of Akkad, the city of Shekhna—with ten thousand residents—was abandoned. A persistent drought, brought on by natural changes in the curtain of rainfall following a volcanic eruption, would not support so dense a population. Refugees fled to the south, straining local resources to the point that the entire Mesopotamian empire eventually collapsed.

There are, of course, any number of speculative scenarios to illustrate the origins of financial exploitation, of urban crowding, and the evolution of warfare, and the rise and fall of past civilizations. From the essentially romantic depicture by Jean Jacques Rousseau of the noble savage being undermined by the brute, unthinking collective, to the more familiar critiques of Edward Gibbon, Karl Marx, or Oswald

Spengler, whose arguments hinged largely upon the exploitation of a labor class by an economically advantaged elite, the results are very much the same: a picture of human avarice, short-sightedness, and ultimate misery. But few historians—George Marsh (*Man and Nature*), Arnold Toynbee (*Mankind and Mother Earth*), Clarence Glacken (*Traces on the Rhodian Shore*), and Roderick Nash (*Wilderness and the American Mind*) being among the compelling exceptions—have focused on the misery unleashed by humans upon nature in the wake of such economic imperialism. And this caesura of scholarship is largely due to a pervasive developmental paradigm that, at least until recent decades, has literally inhibited our ability to see clearly, to acknowledge a world larger than ourselves.

Not only among historians, but in certain cultures, past and present, there are qualified exceptions to the theory of economic myopia. Jain, Bhutanese, Bishnoi, Tasaday, Yahi, and a host of other indigenous peoples and religions have each promoted nonviolence and sustainability. (See chapter 8.) Early Japanese civilization, beginning with the Fujiwara Dynasty in the seventh century A.D., was among those that sought a subdued balance between ecological manipulation (intense terracing, rice farming, construction, and appropriation) and the art of subtlety. This approach to survival could not have differed more radically from the coeval cultures of Europe, the warring Anglo, Gallic, and Teutonic tribes that had usurped the Roman niche. Japanese society made a poetic point of integrating a thorough appreciation of nature—of diverse plants, animals, and tranquil views—with the realities of communal and household life. The result, at least until the middle of the nineteenth century, was a remarkably restrained and homogeneous group of people, in spite of the hideous excesses of certain shoguns and their samurai. No other nation in history has insisted with such vigor that paradise is here and now; that the evident joys and beauties of the creation do not merely intimate, or reflect, but are the very manifestations of the divine.[3] The Japanese pursued spiritual Being, and this was to temper their conquest of Japanese Nature.

Not only did the Japanese excel in the aesthetic cultivation and accommodation of the natural world, but, intrinsically, Japan's populace also understood the perils of overpopulation long before it ever surfaced as an issue elsewhere in the world. During the Edo Period, for example (1721–1846), Japan maintained a steady-state population of 30 million. By contrast, China's population doubled during that same period. Edo (Tokyo) numbered about one million people, and was the largest city in the world, but the Japanese leadership was persistent in its efforts to counteract such crowding. For 125 years, the country deliberately worked at stabilizing its numbers. This effort resulted from a determined Tokugawa shogunate (that also banned all armaments from Japan) and a peasant class that evidently recognized the wisdom in inhibiting large family size. Writes William LaFleur, "It becomes clear that many people in Edo, Japan were limiting their children in order to enhance the quality of their lives."[4] Part of the strategy of limitation—indeed, two key elements, according to LaFleur—which saved the country from outstripping its scant resources, were infanticide and abortion, or *mabiki*—literally the culling of seedlings. While some Shinto scholars of the nineteenth century argued that the more people, the wealthier the country would be (in part because of more "nightsoil," and hence richer harvests[5]), most Japanese, and most Buddhists, saw population control as a means toward achieving nirvana. (Interestingly, Orthodox Christians of the first few centuries also considered the world too full of people. The Messiah, said they, would come; he did not need more souls to greet him.) This is not to say that abortion was treated casually among the Japanese. Even today, the so-called *mizuko* rituals honor the dead fetus publicly. Families gather together at special temples and graveyards where the fetuses are buried. They recite prayers and engrave the recitations on stones. *Jizo,* or protector spirits, are invoked to watch over the aborted fetuses during their journey into the realm of departed souls.

Today, there are approximately 126 million Japanese, roughly 21 million of whom are boys and girls under the age of fifteen. While there was a baby boom after World War II, there was also a dramatic increase in *mabiki*. In 1948, the "health of the mother" was advanced as the legal justification for abortion.[6] With a current total fertility rate (TFR) of 1.55, Japan is projected to begin experiencing a declining population growth after the year 2015. Indeed, toward the end of the twenty-first century, the country is expected to number 12 million fewer than at present. Only a handful of other nations in the world will be able to claim such fertility restraint.[7] Furthermore, there is the lowest infant mortality rate anywhere, at 5 per 1000. A Japanese baby is thirty times more likely to live than his counterpart in the West African country of Guinea-Bissau. If the least harm and greatest good to the largest number of people were one criterion for assessing the worth of an ethical system, Japanese abortions would have to be considered a form of long-term empathy—much like long-term thinking and long-term economics—the eschewal of short-term gains and gratifications (i.e., the pleasures of a baby) for a higher, more lasting purpose (the pleasures of an entire society). Until this century at any rate, Japan's artistic soul had extended the logic of this expediency to the natural world. One need only glance through the country's vast pictorial and poetic archives to appreciate the depth of Japanese nature religion, an astonishing harmony that once infiltrated most details of everyday life. (This generalization is still born out, in microcosm, by a casual stroll, around Kyoto's peripheral greenbelt.) Corresponding to this aesthetic relish was the concerted, even penitent, recognition that overpopulation is harmful to the quality of the experienced world.

Until the "opening" of Japan by the West in the 1850s, the country in many respects was swayed by such potent affiliations. But much has changed now. The age of reverie has been mutilated. Tokyo, still the largest city in the world, boasts nearly 30 million residents. It virtually connects with megacity Osaka, three hundred miles away, their respective city limits separated by an inorganic expanse that cannot possibly suggest anything like the paradise in which earlier Japanese so sincerely believed.

Japan exhibits all the customary ills of human concentrations: near total human appropriation, heavy metal wastes from mining operations leaching into cultivated fields, toxic bioaccumulation in rivers, an enormous increase in oil spills and oil imports, a doubling of nuclear power generators, a global pattern of energy scavenging—be it plutonium from France or tropical hardwoods from Malaysia—and the decimation of wildlife. Tokyo Bay, not surprisingly, is gravely ill.

In 1990, a seasonal armada comprising 10 percent of Japan's squid fishermen combed the choppy waters of the North Pacific, as they have year after year, stringing enormous driftnets over many square kilometers of ocean surface. This tactic of the hunter, in sunlight and in storm, through calm and burdened seas, managed to kill vast quantities (metric tons) of living beings. To be precise—in addition to countless squid, the fishermen caught 1758 whales and dolphins, 30,464 seabirds, 81,956 blue sharks, 253,288 tuna, and over 3 million pomfret. This tangle of death, an infinitesimal fraction of far greater demolitions, was but one clue, virtually unnoticed, to Japan's and the industrialized world's so-called economic *success*.[8] To contrast those few human organisms in the North Pacific with the only other "driftnet" mechanism in nature, namely, the siphonophores (the Portuguese man-of-war best known among them): the former collectively preys upon millions of creatures, the latter, probably in the hundreds, over an equivalent region and duration. Jaded by humanity's spate of large numbers, one is prompted to question whether there is a significant difference between millions and hundreds? The answer is, Absolutely!

Japan consumed 11,455,000 metric tons of fish in 1992, twice that of either the U.S. or China, and six times that of India and Indonesia. In addition, its countrymen cannibalized 2000 tons of whale meat as well as 3.2 million metric tons of chicken, pork, and beef, paying as much as $360 per pound for champion beef (called *wagyu*). The level of the ocean catch is dropping precipitously every year. The biodiversity of the open seas and more heavily populated coastal waters around Japan has been drastically reduced. Only the six-hundred-mile chain of Nambo Islands, (due South of Tokyo Bay and actually part of the greater Tokyo Municipality), retains any sense of natural integrity, in part because some of the islands have been elevated to National Park status.

Most rivers in Japan have been heavily polluted, diverted, and built over. Sewage treatment in the country lags behind that of nearly all other industrialized nations, while the heavy air pollution is as acrid and enervating to the human organism as in places like Mexico City, Athens, and Calcutta.

Even Japan's financial citadel—the envy of the world's money managers throughout the 1980s—seems in 1994 to be moribund and recessionary. There is unemployment, the country's stock market has crashed, and a crisis of lay-offs has infiltrated the once impermeable system of long-term hiring. Confidence in Japan is waning among investors.

In short, Japan's modern experience—fortified no less by a unique tradition of responsible family planning, a sophisticated ecological consciousness, and great wealth—has been to repudiate most of the aesthetic dreams of its history. This appears to be the contradiction, or the price, inherent to all development. But there are other contradictions, as well: despite economic hard times, Japan in 1993 contributed the largest single grant to the Third World of any country, much of it environmentally targeted. There is no rule that money ill-conceived need taint money well-intended.

What the foregoing implies—from ancient Mesopotamia to nineteenth century Japan—is that the human propensity for violence, or for nonviolence, cannot be easily elucidated. Violence comes in moments, events, and trends. It does not preclude peaceful interstices. While the techniques for peace and sustainability—and the justifications for or against violence—vary, the human organism has shown dual dispositions throughout time. Our behavioral ambivalence may actually define our kind, making for little consensus or philosophical certitude. What is even less unclear is the fact that economic development has fueled that ambivalence, aggravating tendencies for better or worse. Just as a disaster will see looters taking advantage, so too will charity flow forth. Prosperity has frequently invoked disparities, neighbors coveting neighbors—nations harboring designs against other nations— but it has also elicited great philanthropy, gentleness, and altruism.

America's economic rebuilding of Europe following World War II marked the zenith of forgiveness and idealism. And while poverty has never proved necessarily pacific, it has universally been shown to enshrine a level of generosity far beyond its means. The response of human beings to hardship can be graceful, even miraculous. Add up these many sides of the human personality and one is confronted by the middle class, a social phenomenon whose birth, following the Renaissance, has signaled an insistence on increasingly higher forms of well-being and security, the presumed precursors of homeostasis. Such economic equipoise has created a population explosion whilst isolating the consumer from the ecological violence inherent to consumerism. Whether forgiving or inflictive, the middle class—fearful of poverty, aspiring after the example of the rich—has proved to be an engine of adversity, the very definition of development. Is there any way out of the syndrome, with its billions of pounds of dead animal flesh, millions of abortions, sprawling megacities of concrete and unbreathable haze, thousands of years of protective and expansionary warfare? Rephrased in Orwellian terms, is there a way to survive en masse, and to remain truly human? The question presumes that we still know what "truly human" even means.

Whether we actually know or not, we are quick to seize upon an answer, namely, the so-called "Developed World," routinely touted as the exemplar, the grand resolution to all such dialectical agitations.

In 1950 the United States had a population of about 150 million. By the end of the millennium, just a few years hence, it will number 272,793,000 persons, plus or minus 3 percent.[9] By the world's standards, most of those people will be middle class. At that time all of the industrial nations together will account for roughly 20 percent of the world's population—about 1.5 billion people, who will be responsible for some 80 percent of all global consumption, not necessarily individual consumption, but mass consumption.[10] The difference between the individual and the mass applies, for example, to a day in the life of a rural African, whose energy intake is essentially one-on-one—a plucked branch or two for cooking fuel, a scooped-up jug of drinking water, hand tillage of a garden patch, or small-scale cattle grazing—versus a North American, like myself. I'm not quite sure how I manage it, but in the space of about sixteen hours I figure that I have conspicuously contributed to global greenhouse gases, to the rape of both temperate and tropical forests, to the death of countless animals (in spite of being a vegetarian), to long-term ocean pollution, acid precipitation, ozone depletion, scandalously inefficient mobilization of energy, the purchase of a stealth bomber or two, yet another unneeded freeway, the government-subsidized butchering of cattle kept on public lands, and any number of other ecologically insane expenditures. By simply being an American, I have conspired with the tax collector, the textile, computer, solvents, plastics, and weapons manufacturers. My clothes, electricity, gasoline, phone calls, mail, travel, and packaged foods all contradict my deepest convictions. I seem to have lost touch with the most basic cause and effect, with the web of life's delicate connections. I have driven seventy miles an hour over nuances at sunset. And I am told that in my own virtual backyard, five endemic California plant species are going extinct, because of people like myself.

Whether in a rich or a poor country, the U.S. or Mali, the net effect of such development and concurrent consumption, when driven by demographic pressures such as we are now confronted with, will invariably spell painful doom for much, or most biodiversity. This is, in my estimation, the central fact of the twentieth century and the crucial factor in considering the whole planet's destiny. This predicament is what I term the *end-loser syndrome.* This ecological no-win pattern has less to do with Developed Worlds, or Developing Worlds, per se, than with *development.*

Again, *recognizing* the perils of development is the difficulty. By and large, individuals are programmed by nature against grasping the large devastation caused by themselves, against internally processing too much shock and bad news. At the same time we are constantly habituating to smaller annoyances and increments of harm. These are possibly neurological adaptations that help us to anticipate and sort out our greatest defensive imperatives from amongst countless lesser dangers and anxieties. These subtle survival strategems do not necessarily curtail our ability to hear the cries, to investigate, to empathize with victims who are actually far removed from our own immediate orbit of vulnerability. Children with whom I've spoken at schools throughout the world know that the ozone layer, penguins, topsoil, and black rhinos are in jeopardy; that their own air and water are polluted;

that pesticides can cause cancer. They are familiar with the oft-quoted "football field per second"-worth of rainforest being destroyed. But this is merely a saying, held up like some colorful poster, an analogy with a familiar shape that conjures up both an aggressive sport—and hence, aggression—and musical bands and cute cheerleaders; three hundred feet of playing field upon which strapping youth play ball; and professionals who are paid large salaries; a sport that releases and dissipates pent-up human energy and is a major entertainment. A football field is thus an imperfect analogy that can scarcely account for the actual tragedy of rainforest destruction. But, given the global scope of that calamity, one "saying" is probably no more off the mark than a whole book, or a library of books, or some bureaucratic conference in Geneva, or Rio, or Cairo. Even "Chico (Mendes) and his *companheiros* were unaware that the devastation they were fighting against, the local deforestation in the municipality of Xapuri, was part of a much larger mosaic of destruction," writes Alex Shoumatoff.[11] We focus on particulars, and wield accessible comparisons.

After all, how could the individual make sense of the fact that, as of 1993, rainforests were down to much less than half their prehistoric amount? Or that, according to a 1991 report of the U.N. Food and Agriculture Organization, tropical forests since 1989 were being lost at the rate of 170,000 square kilometers per year, a 20 percent escalation over even a few years ago? Never before in geological history have different floristic species in vast quantities been driven to extinction all at once.[12]

Because so much of the destruction is occuring in small, local outbreaks (what E.O. Wilson terms "a world peppered by miniature holocausts"), adding up in one's mind the cumulative and irreparable damage to species and habitats is a desperately imprecise task, bound to underestimate the true picture of horror that is emerging. An overall estimate, conversely, cannot begin to intimate all those countless small-scale disasters.

Of the eighty-seven major tropical regions in the world, with a combined total forest area of 1,884,100,000 hectares in 1980, the yearly destruction, by region, is causing the forests to shrink at a speed that most nonspecialists probably would not recognize, or not day by day. Our ability to clearly discriminate among large numbers, additions or substractions, is severely limited, both perceptually and psychologically. In the decade of the 1980s, Africa's forests shrank from 650,300,000 to 600,100,000 hectares; Asia's from 310,800,000 to 274,900,000; and Latin America's from 923,000,000 to 839,000,000.[13] Every year the rainforests are being lost at a rate of 1 to 2 percent, or some 17 million hectares. In fifty years or less, they will be gone. What can that mean to the average American?

Fishermen in Finland, or the Caribbean, are probably unaware that half of all ocean fisheries are being exploited beyond the estimated sustainable yield. A trout fisherman in Colorado is probably not aware that 20 percent of all freshwater fish are extinct or nearly so. A chemist in Louisiana is doubtless aware, but not overly worked up by the fact, that over two hundred major (and countless lesser) chemical accidents take place each year in industrialized nations.[14] A shoe salesman in Moscow and his clients are probably not remotely concerned that billions of animals will be slaughtered in this decade to meet consumer demand for an estimated 12 billion pairs of new leather footwear. A supplier of birds for pet stores in Buenos Aires is likely to feel all the more justified in his or her profession when informed that over 80 percent of all living birds are threatened with extinction, or are declining rapidly.[15] The painful description of the last Spix's macaw in Brazil "desperate to breed," or of the final remaining Bachman's warbler in Virginia, a lone male frantically searching for a female, singing out every spring from the same spot on a branch, hour after hour, but never finding a mate, are increasingly common stories surprisingly incapable of galvanizing an effective human response.[16] Such tales of attrition are fused in a blurred image of destruction that causes us to turn inward, but probably not to cease activities that directly hasten the demise of such creatures.

The World Conservation Union publishes its Red Data Books noting which species have gone extinct, are endangered, vulnerable, and rare. Throughout Europe more than one third of all insects and invertebrate species are threatened or endangered. Largely on account of collectors, the fungi are witnessing an even more devastating loss, potentially a biological nightmare. From the killing of freshwater fish in Africa's Lake Victoria, to the extirpation of profuse plant taxa in Turkey, to the eradication of useful tree snails in Tahiti—the mindless vandalism wreaked by humans is on the scale of a planetary cancer.[17]

Yet who can blame a farmer in Iowa, trying to make up for his family's terrible losses in the floods of 1993, if he is not so interested in hearing about planetary cancers or that a third of all croplands throughout the world are losing productivity because of agricultural abuse? Eighty thousand square miles of previously biologically diverse regions have been reduced to desert through human waywardness. How could such statistics be expected to impress, say, a real estate developer in Palm Springs, California, suffering from the recession, who, if anything, finds the news of more desert a form of competition? Conversely, the tourist from Hamburg who loves to get away to deserts for his asthma, or for a sunny vacation, is likely to welcome the news.

To live on a finite island like Madagascar is to encounter first-hand the sound and the fury as eleven thousand acres of uniquely prolific rainforest that are being cleared, on average, every year, as the population bristles and expands. To inhabit either the state of Sao Paulo or of Parana, two of Brazil's Atlantic coastal regions where only six out of one hundred trees are now left standing, must invite certain unavoidable observations. As will soon be discussed, a recent poll in Los Angeles indicated that as many 70 percent of residents questioned said they would like to emigrate if they could afford it. That was after the riots of 1992 and the fires of 1993, but before the earthquake of January 1994.

An observant resident of Burkina Faso relates how "the region used to be full of wild and ferocious animals such as lions and panthers, buffaloes, hyenas, jackals—and less aggressive animals such as does and gazelles. We were graced with every species of bird on the planet including wild ducks, ostriches, bustards, and the crowned crane. Now these times have become something of a legend, and the animals have disappeared as if under a spell."[18]

But even local crisis is no guarantee of the human response. In wartorn, famine-plagued Sudan, a ninety-six-year-old optimist argues that in his lifetime "the quality of life has changed enormously. In the past, our life was simple in every respect. Today, by comparison, people are living in paradise."[19] In Botswana, where there are now some seventy thousand elephants as a result of the modestly successful ban on ivory imports and exports since 1989, the government has decided, (in 1993) that the country is not big enough for seventy thousand elephants and has recommended "culling" 3 to 5 percent of the herd annually. Five other impoverished countries in Africa want to join in the culling to help make ivory products exportable and "respectable." To many conservationists, such supposedly sustainable hunting represents the height of shortsightedness and wrong-headed cruelty. Yet, confronted with the same population and economic issues as many of those throughout Africa, what postcolonialist would have the right to blame them for seeking solutions?

Twenty-one countries have no protected wildlife areas whatsoever, not for any lack of interest, but from the sheer exigencies of national survival.[20] Seventy-six countries now possess no region that can be biologically described as wilderness, according to the World Resources definition of the term (four thousand square kilometers of totally unadulterated terrain). That definition is the best working understanding of the undefiled habitat size necessary to ensure the largest average of plants and animals with healthy genetic populations. Ironically, the two countries with the least human population density and largest percentage of wilderness (relative to those countries' sizes) are Lesotho and Mauritania. In fact, of the top ten wilderness, by percentage, nations in the world, seven of them are in impoverished Africa, reaffirming that disproportionate environmental damage is wielded by the wealthy nations.[21]

What these sometimes contradictory assessments suggest is that the world is hemorrhaging and must be surgically viewed by region (by body), inasmuch as most species are regionally oriented in their evolutionary adaptation. And because ecological boundaries, not cultural or political ones, are the crucial frontiers, only a transnational analysis of, and commitment to, data and subsequent remediation can effectively soften the burden of global abuse.

Ninety percent of all species exist on land, it is believed, which is where the human population dynamic has enacted its colossal appropriations. As the human population doubles in the coming decades, most analysts concur that there will be a five- to tenfold increase in the global economy. Probably more than any other sign, the dramatically accelerated rate of species extinction over the natural background level argues for humanity's extreme dysfunctionalism. Add to that the momentum of human population growth, which systematically fuels that extinction rate, and one begins to recognize a malignancy unique to the annals of evolution.

From Tonga to the Mauritius, from Jordan to the Republic of Georgia, a biological system (in this case, *Homo sapiens*) that gains too much power over its plant and animal neighbors undermines the whole balance of a given region. To undermine is to *destroy*. There are those who—confronted with human conquest and massacre, the total population dynamic, that whole spectrum of development—might seek refuge from the truth by citing certain physical properties in nature, such as the principle of thermodynamics that holds that nothing, in truth, is lost. Such laws are no comfort in view of the despair, the absolute loss of precious lives, individual personalities, and intrinsically valuable souls that have waited millions, even billions, of years to declare themselves, only to be snuffed out. We know that there is grave disorder among our kind by the nature of what we deem "news," which so hangs upon, and psychically perpetuates, each disaster, act of violence, and posture of war; and by the International Human Suffering Index, a statistical portrait of countries according to the overall quality of life those countries afford, based upon medical and sanitation access, noise, pollution, hunger, crime, poverty, and space. But however bad it may be for our species, it is far worse for all others.[22] One of the greatest of needless paradoxes has been the fact that the less human beings suffer, the more the rest of nature tends to be adversely affected.

Human power leaves its traces in the form of billions of corpses, dead zones, burned out and bulldozed terrain, encroaching deserts, algal blooms, in the alteration of biological cycles, the proliferation of toxins, and the skewing of genes by technology and hybridization. As one scientist in Los Angeles put it, "The Lord made chromium to be eaten, not breathed. . . . It is an environmental sin to put chlorine atoms on a carbon atom. Nature does it only at the bottom of the ocean. Do it on land and it creates freons, dioxins, and hydrocarbons, all of our worst ills."[23]

To turn away from the global forces of nature that we have set in opposition to each other and instead focus piecemeal on the imbalance, region by region, is one way to recognize the tragedy and try to concentrate on and manage a healing process. Such focus is obviously made tenuous given several thousand years of human hubris, of economic self-interest that sees itself disconnected from nature, free and unstoppable. Economic status has redefined our nature. This situation is particularly distressing in the developing world where the largest concentrations of biodiversity and human population size are pitted against each another. While the damage incurred by the developed countries is still far more interregional and extreme than in the developing countries, that relationship may be reversed a few decades from now for a variety of reasons that will be examined in chapter 8. Just at the time when the Earth's immune system may well have been totally exhausted, we can look forward to a new wave of exploitation unleashed not by the rich (who may well have fashioned remedial schemes by then, such as have been outlined in Japan's *New World Agenda* blueprint for the next century), but by the countless billions of poor, hungry, and angry people, most not yet even born, who are clinging to the development ideal as their last hope.

The inherent peril of human population momentum, in concert with this penchant to corrupt all preexisting biological homeostasis, places in doubt our ability to project anything about life in the future, or to actually gauge the full extent of our impact in the present. All that can be said with certainty is that human power is rarely benign, more often destructive and, given current trends and behavioral averages, inevitable. The portrait of that power, and those averages, in country after country, is a study in physical, mental, and moral degradation.

The sum of destruction transcends its parts, just as human actions and awareness fail to connect with the larger picture. Accountability and the motivation to rectify injury are obscured.

## THE GLOBAL PATTERN

Take troubled, determined Mexico, whose well-intentioned populace is currently combing its landmass and offshore regions—as the invading Spanish once did—knowingly driving countless flora and fauna to extinction (such as five out of six remaining sea turtle species in the world.)[24] Or consider Germany, a country that swears it will never repeat the sins of its own dogged history, to paraphrase Martin Heidegger. Yet in 1953, eight years after the German national conscience had time to mull over its debauched and unspeakable past, the firm of J.A. Topf und Sohne of Wiesbaden received from the government a formal patent (No. 861,731) called "Process and Apparatus for Incineration of Carcasses, Cadavers and Parts Thereof" for its design of the Holocaust crematoriums. The company was thus rewarded for having discovered a kind of self-kindling oven, fueled by the burning fats of its millions of victims. Perhaps this resounding perversity of bureacracy was an accident of paperwork, perhaps not. More recently, in 1994, the commandant of the World War II concentration camp, Treblinka, was set free by the German authorities, having served his (full) fifteen-year sentence. This man was responsible, in part, for killing some 850,000 people, mostly of the Jewish faith, in addition to 193 individuals by his own hands. Ten months earlier, Petra Kelly and Gert Bastian, co-founders of the Green Party in Germany—the first congressionally-seated environmentalists in history—were murdered, in the wake of multiple assaults on refugees to Germany. And during 1993, the German courts imposed a chilling legislative hurdle in the path of free abortions.

These are mere highlights of the strange, interconnected, human story—from among many—that give some vague, emergent sense of the pathology at work, a war waged by the very nature of our species, and the dark, inextricable imbroglios of our spirit. In Bosnia, writes poet Joseph Brodsky, "One always pulls the trigger out of self-interest and quotes history to avoid responsibility or pangs of conscience . . . the Balkan bloodshed is essentially a short-term project. . . . For want of any binding issue (economic or ideological), it is prosecuted under the banner of a retroactive utopia called nationalism."[25] The warfare of our past was endless, as are the many, ongoing regional skirmishes around the world—human beings locked in conflict to the point of murder. But all of these battlefields combined do not begin to match the level of destruction, the megatonnage of harm meted out by the more silent, gun-free wars of *progress* and *development*. These two words sail through our consciousness, breezy, proactive, full of promise and supposed comfort. They are the watchwords of every government policy, town meeting, corporate incentive, and social justification. Some would even argue that they are human destiny in an otherwise merciless, unfeeling universe; that without progress, there is . . . nothing! We speak of "underdeveloped" fetuses, capacities, ideas, intellects, and nations, with a mixture of cold detachment, disdain, and caution, whereas we rarely begrudge, or even acknowledge, that which is overdeveloped, except in certain medical cases pertaining to the thyroid or hydro-cephaly.

Yet, it is precisely *overdevelopment* that has distorted everything good and potentially humane about our species. Because the number of people is so huge, and the instruments of its progress so potent, every new condominium complex, golf course, or air conditioner conceals even a far more devastating pain upon the world than a triggerfinger in Bosnia, though they all stem from the same confusion. The tragedy, as Stephen Carter has written (in an essay on Ronald Dworkin's book about abortion, [*Life's Dominion*]), is that "Compromises, by their nature, possess the internal inconsistencies and contradictions that scholars, by their nature, abhor. Scholars want arguments to *make sense*; but politicians know that arguments have to work—which means, in the long run, that they must form the basis for a stable consensus."[26] What I have learned about World War III is that it represents the sum total of human inconsistency and contradiction. While people, turtles, and dolphins, Texas cows and Philippine dogs, are deliberately slaughtered, this war is largely about the very contradiction of Being, of being human, of fighting for a piece of turf somewhere between nothingness and infinity. And truly, as Norwegian philosopher Arne Naess describes it, "it is painful to think" about this self-indicting situation, the ramifications of our presence, our biology, our very name.[27]

There was a time when Being did not mean development; when certain of our ancestors lived truly harmoniously. The notion is bizarre, a hopeless sentiment, because to actually envision this harmony in all its guises, day after day, is to repudiate virtually everything we know to be the truth about our contemporary selves. More to the point, even if we should manage to imagine a plausible sequence of events, of our new life in ancient Arcadia, we nonetheless persist in the way we are, not the way we were. Even if individuals know better—and quite a few of them do—as a species we do not know how to reconcile the imagined, better past with the imperfect present. We cannot stop ourselves. And because we are so many, and multiplying so rapidly, Homo sapiens have fled into the future, frantically squandering the visible resources around them, stealing from their children. The argument of survival has been inverted. Rather than searching for solutions in our anthropological and spiritual beginning—an ontologically sensible but woefully impractical course—we are fixated on upward mobility, more money, more security, immediate returns, and a host of synonyms connoting gratification. Development is the school of thought that enshrines these many activities.

In terms of understanding human psychology, a new 700-room hotel in Jakarta, or a skyscraper in Indianapolis, can only refer to someone's selfishness, however many jobs it may provide. Society applauds skyscrapers, commends new hotels, and rewards the obfuscating of nature if it provides paychecks. Intrinsic merits of the natural world, the sanctity of all life, even the most general ethical considerations, are obsolete when placed beside those rewards. The human impulse to develop—to build cities, walls, towers, to excavate and bulldoze, to reconfigure nature with machines as large as the Golden Gate Bridge, to build weapons of mass destruction—has transcended whatever primeval satisfactions once sustained our kind. Now that there are nearly 6 billion of us in a state of feverish development, it is impossible to avoid the realization that *development is tantamount to war.* This is ironic, for it will invariably drive people away from acknowledgement.

This war is all in the cumulative weight of numbers. According to calculations worked out by Paul and Anne Ehrlich, a child in the United States will yield a destructive force 280 times that of its counterpart in Nepal, Haiti, or much of Africa, 140 times that of a Bangladeshi, 35 times that of an Indian child, 13 times that of a Brazilian.[28] What can such data mean to an American child, let alone an immigrant to Los Angeles (where—ironically—22 percent of all children under five are poverty-stricken)? In South Central Los Angeles, 44.1 percent of all children under eighteen are in poverty.[29] Is it a more destructive poverty than in India?

United States human poverty, the homeless persons on our streets, are murky reflections of the infinitely broader ecological impoverishment that the "American dream" and the habit of western-style economies have globally unleashed. Comparing poverties is not possible, though the homeless I have spent time with in Los Angeles are, in some ways, more forlorn, angry, and defeated, than those in India. If for no other reason, the income gap separating haves from have-nots in developed countries is dramatically greater than in lesser developed countries. Poverty can also incite cruel retribution. Brazilian street waifs—there are 40 million such children throughout Latin America—are increasingly under fire by vigilantes who simply murder them in the night in back alleys, a kind of social extermination. The human fallout from overpopulation, poverty, and environmental corruption is everywhere manifest. Few countries or regions have remained immune to it.

One does not need to go far from home to at least begin to sense the truth of biological damage, and the escalating blur of psychological and perceptual habituation our very presence incurs. Beneath the textual discussions, policy debates, and economic warfare, exists this vast human intransigence, the result of countless genetic eons, the brevity of our lifespans, and the essential selfishness of individual organisms. Can late twentieth-century consciousness modify this biological hierarchy or imperatives? Practically speaking, but the underlying question is whether income generation, the substitution of wealth for poverty, has anything to do with a nation's ability to prevent such ecological damage. "The fact is that nations with lots of resources (capital, scientists, engineers, technology, a per capita GNP of over $4000) are better able to deal with environmental threats than those without monies, tools, and personnel," writes Paul Kennedy.[30] Kennedy goes on to allege that there is a "feedback loop," which presupposes that those who are educated will have an "enhanced ecological consciousness, and a willingness to prevent environmental damage." But I contend that the spotted owl would fare no worse in Ethiopia than in the state of Washington.[31]

Yet, the illogic of growth persists. Reinforcing Kennedy's claim, according to Jaime Serra Puche, head of the Mexican negotiating team for NAFTA (the North American Free Trade Agreement), U.S. $4000 annual per capita income is the miracle cure, the supposedly predictable threshold of prevention. Once a country has reached that economic level, it is able to better invest in cleaning up its environment.[32] This, then, is the illusion clearly stated. Extending the thesis, because per capita income exceeds the $4000 threshold by five times, the U.S. must be environmentally near pristine, which it assuredly is not. The presumption also suffers from three other chilling reality checks. First, out of nearly two hundred countries in the world, only thirty-seven of them have actually reached that threshold. Most nations are far from it, and, in many countries—poor and wealthy alike— per capita income and the dollar value of the so-called quality of life package are dropping, not rising. Second, many of those countries exceeding the $4000 per capita level are running colossal deficits, most disturbingly the United States, which faces a $6 trillion national debt. And, finally, the industrialized world is the last paragon of supposed ecological virtue one would ever want to emulate.

However triumphant a nation's economic progress, there will always be human losers. This was true during the Renaissance, and it is especially true today. But the real *end-loser* will always be biodiversity, whether the country is rich or poor. This appears to be the brutal and bewildering lesson of our age. All the best population controls are being easily undermined by increasing numbers and increasingly concentrated wealth and power. In Los Angeles, a developer pressed to obtain the permits to build a large condo complex on the last remaining coastal wetland in the city. In a supposedly enlightened gesture, his company made it known that 25 percent of the condos would classify as "affordable housing." This was somehow intended to counter the criticism of environmentalists. It characterizes one aspect of the fundamental blind spot in *Homo sapiens.*

Consider what that blind spot, that triumph, the "American dream," has meant, ecologically speaking, in other countries that are either among the elite club of wealthy nations, or struggling to get there. I've chosen to describe, briefly, a handful of such nations that I consider representative of nearly all such countries.

In the Netherlands, a nation tied mercilessly to the vicissitudes of the German economy, a nation that supports 1031 people per square mile only by importing vast amounts of food and resources from elsewhere—millions of tons of cereals, oils, pulses, and minerals, even fresh water—the very earthworms are dying throughout the southern part of the country because of the intense build-ups of chemical runoff and sludge pollution from poultry farms. Ecologists have devised a term, the *Netherlands Fallacy,* to indicate a densely populated region that *appears* to be prosperous, but, in fact, has utterly exceeded its environmental carrying capacity. In the case of the Netherlands, that has meant that the country is dependent upon external inputs.

In Saudi Arabia the rapidly expanding population has been exhausting its nonrenewable groundwater reserves as fast as Brazil destroys its rainforests, or 2 percent per year, in order to irrigate twenty times more farm land today than existed in that country as recently as 1975.[33] Saudis have not gotten hungrier for food, of course, but for money earned through agricultural exports. Their current per capita income is $6230, far behind the United Arab Emirates, Kuwait, Israel, and Qatar, but dramatically more upscale than poor, neighboring Jordan, whose per capita income is lower than $1800 per year. The Saudi population is growing faster than almost any other in the world. With a current total fertility rate (TFR) of nearly 7, the country's 18 million will surpass 70 million late in the twenty-first century. Their country's oil will be long gone by then. And experts believe Saudi Arabia's water will be exhausted by the year 2007. When the oil is gone, the country will no longer have the necessary funds to perpetuate expensive desalination, or to (theoretically) haul in Antarctic icebergs. How will it maintain its intensive agricultural infrastructure in a land where crops can die in one day from the heat if not faithfully irrigated?

Italy should be an exquisite example of the best that money and progress have achieved. The nation's ironic fertility discipline has resulted in near fertility replacement (NRR=1). With 57 million people, a TFR of 1.4, and a per capita income of over $16,000, it boasts the fifth largest GNP, and one of the loveliest legacies of art and cultural sensibility the world has ever known. Yet, by many estimates Italy is the most polluted country in Western Europe. Pursuing industrial stardom, the nation has brazenly eschewed most environmental regulation or enforcement. The Plano, that vast Italian plain stretching along the southern side of the Alps, is now perpetually cloaked in smog, for hundreds of miles. The air pollution intimates other, unseen horrors, such as the countless chemical spills, and the largest garbage dumps in the world, their uncontained toxins percolating into acquifers and the drinking water of Italy's urban populations.[34] The country's topsoil degradation has been likened to that of Somalia, so severe that any kind of restoration is impossible.[35] And, as of 1975, Italy had already lost 95 percent of its wetlands and Mediterranean littorals.[36]

Throughout Europe, at least 50 million acres have been directly despoiled by industrialization. The near total absence of the ancient temperate rainforests previously found in England, Ireland, Scandinavia, and even Iceland, is one clue to this pall of development. (In addition, since World War II, England has lost 98 percent of its old pasture.) It has been estimated that at least $30 billion per year would be required to even begin to redress the damage to European forests from pollution.[37] But biodiversity loss is beyond compensation. In Germany, of the 933 plant species at one time identified, 14 went extinct between 1870 and 1950. But since 1950, following the U.S. Marshall Plan to rebuild the country, 130 more plants have gone extinct, 50 others are threatened, 74 seriously endangered, and 108 in decline. These are astonishing statistics that reflect a continent-wide despoliation.[38] The largest number of endangered species—proportionate to the number of species within a region—are to be found throughout Europe. This fact alone should immediately call into question the assumption that wealth translates into environmental restraint. The long, unflinching history of Western human persecution of other plant and animal species is almost implausible but for the recognizable consistency of its cruelties, which so resemble our own killing of one another, most notably in two world wars, both fostered on European topsoil.

But there is ample historical precedent for such cruelties. The earliest examples of European warfare against other species can perhaps best be sized up in the records of the *venationes*, the Roman animal slaughters, precursors of the later gladiator spectacles that were staged as entertainment for a culture steeped in global conquest. As audiences cheered, hundreds of large mammals were dragged out in nets or taunted into the ring, and there brutally tortured and slaughtered, or predators were unleashed upon them. In one such afternoon's entertainment held in Pompeii in 55 B.C., in the shadows of lurking Mt. Vesuvius, thousands of lynx, monkeys, lions, elephants, and female leopards were massacred.

While King Henry IV used the best skins from twelve thousand squirrels, plus eighty ermine pelts, for one of his robes, a czar was known to have had a quarter of a million ermines killed for his coronation garments. Parisians wore beaverskin coats, as well as wolf. The last wolf was seen foraging along the streets of Paris in 1419. By 1526 the last British beaver went extinct. Whether along the coasts of Nova Scotia, or Brazil, European exploration was much motivated by the insatiable appetite for soft fur. In just one year (1895) approximately ten million furs passed through the hands of dealers in London. In the 1870s, ten million birds a year were shot in the U.S. mid-West. Whole colonies were extinguished, filling hundreds of freight cars. In Paris ten thousand workers artfully transformed the feathers and bird skins into ornamental goods. Much of European cultural history can be characterized by this particular obsession, in which always figured the skins of sable, marten, fox, otter, wolverine, skunk, lynx, ermine, bear, and tiger, and the feathers of countless bird species.[39]

But if Western Europe's environmental track record is deplorable, today's Central European battlefields are even worse. These are the RICs, or Rapidly Industrializing Countries, which have their hearts set on achieving that $4000 magic number. The ghastly environmental legacy of places like Copsa Mica, Romania, and the Katowice–Krakow areas of Poland, and the Bohemia region of Czech make a mockery of World Health Organization pollution standards.[40] In one part of Poland, levels of lead in the soil were recently found to be fifty times the WHO limit, or 19,000 parts per million. Poland's Bay of Gdansk is considered the most polluted body of water in the world. Lead and cadmium pollution in the Upper Silesian soils around the towns of Olkusz and Slawkow are the highest levels of contamination ever recorded anywhere.[41] Contact through dust, food, water, and soil has resulted in IQ declines by as much as thirteen points in exposed children.[42] Yet, the country, desperate for hard currency, accepts additional hazardous wastes from the West. More and more vehicular travel, leaded petroleum manufacture, the increased use of so-called unclassed (filthy) water for irrigation and industrial purposes, increasing toxic discharges of phenol, ammonia, and heavy metals from coal plants, and the abundant untreated animal wastes from agriculture and slaughterhouses are creating ongoing havoc with Central Europe's environment. In parts of Czech, 80 percent of the surface water is mired with such pollutants. Across Bulgaria, 66 percent of all silver firs are dying. In the Russian industrial town of Sterlitamak, 84 percent of all births have been deemed abnormal and high infant mortality rates corrolated with high levels of toxic waste in the air and water. Government officials throughout the Commonwealth of Independent States have revealed a veritable Armageddon of ecological statistics, such as that more than a third of all food (two thirds of all fish) and nearly three-quarters of all water in many of the states are seriously contaminated.[43] Throughout immense Siberia, the wasteful burn-off of natural gas flares are said to pollute the atmosphere as severely as the fires of the Persian Gulf War. In addition, millions of tons of uncurtailed Russian spills and leaks—at least

10 percent of Russia's total oil production—have damaged the soil and plant ecology of much of Western Siberia.[44]

Each of the RICs throughout the world has shown the same economically driven devastation, the nearly identical, malevolent destiny written without distinguishing punctuation marks. By definition, no city today is anything but a brute, negative energy flow. Outside the urban environment, the RICs have spread their ruinous traits. Forest losses have quadrupled in these countries during the last twenty-three years. In Thailand, with a per capita income double that of Indonesia, forests are less than 28 percent of the country total, down from 55 percent in the early 1960s. Though the Thai government banned all logging in 1989, the very next year forty thousand hectares were destroyed secretly and Thai environmentalists say the law is at the mercy of industrial aspirations. In addition, the increase in prawn cultivation for export has destroyed mangrove forests in the southeast, while the fishing fleet, by resorting to ever-larger vessels, has completely upset the marine balance in the Gulf of Thailand, a syndrome reminiscent of that in Kerala. In Bangkok, there is no sewage treatment for some ten thousand metric tons every day, which simply are flushed into the four rivers that flow through the heavily traveled center of the city to the sea. The words used to describe such deleterious effects as "downstream nutrient enrichment of rivers," or "deforestation and sedimentation," fail completely to account for the actual degeneration and demise inherent to these slovenly human practices. On land, the country's energy extraction and consumption rate has been escalating at an astounding 300 percent per year, twice what had been predicted by officials in the 1980s.[45]

In rapidly developing Malaysia, whose per capita income is nearly four times that of Indonesia and double that of Thailand, the *National Forestry Policy* of 1978 is touted as being the closest to so-called sustainable logging in any tropical rainforest in the world. But only along the Malay Peninsula. In fact, throughout most of Sabah and Sarawak, where at least 50 percent of all hardwoods in the world originate, the destruction, and the profits, are out of control. The International Tropical Timber Organization has stated that the old primary forests of Sarawak will be gone by the end of the decade. At the same time, the country is predicting a doubling of its electrical capacity during this decade, which some predict will "tax Malaysia's reserves of fossil fuels and its fragile ecosystem of rivers and rainforests as never before." In Sarawak, such development has translated into the nearly $4 billion Bakun hydro facility, which will displace 695 square kilometers of forest with water, as well as forcing out six tribes along the Balui River. The World Bank has analyzed infrastructure constraints to such economic growth throughout countries like Malaysia and has found that "manufacturing has outpaced infrastructure" in nearly every energy-capacity domain, whether power plants, freeways, waste disposal, water supplies, or public housing.[46] Most troubling of all, the president of Malaysia has expressed his desire for a Malaysian-produced motor vehicle, prompting a pronatalist stance based on the presumption that you need a more sizable population than, say, 20 million people to effectively sell a new line of automobile.

When the United States suggested targets for controlling population at the 1974 Bucharest Conference on Population, many Third World nations responded by calling for a "new international economic order" that would assure them access to world markets. The Green Revolution was part of that new order. "Development is the best contraceptive" was the motto at work in both the developed and developing worlds. Development, it was universally felt, would ultimately provide the human race a way out of overpopulation. But, in Mexico it didn't happen. Development, and the country's Green Revolution successes, have been negated by the rapid growth in human numbers. Throughout Latin America, a 10 percent decline in per capita food production has coincided with increasing wealth for the few, poverty for the many, and rapid population growth of 2 percent per year. While Mexico's GNP grew by over 9 percent annually throughout the 1970s, boosted by oil development, the hidden costs included the loss of some 2.4 million acres of forest each year. Considering that Mexico hosts one of the largest diversities of mammalian species in the world (larger even than Brazil), such forest attrition is disastrous. In Mexico City, nearly forty thousand factories and over 3 million motor vehicles discharge their uncurtailed contaminants almost continuously. The city's air pollution levels in December 1992 were comparable to the killer smog of London that resulted in four thousand deaths in one week during 1952.

The standard demographic equation that holds that rapid development should eventually result in unambiguous population decrease, has not proved to be the case in either Mexico or Brazil, thus far, where 1.8 and over 3 million newborns, respectively, need to be fed and housed each year. Both countries have TFRs of over 3.0. Mexico's population will exceed 104 million by the year 2000, up from a current 90 million. Brazil will harbor 175 million people within five years. Industrialization in both countries has merely widened the gap between the rich and the poor, exacerbating preexisting environmental ills. More than half of all Mexicans and 58 percent of all Brazilians are below the poverty line. Mexico insists that it has three times more environmental inspectors per business enterprise (of which there are some six hundred thousand in the country) than in the United States, and that increasing wealth will invariably result in tougher environmental legislation and enforcement. But, so far, this has not proved to be the case.[47]

Brazil, meanwhile, has come to typify absolutely everything that's wrong with economic development. As of this writing, 235,000 square miles of rain forest have been destroyed in that country, yet the population is 75 percent urban, as in the United States. This suggests that the concentration of wealth and power will not differ substantially in its destructive manifestations from an essentially rural population, such as India, Indonesia, and China. Brazil's mean per capita income (applicable, as mentioned above, to about 42 percent of the population) is $2550 per annum, slightly below that of Uruguay and Suriname, but higher than any other country in South America. Many of Brazil's boomtowns resemble Cubatao, known as "the valley of death." A virtual suburb of Sao Paulo, sprawling along "four lifeless rivers," the city is one of the largest petrochemical centers in all of Latin America. Data by the World Health Organization indicate that, at least statistically, Cubatao's enormous levels of air, water, and soil pollution should not support human life. Indeed, according to Jacques Cousteau's staff, writing in 1980, "There are no birds, no butterflies, and no insects of any kind, and when it rains on particularly windless days, the drops burn the skin." In the 1970s, there was, on average, one emergency medical call for every two residents throughout the year.[48] Yet, as of the late 1980s, Brazilians were flocking to Cubatao because it boasted some of the highest per capita wages in the country. Most of the laborers there, however, live in filthy, ruinous shantytowns.[49] Cubatao is one more example of what the promise of the United States can look like to those elsewhere who emphatically aspire toward it. And despite Brazil's higher income levels (relative to most of Latin America), most residents of the country have not chosen to reduce their childbearing zeal. The TFR is still over 3, and if the country were to income-average its colossal depreciations for loss of biological integrity, every Brazilian, and every Brazilian's descendent for the foreseeable future, would be permanently and irretrievably in debt. In Brazil's interior, many women have already given birth eighteen times by the age of forty-two. According to Julia Preston, that is "not an unusual number. . . . Burying children is a routine function of motherhood. Maternal affection is an acquired luxury that mothers rarely experienced when they were young and cannot consistently give to their families."[50]

Unlike Brazil, in Chile demographic transition has taken place, yet the war against nature—impelled by the desire for wealth—continues. The country's infant mortality rate has gone down almost as dramatically as in Kerala, India (from 79 to 23 per thousand) and the TFR is 2.4, virtually the lowest in all of Latin America. Yet ecological transition has not occurred. Slums and rapid urbanization have caused countless endemic problems. All sewage is untreated and goes directly into the same rivers utilized for irrigation. This has resulted in major typhoid epidemics. Conservation is low on Chile's list of priorities; 57 percent of allegedly protected trees have been cut down.

In Colombia, the fecal bacteria count downstream from Bogota is an astonishing 7.3 million per 100 milliliters. Safe drinking water should not exceed a count of 2000. The country is now witnessing periodic outbreaks of cholera as are a dozen other nations throughout Latin America where there were 400,000 cases in 1991.

While democratic Costa Rica points to its high caliber investment opportunities and high proportion of protected wildlife areas (roughly 12 percent), in fact, the country's deliberate destruction of much of its forest lands has been calculated at a cost to each of the nation's residents of $69 per year in natural resource depreciation. In addition, over-exploitation of the country's marine resources has meant that the personal income of the fishermen actually fell below the level of welfare payments to the destitute. Overall, Costa Rica's accounting framework has ignored the country's impact on natural resources such that in 1989, according to Robert Repetto, the government overstated "net capital formation . . . by more than 70 percent."[51]

And, in poor Guatemala, whose TFR exceeds 5.4 and whose per capita income, like that of Honduras and Nicaragua, is less than half that of most other nations in Latin America, the people are frantically looking for ways to catch up. Once again, destruction of the forests—at one time profusely populated with wildlife—seems to be the imagined secret code. The mountain people are deserting their traditional lifestyles, roaming the lowlands, burning down the trees, planting corn (milpas) and beans, quickly exhausting the soil, and moving on. The Peten region is now covered by a pervasive fire haze indicative of the same kind of war zone that is engulfing the Amazon.

In Uruguay, desertion of much of the rural areas has meant that the majority of the country's residents is now situated in the capital, Montevideo, where poverty and the overconsumption of a limited resource base is endemic. The utter imbalance of population distribution in that country, and the jarring impact such disproportions wield, is suggestive of another glaring problem associated with overpopulation, namely, the breakdown in the rational distribution of goods.

Between 50 and 90 percent of the rapidly industrializing countries' exports are primary products—like forests, fuels, and minerals. This means that the adoption of American-style consumerism is not increasing service industries (whether consultants, plumbers, lawyers, or educators), but rather, directly brutalizing all life forms for primary resource extraction at giveaway prices. At the same time, much of the revenue from that encroachment is earmarked, not for recompensating nature, but for additionally inflictive public works projects. In Southeast Asia, for example, $600 billion has been collectively budgeted for new roads, telephone lines, bridges, and viaducts. In the developing countries the game of catch-up has engendered military spending at a pace three times that of the industrialized world. And the $173 billion expended on the military in developing countries in 1987 was two to three times more than those same countries spent on health and education.[52] The quest for what human beings imagine to be economic security is simply bankrupting our capacity to restore ecological and moral balance to the world.

It can certainly be argued that human economics has replaced the forces of human evolution. Economic impulses are destroying faster than nature can replenish. The law of averages knows no biological analogy, and thus offers no theoretical resolution. While our life and death spans are fixed, more or less, it seems plausible that we might be capable of destroying everything around us, sustained by nothing more substantial than that conceptual big bang, an explosion of ideas during the past 15,000 years. Physically, we have scarcely changed. Mentally, we are a new species, unrecognizable to our late-Paleolithic forebears who lavished such splendid images on the cavern walls at Lascaux. Mental speciation, without a corresponding physical accommodation, implies a disease, a form of chaos. So deeply embedded is this developmental paradigm, this *Weltanschauung*, or world view of economic construction and destruction, that today's human organism seems lost (though not unimaginable) without it. A species sustained by mentation, and increasingly global, technological homogenization, prefigures a true monoculture, correspondingly vulnerable to its own biological downfall. When does the biological bottom line engulf us, like a mythical sky falling on our heads? We speak of an ecological price to pay, but is any price relevant to a civilization that has already been bankrupt for many years? Whether the debt is 6 trillion, or 100 trillion dollars, we would not out-populate and out-produce our attrition. A billion babies may starve to death, but does that matter, in biological terms, if 4 billion newborns replace them?

A few have taken seriously the question of redemption, and sought ways as individuals and communities to combat progress, in the manner, say, of the Amish in such places as Lancaster County, Pennsylvania. As individuals, this is possible, to varying degrees, of course. But for the most part, the last remaining minimalist, anti- or preindustrial societies, such as those of Brazil's Kayapo or Yanomami, are being systematically eliminated or assimilated. Soon, their modest example of low population impact is likely to be lost.[53] It is left to governments to steer population policy. Individuals, acting on their own, cannot do it.

In contemplating Paul Kennedy's book, Robert Heilbroner posed the challenge and the risks in this way:

"What form of political leadership will suffice to halt the juggernauts of demographic, economic and ecological change? Can demographic explosions be halted without recourse to severe, even repressive population policies? Can an allocation of carbon emission rights be instituted or enforced without military force? Can the imperatives of capitalism be permitted to endanger the very continuity of the Western world (not to mention the underdeveloped continents into which its enormous energies are increasingly directed)? And if not, what socioeconomic system would replace it?"[54]

The World Resources Institute has stated that if the engineering and global proliferation of sustainable technology does not precede the anticipated $50 trillion dollar world economy in the next century, then "the environmental impact could be devastating."[55]

The devastation is already upon us, obviously. We inhibit our response capacity by habitually postponing the day of reckoning. If there were no future, we could postpone nothing. All the indicators of environmental decay are such that the future of biological complexity is finite. One can calculate the demise of species in years, just as it is possible to record the disappearance of trees, and the increasing chemical imbalances of water, air, and soil. We do not have years. The global economic recession reflects a vastly more paralyzing biological deficit. There is no time, none. Nor is there any better means of calibrating the level of biological attrition than by focusing upon the urban laboratory, where destruction is so massive, death so unabashed, that we can literally see it, feel it, hear it. Out beyond, in the wilderness, such human impact is less often perceptible.

That laboratory of death encompasses not only the RICs (rapidly industrializing countries) but also the NICs, or newly industrialized countries, including Hong Kong, Singapore, South Korea, and Taiwan, countries whose spectacular growth has become the envy of 150 other nation states. Hence, the terrible, ethical predicament that few are willing to discuss openly: in our proudest moments of city life, we actually stand on the brink of complete obliteration of nature. We might as well have engineered a Martian landscape, replete with sentimental relics of biological memory, nature documentaries, and domed arboretums.

For twenty years, beginning in 1965, annual economies in the NICs grew by 6 percent each year (far more exhuberant, even, than Japan's 4.3 percent and the U.S.'s 1.6 percent growth). But like the U.S. and Japan, the NICs have paid a terrible price.

Taiwan, for example, is one of the world's most densely populated countries. It has the largest volume of daily trading on its stock market in the world ($1.5 billion), but it is the most volatile because investors appear to be interested strictly in short-term profits. It is, from the investor's point of view, a "casino"; and that's how the country's biological heritage has also been viewed. According to the manager of one Taiwan fund traded on the New York Stock Exchange, "They (the Taiwanese investors) act on rumors and tips rather than on fundamental analysis." It means that this country of 21 million people (and a low TFR of 1.7) is a long way from "green" business practices.

Taiwan, along with South Korea, uses more pesticides per hectare than any other country in the world, while treating less than 1 percent of its human waste. Consequently, the country has a higher number of hepatitis B cases than any other nation. Its entire west coast is an ecological disaster, its rivers and air are severely polluted, and associated health problems, like asthma, are raging. Estimates on clean-up have exceeded $300 billion. Taiwan's leadership has owned up to these costs, but whether the money will actually be allocated and spent is yet to be seen.[56]

In South Korea, there are more engineering students graduating each year than in all of Great Britain, Germany, and Sweden combined.[57] Yet all of that expertise is not being oriented toward environmental sustainability, but rather to uninhibited, unmediated profit. South Korea has just recently topped the $4000 per capita magic threshold, yet 75 percent of its sewage remains untreated, pesticide and fertilizer poisoning of the soil and water is extreme, and much of the tap water is deemed undrinkable. At the same time, the country consumes energy much more intensely and inefficiently than most other industrialized nations—three times that of Japan, and roughly 17 percent more than that of the U.S. While NICs such as South Korea have made enviable strides in lowering their total fertility rates providing access to health, education, and land, their dismal environmental records (and unfair labor practices, particularly with regard to women) require a sober rethinking of the premise upon which these gains have been won.

It is estimated that current trends in energy use, if unabated, will result in a quadrupling of the annual global energy demand by the middle of the twenty-first century as a result of population and economic growth.[58] For places like Siberia, that can only mean a point of no return. Siberia's degradation will eventually alter an enormous biological heritage of swamps and temperate forests and tundra, fragile systems the size of Europe that will never be the same. Ironically, given the widely understood alternative solutions to non-renewable energy exploitation, such damage is avoidable, technically speaking. Yet, energy conservation and restitution, by themselves, are not enough to curtail population-induced ecological warfare.

In California, the technical acumen, enlightened public utilities, sophisticated city and county management, and the general public, are as energy conscious as nearly any in the world. Furthermore, just between San Francisco and Los Angeles, there are more environmental watchdog organizations than can be found anywhere else in the United States. But California, particularly Southern California, is an environmental paradox of grim proportions, despite its good looks. Imperceptibly, over the course of a mere century, California has succumbed to the same irreversibility as Western Siberia's ecodisaster, the dramatic human trespass witnessed in the state of Kerala in southernmost India, and the transmigrational tragedies afflicting a city like Bandung, Indonesia or Shanghai, China. Remarkably—and it says much about the American personality—people in California are talking about three things: taxes, crime, and health care. But few are referring to the overtaxed biology, the crimes against nature, and the awesome breakdown in the health of California's total ecosystem—all symptoms of an unsustainable population boom that rivals that of any developing country. As the word applied to the Netherlands, one might similarly call this the *California Fallacy*.

## THE CALIFORNIA FALLACY

California has always been understood in the light of beauty and freedom, of behavioral unrestraint and psychological deliverance. And without belaboring its myriad gratifications and braggadocio, the *dream* of California has exercised the same persuasiveness over much of the human species as the New World itself once held for Europeans. England pursued an industrial revolution; America searched for California. The gold rush has never ceased. In pursuit of pleasure, California has coined three global reminders, among others, of her stunning bounty and indifference: "Surf is up," "Frankly, my dear, I don't give a damn," and "The Wonderful World of Disney." But the cockiness and romance of the place are wearing thin. There are too many people. Each natural disaster thereby inflicts more harm. Inquire of any local— whether Chumash Indian, psychiatrist, California Condor, policeman, a Braunton's milk vetch, a commuter from Orange County, a school teacher, a utilities official, a little kid in South Central, a surfer, or a family planner—and you are likely to get the same answer: "Overpopulation." It is this numbers dilemma that accounts for the severity of environmental destruction throughout California.

More specifically, consider that in the two most teeming nations— China and India, the birthrate is approximately 21 per thousand and

31 per thousand, respectively. But in California, particularly Southern California, the birthrate in 1990 was an astronomical 84.6 per thousand, a 2.1 percent per annum growth rate overall. California's population is growing at a rate of 646,000 a year, not including illegal immigration. That fertility penchant is three and a half times greater than the average population growth rate for the world's more developed countries; it is even higher than the world average of lesser developed nations. One demographic research analyst I interviewed in the Planning Department for the city of Los Angeles told me with a heavy sigh that "fertility numbers have tripled." And he went on to say, "I can't even publish them (the data) they're so outrageous."[59]

Why isn't this major news? Why aren't congressmen in a state of panic? Why aren't Americans in an uproar? Because, in rich, spacious America, few people care or understand what these numbers bode. The U.S. has no official population policy.

On the twentieth anniversary of the Supreme Court's *Roe v. Wade* decision, I spent part of the morning with Dr. Joan Babbit, then executive director of Planned Parenthood for the city of Los Angeles. She's a veteran of what, in L.A., might best be termed the *demographic wars,* so I expected to meet a hardened cynic. Instead, I found in her enormous humor, clarity, and compassion. She had started out in pediatrics, graduated from medical school at the precocious age of twenty-three, only to be appalled by what she discovered in the Vermont community in which she went to work—namely, four thousand unwanted children—castaways and wards of a poor state. That's a far cry from the estimated 40 million unwanted children in Latin America, but it was a powerful enough revelation to back her into a public health career, which she soon discovered equated with population and ecology. "There are four things at the end of this century that are unprecedented in history," she told me. "Limited resources, pollution, nuclear energy, and overpopulation. Overpopulation is the key to everything else." In a democratic country, resource, pollution, and energy issues can be tackled, albeit inadequately. Health officials have a much harder time counteracting overpopulation.

Dr. Babbit described one of her first encounters with the predicament. "I can remember cringing inside when an impoverished four-teen year old with gonorrhea insisted on retaining her pregnancy because it would mean she could then become independent, which is to say, go on welfare." And then she reflected out loud, "Who am I, or who are you, to say you're not worthy to have kids? . . . We believe in reproductive freedom. You have to think it through." And she described someone she knew who had fourteen kids. "They wanted four-teen kids. That's their choice. If you believe in freedom of choice and privacy and the government staying out of the bedroom then you have to put across the fact that choice involves not only being able to choose an abortion but it also means I can choose to retain a pregnancy or to have as many pregnancies as I want. . . . You will never find me in public talking about population control or zero population growth. What I do personally is something else."

"And when you retire?" I asked.

"When I'm through here I'll probably belong to ZPG [the zero population growth organization]," she said quietly.

"You respect personal freedom. But this means, in reality, that the one organization that everyone assumes is systematically *curbing* the much-touted population boom, is, in fact, perpetuating the status quo," I said, by way of conjecture.

"It's very distressing," Dr. Babbit acknowledged. "The planet goes to hell. I've thought about that a lot. I'd love to sit here and rule because I'd change a lot of things. If you'd like to set us up as benevolent dic-tators, I'd love that. You have to be professional enough to understand the difference between counseling and manipulation. There is a fine balance. I counsel. If I see somebody who is pregnant, has too many kids, is on welfare and has all these other problems, I have no qualms, at all, trying to help her make a wise decision for herself as skillfully as I can. (But) it still has to be her decision."

The inevitability of contradiction—the realization of an over-populated world and an allegiance to freedom of choice—is as much in the phrase "family planning" as in the truth. Family planners in this country do just that, they *plan* out a pregnancy. They do not, as a rule, discourage pregnancies, even though the economics of pre-natal care and delivery are becoming more grim every day.

In constant dollars, Congressional "Title X (Ten)" funds (those specifically designated for family planning) have steadily lost value because of inflation. Every year Planned Parenthood has had less and less money to work with. In 1992, $13 million Title X dollars were allocated to California. In addition, the state itself coughed up an additional $61 million for family planning. Eighty to ninety percent of clients are also eligible for MediCal subsidies, paid for, of course, by the taxpayer. A woman walks in off the street, wants to have a baby, and is likely to build up $6200 in costs to the system, assuming she is not high-risk; if there are complications the bill can be much higher. There is never anywhere near enough government funding support. Dr. Babbit would like to see the funding levels immediately quadrupled.

In addition to women off the street, Planned Parenthood sees countless illegals who do not want to be identified, and for whom the clinics will never be reimbursed. The organization is also treating more and more "multiple pathologies" among the homeless and skid row victims, conducting testing for HIV and STDs, as well as handing out contraceptives.

Unit cost to the manufacturer for birth control pills is about a third of a cent. They are retailed for about sixty cents each. Planned Parenthood gets them for half that price. But the whole system of healthcare in this country has meant that pregnancy, and children, is virtually unaffordable. "We need to shift from heart and lung transplants to doing basic and preventative care for little kids," said Dr. Babbit. "The vested interests prevent us from doing this. In America, the rules have stifled us: you do not give contraceptives until you've given the patient a physical, a blood test, a hemoglobin, a urinalysis, they've been screened for cancer, sexually transmitted diseases, et cetera. That's where you get into the $80 to $100 costs before some birth control pills are even given out. There is no technical, scientific reason why pills should not be given over the counter."

During the Reagan/Bush years, Congressional funding for family planning could not be used for any abortion-related procedure, or even for counseling. You could not so much as mention the word without the risk of losing your funding. The very morning on which I met with Dr. Babbit, Clinton had rescinded that Reagan/Bush gag order by sanctioning abortion counseling at four thousand federally funded clinics, ending the ban on fetal tissue research, opening the way for the importation of the abortifacient RU486, allowing for abortions at U.S. military hospitals abroad, and renewing funding for international organizations that performed or promoted abortion. "Many believe that this is one of the most important environmental steps we can take," said Clinton.[60] Indeed, according to population expert Sharon Camp, "Nothing, nothing, NOTHING" would bring down fertility rates more effectively than providing abortion services for women worldwide.[61]

But, some six months later, when it came time for the House of Representatives to vote on lifting the seventeen-year-old ban on actual federal funding for abortion, the backlash was intense. They voted against it, thus condemning poor women to fertility enslavement. The author of that ban, Representative Henry J. Hyde, a Republican from Illinois, stated that opponents of the ban were seeking to "refine the breed," in other words, to cull the herd of poor people and minorities. Members of the House warned that they would never stand for an abortion provision being incorporated in the nation's health care package. This sentiment was conveyed inside Congress, while outside abortion foes were becoming perversely emboldened following the March 1993 killing of Dr. David Gunn, a man who had performed many abortions.

More recently, Democrat Representative Anthony Beilenson from California has been pressing Congress to increase funding for population control. America's TFR is 2.0, its annual population growth rate, 0.93. But, in Beilenson's home state, the numbers are frightening: 30 million residents, an annual population increase in Southern California of over 4 percent (twice the rate for the rest of the state), and projections of future net immigration destined to double the fertility momentum in the coming decades. What makes that so troubling is that the State of California is the eighth largest economy—after the G-7 nations—in the world. It produces 50 percent of all fruit and vegetables for the United States. Its energy needs and global impact are intense. By perpetrating the American dream around the planet, California's population growth adds an enormous amount of kindling to World War III. How can such allegedly good times be turned back? Clichés do not die easily. Family planners and ecologists know that, short of a revolution in consciousness, a holding pattern needs to be fixed, and that means creating legal barriers to development and levees to hold back the flood of fertility. It also implies some critical points of view with regard to immigration.[62]

Combating not only the population crisis in California, but a global TFR hovering around 3.2, Beilenson is trying to wrest a commitment from the United States of $725 million for family planning programs in 1994, $800 million in 1995, and $1.4 billion (in 1990 dollars) by the year 2000. In 1989, the U.S. gave less than $200 million for international population assistance.[63] But according to polls, 57 percent of the American public support such increases, as do at least 160 members of Congress who signed a letter of endorsement for Beilenson addressed to the chair of the House Foreign Operations Subcommittee of Appropriations, which must decide on such matters as international family planning funds. That increased budget share would help provide fuller access to contraception for every person on the planet by the end of the millennium, and that includes high school students in Los Angeles, who are reproducing faster than teens anywhere in the world. More than half of all those on welfare in Los Angeles are teen mothers. Forty-four percent of the high school dropouts in the city are girls, 80 percent of whom are pregnant. In 1990, Los Angeles witnessed twenty-four thousand teen babies, more than any other region in the U.S.

A century ago, the onset of puberty occurred on average at age fifteen. Today, possibly as a result of the very exposure to omnipresent sexuality, puberty begins around the age of eleven. It's as if the body is meeting the demand on its own. It has been pointed out that an average adolescent is exposed to fourteen thousand instances or intimations of sex on television every year, of which only 1 percent has anything to do with education or contraception. The so-called "Don Juan" of L.A.'s Lakewood High School "Spur Posse" claimed to have achieved "70 points" (sex with seventy different teenage girls). Data from throughout the U.S. does in fact show that nearly a third of all teens between the ages of fifteen and nineteen have had at least six sexual partners.

For nearly seventy years, the American public has recognized the inherent economic and social problems associated with overpopulation. Even when their lawmakers vacillated on the subject, Americans have consistently upheld the concept of population control. The condom (first introduced in Gabriele Fallopio's *De Morbe Gallico*, a treatise on venereal disease published in 1564) was popularized in this country around 1914, when Margaret Sanger, a New York nurse, began publishing a monthly magazine, *The Woman Rebel*. She was arrested and indicted under the stubborn Comstock law of 1873, which made it illegal to disseminate by mail information pertaining to contraceptives, declaring such material "obscene." Sanger fled to Europe, where the public was considerably more sophisticated and practical about these matters, but she insisted on returning to Brooklyn in 1916 where she opened the first birth control clinic in the United States. Police promptly closed its doors. But eventually, Americans sided with birth control; in step fashion, in 1925, in 1931, finally in 1937, the religious and medical communities came to accept family planning as fundamental to medical practice.[64] (At the same time, many U.S. scientists were promoting the concept of eugenics, which the emergent German Nazi party employed to initiate compulsory sterilizations.) In 1965, President Johnson advocated U.S. aid for birth control projects abroad. In 1968, however, Pope Paul VI's encyclical, *Humanae Vitae*, banned artificial contraception. That same year, Paul Ehrlich countered the Pope with his astonishing book, *The Population Bomb*. A year later the State of Virginia and the country of Singapore both passed the first set of non-restrictive sterilization laws, and President Nixon allocated funds for birth control and family planning within the U.S. In 1973, the U.S. Supreme Court upheld *Roe v. Wade*, which was intended to safeguard a woman's right to an abortion.

In November, 1970, Pope Paul VI again rejected birth control programs, stating at a United Nations conference on food and agriculture, "There is a great temptation to use one's authority to diminish the number of guests rather than to multiply the bread that is to be shared."

This, among other effects, will undoubtedly favor a rational control of birth by couples who are capable of freely assuming their destiny. The careful use of language tends to soften what was, in fact, an implacable, desperately unrealistic position, and one that dreamily, sadly that assumed people are rational.[65] In 1984 in Mexico City, the U.S. under Reagan reversed its population policy, claiming that population growth was a "neutral phenomenon." A year later, the U.S. Congress dropped its support for the United Nations Fund for Population Activities, as well as the International Planned Parenthood Federation, and suspended all participation in those programs that in any way supported abortion.[66]

And yet, it has long been understood that there is no better investment, nor more powerful technology, than contraception. According to the Alan Guttmacher Institute, for every $1 spent on family planning, California taxpayers save nearly $12 in welfare costs. In fact, each unintended pregnancy costs California $3317, multiplied by nearly 100,000 unintended pregnancies a year since 1990. Teen births noticeably started rising across the U.S., particularly in Southern California, in 1987. Nearly seventy-one thousand of all births in California in 1990 were to women under twenty.[67] According to the Surgeon General, Dr. Joycelyn Elders, America does "the sorriest job of any country in the world providing family planning. We're always running around hollering and screaming about abortion—and abortion is not the issue. The issue is providing family planning services for all women who need them. Right now, the rich have them, but we don't care about the poor."[68]

President Lyndon Johnson had subsidized family planning in 1964 as part of his War on Poverty. But when it was initially introduced into Los Angeles through the newly created Office of Economic Opportunity, African-Americans perceived it as a form of genocide, the wealthy whites trying to control the population of the poor blacks. Planned Parenthood was essentially kept out of the poor communities on racial grounds. Community doctors were timid before such controversy, much as politicians were in India after the so-called "Emergency" under Indira Gandi. With the Reagan years, family planning came to a virtual halt in the poorer communities of all the big cities.

Under twelve years of the Republican White House, over one thousand health clinics specializing in birth control were closed, creating a social time bomb of unwed teen mothers in neighborhoods of soaring illiteracy and unemployment—"greenhouse conditions" for a baby boom. Along with the Philippines, Saudi Arabia, Malawi, and Haiti, the U.S. was recognized in 1991 by Population Action International as among the least progressive countries in the world in terms of expanding access to family planning services. Conversely, India, Thailand, Colombia, Morocco, and Kenya were hailed for their efforts.[69] Today, there is but one organization in South Central L.A. that receives family planning funds through the Title X federal program: a women's health center.

Less than half a dozen states in the U.S. have populations larger than Los Angeles', yet there are only ten Planned Parenthood clinics in the whole metropolitan region of L.A. Serving a population base of at least 12 million people, Planned Parenthood is calling for twenty new clinics throughout Los Angeles in the 1990s. At present, they serve thirty-nine thousand patients a year, who visit twice on average. Nationwide, 3 million people are served each year by Planned Parenthood. The organization would like to think that they are at the bottom of a hopefully very steep growth curve. If patients say they can't pay, they are taken at their word. Once in the door, a woman (it's almost always a woman, not a man) will get assistance. In addition to the ten primary Planned Parenthood clinics (including a mobile unit), there are another forty to fifty private agencies and clinics throughout L.A., as well as the county health services—a total of some sixty clinics in the metropolitan region. Yet, two out of three women in need of help, or half-a-million females, never get into the system. Many women are not assertive enough to walk through that door. They are afraid. To go to a clinic, says Dr. Babbit, a woman has to have taken control of her body and her situation. Few cultures support such female empowerment, whether in China or East Los Angeles.

Net immigration now comprises about a quarter of all population growth in the U.S.—Third World consumers suddenly stepping into the consumptive cornucopia of the First World. Among them, Latinos have the highest birthrate in L.A. County, in the whole United States for that matter. According to many Latina womens' perception of their "significant other," the male demands control over his mate. He wants his wife to be stuck at home caring for babies because that will reinforce his power and privilege. It may not be an exaggeration: Latina women are having four to five children each. California's fertility rate has increased 17.5 percent in the past decade, making for an overall 26 percent increase in actual population—from 23,668,000 to 29,760,000 as of 1990. Half of those were Latinos. In L.A., Latinos make up 37.8 percent of the population, and utilize 47 percent of all family planning services. In 1990, the city of Los Angeles saw 204,124 births, of which 113,786 were to Latinas.[70] The Latino population of Los Angeles is expected to reach 4 million by the year 2000.

According to Melinda Cordero, the coordinator for the Promotoras Communitarias, a Mexican-originated family planning outreach program that now operates in Los Angeles, Latina women tend to believe that it makes them "more of a woman to have more children," or at least that's what they've been told.

In 1990, of the teenage women between the ages of fifteen and seventeen who gave birth in L.A. County 6329 were Latinas, or nearly 35 percent of total teen births. Their birthrate is higher in Los Angeles than in Mexico City. Their plight—as it needs to be viewed—is such that they have often become the victims of male sexual gang members (the "homeboys,") who molest them and who refer to them as Ho's (whores) or OPP (Other People's Property), or "breeders."[71] Typically, these young women are 200 percent below the poverty level. Afraid of the county hospitals, of officials in general, they are alone in their pregnancies.

Cordero and I drove out past the housing projects, through the smogbound southeast portions of the city—Maywood, Southgate, Boyle Heights—all heavily Latino areas. At a community center in Montebello, she spent an hour with fifteen or so mothers and two fathers—

and countless children—describing contracepive options. She sat casually on a table before the crowd, speaking in Spanish, using a pointer and anatomical chart and video animation to expertly illustrate aspects of a woman's monthly cycle, talking about the female body and self-esteem, and offering cartons of free condoms and foam for the women. After the talk, most of those in the audience swept up the supplies into their handbags.

Some of those women had eight children. But family planning was never too late. About six hundred people have been introduced to the Promotoras program so far. Forty-three instructors, all but one of them female, have been initiated. The meetings have the look of a Tupperware party. Trainers get $50 for going through the course and $25 for each session they conduct. Each trainer will recruit women and try to have six meetings a year.

I asked one twenty-six-year-old, somewhat frazzled-looking Latina mother of four whether her childbearing days were over. "If God wants me to have another child what can I do?" It was a response I've heard all over the world, irrespective of the religion or culture. These women were all low income or poor, unlikely to ever get an abortion, afraid of pills, wary of the IUD. And while statistics show that nearly one in three American women will eventually be sterilized, the Latina women have not followed the national norms.

Men discourage the women in Latino cultures from even going into the clinics. The fathers don't believe in birth control at all. It is not religion that motivates this male-dominated ethos, but the quest for sexual control. So it is up to people like Melinda Cordero and Dr. Babbit to guide these ambivalent mothers, assuring them of free choice, offering mature, womanly, and logical advice when called upon. "You can't go directly at it. You go indirectly," says Cordero. Often for the teen mothers the Promotoras instructors are like surrogate parents, encouraging the young women to go to school, to get out there and rise up as individuals. Dr. Babbit believes that education is the key; that the women have a life that is separate from those of their children; that they own their own bodies. After that, the hurdles include convincing them that the pill will not give them cancer. And that abortion is not an evil word.

"People are not stupid," she told me. "We've had two hundred years of democracy. We know that educated women are not going to have a dozen children. . . . People are smart enough to realize we're cutting down too many trees, that the aquifer under the United States (sic) is greatly endangered by too many people," and so on. But, according to Lisa Kaeser, a senior public policy associate at the Alan Guttmacher Institute in New York, "Back at the start of the family planning movement, our attitude was that all you needed to do was to get the information out. Then people would make the best decisions and there would be no more unintended teen pregnancies." But it didn't happen that way, or not in Los Angeles.[72] Women in California had 321,072 known abortions in 1990, 83,500 of them teenagers; this is the highest abortion rate in the United States, over 46 per thousand women, compared with 27.3 elsewhere in the country. Yet California has no legislative mandate to teach sex education in the schools. In fact, the state has ordered that "no student may be required to attend sex education classes" and "parents must be notified that sex education material is going to be taught and given the opportunity to review it, and to refuse permission for a child to participate."

It has been argued that among the major industrialized nations, the U.S. has the highest rate of teenage pregnancies precisely because of the taboo that persists here against sex education in schools or at home. Only with the advent of AIDS could pregnancy prevention workers even begin to make headway getting into the schools. In the minds of many, for a high school girl to even ask about contraceptives must translate into the fact that she is promiscuous, even though every statistic now shows that teenage girls, like boys, are freely exercising their normal sexual curiosity. Surgeon General Elders has advocated the distribution of free condoms throughout U.S. schools. In New York, Philadelphia, San Francisco, and Los Angeles, superintendents and school boards are taking that call to heart. The L.A. Unified School District has thus far given out thirty-eight thousand free condoms to approximately 15 percent of its student body. The costs, in the context of the annual education budget, are absolutely trivial. Only 5 percent of all parents of students have rejected the program.

Part of the problem with controlling population in L.A., whether one is speaking of teens or older women, is the fact that more than one hundred languages are spoken, and one hundred separate cultures are at work in Southern California.

Consider the 125,000 Hmong who have settled in California (and to a lesser extent in Minnesota). They have brought with them the highest fertility rate in the world—9.5 children per woman—the earliest marriages, and the highest welfare rate of any refugee group in the country. Sixty-two percent of them rely upon public assistance (versus a 7 percent average for the rest of the U.S.). Nobody anticipated that the second generation of immigrants would also be on welfare. It is estimated that between 50 and 70 percent of all Hmong brides are younger than seventeen. Sex with anyone under fifteen is a felony in California, and marriage under eighteen requires the court's permission, so the Hmong act in secret. Why do they have so many children? One explanation is their legacy of suffering. The Hmong fled persecution in China in the 1800s to the highlands of Laos. Later on, they were recruited by the CIA to fight the Vietcong. They died in scores but had scores of children to make up for it. Or that's one explanation.[73]

Whether Hmong or Filipino, Mexican or Korean, Chinese or Polynesian, East Indian or Pakistani, Irish or Guatemalan, more than half of all immigrants to the U.S. find their way to California, most to the L.A. area.[74] Immigration is expected to reach a million or more annually in the coming years. Every week six thousand legals enter California, in addition to thousands of illegals; and it is unlikely that earthquakes will dissuade this avalanche. Even the majority of Latino groups within the U.S. say ethnic migration is excessive. In the past decade, 7.3 million legal immigrants have settled in the U.S. The new amnesty program and the *Immigration Act* of 1990 have increased this flow. In the next decade, as many as 10 million legal and illegal immigrants will enter, with at least half of them coming to California for permanent residence. In addition, 3 million applications for undocumented immigrant amnesty were submitted in 1987 and 1988.

The majority will become permanent immigrants, then U.S. citizens, and will then be able to legally bring additional family members into the country.[75]

By the year 2040, some two generations from now, California is expected to more than double in size, to over 63 million.[76] Southern California's seven major counties are expected to account for approximately 37 million of that total.

In planning for so many people, the Southern California Association of Governments has cited a number of critical shortfalls and anticipated ecological deadlocks. For example, there will be five times more traffic congestion, diminishing average commuter speed from forty to ten mph. Presently 348,000 vehicles per day plod past Vermont Avenue on the I-10 Freeway.[77] Millions of new homes—800 every day—and offices and at least one school a day for 640 new students, will be built upon 650,000 acres of already scarce open space. (Other data suggests that urbanization will convert three times that much acreage.) Many new roads will be built, adding to the existing 21,086.5 paved miles in the city of Los Angeles (an amount equivalent to nearly two-thirds of all city property!)[78] One out of 12.6 households will be without water, while wastewater treatment capabilities will be overwhelmed throughout the Southern California counties—as much as 51 percent in San Bernardino, 45 percent in Riverside, and 19 percent in L.A. By 1996, available landfill sites will also be exceeded throughout Los Angeles city proper, as the discharge and dumping of hazardous wastes increases from approximately 18,250 tons to as much as 1.48 million tons per year. An additional 60,740 gigawatt-hours of electricity will be needed, in spite of existing enhanced energy efficiency programs and Southern Cal Edison's projection that future innovative electrical consumption will save enough energy for an additional 1.5 million homes every year.[79] As for air quality, every major pollutant type— reactive organic gases, carbon monoxide, nitrogen, and sulfur oxides, surface ozone, and particulate matter—will increase, as much as 500 percent in some cases.[80]

Population density throughout the State of California is about 184 per square mile. But in most of Southern California it rises to over

2000 and in Orange County is nearly 3000 per square mile. With a doubling or tripling of density, what is the impact on the environment likely to be?

To answer that, it is crucial to acknowledge that California and India's southern state of Kerala, share a remarkably similar history of ecological abuse. Their population sizes, multicultural immigrant mix, and revenues from elsewhere, invite comparisons. But what most dramatically distinguishes one state from the other is the fact that while Kerala is one of the poorest states in India, and thus in the world, its family planning initiatives are among the most successful; whereas California is the richest state in the U.S., but its ability to curb population growth is the least effective. There are 30 million people, more or less, in each southern, coastal state. In Kerala, more attention has been paid to maintaining a standard of health and contraceptive awareness that would result in a lowering of the total fertility rate. There have been only marginal efforts at diminishing human impact on the environment. In California, despite the valiant efforts of organizations like Planned Parenthood, family planning has had a minimal impact. And, while environmentalists have won numerous victories, those do not begin to counter the all-embracing destructiveness of a population boom. California's overall loss of biodiversity and habitat is a chilling reminder of the total population dynamic. Both regions, then, comply with the end-loser syndrome scenario. All that can be hypothesized is that in absence of existing family planning, the environmental damage would be even worse than it already is.

There are areas remaining in California that recall an earlier time of biological majesty. From the northern Trinity Alps to the high sequoia toward the south, along the upper creeks of the Big Sur, in the least frequented valleys off the John Muir Trail, or among some of the Channel Islands of Southern California, are to be found micro-worlds that seem to predate the arrival of man.

For example, one can lounge in early summer among hundreds of sea lions that come to mate on the dunes and beaches of remote San Miguel Island. Or, swimming along the four-hundred-foot cliffs of Santa Cruz Island, just forty miles northwest of Los Angeles, one can enter a tranquil world of drifting *Macrocystis*, the lush groves of kelp. The light penetrates these vertical columns and transforms the pellucid azure waters into gold and emerald havens of life. In the deep waters, along the intertidal pools, and carousing in the hypnotic kelp jungles and eel grass are as many as twenty-five phyla of marine animals—sharks, rays, fish of numerous varieties, cetaceans, lobsters, and pinnipeds. Out away from the cliffs, the minke, blue whales and dolphins breach for the sheer sport of it, and large flocks of cormorants and gulls, brown pelicans and terns follow their antics, swooping after scraps, riding comfortably on the bunches of surface kelp. Enormous grottoes punctuate the rocky fortifications of Santa Cruz Island, alive with the echoes and aerial maneuvers of the little cliff swallows. Up above, along the knife-edge rims, nonindigenous sheep introduced by ranchers at the turn of the century have gone wild and run riot over the high promontories of grass. Tough old boars and a few feral horses also inhabit these island grasslands. Once home to a large number of Indians, today the soft archaeological middens are the only signs that humans ever slept and ate and dreamt here. When Juan Rodriguez Cabrillo landed on nearby Santa Catalina Island in 1542, about three thousand Chumash Indians were living there. Today there are only the most benign traces that they ever even existed. Whole cultures came and went without the slightest disturbance. Santa Cruz Island is virtually uninhabited. Few visitors ever venture there, though hunters come now and then from the mainland to kill the boars and the prize rams. Most of the islands have been designated a national park, one of those aforementioned victories.

These last remaining oases of ecological integrity offer an intense contrast to most of the coast, where population density has increased 270 percent in the past fifty years. Spend a few days at a place like San Miguel Island, or in the heart of Death Valley, or amid the high Evolution Lakes, then head back into the confluence of Los Angeles—you will feel the insane sting and morbid awakening of World War III, if only for a few minutes prior to rehabituating to the city, which we have grown up accepting as intrinsic to the world.

During those acute seconds of transition from the wilderness to the urban, Los Angeles explodes in delusional directions, a kind of Hiroshima of the senses that some people actually relish. We accept the devastation as prima facie evidence of humanity and its place in the world. We call it human.

Consider what "to be human" actually means, in the case of Southern California. Over fifty thousand humans die in the Los Angeles metropolitan region every year, though that scarcely touches the net gain in population growth. People die all the time, everywhere, and so such mortality is considered normal. In fact we are so accustomed to what is, in essence, biological reincarnation that we scarcely even question it. That habituation to—and national denial of—death is equivalent to our *total blindness* when it comes to the city and its devastating appropriation of nature. Because we are part of the avalanche, we do not know it to be so. In fact, historically humankind has relished that appropriation, more at peace with an inorganic bathtub or a bookcase than with a cockroach or a raccoon. Arrogantly, we might argue that these preferences are destiny—the basis upon which our complex cerebellum has expressed itself in contradistinction to the rest of nature. Wilderness, after all, is etymologically a denegation, meaning, *place of wild beasts*. Yet the fact remains that our neurological hardware and its sensitivities are essentially the same as those of a rat. Nevermind that society hates rats. There is no law that protects them from our wildest, most sadistic behavior. This reversion from the truth of ourselves, taken to its most "civilized" extremes, accounts for our scientific paradigms and the way we observe, the conclusions we predictably draw. Psychologically we are almost incapable of confronting the nature of human appropriation.

Thus, the leading causes of death in Los Angeles are described in such a way that they give little or no clue to environmental causation. Most deaths are associated with heart disease (32 percent), are malignant neoplasms (22 percent), are cerebrovascular (7 percent), or the result of chronic obstructive pulmonary disease (5 percent). In defining death in humans, and in our approach to most medicine, we have dealt in effects, not causes.[81]

However, mounting evidence and hundreds of research papers have begun to link at least some of those deaths to the area's chronic pollution problems.[82] As early as 1960, studies in 117 U.S. cities showed several definitive relationships between sulfates, particulates, and human mortality. In addition, investigators have found conclusive causal relationships between stomach, prostate, esophagus, and bladder cancers, and ambient concentrations of total suspended particulate matter.[83] In one recent investigation, six thousand nonsmoking, vegetarian Seventh Day Adventists in Southern California were monitored for six years and it was demonstrated that the "risk of malignant neoplasms in females increased concurrently" with higher total suspended particulates, the stuff of countless industries and the automobile. In addition, an association between ozone and respiratory cancers was noted.[84]

While an estimated ten thousand human deaths come from cigarette smoke, countless others result from reactive organic gases like benzene, methane, and petroleum, butane and propane; from suspended particulate matter in the form of dust, soot, smoke, and chemical droplets all in the submicron size that reach the lung's vulnerable lower alveoli. Some air toxics, like hexavalent chromium, exceed the high risk levels (5 nanograms per cubic meter) four to six times in many parts of Los Angeles. Other trace metals from such materials as paint stripper, carcinogenic pesticides in food, and occasional cancers picked up from drinking water add to the fatal puzzle. The number of environmentally related diseases in L.A. affect the eye, nose, throat, brain, stomach, heart, and lungs. It was recently noted, for example, that asthma among African-American children has doubled. The Agency for Toxic Substances and Disease Registry (of the Department of Health and Human Services in Atlanta) has cited seven definitive health risks to humans living near hazardous waste sites: birth defects and reproductive disorders, cancers, immune function disorders, kidney dysfunction, liver dysfunction, lung and respiratory diseases, and neurotoxic disorders.

Among L.A.'s refinery workers, hydrogen sulfide exposure has been shown to impair the nervous system, the ability to stand upright, and balance and reaction time. Chemicals from various dumpsites in Los Angeles have impacted residents during storms, when there is overflow of holding areas, or when watering tanks evaporate and those persons downwind are exposed. The oldest human functions, the brain and nervous system, may be the first impaired by chemical wastes. It has been suggesed by some researchers that this generation of children throughout much of L.A. county may be intellectually impaired by five IQ points due to the high levels of lead in their bodies. In 1989, a USC Medical School study showed 80 percent of those young homicide victims examined to have "notable" lung abnormalities and 20 percent exhibited severe lesions in the lungs. In some parts of L.A., children—who respire as much as seven times more than adults—were shown to have 75 percent of their lung capacity reduced because of the heavy pollution.[85]

Los Angeles has the worst air pollution in the United States. In 1991, *each* of over 8 million automobiles (making 11 million separate trips) churned out 13.2 pounds of hydrocarbons, 231 pounds of carbon monoxide, 23 pounds of nitrogen oxides, and 14,500 pounds of carbon dioxide. In addition, there are over 60,000 major sources of air pollution on 23,000 sites in the city putting out approximately 9000 *tons of air toxins every day*! Rubber specks, oil leaks, methane-producing road surfaces, paint and other solvents from point sources, all contribute overwhelmingly to the hydrocarbon problem. Meanwhile, Los Angeles also has the most dangerous ground-level ozone burden in the country. More than 2 percent of the ozone layer overhead in the stratosphere has been depleted—also a U.S. record—while the city's atmosphere contains at least 25 percent more $CO_2$ than it did a century ago. On the macrolevel, California ranks with the ten industrial countries of the world for its contribution to global warming and ozone depletion. Even the micrometeorology has been usurped by humans. For example, the urban heat effect means that Los Angeles at night is ten to fifteen degrees warmer than the surrounding countryside.[86] A resident of Los Angeles literally warms the planet nearly 9 times more than a Chinese and 14.3 times more than an Indian.[87]

Coupled with air pollution is serious ground pollution. Los Angeles county produces fifty thousand tons of garbage every day, more than twice that of Tokyo. The seventeen landfills are running out of room, even though one of them is a monstrous two square miles in size. Of the 100 million pounds per day of waste, at least one-hundred thousand pounds are toxic and hazardous.[88] The amount of this waste is amplified when one adds to it all of the sewage produced. From 1925 to 1950, beaches south of Santa Monica were frequently quarantined because raw sewage, virtually untreated, was flowing directly into the Pacific. If you wanted to walk on the beach in Venice, you often had to wear high boots because the fecal matter was ankle deep. Even after World War II, two hundred fifty tons of solid waste and millions of gallons of wastewater continued to flow directly into the ocean. Wastewater was eventually treated and converted to sludge—a toxic mass that was diverted into a veritable "Grand Canyon" of wastes a few miles out in the Pacific. Marine life was wiped out in the vicinity. Today, the city continues to hemorrhage forth hundreds of millions of gallons of half-treated sewage and two-hundred thousand pounds of toxic-laced solid matter into the ocean every day. And whenever a bad storm hits, the 6500 miles of sewer pipes and the wastewater facilities are overwhelmed and the runoff goes directly into the ocean. In 1989, "150,000 pounds of lead, 4.5 million pounds of oil and grease, and a slurry of garbage and pet feces (washed) down to the beaches."[89] In 1991 60 million gallons of chlorinated water entered the ocean during a week of storms—six times the amount of killer liquid as that which flowed from the Exxon *Valdez*. In January 1992 another routine storm produced four million gallons of runoff which entered the ocean here.[90] I remember it clearly because I was writing this chapter while looking out at all that sludge. And it was disclosed in early 1994 that over a period of fifteen years, a single oil company north of Los Angeles knowingly leaked 8.5 million gallons of petroleum thinner into the ocean, making it the largest such spill in California history.

The result of such accumulated wastes is the killing of U.S. waters, whether in Puget Sound, New York Harbor, the Mississippi River, Chesapeake Bay, the Great Lakes, or Southern California's coastline. The last salmon spawned in the Los Angeles River in 1910. Ninety percent of the city's wetlands have been destroyed. Shellfish vanished, bacteria species proliferated, kelp died out, and the bay—by various estimates—today is dying. The power plants discharge hot water into the ocean—1.5 billion gallons of it every day, covering hundreds of square miles of ocean. DDT concentrations off the Palos Verdes Peninsula just south of Los Angeles, are the highest in the U.S. Correspondingly, Southern California marine life has the highest bioaccumulation of toxic chemicals of any fish in North America. The Food and Drug Administration (FDA) has declared much of it unfit for human consumption, particularly the bottom-dwelling species. These are just a few of the many legacies of "successful" development and the American way of life, which so many political leaders and administrations have proudly hailed.

Add to this the so-called "sleeper" issue, namely, nonpoint (difficult to analyze) pollution sources in the state. The EPA lists 295 California waterways as polluted. Of these, more than 94 percent are from nonpoint sources—diffuse causes like "street runoff or cow pastures."[91] Other difficult to trace pollution sources include ozone. As of 1977, L.A. basin ozone emissions accounted for at least a million damaged ponderosa and Jeffery pine trees as well as degradation of coastal sage scrub communities. Seedlings of the giant sequoias hundreds of miles away are showing reductions in root growth, while a number of other pine and cedar and black oak species are showing definitive signs of vulnerability to increasing ozone pollution.[92]

Plastic and synthetic fiber pollution continues to kill tens-of-thousands of California marine animals every year, either through ingestion or strangulation.

And pesticides pose an even greater hazard. There are still no environmental impact records kept on countless numbers of unrestricted pesticides. As for the restricted ones—85 million pounds were applied in 1986 in California. Considering the amount of food produced in California for the United States, it is thus not surprising that more than 36 percent of all foods in the U.S. were found to have pesticide residues in 1991. Little progress has been made in the state to regulate these toxins. According to the authors of the California Policy Seminar Research Report (In Our Own Hands), the state department mandated with protecting the environment and the public from pesticide poisoning has been unsuccessful thus far in implementing compliance with safety standards.[93]

In addition to the ground-based wastes, runoff, air pollution, pesticides, and marine contamination, countless other fallouts from Southern California urban life add to this perplexing picture. Electromagnetic pollution, coming from the city's high-voltage wires, is suggested to correlate with nervous system tumors in children.[94] Environmental factors have been blamed for a remarkable drop in male fertility and the number of spermatazoa, whose volume has plummeted from an average of 60 million per ejaculation to 20 million, not just in Los Angeles but in cities throughout the developed world.[95] Whether indoors or outdoors, there is no escaping the enormous range of adulterants that a city like Los Angeles imposes on human life.[96] The so-called "sick building syndrome" has been studied extensively in Los Angeles, where radon concentrations and a number of other indoor-specific pollutants have been traced to the preponderance of morbidity, given the amount of time that individuals spend indoors. There is a large research literature on these many ill-health effects.

In the mental health realm, it is known that there are more psychiatrists per square mile in Los Angeles (some two thousand) than anywhere in the United States, and that does not include L.A.'s additional arsenal of psychologists and counselors, a huge reservoir of mental pacification in a sea of tumult.[97] Add to psychic pollution the myriad other forms of stress occasioned by city life, and an even more baffling statistical saga emerges—whether in the form of crime, suicide, countless mental disorders pertaining to sheer population pressure, sensory annoyance, mental pain and anguish, and the feeling of vulnerability, uselessness, isolation, and loneliness.[98] In 1991 there were a total of 67,820 known mental patients who filed for insurance in Los Angeles but it is estimated that as many as half of all mental patients do not file for insurance claims.[99]

A record number of homicides were committed in L.A. County in 1992—2589 killings, an 8 percent increase over the previous year. At the same time, nearly 300,000 hunting licenses were issued in the state of California that year. Crime in the U.S. is forty times greater than in Japan, twenty-one times higher than in all other industrial countries. Within the U.S., Los Angeles shows the highest number of such crimes. Some school supervisors in the U.S. have now sought burial insurance as part of their school packages—so prevalent is the violence in schools. At one high school in Hollywood, students are periodically searched for weapons before entering the classroom. Recent polls showed Salvadoran war refugees emigrating from L.A. back home because they found Los Angeles too dangerous and too poor. Hate crimes in Los Angeles rose by 11 percent in 1992, typified by the beating of Rodney King, the killing of nearly sixty people, and damage exceeding $1 billion, in the L.A. riots.[100]

Hundreds of sirens can be heard throughout the city every twenty-four hours. And every day 4696 airplanes take off or land somewhere in the Los Angeles basin—17 percent of the nation's total and the largest combined source of aerial noise and pollution in the world. Most of these planes are known as Stage 2 aircraft—like 727s, 737s, and 747s—meaning that they emit noise in the 120- to 130-decibel range. Very loud! The noise comes from the tip of the jet engine propeller, which turns at supersonic speed. Military planes, like the AT6 warbird and the incessant helicopter fly-overs throughout the day and night, are even louder. Indeed, there are many hours of the day when Los Angeles seems to be putting on a perpetual airshow, or mobilizing for war.[101]

The colossal combination of all these physical, chemical, and psychological ills and assaults has not dampened the enthusiasm of millions of pilgrims who make their way to Southern California each year, whether to settle here and look for work, to visit Disneyland, to thrill to the horrors of an 8-point plus earthquake at Universal Studios, or to simply "hang out" at Venice Beach. Despite an exodus of Los Angelinos to New Mexico, Oregon, and Colorado, many remain in the state, and many more arrive here or are born here. Between its grateful immigrants, starry-eyed tourists, 14 million heterogenous locals, and 130,000 gang members, its rich and famous, its millions of poor, and 40,000 homeless, L.A. utterly challenges the doctrine of progress. As the foregoing has endeavored to suggest, "progress" is costly. Many people are mesmerized by Los Angeles. Millions of others obtain intermittent pleasure from the experience of living in the city. That erratic joy seems to justify the ignoring of the city's dark side. But the real victims of this two hundred-year-old spasm of development are not merely the millions of other people who do not particularly relish the urban experience but have no easy way to alleviate their situation, but California's biological heritage. And this is where Los Angeles, and much of the state, has set a melancholy example of irresponsibility that the rest of the world appears bent upon following.

## THE END LOSERS

While the human fiasco in Southern California is veiled beneath the illusion of success, its impact upon other life forms is less concealable.

I set out one day to calculate the amount of "green territory" left in the metropolitan region of Los Angeles. The city *proper* contains 470 square miles, or 300,800 acres. The *county* officially lists 4025 square miles, or 2,576,000 acres, within its boundaries. In the fifteenth century, several thousand Native Americans lived scattered throughout what is now Los Angeles. Even then, the L.A. basin was referred to by locals as the "Bay of Smokes" because of the natural low inversion syndrome, which trapped the accumulation of smoke into a persistent haze. The vertical shifting, or atmospheric carrying capacity, was always limited by dint of geography. And the same limiting situation prevailed in terms of water availability. It has been estimated that the inherent water and land resources could reasonably support no more than two hundred thousand people in the L.A. basin. By 1900 Los Angeles city had expanded to 44.35 square miles of inhabited territory and its population to 102,479.[102] At that time residents were consuming 26 million gallons of water every day. Now, Los Angelinos consume far more than that in one hour. In fact, it has been calculated that a staggering 1.5 million gallons of water per Californian are taken up for food production every year. To produce all that nourishment, nearly 30 percent of the State of California has been irrigated. Most of the water comes from outside the state. As in the Netherlands, 30 million Californians survive only by scavenging their most precious substance from elsewhere.[103]

According to Paul and Anne Ehrlich, a million people in a city need four hundred square miles of surrounding land in crops for even a vegetarian diet under theoretically maximally efficient utilization. Multiply that by an inefficiency factor of probably five-to-ten (based upon current modes and standards of energy extraction, and the small number of vegetarians within the overall population), and the more realistic picture emerges: to ecologically accommodate a million people, their impact *should* be diffused over a region between two thousand and four thousand square miles in size. It means that Southern Californians need something like thirty thousand to sixty thousand square miles of open space. In a sense, that's one way to define the size of the city and it means that Los Angeles county is not four thousand square miles but tens-of-thousands of square miles in extent.[104] But if one compares this illusory hegemony of consumption with the known truth of a large carnivore's needed territory, sixty thousand square miles might only serve a few thousand animals, not ten or fifteen million. Consider the predators of Serengeti (referred to in chapter 5) or the fact that a single Alaskan grizzly needs approximately one hundred square miles to survive. Human "ingenuity" may have maximized and consolidated its extraordinary exponential powers of extraction through invention and engineering, but the costs, while hidden, have not diminished. Efficiency does not reduce the number of equivalent consumer burden, except in relative terms. To look at a large human population is to peer through a veil of deftly concealed ecological costs, a dense conurbation of bills and unpaid-for destruction.

Los Angeles easily exploits the aforementioned sixty thousand square miles of open space, but they are not anywhere in Los Angeles county. In essence, it has vastly exceeded its carrying capacity. In fact, as of 1986, a mere eighty-three hundred so-called "agricultural acres" remained in L.A. county—but by definition that includes barren flood plains and golf courses. (Add to that meager figure the roughly seven acre garden patch given by the city in 1993 to the residents of South Central L.A. following the riots.) As late as the 1960s, some one hundred farmers cultivated about thirty-five hundred acres of vegetables between the South Bay and Santa Monica. Today only sixty acres of actual cultivated area are left. Spinach fields were turned into housing tracts, rows of celery in Venice became dry cleaners and liquor stores. Farming in the city environment is virtually cost-prohibitive. The would-be farmer is assailed on all sides by high water bills, building codes (greenhouses with fiberglass roofing are illegal, for example), required public parking spaces, special ramps and bathrooms for the disabled, and actual zoning regulations that specifically forbid "farming"! It is ironic that where the most people live, the production of food is virtually illegal.[105]

The twenty-eight thousand existing public acres contain many buildings, whether police stations, public utilities, city halls, community centers, or fire stations. Even the parks could hardly be described as "green." Most have been retrofitted with public bathrooms, parking lots, tennis courts, basketball courts, artificial lakes, trails, park benches, and gardens shorn of all diversity save for grass and a few species of flowers. The last stretch of marshland in Los Angeles has been covered over. There are a few hundred acres remaining of the Ballona Wetlands on the ocean, which are frequently inundated with polluted runoff. Aerial studies of L.A. go down to only the 2½-acre resolution, which means that every backyard is excluded from official tallies. But if you examine the average neighborhood south of Wilshire Boulevard—a somewhat consistent, at times ambiguous, socioeconomic and ecological dividing line from the ocean to downtown Los Angeles, a distance of nearly twenty-five miles—approximately 80 percent, on average, of any property is given over to construction, not biodiversity. Add to this the industrial sections of the city—south and southeast of downtown—in addition to all the highways, and the 80 percent Total Human Appropriation factor escalates to about 90 percent.

In the 1950s the city, excluding the San Fernando Valley, commissioned a map of so-called "building footprints" and even then it was discovered that there was very little that had not already been touched, or converted (as the Indonesians understand the term). Today nobody in government, or in research, actually knows how much of nature has been appropriated in Los Angeles, and few people seem to care. After all, a city is a city. Furthermore, Southern California is large. There is plenty of surrounding desert in which to expand, as the growing congestion out toward Palm Springs, one hundred miles east of downtown Los Angeles, indicates.

Los Angeles possesses America's only urban mountain range, a rock-strewn cordillera mostly lower than two thousand feet in altitude, that stretches from the Ventura county line all the way into Pasadena. Beyond this the uplift becomes known as the San Gabriels, real mountains deep in snow during winter, ascending to nearly eleven thousand feet out above Palm Springs, highlighted by many thousand-foot vertical cliffs. Within Los Angeles county the mountains comprise 150,000 acres with over thirteen hundred plant and animal species. Oddly enough, Los Angeles city proper has less open space per thousand people than New York, only 1.2 acres.[106] The pressure on the Santa Monica mountain ecosystem from real estate development and human recreation is intense, though the impact is far less demonstrable than upon surrounding flatlands, which are so much more accessible.

Beneath the dense metropolitan shroud is the all-giving, all-receiving soil. I spoke with several soil scientists in Los Angeles who agreed that, as far as microbial activity was concerned, the dryness beneath all those homes and buildings meant that most bacteria were probably in a steady state, awaiting moisture, weather, and colonization by plants. While toxins will alter individual microbes, microbial populations are highly dynamic. They will wait it out, or readjust. The same cannot be said of most higher organisms.

California hosts a known 7850 plant species, a third of them found nowhere else on the planet. In addition, there are 742 species of vertebrates, 540 bird species, 214 terrestrial and aquatic mammalian species, an estimated 28,000 species of insects, 47 amphibian species, 82 species of reptile, 110 different species of fish, 1200 species of lichens, nearly 5000 different types of gilled fungi, 300 to 400 slime mold types, and 660 species of mosses and liverworts.[107] This profusion places the state among the most biologically diverse of regions in the world, and renders it exceedingly vulnerable to the above-mentioned human activity.

At least seventy-three major plant and animal species have gone extinct in California in the last one hundred fifty years, including the grizzly bear, the state's official mascot. The last one was killed in the mountains of Tulare County in 1922. In fifteen years, the sturgeon went from being healthy and widely manifested in California waters to near extinction. The sharp-tailed grouse, last seen in 1915, went extinct in less than thirty years from the commencement of its habitat loss. A single dam on the San Joaquin River wiped out an entire species of chinook salmon. Most native frogs in Southern California have vanished.[108] The California condor was a common sight, at least until 1870. By 1943 only one hundred were left. As of 1976, only two had been seen in the wild. On April 19, 1987 the last wild California condor was captured. Seventy-six of the animals have been kept in an isolated captive breeding environment. Eggs have been inadvertently crushed by researchers, and of those few released back into the wild, at least three have died as a result of flying into electrical wires.[109]

Of California's known extinct species, twenty-four were endemic—in other words, they existed only in California. But the situation is much more dire than the number twenty-four. Countless other species are in sure decline, though perhaps not quite at the point of immediate extermination. In the findings of the 1987 report requested by the California Senate Committee on Natural Resources and Wildlife, an astonishing 33 percent of all California mammals, 25 percent of all birds, 33 percent of all reptiles and amphibians, 40 percent of freshwater fish, and 12 percent of native tree, shrub, and wildflower species, were deemed threatened with extinction, based upon current trends. The report states that 178 natural communities (47 percent of California's wildlands, or more than 47 million acres) are "imperiled." And that "all or nearly all marsh, riparian and bottomland, and sand dune habitats and four-fifths of herbaceous, estuary, lagoon, and bay communities are rare or threatened. A quarter to half of the scrub and chaparral, woodland, forest, lake and pond, and stream communities are uncommon or threatened."[110]

Over 17 million acres of wilderness have been completely destroyed in California.[111] Along with the extreme surface losses, aquifer supplies underlying the San Joaquin Valley are being depleted by an estimated 500 billion gallons or more every year. Hundreds of square miles of the valley—America's premier agricultural production area—are now brackish and turning to salt. Many have predicted that at current population and agricultural growth rates, a startling proportion of California could become desert during the twenty-first century. This is all the more devastating when one considers that California contains 25 percent of all plant species in the United States and Canada combined, with 2140 of those species unique to the state.[112]

Concluding from their many case studies, the authors of In Our Own Hands admit that "there is as yet no effective means of protecting these habitats." Using their own urbanization statistics and population projections, they predict that "between 1 and 2 million acres will be urbanized in the next decade to accommodate population growth."[113] At the high end, that will add 40 percent of converted area to the existing urban acreage in the state. In addition, there are currently thirty new proposals for large dams in California being studied, while California's forests continue to be logged "three times faster than the national average."[114]

With the exception of the Mugu Lagoon, to which public access is strictly controlled by the U.S. Navy, all other estuarine areas in Southern California have been disturbed or destroyed. A groundbreaking 1972 study by Rimmon Fay and colleagues at the Center for California Public Affairs in Claremont, California, highlighted the specific threats to the hundreds of species depending on the estuaries. Pacific herring, for example, deposit their eggs on the eel grass and the young spend their first year living in the immediate vicinity. At least a dozen other species including barracuda, halibut, perch, and spotted sand bass use the grass as a nursery. Many of the more than one hundred fifty avian species and more than 2.5 million wildfowl that come to Southern California each year along the Pacific Flyway between Alaska and South America, over-winter among the estuaries. The dredging of these fragile ecosystems to build harbors removes the mudflats, disabling the ecosystem so that it cannot regenerate crucial fertilizers.[115] In addition, the authors revealed that an individual walking across a mudflat can annihilate all ghost shrimp burrows and 50 percent of all clams along the path taken.

At the Santa Monica pier, the fish are largely toxic now, the benthic algae gone, though the breakwater is still inhabited by occasional starfish and sea urchins and barnacles. The Venice Canals possess some clam beds and occasional fish, heavy algae and ditch weed (the resident duck populations were found to be disease carriers and condemned to death), while Marina del Rey supports a "marginal marine biota" despite extremely dangerous industrial wastes from Ballona Creek, the normal storm drain runoff, and high levels of antifouling paints from the hulls of thousands of recreational boats. At Dana Marina, submerged sewer outfalls and an artificial habitat have all but destroyed the previously healthy biology of the harbor. During the 1950s in Long Beach Harbor, the apparent policy of the city was to drive to extinction all known forms of marine life there so as to protect wooden ship structures from the destructive boring of shipworms and gribbles.[116] This shortsighted strategy had been "successful" up to the early 1970s, when it was decided that water quality should, in fact, be upgraded. Thus, a few biotic components were "allowed" to return, namely, some invertebrates and pollution tolerant algae. This "return" indicates that where even minute portions of the original ecosystem remain more or less intact, or where natural conditions are restored even minimally, a remarkable biotic profusion can be found, including 158 species of birds and many invertebrates. Policy changes can reverse, over time, much of the damage that has been wreaked by humankind. Nature could make a comeback if given half a chance.

All that one can predictably say about the beaches of Los Angeles is that they tend to be white and broad. Otherwise, they represent the zenith of human modification. All the dredging, construction of piers, jetties, pipelines, harbors, and the congestion of urban life, in addition to storm runoff and daily wastes have poisoned and dislodged the entire biomass of the last several thousand years. While the occasional worms and clams, sand dollars and sand crabs can be found, most tidepools have been scavenged and shorn of all but the sparsest biological communities. Studies of impact on tidepools by curiosity seekers have generated certain predictions: every rock weighing less than four hundred pounds will eventually be turned over, its clinging inhabitants destroyed. The intertidal communities, the integrity of the former salt marshes, the protected embayments utilized by migratory fowl, are mostly gone. Marine life has been essentially smothered by chemical and physical discharges, by sewage sludge, by trace metals of mercury, lead, cadmium, and chromium, by acids and alkali compounds, particulate organic materials, chlorinated hydrocarbons, and thermal heat from utility plants. This latter process sucks vast amounts of water from the ocean—with everything in it, such as larval stage organisms—and passes it through heated power plant condensers to cool them before reversing the flow back out into the ocean.

The California Department of Fish and Game began record keeping in 1916. Fish production peaked during the 1930s, and then began a near constant decline. In the 1920s and 1930s, beaches in Los Angeles were inundated with oil, which may well have contributed to the loss of fertility in the oceans. The Los Angeles housing boom and an oil bonanza coincided. House owners often drilled three or four oil wells in their backyards a few hundred feet from the water's edge. There were thousands of such wells by the time of World War II. Most recently, the Chevron Corporation got the go-ahead for routine transport of oil in shore-hugging tankers from Santa Barbara to Los Angeles, and for increased production by over 50 percent, to eighty-five thousand barrels a day. California's governor hailed the Coastal Commission's go-ahead a "courageous stand." While local environmentalists have pointed out the many calamitous oil spills and oil leaks in Southern California's history, and the statistical likelihood of others, they do not have the financial support to fight the Coastal Commission's decision. (Chevron recently opted not to initiate the production increase go-ahead, but others are queuing up to do so.)

All of this carnage—most of it largely unappreciated by Californians themselves, who are unwitting agents in the destruction—translates into an estimated 663 plants and 220 vertebrates that are on the brink of extinction throughout the state. Yet, between the federal and state governments, only 128 animals, 213 plants, and 103 insects are listed as endangered, or candidates for listing. In metropolitan Los Angeles, there are 174 species *officially* listed as endangered, rare, threatened, or extinct. Those organisms include the Guadalupe fur seal and the California bighorn sheep (can anyone even imagine the preconditions necessary in the 1990s for a wild bighorn sheep in Los Angeles?), the dwarf golden star, the short-eared owl (last seen in 1932), the California black rail (which disappeared as of 1928), the Coopers hawk (gone since 1921), the American peregrine falcon, the island night lizard and tiger beetle, the El Segundo blue butterfly, seven species of salamander, the black toad, and California freshwater shrimp.[117]

Official restraint in naming and protecting species typically hinges upon political compromises effected under regimes of economic pressure by special-interest groups. These in turn are linked to, or derive directly from, the implacable realities of California's population boom—the number of unemployed, the state's budget committees with their own vested interests, a noncohesive patchwork of human values, the skewed self-interested hierarchy of social priorities, and ultimately, California's fundamental ignorance of and lack of sensitivity toward, its own astonishingly privileged nature. All of these characteristics work to catapult development forward. Add to this mechanism the fastest growth, and largest financial stakes in all of the U.S., and the true image of a California in crisis begins to emerge.

Only 6 percent of land in the state is biologically protected. That would include a place like the center of Yosemite Valley, with its hotdog stands and summer haze from congested camping—ten thousand people a day during the late spring and early summer—a region readymade for the incoming avalanche of recreational vehicles, motorcycle caravans, and humanity in general. The first automobile entered Yosemite in 1900. Under newly established Department of Interior jurisdiction, two regiments of the army assumed permanent residency. A power plant and telephone lines were installed and a railroad track was laid to the entrance of the park. Under the National Park Service's first director, Stephen Mather, automobiles were officially welcomed.

"Our chief duty is to make our National Parks comfortable for all the people. . . . It is the right of the hundred million owners of the Yosemite to visit that incomparable shrine in as much luxury as each can afford, and it is the duty of the Government representing these millions to provide each one who comes as nearly as possible, all things considered, the degree of comfort, even of luxury, that he requires . . ."[118] By 1920 not a single Southern Miwok Native American was left in Yosemite. And by 1928, nearly 1.5 million automobiles had come into the sanctuary of sanctuaries.[119] Yosemite is but one prime example of how difficult it is to ensure the basic principles of conservation within a democracy.

California's many desert areas pose another level of compromise. The Bureau of Land Management estimated in 1980 that 47,900 acres of desert in California, some of it allegedly protected, had been disturbed. But, with thirty-five thousand miles of roads and trails in the desert affording access for destructive desert bikes and cactus collectors, other researchers at the University of California have estimated twenty times that much impact and damage, or nearly one million degraded acres.[120] Discrepancies in preservation figures begin to resemble similarly frustrating information gaps in many developing countries.

There are only 265 game wardens to protect the entire State of California from poachers, of whom 95 percent go unpunished. The wildlife protection budget for the state has dropped to a mere $22.9 million, gigantic by Indonesian or Indian standards, but a paltry sum for the eighth largest economy in the world.[121]

Poachers clandestinely lay miles of drift lines studded with thousands of hooks just offshore, by dark. And, as journalist Paul Dean writes, they "mow down elk, including their calves, with AK-47 assault rifles. Or run over wounded deer when bullets run out. They harvest organs and paws from bears and leave the carcasses to rot. They bring to this war forged permits, spotlights, diving gear to take lobster and abalone from protected areas, off-road vehicles, police radio scanners, infrared spotting scopes, military weapons, illegal lines that fish a mile of ocean. . . . And, an understaffed and outgunned state Department of Fish and Game can't slow the slaughter."[122]

Further complicating the scenario, some Southern California land-owners cut down their coastal shrub to discourage endangered species—such as the gnatcatcher bird—from nesting in the area, and to prevent such developable lands from being contested by environmentalists. Farmers in Riverside county have actually destroyed their fallow fields in order to eliminate endangered rodents like the kangaroo rat, which some have idiotically blamed for the fires of 1993. "Private landowners across the United States are quietly waging an underground war against endangered species, killing them off or destroying their native surroundings to avoid costly government restrictions intended for their protection." Elsewhere in the country, bald eagle nests have been burned, owls killed, sea turtles bludgeoned.[123]

While its exercise of policing power, protective custody, and formal, scientific definitions may be hamstrung to a certain extent, the Department of Fish and Game does not mince words when it comes to listing the human causes of extermination, the very verbs of warfare. The following activities are deemed responsible for the demise of plants, in order of magnitude: development, livestock grazing, off-road vehicles, roads, agriculture, other exotic plants, water projects, human and equestrian trampling, fire management, feral animals, mining, landfills and garbage dumping, collecting, logging, flood control activities, energy development (pipeline and powerlines), water quality degradation, hybridization, vandalism, disease, and finally, climate. For animals, the course of degradations is as follows: development, water projects, newly introduced predators, agriculture, livestock grazing, off-road vehicles, other human disturbance, pesticides and other poisons, flood control, exotic plant imports, climate, energy and mineral development, roads, logging, disease, habitat fragmentation, extinction of habitat, hybridization, water pollution, and collecting. If one were to catalogue the behavior of *Homo sapiens* throughout history, many of these activities, in one form or another, would be recognizable about twelve thousand years ago, a period noted for the beginnings of agriculture in what are Iraq, Thailand, and Israel. But what distinguishes today's peril, of course, is the sheer *number* of people engaged in these enterprises.[124] Remember, Southern California's seven counties are verging on a population size of 37 million by the year 2040.

In trying to protect what's left of Southern California's biological integrity, the Department of Interior in Washington has suggested a compromise model, beginning with the rare gnatcatcher. Interior has agreed to "minimal development" in the area of the bird's nesting grounds, one of the only times in the twenty-year history of the *Endangered Species Act* that the Interior Department has compromised with respect to allowing any development whatsoever. There could be no clearer sign of the escalating war against nature than a compromise from the one public trust mandated to protect America's wildlife from greed, when all else fails. But this incident of compromise is much larger than the Southern California gnatcatcher; it is a codified pattern. As of December 1992, the Fish and Wildlife Service had classified a mere 755 species of plants and animals in the U.S. as threatened or endangered. Worldwide, according to the Threatened and Endangered Species lists, candidates number 1284 (for plants) and 3964 (for animals). What these dramatic understatements belie is the extent of the homocentric, economic imperative; the appalling arrogance and ignorance and political inertia that would obfuscate and deny the scientific truth. To argue that fewer than fifty-three hundred species out of a possible 100 million on the planet are currently threatened or endangered is akin to denying that the Holocaust ever happened. More infuriating is the fact that the government's very assessment criteria necessitate such blatant falsehoods. Is it possible that all the excellent science has had so little impact on the convictions of our political administrators?

Further eroding one's confidence in such official statistics is the wording (and practice) of the *Endangered Species Act* itself, which has been described as "a landmark decision." Under the provisions of the act, it is legal and allowable to "take" a species, even though it is listed as endangered or threatened. Hence, the true groundrules of compromise. And what does "take" really mean? It means that one can "harass, harm, pursue, hunt, shoot, wound, kill, trap, capture, or collect, or attempt to engage" if the activity is "incidental to a legal activity"—that is, if it is not the primary activity—but is somehow the by-product of a legal activity. In addition, such killing and so forth must meet the criteria of a "habitat conservation plan" such that the activity does not jeopardize the overall survival or recovery of the species.[125] Such wording and intent are classic "doublespeak," Orwell's term for linguistic ambiguities deliberately contrived in order to conceal evil or perpetuate the status quo. Such grammar fosters destruction under the guise of good-intentions. From 40 to as many as 59 percent of all species protected under the *Endangered Species Act* in the U.S. are declining in number. Only five species out of more than six hundred have made recoveries. Less than seven hundred thousand acres nationwide have been set aside for endangered species. Habitat destruction is the key to their demise and virtually nothing is being done to curb America's trespass. Yet the U.S. Inspector General's office estimated in 1990 that it would cost no more than $4.6 billion to save the six hundred or so most endangered domestic species. The figure is difficult to assess considering a pattern that seems to authorize as much or more budget allocation to lawyers and additional studies as to actual restoration of habitat. In addition, the U.S. Fish and Wildlife Service's total budget is only $8.4 million, which means it did not have the clout, between 1987 and 1991, to stop 1966 federal construction projects that directly and knowingly accelerated the demise of existing endangered species, all in the spirit of so-called "compromise."[126] These bureaucratic labyrinths and denials are desperately discouraging. Pessimism prevails throughout the conservationist community. Humanity appears almost incapable of marshalling the necessary reserves of intelligence and compassion to forge an end to the war, even in a place as blessed, and seemingly aware, as California.

## AMERICA'S CHALLENGE

At the beginning of the EPA Superfund project, there were thought to be about five hundred sites that could be cleaned up in five years at a cost to taxpayers of $1.6 billion. Now the list is projected to hit four thousand sites, and the cost of cleanup to exceed $1 trillion, including Defense and Energy Department waste sites. For even partial cleanup, a price tag of $135 billion dollars and a time frame of thirty-five years has been fixed.[127] It's not going to happen, in my estimation. The U.S. is going to walk away from these sites in the belief that, 1) Americans cannot afford to clean up their mess and, 2) it doesn't matter, considering how large the country is. We will put up signs in region after region that read, simply: "Poisonous. No Trespassing." Forever. Paul Kennedy points out that in 1988, 65 percent of federal research and development monies went to defense, but only 0.5 percent to environmental protection.[128] In the U.S., incomes in constant dollars have fallen steadily since 1973 making it all the more unlikely the country can sustain even the past levels of international largesse, let alone future environmental stabilization. At present there are 1275 National Priority toxic sites in the United States, of which only sixty have been cleaned up. According to President Clinton, "the Superfund has been a disaster." As with the *Exxon Valdez* spill, nobody has been able to determine "when clean is clean." Given the U.S. culture of ecological compromise, we have totally forgotten what it used to be like. There is virtually nobody alive today who was born in a tepee, or who can recall a life of preindustrial (relative) harmony. And so, all is habituation, or impotent sentiment.[129] Furthermore, there has been a flood of litigation involving individuals, states, the federal government, and whole communities over who is responsible, and who is to pay for remediation. Cleanup standards and liability provisions have dogged the EPA Superfund, which first emerged as part of the *1980 Comprehensive Environmental Response, Compensation and Liability Act,* CERCLA. "Joint and several liability arrangement" is part of that law, meaning that several parties can be held responsible for cleanup. One litigation in New York involved 257 parties, which hired 136 law firms suing 442 insurance companies, who in turn hired 72 other law firms to defend them. An entire Girl Scout troop was implicated in another suit, and seemingly innocent town residents whose garbage had simply been taken to a particular landfill were named in yet another superfund lawsuit.[130] According to the EPA, by the end of the millennium 75 percent of the forty-five hundred landfills in the U.S. will be overflowing.

Gains in the 1970s from municipal wastewater discharge treatment have been outstripped in the 1980s by population growth and the problems are fast increasing. Seven hundred thousand tons of hazardous waste are cast off every day in this country, and 11 billion tons of solid waste every year. Population growth is expected to increase municipal solid waste by another 35 million tons a year throughout the 1990s, while only 20 million tons of new landfill capacity are being added annually to the system. It means that nearly all Americans are co-conspirators. No one is free of sin. No one has the right to cast the first stone. The U.S. has socialized culpability. Since the U.S. system is at fault, and no other highly populated system is any better, the individual is actually relieved of the burden of guilt. Both the sting and substance of criticism are invalidated. The system is to blame. It seems to make the problem—from the individual point of view—simply go away.

And thus, because of this convenient, democratic deferral of guilt, nature in the U.S. is in a state of calamity. Outside of the cities, this is not necessarily obvious. To drive, for example, across New Mexico, Utah, and Nevada, or through central Nebraska, or much of Maine—in fact, through almost any region of the U.S.—is to marvel at the size, the grandeur, and the diversity of this country. Our wilderness areas and national parks set precedents of conservation throughout the world, as do statistics pertaining to the number of visits each year to these preserves by Americans. But this apparently attuned consciousness, what some might be prompted to call a late-twentieth-century renaissance of ecological awareness, is burdened by much lip service that atones for without remedying; that masks a pervasive and escalating deterioration of the natural heritage of the U.S.

Barbed wire fencing and freeways, which kill and impede freedom of movement for countless species, are often the only two clear and present insults in an otherwise seemingly unfettered wilderness. But train tracks are another. In 1993 countless endangered grizzly bears were known to have been struck down as they fed on corn that had been spilled along the Burlington Northern tracks in Montana. Over 131 million acres have been given over to railroad tracks in the U.S.

Like the exploded corpses of train-killed grizzlies, their bodies scattered beside railroad ties in distant forests, pollution is not normally visible either, be it in the soil, the rivers, or the air. Most strip mining and clear-cutting operations are deliberately concealed from passersby. For example, Iron Mountain on the outskirts of Redding, California, hosts the largest toxic waste site in the western United States. Surrounding streams are poisoned or dead, hundreds of thousands of fish have been killed, and for ninety years the copper mining has been known to produce a steady flow of contaminants into the Sacramento River—seven hundred pounds each day of copper and zinc. But all that appears oddly luminescent when the light hits the river in early morning or late afternoon, much like a late-nineteenth–century landscape painting. The mine was closed down in 1963, but rain percolating through the exposed ninety-year-old rifts and caverns have engendered a chemical reaction that will remain active for three thousand years, continuing to liberate trapped toxic metals from the mountain walls into the soil and water. Ten miles of creek below the mine are permanently sterilized. In the 1980s, acid residues from a second mine in the region killed some four hundred thousand steelhead and salmon at the nearby Mokelumne River. In 1991 a train derailment added to the Sacramento River's woes some nineteen thousand gallons of pesticide and took out forty miles of aquatic life. None of this is evident to most locals, let alone passers-by.[131]

This inobvious carnage is multiplied inexorably throughout the United States. In 1990 the U.S. produced 614 thousand tons of pesticides, 982 thousand tons of plasticizers, 57 million tons of petroleum, and 33 million tons of plastic and resin. Paper, tobacco, printing, primary and fabricated metals, leather, and electrical processes release more than 1.5 million additional tons of toxics into the environment every year.

Add to this systemic refuse the realities of over seven thousand major oil spills in and around U.S. waters, as well as the tens-of-thousands of inland spills and the perpetual problem of leaking pipelines. The beneficiaries of that oil—you and I—traveled over 1.5 trillion automobile miles in the U.S. in 1990, and consumed 70.73 billion gallons of gasoline doing so.[132] Sixty thousand square miles of the United States have been paved over, an area equivalent in size to the state of Georgia, in order to provide roads for all that travel and gas consumption. Imagine an entire country, the size of Georgia, transformed into tarmac in the name of progress, where not a single blade of grass could ever grow.

In 1990 a total of 27.3 million curies of high-level radioactive waste was spread across the combined commercial sites in the U.S., and another 1.02 billion curies—400,000 cubic meters of high-level radioactive wastes—at Department of Energy and Defense sites.[133] Once again, most Americans are completely unaware of these poisons in the environment.

Rural population density in the United State is around twenty per square mile, relatively low on a perceptual scale, though sufficient to interrupt the normal ecology of many wild populations. But again, that is not a readily discernable fact. We do not miss the New Mexico jaguar, the last of which was hunted down in the early 1800s; nor do most Americans know or care that several thousand species of apple and pear trees had gone extinct in this country by the year 1900.[134]

Hunting takes a more immediate toll on the biodiversity around us—some 250 million animals are killed every year in the United States, according to philosopher/activist Peter Singer[135]—but most people will never experience that toll, or bother themselves about the plethora of angry, sanctimonious, and anachronistic hunters stalking the wilderness at dawn.[136] And the same oblivion, or ethical depravity, concerns the U.S. addiction to dead animal products of all kinds. That includes 6 billion chickens, hundreds of millions of pigs, cows, horses, sheep, and goats, and nearly 14 million animals killed by traps. Only upon closer or dedicated inspection, often with the eyes of a biologist, the stealth of an informed environmentalist, or simply the sheer stamina of someone with a diligent conscience, are the signs of this professionally hidden warfare revealed.

But the facts cannot be obscured. The U.S. is the largest producer of forest products in the world, and the second largest exporter of raw timber.[137] The forest summit in Portland, Oregon on April 2, 1993, invited the brazen reaffirmation of this mentality. The much-vaunted cries of that summit were over jobs, possibly concerning even fewer employees than were recently layed off by IBM with not so much as a blink from the rest of the country.[138]

What was really at stake at the Oregon summit were the last remaining mature forests in North America. Where was talk of retraining? Of banning exports of logs, of shifting the emphasis from the northwest to the fast-growing plantations of the southeast? Of managing demand to the extent of imposing a moratorium on housing starts? Of refocusing, in other words, a country's energies away from destruction and toward creation? These shifts in thinking and management had been understood by scientists and planners at least since 1909. There were 800 million acres of forest in America at the time of the Pilgrims. By 1900, 75 percent of that had been stripped. In 1909, President Theodore Roosevelt instigated a national conservation conference to address his nation's rapidly dwindling natural resource base, as it was even then perceived. Seven years later, the National Park Service came into being and with it the first stirrings of an international conservation movement. Yet, in just the past ten years, the federal and state governments have allowed two-thirds of the remaining old forest in this country to be cut down. At the summit, it was pointed out that we should err on the side of ignorance, in the modest recognition that we still do not understand the complexities of forests; indeed, for all we know the trees are like the whales, except that we have not yet learned how to look a tree in the eye honestly or humbly, let alone listen to one. And the same can be said of the wild salmon of the Pacific northwest, all five species of which are now endangered because of aggressive human overconsumption.

Shift from the northwest to the southeast. There, in south Florida, population growth is equivalent to that of Bangladesh. Nearly 18 million people will reside there in the year 2010. Ninety percent of the Everglades' birdlife has vanished and many other species—like the Florida panther and manatee—are on the brink of extinction. Natural food sources for Floridean fauna are competing with the second fastest population boom in the United States and have been consumed or destroyed by humans, at an unbelievably malevolent fever pitch. The wetlands ecosystem is collapsing; the Ehrlichs have speculated that the Everglades National Park system might be forced to close its doors by the year 2000.[139] The U.S. has lost more than half of its coastal wetlands. As described in the World Resources report, "The volume of nutrients entering coastal waters can generally be correlated with the number of people in the watershed. As the population increases, so does the pollution. In 1991, a study of forty-two rivers directly linked the population density of a watershed and the amount of nitrogen pollutants in the watershed's coastal area, a relationship long understood by biologists but previously unquantified."[140]

In 1700 less than 10 percent of the world's population lived in cities. By 1950 the world figure was 30 percent, but in the United States and Europe, urban populations had swelled to 64 percent and 56 percent respectively. In theory, this presumes less and less impact on more and more open space, a trend that might ultimately lead to the liberation of nature. Unfortunately, the data is not supporting that hope. In the United States, metropolitan populations have doubled since 1950, from 85 to 180 million residents. But all that concentration of energy capacity and jobs has only served to build up a far-reaching infrastructure of exploitation: to defoliate tens-of-thousands of square miles; to accelerate the proliferation of toxins around the world; to exacerbate the domestic gap between the rich and the poor. We have an increasingly unaffordable welfare state, uncertainty and rancor regarding the rights of immigrants, a national paroxysm of crime, depression, and boredom, and a breakdown in education that has left as many as 84 million Americans—a third of the nation—functionally illiterate. According to Paul Kennedy, "one in seven adult Americans tested recently could not locate his or her own country on a world map, and 75 percent could not place the Persian Gulf—even though in the late 1980s many of them favored dispatching U.S. forces to that region."[141] Thirty million people are hungry in the United States. Ten percent of all Americans are dependent on food stamps. The U.S. placed last among industrialized nations in child mortality and life expectancy, while 375,000 Americans have been born preaddicted to cocaine and heroin.

Perhaps other, equally galling statistics could be cited for every country, throughout history. The periods of the Hundred Years War, the Thirty Years War, the Crusades, the wars between England, France, Spain, and the Netherlands, the two hundred fifty thousand known battles since the time of the Renaissance, all indicate a human propensity for barbarism—lawlessness, beheadings, cannibalism, torture, pandemonium, despoliation. What have we learned? Only that the more people there are, the more possibilities for injury, pollution, and aggression.

These physical and cultural aspects of economic urbanization have backfired upon nature. Among industrialized nations, the U.S. has by far the highest number of endangered species.[142] Where there is poverty, illiteracy, and crime, there is less money or sensibility that can be mobilized in defense of the environment. Ironically, where there is wealth and a relatively high level of universal secondary school education, there is an epidemic of hedonism. Like ants, human beings are utterly focused upon their own problems, which they do not, will not, perhaps cannot grasp as part of an ecosystem.

The numbers, the death tolls, the production figures, miles of paved road, tons of toxins, decibels of noise, percentages of growth, are all intangible, relative to nothing. They defy viable definitions and transcend our ability to exert ethical controls, or even to think clearly. Take the most blatant case of this statistical juggernaut: the Holocaust. While society as a whole condemns with all its heart what happened in Germany (and Austria, and Poland, etc.), the sheer numbers of dead souls can never be assimilated or reconciled. The killing of between 6 and 8 million Jews is unimaginable (though not to the survivors whose parents and whole families were melted down in the ovens). Given this humanly "unimaginable" situation, most people, other than Jews, would probably like to forget the Holocaust, to move on. And so they have (notwithstanding Stephen Spielberg's sensitive epic film, Schindler's List, and Jean Paul Sartre's insight that even if there were no Jews, there would be anti-Semitism). Would the world pause and reflect a little longer perhaps, if not 6 to 8, but 12 million Jews had been slaughtered? In fact, according to official French documents released after the war, between

18 and 26 million men, women, and children had been killed by the Germans through hunger, cold, pestilance, torture, medical experimentation, and other forms of murder.[143] Is there a numeric threshold where ethical thinking undergoes epiphany? Where we are suddenly transformed by the data? What if all the Jews had been exterminated, like other, smaller groups of humanity in the past—the Tasmanians, or the Caucasian Pygmies in northern Burma, once known as the Daru? Would that level of attrition force us to reevaluate the causes and consequences of human nature and do something about it? Or do the larger numbers and disasters actually dissociate truth and meaning from the human capacity to comprehend, to be moved by feelings, to act upon important information?

It is said that one truth is equal to all truths. But can we actually distinguish one inconceivable number from another, let alone attach a hierarchy of ethical determinations to the varying quantitative levels and degrees of disaster? Are we any more distressed by the knowledge of a billion hungry people in the world than we would be by, say, 900 million? Or is the prospect of saving 100 million children from starvation even enough to elicit a twenty-dollar donation, a call-in to an 800-number on TV, or an incentive for a congressional resolution? If it is not, then how can there ever be a serious understanding and application of family planning in any country? How can Americans intelligently grapple with the fact that there are well over 200 million firearms in the U.S., or that in America, men rape women seven times more frequently than anywhere else in the world? How can we fathom the thousands of superfund sites, or billions of slaughtered animals, or hundreds of thousands of homeless people, or billions of curies of radiation in the soil?

We are good at rallying around the plight of a single individual, or certain slogans—"family values" or the "war on poverty". All of America clung to the radio for a week once, when a little girl named Kathy Fiscus fell into a pipe in Texas. The country put all of its resources into extricating her. But in 1994, six dead Bosnian children, shelled while innocently sledding, elicit no more than a few moments of news. The fact is, the more people, the more quantities of problems, the less likely we are to deal effectively with them. Specifically, how can we formulate and execute any policy involving mass numbers of people, when Darwinian behavior, the ethics of personal survival at any cost, appear to be enshrined in our genes? America's rejection of a few cents added to the price of gasoline to help preserve the environment is simply one example of our inflexibility and astonishing short-sightedness. We seem incapable of collective altruism, preferring personal escapism in the hope that we will somehow muddle through.

Economic development, the outgrowth of overpopulation, is the policy by which this biological contradiction is furthered. Aspiring for aspiration's sake, we have as human animals so long insisted on the acquisition of wealth, there seems now little hope of turning back. The rise of capitalism in Moscow and Beijing reaffirms this. Indeed, having lost sight of the original world—that Paradise envisioned, for example, by the Pure Land Buddhists in Kyoto, Vermeer in Delft, Percy Shelley in Chamonix, by Sir Thomas More in his *Utopia*, or Sannazaro in his *Arcadia*—we are left with nostalgic consolations. I dote upon certain traces, an island I know in French Polynesia, a mountain in Antarctica, a painting at the Mauritshuis in the Netherlands, a musical score written by Vivaldi in Venice, and so forth. Those who crave a better future are left with two choices: retreat, a kind of desperate, personal solipsism; or the endless, anonymous, thankless quest to "change society." America incites and allows for both impulses, mixing them up without regard. Great cities offer great consoling visions. Los Angeles can be a fabulous place, when the hills are not burning, or the Earth not upward thrusting, transforming homes into deathtraps. Some days the city seems perfectly tolerable, the Southern California that so many the world over have dreamt of; a California that altogether contests this admittedly cynical inventory and unrelenting spasm of complaints that I have endeavored to lodge for the record. On such days, there is no apparent pollution, suicide, crime, animal slaughter, skid row, AIDS, extinction, or even future, for that matter; only the ephemeral moment of wind, sea, blue sky, luxuriant beauty, and lunch at the Hotel Bel Air, if you happen to be in that part of town and can afford it.

Whether one lunches at the Hotel Bel Air or the UCLA student cafeteria a mile away, there is an additional sense of optimism that pervades the professional communities in Los Angeles, a vigor and hope that the condition of the environment can, and will be ameliorated; that family planners will have increasing success; that the greed of individuals will be tempered by sound governance and ingenious tax schemes; and finally, that the human pluralism of Southern California, with its artistic and intellectual riches, will lend a unique vitality to all future humanitarian endeavors. Every city has this burning upside of hope, this rationale for going on, whether Santa Monica or Calcutta.

Thus, two distinct impressions are likely to emerge from the melee of late-twentieth-century development, the *opposing Californias* of economic theory. On the one hand, there is the justified perception that the aching, urban goal of achieving greater and greater "wealth" is, in fact, bankrupting the planet—a formidable paradox drastically aggravated by nearly 100 million newborns every year; by nearly two hundred countries each struggling with no other purpose than economic prosperity, at whatever cost to nature and that huge parliament of disenfranchised human beings. On the other hand, there are millions of caring, knowing professionals who want to make a difference and are determined to make that difference, though many of them realize that there is little time left.

How does the impatient individual, increasingly aware of the seriousness of the human predicament and the many frustrations belonging to compromise, go on living a productive life? How is any progress or sustainable development possible, given the dark tenor of the definitions and semantics, as I have here employed the terms?

For purposes of developing an internal compass reading, I start with the principle of absolute zero impact: what would it require? What logical number of people might be sustained, and by what means? Aristotle believed that the ideal city could not exceed a population of fifty thousand. Plato reckoned on a mere five thousand inhabitants. To what extent is human decency, kindness, joy, dignity, and love— and for that matter, art—dependent upon the advances of civilization, technology, and cities? Are trade-offs ever possible? In some theoretical future—tomorrow, for instance—would it be possible to say "No" to a Mars mission, and "Yes" to universal literacy? Many will argue that the two are by no means mutually exclusive. Yet the global recession and the historical finitude of wealth makes it readily clear that a culture cannot undertake every expedition of which it might be capable. The Soviets built a space station, but could not guarantee fruit or protein to their citizens. The Japanese are doubling their nuclear power capacity, but have not yet managed to provide adequate sanitation services for over a third of their population. India fashioned the Taj Mahal, as well as the largest democracy in the world, yet nowhere in the country is drinking water unpolluted. For that matter, nowhere in the United States is a river or a creek entirely unpolluted, says the EPA. Americans have not found the means to pay countless prices: an AIDS or cancer cure, or enough shelters for the homeless. How shall we finance a space station or assume the costs of prenatal care for ten million new immigrants in the coming years when already over thirty million Americans are not getting square meals? A balanced budget would surely provide more essential humanitarian elbow room. But what no one has been talking about is the root cause of U.S. economic imbalances, and that is our population explosion at home. Of the developed nations, only France and the Netherlands have official population policies, and only the Dutch have advocated negative growth. The U.S. has no such policy.

We live in a world more fraught with disaffirmation and the antithetical than ever before, it would seem. Yet every disparity declares a priority, invokes a conviction. Our convictions are being tested, today, by new amplitudes and proportions, a crisis of harm that must force us to requestion all the fundamentals of our existence: who are we, and what are we here for? And while these conjectures have always been current, forever relevant, the massive, unchecked multiplication of our *presence* now challenges the core of our humanity and ingenuity.

Consider what, if anything, humanity has accomplished, judging not by its own standards, which are necessarily self-serving, but according to those of the Earth? By simply shifting this crucial point of view—the goal of all ecology—we cannot fail to register a certain regret about the industrialized nations; to recognize everywhere the calamity of our presence. The inventory of damage has left traces in our consciousness, to be sure; and in our lungs and mammary glands and children. Moment by moment the list of atrocities, of incoming data, escalates, and we are increasingly unsure of ourselves. We are less capable of justifying our very presence, which, in turn, incites a host of ontological difficulties. For example, some may question the notion of a divinity because it does not seem reasonable that any God would condone the ecological blasphemies of one species out of so many hundreds of millions of others that have gone about their business without leaving so much as a footprint. Why Homo sapiens? Friedrich Nietzsche had remarked that we, alone, are the "sick species." Our discouragement, even self-disgust, leads to apathy, or cruelty. Family planners cannot do their job if they are awash in recrimination and gloom. Surgeons, in the act of saving a life, will most likely falter if they cast their thoughts to Ecclesiastes and focus on the inevitable fate of all beings (dust, in other words), rather than the task at hand.

This practicality may seem obvious, but it tends to blur at the edges of the late twentieth century for reasons that are both ecological and psychological. We are assailed not just by terrible choices, but by circumstances whose magnitude offers no earlier analogies. The full force of this "logic jam" is, perhaps, best appreciated in terms of human demographic paradox, the subject of the succeeding chapter. I call it paradoxical, because, in assessing the coming decades demographically, I am struck by a remorseless conundrum that nearly has a mind of its own: an already desperately overpopulated species furiously breeding, poised to double, triple, or even quadruple its numbers; an organism that claims to desire peace and tranquility, but whose collective violence is fast escalating in all ways; a genetic population whose territorial imperative has utterly condemned that thin veil of other lifeforms gently covering the planet. This condition of knowing abuse and chaos and pain we still proudly hail as "the civilized world."

The crude mathematics of our situation have persistently knocked at the door of all biological laws. Should we actually open that door, as we appear to have done in cities like Los Angeles, Shanghai, Bombay, and Tokyo, Mexico City, Sao Paulo, and elsewhere, there is simply no telling what might happen next. Some theorists believe that nothing will happen. That life will continue to drag along in depressing fits, some positive, some negative, a severely asthmatic life, wan and sapped of vitality. Others argue our fertility momentum has already made it sadly clear that every expression of misery, from Pandora's Box, to the Apocalypse, to ecological doomsday, is now operative. Whether one is a firm believer in the power of the human spirit or not, does not alter one irrevocable fact: that unless governments, and those billions of human beings they supposedly represent, act now, in a globally coordinated fashion, to reduce our numbers and refine our economic impulses, all other "solutions" are likely to be useless.

Hundreds of scientific Nobel Prizes, and a century of vast technical acumen, have only accelerated the present crisis. Life expectancy—for many—has surely increased. But for just as many others, life has gotten much worse. Writes John Bongaarts, "Providing a diet equivalent to the First World diet in 1990 would require that each hectare increase it yield more than six times. Such an event in the developing world must be considered virtually impossible, barring a major breakthrough in the biotechnology of food production. . . . A more troublesome problem is how to achieve this technological enhancement at acceptable environmental costs. It is here that the arguments of those experts who forecast a catastrophe carry considerable weight."[144] In Russia, that catastrophe is already underway, as witnessed by an actual "population implosion" characterized by male life expectancies which have fallen to a mere sixty years—less than in India—the death rate having increased by 20 percent between 1992 and 1993. In fact, in 1993 there were almost eight hundred thousand more Russian deaths than births. Infant mortality, the rate of abortion and of unnatural deaths (like suicide) have risen rapidly, while the Russian psyche seems unequipped to deal with so much medical, ecological, and economic adversity. A sign of the times? Demographers are now likening the new Russian data to what one would expect from a country like Bangladesh.[145] Such social chaos, hardship, and mortality bodes of a lower Russian population in the future, but signals the human price that all nations can similarly expect to pay if population increases are ever to be reversed. There are ways to soften the reversal, of course. Family planning, high contraceptive use, and adamant laws and policies of ecological sustainability. Short of these, humans, like most other life forms we have displaced, are going to be caught out in the storm of development.

Whatever we hope, or say, or do, our volition will most likely be pulverized by the uncontrolled flood of newborns at the heart of the population explosion. To ponder this virtual double bind is to taste a form of insanity. "When elephants make war," says an African proverb, "the grass gets trampled. When elephants make love, the grass also gets trampled."[146]

# Demographic Madness

Anyone who believes exponential growth
can go on forever in a finite world is either
a madman or an economist.

*Kenneth Boulding[1]*

Midway the path of life that men pursue
I found me in a darkling wood astray,
For the direct way had been lost to view.

*Dante[2]*

## CHOMOLUNGMA

It is estimated that every second more than twenty-eight people
are born and ten die; that every hour, more than eleven thousand
newborns cry out. Each day more than 1 million human conceptions
are believed to come about, resulting in some 350,000 new cases of
venereal disease, and more than 150,000 abortions. Among those
newborns, every day 35,000 will die by starvation, 26,000 of them
children. Meanwhile, each twenty-four hours, the pace of war against
the planet increases, sometimes in major affronts, other times imperceptibly, at least by our limited perceptual standards. That war includes
the loss of 57 million tons of topsoil and eighty square miles of tropical
forest and the creation of seventy square miles of virtually lifeless
desert—every day. At current birth and death rates, the world is
adding a Los Angeles every three weeks. If average human growth rates
were to continue at their present course (the so-called "constant fertility variant") the world's population would reach at least 10 billion
by the year 2030, 20 billion by 2070, 40 billion by 2110, and 80 billion
by the year 2150.[3] Most social scientists believe the figure will be,
at worst, between 12 and 15 billion. There are many reasons—some

delusional—for such "optimism," and I shall take up the pros, and cons, of each of them.

The word *demography* has its origins in ancient Sanskrit and pertains to the division of a country or tribe. In Greek, the addition of *cracy*—meaning power—to *demos*, meaning people, lent moral and theocratic legitimacy to the primeval notion that human beings controlled the country in which they lived. Assisted by a whole new menagerie of tools and motives not found, previously, in nature, human meaning—the nature of our hominid self-interest—campaigned early on to overpower whatever facets of our biological selves, and of biology in general, might dampen our hedonism.

While I have felt the oppressive weight of people—whether I envision them raging across the countryside, teeming over the concrete, slaughtering other species, burning down forests, or raping one another—I am still not quite conversant with the paradox of my own complicity. That, by association, as one more member of that species, I am as guilty, as out of control and blind as the next man. While my wife and I as yet have had no child, we are two Westerners, equal in ecologically destructive impact, I imagine, to at least four hundred of our brethren in the developing world. My response to that estimated ratio should be colossal guilt. Yet, in truth, I cannot trace, or visualize the known burden of myself upon nature. The quantum leap from my personal conduct and lifestyle to an imperiled world is rather academic. Nor, for that matter, do I even grasp the real meaning of an entire planet in crisis. I have surveyed many of the battlefields, counting corpses, enumerating misery; have felt a slack stomach, picric eyes, and paralyzed throat amid the stench of blood or acrid water; and been chronically depressed by the horror of animal murder, the foul urban squalors and felled tropical forests. And yet there has usually been some oasis of sanity, far from that madding crowd, to which I could retreat: a spiritual trust, or aesthetic delectation steeped in associations and memory; the many heartfelt exchanges of the soul; or, finally, some distant archipelago of tranquil isolation central to the safeguard of spiritual being, to which I could always flee. Adding to this escape route, at least until recently, the Earth seemed infinite, perpetually in renewal, even forgiving of our behavior.

Though wars and disasters have long engendered refugee populations, today's melancholic masses suffer from a more insoluble dilemma than mere tragedy, namely, an incessant, unavoidable proximity to more and more members of their own kind. This biological configuration, unique in the annals of societal experience, outweighs all demographic theory or complacency. Even the theological struggle to explain, and resist, evil falls short of the psychological mayhem, the omnipresent paradox of so many *Homo sapiens*. It is a new order of double-bindness, of having no place to turn, or to run. "There is a concept which corrupts and upsets all the others," writes Jorge Luis Borges. "I refer not to Evil whose limited realm is that of ethics, I refer to the infinite."[4] A billion people, and trillions of other individual creatures, are casualties of the overall human presence. That presence induces not only a scarcity of air, water, and soil—the bases of our life—but everdwindling privacy, the volatility of competing demands, desires, and information, and the malaise of *contagion*. The psyche is overrun by these many components of congestion. Our life is a love/hate relationship. We must constantly ask, "Why?" and are stricken by implacable doubts pertaining to our very nature. We may complain of rising crime, increasing noise, filth, traffic, and troubles, yet cities are growing in size, not shrinking. Awareness of these problems is having no countervailing effect. The world is indisputably becoming urban, not rural. Ardent consumers, people continue to be propelled toward cities and megacities, where consumption is offered the widest range of possibilities. Overall, 72 percent of all new urban households in developing countries are located in slums.[5]

Yet, there is an insatiable nostalgia for rural calm, a legacy like that in parts of Provence, with its rocky garrigues, wild vines, and edible landscaping, a zone of odoriferous gardens and rough grasses somewhere between the heart, the stomach, and the imagination. If every human being on the planet presently, moved to that mythical country, population density across the terrestrial world would average 0.4 persons per acre. But considering how human beings live, what we do, our restlessness to build, extract, exploit, and mobilize, even that is a *very* crowded world. Moreover, that figure has nearly *doubled* in just the past generation. Every myth seems to be contradicted by current data. Note this following item from the Hong Kong–based China news services (CNS) published July 23rd, 1992, and titled "Deathlessness in Tibet Village": "Nobody had died in the remote village of Dongzi in the mountainous region of Tibet for 50 years. . . . The village, which has not reported any deaths since 1942, has one 142-year-old and 186 people aged over 130 . . . All of them are healthy and free of ailments," CNS said, adding, "the village's population has soared from 680 to 6234 as a result of the residents' longevity." Even Shangri-la is bound by an ecological carrying capacity.

When I was born, in 1951, only five urban areas in the world had populations of 5 million or more. Today, there are nearly fifty cities that size. When I was a child, Mexico City contained about 3 million people. It now has closer to 25 million. In 1951, there were fewer than 300 million urban dwellers in the Third World. In another generation, there will be roughly 4.5 billion such people. In Africa, over 90 percent of them will be living in shanty towns. From our first human ancestors to the early 1950s, the population of *Homo sapiens* had grown to 2.5 billion. In a mere forty years since then, in the span of my life, we have more than doubled that number. Over 90 percent of the growth of the human population has taken place in less than one tenth of 1 percent of our whole species' history. And because half the world's population is now under the age of twenty-four, a third younger than fifteen, it is clear that we have currently reached the base—not the top—of a demographic Mount Everest, or Chomolungma, as those Buddhists dwelling directly beneath the mountain call it.

Though this century has already witnessed an unprecedented human population explosion, we are presently poised to proliferate as never before. Three billion young people will enter their reproductive years in the next twenty years. Population scientists have devised countless graphs to put exponential growth in perspective: sinuous curves that vacillate, alternate, or convulse. Others have balanced human growth against the loss of keystone species, or prime agricultural lands, or the volume of depleted aquifer reserves, or the rise of new-comer "trash species" in the oceans (such as spiny dogfish that have, in their turn, usurped the territory of previously stable indigenous marine residents). Others have singled out trends in energy intensity and consumption as a barometer for gauging optimal population size; or the amount of manufactured or experimentally detonated kilotons of fissionable material, of mass immigration, the number of environmental refugees, or, simply, attrition from hunger and disease.[6] Many have analyzed existing food stocks and agrotechnologies to determine the world's human carrying capacity. According to one study, it so happens that the current global population of 5.5 billion—if strictly vegetarian—might well be supported. However, at even limited levels of meat consumption, the carrying capacity plunges. If 25 percent of the total human diet were animal-based, as is the case among most North Americans, half of the world's human population would starve.[7] And as it turns out, according to the British Intergovernmental Panel on Climate Change whose findings were released in 1990, meat-eating and dairy consumption are projected to increase by 45 percent worldwide in the coming generation, while 90 percent of all encroachment on tropical forests is now induced by the hunger for new cropland. Taking into account the moral and ecological global imperative for vegetarianism, many agroscientists have argued that North America's "ideal population" should be somewhere between 40 million and 150 million. On average, to sustain a nonpetroleum-based, well-fed economy at current levels of affluence, it has been pointed out that the U.S. population should exceed no more than 100 million people. But since the populations of the U.S. and Canada are projected to continue increasing (even minute annual growth can result in a doubling of populations within 64 years), and since current domestic legislation has perpetuated virtual public land giveaways and outrageous tax write-offs to the cattle industry, as well as a whole system of subsidies, there appears to be little hope for the near future that Western policies will lead the way toward global agricultural stabilization.

I can think of no better image of this steadfast pyramid of human consumers, this Total Population Dynamic, than that of a mountain like Chomolungma (29,032 feet), cresting above the surrounding Himalayan roof of the world. Where the analogy abruptly ends, however, is with the adjectives appropriate to such a sacred mountain as Chomolungma: wild, free, and magnificent, descriptions that no longer apply to most of the world, and which certainly contradict a future planet whose human population might be 12 billion, 15 billion, even 20 billion carnivores. Ironically, already Chomolungma herself is showing signs of wear. The effects of overpopulation are starkly clear. Since 1951, over four hundred climbers to the mountain have left some forty thousand pounds of nondecomposing refuse strewn upon what is, in fact, one of the most fragile of ecosystems, an eight thousand meter peak.[8]

Some social scientists continue to ask, "At what point does the real population crisis set in? Or does it ever set in?" In parts of Egypt, that crisis has meant that school children obtain no more than three hours per day of instruction, so as to make way for incoming shifts of other students; in the Pacific Caroline Islands, the Yapese people were said to have been so overpopulated, and consequently destitute, that "families lived miserably on rafts in the mangrove swamps . . . and that sometimes four hungry men had to make a meal from a single coconut."[9] In 1722, 111 people were left on Easter Island. Polynesians had settled there as early as the third century A.D., and had attained a stable population size of seven thousand. But excessive deforestation meant that no more fishing boats could be constructed and the eventually malnourished population split into two warring tribes that only succeeded, ultimately, in wiping one another out. In that case, overpopulation and ecological crisis were the classic prelude to human extinction.[10] In the Philippines, once the lowland agricultural areas were utterly saturated with people, highland encroachment became suddenly serious. Water scarcity in Egypt and Pakistan, the absolute lack of new arable land in Kenya and Madagascar, are each crisis-level circumstances.

The word "crisis" is not entirely useful, however. For what does it ultimately explain? Even in the densest, most gnarled crannies of urban experience, the human being may discover love or pleasure, and hence, habituates to a city's neurological rigors. Out-migration from rural deathtraps—as witnessed in Ethiopia, Somalia, the Sudan, and throughout much of West Africa—has always presented new opportunities for salvation. There were nearly 10 million refugees following World War II. Today, there are at least seven times that many environmental refugees and little undisturbed land left to newly inhabit. Indeed the three hundred thousand Burundians who recently fled from one civil war into another in Rwanda, are truly stranded, like the several hundred Palestinians caught out in the snow, in 1993, between Lebanon and Israel. Nevertheless, biologists know that evolution, in theory, occurs in "harsh places." Thus, some apologists—political, scientific, and religious—have sought to explain away all human dilemmas according to poverty or morbidity, not overpopulation. Indeed, tax exemptions for children in many countries, including the U.S., have only served to perpetuate the notion that the more people, the more consumers, the healthier an economy.

Yet there is absolutely no doubting the slavish links between human population growth—the daunting number of consumers—and environmental catastrophe. For example, even if each U.S. household were to replace just three incandescent bulbs with compact fluorescents, U.S. energy savings over the seven-year life of the bulbs (1 percent of the total American annual electrical budget) would be completely nullified by far greater population/consumer increases during that same period, according to the Population–Environment Balance organization in Washington, D.C. It is known that, at a minimum, 75 percent of all global demand for fuelwood is the specific function of increasing numbers of consumers.[11] By the year 2030, at least 8 billion people will reside in countries containing, or previously containing, tropical rainforests. But no longer is fuelwood likely to be available to these multitudes, who will have cut down the trees for fuel and replanted with fast-growing agricultural crops. Reforestation is occurring at only a fraction of what would be necessary to ensure long-term fuelwood supplies. And for many of these transmigrants, fresh water will no longer be available, not even industrial water for agricultural purposes. Most resources, in fact, will have been depleted as 94 percent of all popula-

tion growth in the coming decades will likely occur in those developing countries with the most impoverished resource base. In Pakistan,
even by the late 1970s there was fuelwood scarcity. Yet Pakistan's population is poised to more than triple in size in the coming two generations,
while only 2 to 3 percent of the country's forest cover remains.[12] By
2030 in Madagascar, Cameroon, and Cote d'Ivoire, countries with
severe biodiversity loss already, respective populations will have leapt
from a present 12 million in each of the three, to 49 million, 67 million,
and 83 million. Vietnam will increase from around 68 million to 168
million; Myanmar from 43 million to almost 100 million. In each of
the eighteen so-called biodiversity "hot-spots," sizable population growth
is condemning all life forms therein. Where population is increasing
most rapidly, ecological deterioration tends to be the worst, infant and
child mortality, maternal mortality, and the extent of malnutrition and
hunger, the highest.

The trends have long declared themselves. At present, for example, while the world population is growing by 1.8 percent per year, net
grain output is growing at less than 1 percent per year. By the end of
this century, the amount of arable land per person on the planet will
be less than half of what it was in 1951. According to one model, because
of the cumulative environmental impacts of overpopulation, global
grain harvests will plummet by as much as 10 percent every forty months,
on average, in the coming decades, and between 50 and 400 million
additional people will die from starvation.[13]

Nor has this net food deficit been mollified by the nearly 8 billion
tons of additional "artificial fertilizer" spewed into the atmosphere each
year in the form of carbon dioxide emissions, at least two-thirds of which
are the explicit result of overpopulation (i.e., more automobiles and
electrification).[14] While future fossil fuel generation in the developed
countries may be modified by gains in efficiency and the implementation of best available technologies, in the developing world per capita
$CO_2$ emissions are projected to double in the coming generation. For
governments in sub-Saharan Africa, where agricultural production has
declined by more than 2 percent annually during the past twenty-four
years, condemning at least 30 million people to starvation and chronic

malnutrition, and 320 million others, or 65 percent of the total sub-Saharan population, to "absolute poverty," such indicators are bewildering.

The sheer habit of finitude plagues any mind that attempts to reconcile seemingly irreconcilable conflict. The paradox stems from the essential quality of ungraspability that characterizes all intractable knots, insoluble dilemmas, and higher mathematics. Even the all-familiar "zero"—the heart of exponential affairs, whether in biology or physics—eludes easy cognition. We use the zero almost as casually as common change. But by 10-to-the-fifth power—or, 100,000—we are no longer capable of visualizing the size. And this is precisely the point at which Aristotle, writing in Book VII of his *Politics*, suggested the emergence of a crisis in numbers. Human population, he said, is only in balance when it is "both self-sufficient and surveyable."[15] Beyond that surveyable, self-sufficient size, Aristotle advocated compulsory abortion, as well as the mercy killing of children born deformed, though he did not stipulate what kind of deformities. But what does "surveyable" mean in the twentieth century? A man surveys his property. But what about a place like Macao, the most densely human-populated corner of the planet, with 61,383 individuals per square mile over a region of ten square miles? (By comparison, the second highest density is to be found in Hong Kong, four hundred square miles and 14,315 people per square mile.) What about a city like Los Angeles, with over four thousand square miles? That city, and most others, has transcended surveyability, except for the most basic views from above, and statistics—the approximate number of people, the birth and death rates, the kilowatt hours of electricity consumed, and the level of hydrocarbons in the air, pesticides in the food, or trace metals in the water. Most other realms of subjective causation, such as "dynamic homeostasis" in ecology, "punctuated equilibrium" in biology, "wave-particle duality" and Heisenberg's Uncertainty Principle in quantum mechanics, "dissipative structures" in chemistry, and "chaos theory" in mathematics, are beyond cognitive accessibility, somewhere between intuition and metaphysics, bolstered by interpolations whose data continues to shift, to be fuzzy and unreliable. And so, I would argue, is the demon of demography. All of these facets of life on Earth share a fundamental unpredictability that is psychological, inflated or deflated according to the particular day, or personality of the analysis. In other words, grounds for much blue sky.[16]

In an essay he named "Despair Viewed Under the Aspects of Finitude and Infinitude," (part of his larger work, *On Fear and Trembling*), Soren Kierkegaard psychoanalyzed an individual at sea amid blue skies—evolutionary forces and vast aggregates beyond his ken or control. Struggling between everything and nothing, the individual is one who seeks shelter, comfort, and security. All else is a blur. When we attempt to meaningfully manage that amalgam of ungraspable facts, we court true existential despair.[17] Often, paradoxical thoughts (i.e., the individual versus the species) are wont to adopt conventional assumptions and are led, by extension, to a kind of insanity.

One such extension, discussed earlier, is the ideal of progress. The appetite for progress is insatiable, unleashing an incalculable avalanche of biological fallout. Another is compassion—the goodness of the human heart—which is, every day, limitless in the countless gestures and deeds it may initiate, thus inviting no end of conceptual possibilities, or conversely, melancholy at the failure of the heart to perform anywhere near its potential. Hope and compassion, like awareness, are constantly on trial before the bar of greater opposition; forever tested by inertia and stark negation. We have been "aware" of the capacity for human cruelty—and ways to curb it—for countless millennia. Yet, in this century more hideous aggression has been perpetrated by human beings than at any other time in our history.

Even within the family planning arena, it is generally recognized that, in and of itself, *awareness* of some kind of contraceptive method does not often translate into contraceptive prevalence. In Kenya, for example, nearly all married couples are aware of some form of contraception. Yet no more than 33 percent actually practice birth control.

There is no more daunting paradox, no greater logic problem, than human demographics. Mathematicians have termed such problems NP-Complete, or, "nondeterministic, polynomial-time-complete," referring to exponential combinations that are so myriad, so irreducible, as to have become positively impenetrable.[18]

Chinese mystics spoke of a similar paradox when attempting to fix the *meaning of life*: "One who knows does not speak; One who speaks does not know," wrote Lao Tzu in the *Tao Te Ching*.[19] By its very nature, all paradox turns endlessly upon itself. Jung argued that reality— the very fullness of life—can only be experienced as paradoxical. Throughout the history of philosophy, religion, and art, paradox has always proven catchy. Consider the following seemingly paradoxical utterances: "When freed from abiding (with thoughts), you are said to be abiding with the non-abiding" *(Prajnaparamita Sutras)*; "Form is matter" *(Upanishads)*; "The first shall be last" (from the *Gospels*); "So shalt thou feed on Death, that feeds on men, And, Death once dead, there's no more dying then" (Shakespeare, *Sonnets*, 146); "Ask yourself whether you are happy and you cease to be so" (John Stuart Mill); "The supreme triumph of reason is to cast doubt upon its own validity" (Miguel Unamuno).[20] In attempting to describe Zeno's paradox, Bertrand Russell referred to it as "the problem(s) of the infinitesimal, the infinite and continuity." The Greek philosopher lets fly an arrow and attempts to calculate when it should reach its mark, only to realize, at least theoretically, that the arrow is nowhere and everywhere. The arrow is never in motion but always occupying one point or another along its trajectory of infinite points. At each instant, the arrow does not move.

This is the essential paradox of population growth, as viewed from the perspective of the individual.[21] And it also explains why, in the absence of a "sexier" crusade, few corporations are willing to donate charitable funds to population control programs. They cannot "see" an outcome. It is, again, unsurveyable. There's no quick payoff. Because the person occupies a point in time, human growth is virtually imperceptible, a living reminder of our species' perceptual limitations. We can't see in the infrared, hear ultrahigh frequencies, nor visually distinguish a cluster of fifty million from fifty-one million.

All of our physical capabilities, desires, and metaphors are conditioned by the need for stability, a precursor of so many of our formulations, general principles, hypotheses, and behavior. Hence, demographers (unlike astronomers and particle physicists, whose gazes uniquely eschew these all too human character traits) have tended to establish arbitrarily a point of *stabilization* for the human population, much in the manner of Hegel, whose "dialectic" presupposed eventual synthesis and resolution in the human psyche. The World Bank and United Nations have each predicated their respective population projections according to this general impulse, appropriating a formulaic optimism as a means of asserting the inevitability of classic demographic transition. Those few antisocial pessimists—ecologists, most of them— who see massive ruination, the human population reduced to a meager coterie of hunter-gatherer survivors in much reduced circumstance, are a minority, politically incorrect, the bane of family planning. In fact, most family planners are convinced that they hold the key; that things will settle down. Life will triumph. The world will once again be able to go back to business as usual, sometime by mid- to late twenty-first century. Sooner than that, in fact as early as 2010, it is believed that rural populations worldwide will even begin to decline in number as the cities fill up with migrants, and this de-escalation is predicted to be a blessing for the remaining wildlife. There will be regional scarcities, intermittent famines and plagues, an immigration problem here and there, but the world is not about to come to an end. Indeed, there are more parks, more food, more museums and forms of entertainment, and vastly more energy extraction and the production of capital goods, than at any other time in human history. Immunizations, CDs, cheap airplane fares, fax machines, home video–dating, the availability of ripe avocados in mid-winter, are all presumed signs, for some, that times are good. But again, the year 2010, the whole mental image of humanity's developmental triumphs, the very hope of family planning, is totally relative, even arbitrary. The current data and trends in no way support such unified enthusiasm. Such are the pitfalls of quantitative analysis.

## GLOBAL SHOCK

Environmental archaeologists and demographic historians have long known of the millennial long waves (MLWs) of human population expansion and collapse. Thought of most simply, these are curves on a graph that fluctuate in accordance with the rise and fall of population size. MLWs reveal two distinct patterns. The first is analogous to the periodic fate of other species, namely, a healthy population resurgence following disaster. For example, in the Rocky Mountains, 90 percent of all those pine trees upon whose bark and cones local squirrels feed, will die every few years. Yet, because the squirrels themselves are periodically culled by the forces and attrition of hunger, the trees are able to perpetually regenerate healthy sylvan populations. In a similar way, business opportunities and prospering families abounded in Europe following devastating population declines during the Black Plagues, which in turn spawned a doubling of overall human numbers.

The second pattern of MLWs is analogous to mutational bacteria, viruses, and pesticide-resistant insects, namely, a shorter and shorter period between undulations of disaster and rebound. What this has meant is that the more epidemics, crime, war, and infant mortality, the more rapid the human population resurges. Tragedy invokes biological success. The more disasters, the more people; the more people, the more disasters. But the irony must not be missed. The logic of all things finite permits catastrophe only up to a point, beyond which a tragedy too vast, one that has totally closed the ecological circle of attrition, will admit to no succession. Such ineluctable confluences of human numbers following in the wake of human self-destruction, must unleash a global Armageddon on a scale that will admit to no swift and easy population resurgence thereafter. There must come a time when one inferno will actually prevent us from repopulating. A nuclear holocaust, for example; or an ecological virus of global proportions. That will mean, of course, eventual extinction. For those few who temporarily survive this vague and distant bang, or whimper—however one views the generic calamity—it might be a blessing. Surely the rest of remaining creation is likely to view it that way. But there is no consolation whatsoever in this post-human scenario. The events of which I speak would undo, within a human generation or less, the dreams and miracles of the most recent several hundred million years of biological activity on Earth.

For our species to achieve long-term stability amid such mathematical perils, we must somehow provide more generously and equally for all people, without destroying the biosphere—a colossal hat trick as yet to be learned. That time bomb is now ticking. There is no one alive who does not hear it, in some form or another. The soul hears, and fears it, but knows not what to do. It is a lurking San Andreas fault to the millionth degree. Every organism dimly perceives the growing shadow, even in a tiny country like British Montserrat, in the Leeward Islands of the Caribbean, with its total population of eleven thousand. By the year 2025, it is expected to hit a whopping fifteen thousand. Already, in Montserrat, the incidence of teenage pregnancies has become a problem. In Oceanic Vanuatu the 1990 population of 151,000 is expected to reach 205,000 by the end of the 1990s. And, with its current high TFR of 5.53, the island population is expected to double by the year 2015, a mere generation away. Ten years later, in 2025 Vanuatu may well contain 348,000 people. To feed and house such an increase in so small a region can only have the most devastating environmental consequences, reminiscent of Easter Island. Vanuatu is a surveyable problem. A dot in the ocean. And yet, even in this one workable microcosm, no solution to the population-environment collision has come to light.

Presently, in the mid-1990s, eight to ten Paris-size cities a year are being added to the world's urban population. There is no way, as yet, to conduct real time population audits of these new megalopolises; no method by which the ecological or psychological impacts may be measured. Too much is happening, too fast. Thought of differently, every eight months there is another Germany, every decade, another South America.[22] And while TFRs throughout the developing world have declined from 6.1 to 3.9 on average, and overall population growth from 2.1 to 1.7-to-1.8 percent per year, getting these numbers down any further is appearing next to impossible. In fact, because the base population is so much huger than ever before, runaway growth is utterly outstripping the aforementioned declines. The numbers, then, are tragically deceptive, like declaring, "We have cut down all the forests to build housing, but at least it's *affordable* housing." For the human race to check its growth even at 16.5 billion, it will have to reach a two-child (replacement) family by the year 2080. In the United States, at present, only 23 percent of the population actually conforms to a two-child norm. And in all but a handful of European countries, Japan, and a couple of cities in China, the two-child norm now appears unlikely.

A global demographic transition assumes that declining fertility will follow declining mortality, an assumption that is proving to be less and less tenable in most low-income countries like Indonesia and India. These countries are stricken with an average sixty-two-year life expectancy, and a mean infant mortality of 94 per 1000, versus 8 per 1000 in the industrialized countries.[23] More than half of the poor populations live in ecological zones that are biodiversity hotspots, those particularly vulnerable to further economically induced disruption. Eighty-five percent of all future megacities—ecological tinderboxes—are in the developing world. Each year, for example, Egypt, Ethiopia, and Nigeria add more people to the world than all of Europe combined. What we are seeing is the breakup of any predictability, just at the time when more and more government agencies and administrations are depending upon that presumed underlying orderliness to fulfill donor-country targets, meet international treaty mandates, pay back loans, and get reelected. But ecological and economic fragmentation are rendering obsolete all classical demographic transition theory. Post-industrial fertility declines required a century in northwest Europe, but only thirty-five years throughout much of the Pacific Rim and Cuba. In Sri Lanka, China, and Costa Rica, diminished fertility did not coincide with noteworthy industrial development, whereas in Brazil and Mexico zealous industrialization has had little impact on continually escalating fertility rates.

By the year 2000, projections show that 1.2 billion women of reproductive age will have three children on average, and the number of those children under fifteen will have risen to 1.6 billion. By 2025, reproductive-age women will number 1.7 billion and their offspring also 1.7 billion. As technology enables women to conceive later and later in life, TFR data—sampled, typically, at one time—will tend to be less and less reliable. Moreover, male TFRs are yet to be accounted for. What this all means is that projections for future stabilization—be they population-oriented, or ecological—are fraught with uncertainty. Even in 1994, demographers are unsure of Mexico City's population. The variance is more than 25 percent.

Yet, according to the U.N., the U.S. Bureau of the Census, and the World Bank, demographic transition will effect a global human population stabilization in the middle of the coming century.[24] The sophistication of demographics today enables family planners, economists, and policymakers to essentially justify any claim. For example, if the human population simply adopted zero population universally, within a century or so there would be no more human beings on the planet. With a globally consistent TFR of 1 (as in parts of Shanghai, Germany, Italy, Portugal, and Spain), we would end up with just over 2.5 billion people in a very few generations—the number of humans in 1951, when the nuclear arms race and the specter of radiological fall-out had attained its apogee, a fact not to be forgotten.

But there are countless, glaring contradictions. According to an executive summary of a forthcoming document from the U.N. Population Division,[25] the range of human stabilization is between 6 billion (clearly impossible) and 19 billion for the year 2100. It is disturbing that just ten years ago, the U.N. calculated that the stabilization spread would be between 7 and 15 billion. The new figures are based upon a replacement fertility rate of 2.06, either sooner or later—that is to say, a two-child woman (who will consequently experience some 450 menstrual cycles in her life).[26] But if fertility stabilization were at 2.17 (a mere 5 percent higher than replacement levels) the world's population would hit nearly 21 billion in the year 2150! What is clearly terrifying to consider is that a 5 percent higher than replacement level seems not only plausible, but dreadfully conservative. And yet few people in government, or in science, dare to speak about a human population of 21 billion. Only the Vatican continues to uphold a belief that the world could carry on, even with 40 billion people. Most scientists consider such convictions absurd. What is certain is the fact that human fertility is more combustible than ever before; its extent, advantage, brain power, carnivorous habits, footprint, hedonism, and malevolence, far in excess of the whole biological past combined.

What if, as outlined earlier, even the 2.17 figure is nothing more than demographic dreaming? What if a global average TFR of 3.3 to 3.4 persists, as it has for the past decade? After all, Africa is projected to remain at a TFR of 6, its population expected to triple to 1.6 billion by 2025. And in 1995, fifty-three countries will still have child mortality rates at or above 100 per thousand. (In Europe, the average is 13.) Twenty countries in 1995 will still be officially considered malnourished, providing less than 90 percent of those respective populations's caloric requirements, engendering "stunting" and "wasting", as the conditions are known. And remarkably, sixty-eight countries containing 18 percent of the world's population are currently either maintaining, or actually encouraging a rise in their fertility rates.[27] The major population centers—India, China, Indonesia, and the U.S., which until recently prided themselves on family planning—are all losing ground to wave after wave of new baby booms. Given the enormity of the world's present fertility base, and the aforementioned (theoretical) undeflected growth figure of 80 billion by the year 2150, the triggers seem to be transcending culture or technology. They are purely mathematical.[28] And this is terrifying. For, in spite of all the talk of environmental mediation, international family planning, and the very best intentions, such mathematics indicate a war that has gotten totally out of our control.

Population projections are based upon a host of assumptions. If the assumptions are off to any extent, so are the projections. Because population momentum is built into the age structure differentiation of any population, growth will continue for many decades before the assumed new norms set in. But many things can happen during those decades to offset or countermand prevailing reproductive behavior, making all such predictions vulnerable to change. As previously indicated, disaster tends to work in favor of more and more people, ironically. Nearly nine million people died in World War I. Yet, within one generation, Western populations soared well beyond their earlier numbers. Over 36 million perished in World War II, and again a huge baby boom followed. Triumphs in health care, lower mortality rates are deemed the basis for a universal sense of family security, and hence, lower fertility. But in some countries, such as China, the reverse is taking place, as one might expect it to: family security translates into an economic comfort level which engenders more, not less, children. Economics, it is argued, determine the number of children. Poverty makes for larger families, wealth for smaller ones. India and Singapore are nearly always cited as examples of this two-pronged theory. But Singapore is going the opposite direction, at least since 1983, as is China. And, as for the much-touted principle of education, in Kerala it has had a major impact on fertility, but less so in California, or Mexico, or Malaysia. And in poor Mali, the literacy rate has actually declined in the past two decades, not risen. As a rule, of course, the rich nations have shown that literacy and wealth do drive down family size.

No U.S. depression

The biggest contradiction inherent to all stabilization projections is the very assumption that the human population will ever stabilize; that the human condition is inevitably headed toward more wealth, more leisure, more security, when in fact all indications suggest the opposite. The standard of living, the so-called quality of life package, is rapidly diminishing for more and more people, billions of people in fact. And if this opposite is true, how can anyone trust that fertility and mortality will eventually fall into some heavenly harmony? Only a fool would buy into that argument. Disaster has always made for even more people in its aftermath. Economic and educational advantages are slipping. On what grounds, then, can demography assume anything about demographic resolution?

According to the World Bank Working Papers (which are strictly that, working papers, without the final sanction of the World Bank) that future "harmony" will result in 12.2 billion people in the year 2150, 88 percent of whom will be living in less developed countries.[29] Ultimate stabilization figures are said to be based upon "current fertility and mortality conditions." Given the fast-declining health of the environment, and our lack of definitive knowledge regarding even the most basic biodiversity inventory of the planet, or even of most countries, how can any conclusions be mathematically, extrapolated, from the "current" situation? Such "fertility and mortality conditions" rest upon several troubling assumptions. For example, it is presumed that the conditions making for the most progressive region within a country will spread to all other regions in due time. In India and China these are doubtful deductions. For example, the replicability of the Tamil Nadu/Kerala TFRs, literacy rates and IMRs may not be possible in any other Indian states, in which case, if India continues to grow it will double, triple, and quadruple its size within a timeframe that is projected, instead, to stabilize. While Bangladesh has been cited by family planners as an exemplar of recent positive change, some fundamentalist Islamic religious leaders are trying to reverse that profile. They have now taken—according to local press reports in the country—to destroying marriages and publicly beating social welfare workers accused of teaching illiterate women to read and write. Evidently, these clerics fear that women, once literate, will convert to Christianity. Zimbabwe has achieved a rather progressive contraceptive prevalence rate—over 30 percent—and yet, its TFR is still 6. The region of Chogoria in central Kenya is a good percentile point below the rest of the country in terms of TFR, but its success does not appear to be spreading elsewhere in the country.

The entire North American growth rate of an annual 0.9 is projected to decrease to 0.3 by 2025 as a result of diminished net migration from Latin America. Yet statistics from California, the country's most populous state, show a doubling of numbers (comparable to Uttar Pradesh), and an enormous increase in migration (not unlike the situation in India's Bihar state). The notion that the world's total fertility rate is projected to decline from 3.3 to 2.9 in seven years, and then to 2.4 by 2025 is based upon nothing more than vague institutional optimism. "For countries with high fertility, the trend is assumed to be downward, and substantial fertility decline is projected."[30] Just like that! But if, as history has shown (in all but a few countries, like Kenya and France), economic turmoil actually induces higher population growth, then this premise of declining numbers is not to be trusted. Fifty countries currently show incomes under $610 per capita per year.[31] Another fifty-five or so are lower-middle income, from $610 to $2399. Still another thirty-three countries are, by any Western standards, poor. These include Czech and Brazil. All of these nations demonstrate the very economic adversity that has traditionally lent itself to higher and higher birthrates, not declines. And if one assumption is incorrect, many probably are. For example, a higher TFR value will skew the anticipated decline in the proportion of individuals under the age of fifteen, meaning a larger percentage of childbearing females, and thus a larger built-in population momentum than expected. Optimism on this stage of life is a house of cards. Thus, the editors of the World Bank Working Papers caution that the data should be read tautologically with the "universal qualifier that population will follow the indicated path if the assumptions prove to be correct."[32] And if the assumptions prove to be incorrect? Add a global recession and the mounting "trigger" of ecological collapse throughout the world, and all such population assumptions and projections go up in smoke.

It would be wrong, for example, to extrapolate current fertility trends from China's population control successes of the 1970s. The country lowered its TFR from 5.8 to 2.5 between 1970 and 1985, but only under very unique circumstances that failed to materialize after one year of like-minded efforts in India. Furthermore, while declining fertility assumptions by the World Bank and others are said to have no predictive relationship to socioeconomic factors, those are precisely the factors that have held sway during the initial fertility transitions in the two most populous countries in the world, and now appear to be doing so, contrary to most expectations, in Kenya.[33]

To appreciate the sense of discrepancy that prevails within the demographic community, note that the World Bank Working Papers project that China will "remain the most populous country until 2120."[34] Yet dozens of independent demographers all predict that India will overtake China within two generations; the Working Papers do not see that happening until 2120. Like some macabre game of "chicken," two conflicting sets of data race toward a brink, crossing over the domains of urban and rural healthcare, ecology, religion, anthropology, politics, the world market economy, the shifting geo-political landscape, the engineering, biotechnology and military sectors, even ecotourism, and telecommunications. The falling market for coconuts, the rising value of tennis shoes, the pope's travel itinerary, new emergent contraceptive technologies, the age of certain political leaders, the extent of groundwater pollution, or the death of mackerel, daily calorie supply, the scope of infant immunization and secondary school enrollment, the rate of inflation, political freedom and civil rights, all enter into the equation of demographic projection. To better appreciate the complexity of prediction, consider just one aspect of that equation: contraception technologies.

Carl Djerassi, who developed the birth control pill, speaks encouragingly of a simple "menses inducer" that eventually will be taken orally once a month just prior to menstruation. If she is pregnant or not, the pill would induce a woman's monthly bleeding. But he also sees future contraceptives being provided by the immune system, not hormones; a contraceptive vaccine that might innoculate a woman against human sperm, or against her own eggs; a vaccine that coats the woman's eggs with antibodies to impede sperm and to short-circuit the moral issue of killing fertilized eggs. Levo-ova (used only in Sweden, currently) is a progestogen-only IUD device that impedes sperm while reducing bleeding tenfold. A vaginal ring is also in the works for the near future. A woman can insert this plastic ring and take it out for her period. It releases progestin into the bloodstream, as does Norplant. Abortifacients like RU-486 will more than likely be introduced in countries everywhere, allowing women in fundamentalist nations to implement birth control without their husbands' knowledge. Dr. Sheldon Segal, who invented Norplant, is now developing a similar male implant that will suppress sperm production on the theory that lower sperm count will suffice to prevent pregnancy. The technology will be available by the end of the 1990s. The American Medical Association, meanwhile, has encouraged the advertising of prescription contraceptives directly to the consumer. And it is likely that most contraceptives will eventually be sold without a prescription needed.

But all of this restraining technology may have a very limited impact on the most populated quadrants of the planet, where it is most needed, for lack of funding or confidence. In the poorest countries, such as Ethiopia, a single condom can now cost the equivalent of ten days salary. Yet economic inaccessibility is the mother of invention. There are over three thousand known indigenous herbal contraceptives. In some parts of Africa, the oldest contraceptive technology of all is coming back into vogue—namely, breastfeeding. In a recent study of women in twelve countries, it was shown that breastfeeding beyond the resumption of menstruation and sexual relations, following a birth had a "considerable" contraceptive effect.[35] UNICEF says that breastfeeding could save 1.5 million lives a year, as well as sparing the 250,000 children who are permanently blinded from vitamin A deficiency disease. Yet, few hospitals or doctors normally advocate breastfeeding as a form of contraception. And the same bias prevails with regard to the "rhythm method" which can be 97.5 percent effective, according to the World Health Organization. Even periodic abstinence has an 80 percent effectiveness rate, better than spermicides or sponges, about the same as a diaphragm, and just slightly less than the condom.

And yet, despite forty years of concentrated family planning worldwide, neither breastfeeding (which of course requires no genius, no literacy, no technology whatsoever, just Mom) nor other natural methods have exerted the kind of population controls that many have hoped for. Experts argue it is because the educational component is lacking. Others attribute this missed opportunity to maternal morbidity, economic fallout, ecological stress, or lack of access to family planning services *affecting at least 300 million couples.*[36] During the decade of the 1990s, another 100 million couples will need family planning services just to maintain the status quo. According to a WHO/UNFPA/UNICEF statement, every year 13 million children die before their fifth birthday and five hundred thousand women die from pregnancy complications, but "with current technology, the majority of these deaths could be prevented."[37] UNICEF has saved 1.9 million children every year just through measles vaccinations. Another eight hundred thousand annual deaths are yet to be averted through an increase in such injections. Immunizations for diphtheria, whooping cough, and tetanus have all effected resounding improvements in child survival rates.

There is no question, then, that new technologies, and the greater application of existing ones, could alter the demographic outlook. By 1998, for example, a malaria vaccine (SPF66) is expected to be made available, and this could have an enormous impact on higher fertility rates. Extended breastfeeding and universal immunizations, by themselves, could have demonstrable impacts. The open question is: will such technologies and techniques increase or decrease the overall size of the human population, keeping in mind that it is precisely the advances in medical technology, for example, that paved the way for India's and Africa's population booms?

On Kenyatta Day, October 20, 1993, residents of Nairobi read in bold headlines in their morning newspaper, "MAJOR ANTI-AIDS BREAKTHROUGH." The report pertained to research by a Kenya–Canadian team that focused on prostitutes in Pumwani, one of Nairobi's shantytowns. Seventeen hundred women were studied for eight years. Twenty-five of the women, though constantly exposed to the virus, escaped infection and it was believed that the women all shared an immunity to to HIV, or an ability to make virus-neutralizing antibodies; "a breakthrough in the discovery of an anti-AIDS vaccine before Christmas" was indicated. Weeks later, the French were also claiming to be on to a "breakthrough."

But the shadows of Africa's epidemiology remain: do away with the tsetse fly, AIDS, and malaria in Africa, and there is no question that the continent's population will increase, inviting more suffering, hunger, brain damage, more poverty, more wars, more biodiversity loss and animal pain. In Pakistan, medical advances have pushed the crude death rate down by 50 percent since that country's independence in 1947. Yet the birth rate has remained stubbornly at nearly 45 per thousand, creating an enormous net population gain. Hence, in one sense, medical progress—like technology in general—may serve to exacerbate the population debacle. As Norman Myers has written, an "operational evaluation of Pakistan's carrying capacity becomes, in essence, a matter of judgement."[38]

In at least one crucial sector—agriculture—technological advances are falling quickly behind population growth. Nearly 15 million tons of grain are lost each year as a result of known environmental destruction, an amount equivalent to 50 percent of all known technologically induced gains in grain output. At the same time, 20 percent of the world's (economically advantaged) population spends over $200 billion a year on agricultural subsidies to fight off exports by 80 percent of the world's (economically disadvantaged) people. The equation should be, if anything, reversed, in any honest and moral economic system.[39]

Such data leaves the question wide open as to whether or not new technologies are likely to have any impact on curtailing fertility in the future. A male implant that reduces sperm count may be redundant and ineffective, for example, inasmuch as sperm counts worldwide since the time of World War II, have already plummeted for reasons not yet understood—perhaps ecological poisons, biochemical stress, or the planet's own counterstrategy?[40]

Every factor influencing demographic projections confronts the same unknowable forces. Disaster breeds disaster. Medical technology devised and applied compassionately can inadvertently invoke even greater pain and suffering by keeping more people alive for longer periods and setting the groundwork for more and more children. Healthier children are understood to be the secret to fewer children. And yet, just the opposite can also be argued. There are no rules. Human nature, like every other force of nature, defies consistency.

No fact of biology is less ascertainable, and more important to population figures, than the death rate. Currently about 9 people per thousand die every year. The extremes are Mali and Japan, where 167 children versus 4 children per 1000, die annually. Will such ratios change with the advent of more and more education, health care, technology, and financial assistance? According to one investigator, a mere $3.2 billion per year, targeted at secondary school enrollment for girls throughout the developing world, would end the "cycle of female deprivation and mortality."[41] But so far, that money has not been found. Like the stalled TFR declines, infant mortality is now declining far more slowly than in previous years, while the number of adult

illiterates and children not enrolled in school is rising rapidly. Heart
disease and many forms of cancer are also everywhere on the increase.
Eight major tropical diseases are expected to kill as many as 4 million
people each year through 2010, a considerable increase over today's
figures.[42]

There are hundreds, possibly thousands, of other serious ecological
viruses and bacteria that may easily be liberated by human trespass, only
to inflict unexpected consequences—like a Black Plague—which
would, possibly, change the whole demographic picture, though it is
impossible to say in which direction. "More than any single carrier,
it is human encroachment that ultimately precipitates the emergence
of killer viruses," writes Stephen S. Morse.[43] In Japan (Japanese encephalitis), Argentina (hemorrhagic fever), U.S. (Seoul virus in
Baltimore), Panama (Venezuelan equine encephalomyelitis), and in
nearly every other country, the connection between increasing ecological destruction and the sudden exposure of humans to long isolated
or dormant bacteria, protozoans, fungi and viruses, has been identified.
A case in point is Lyme disease, the work of the spirochetal bacterium
known as Borrelia burgdorferi, "directly linked to changes in land use
patterns" on the east coast of the U.S., where rapid reforestation, new-
growth forest, and intense residential suburban development, placed
human hosts, white-footed mice, deer, and Ixodes ticks in the same
neighborhoods.[44]

Never before has there been such an avid exchange of "information" between large species (Homo sapiens) and microbes. Of the more
than 100 zoonoses (animal infections transmittable to humans) and
over 520 known arthropod-born "arboviruses," at least 100 have been
shown to cause diseases in people. Some are among the deadliest diseases
ever encountered. In 1976 in Zaire and Sudan, as many as 90 percent
of those infected with Ebola virus died horribly within weeks. The
disease is believed to have been picked up by contact with monkeys.
Richard Preston's detailed description of death by Ebola is worth quoting
at length, if only to hint at the real agony inherent to ecological viruses and the population explosion. What begins with a mild headache

and a certain unhappy expression in one's eye, soon takes on night-marish complications.

> The patient's bloodstream throws clots, and the clots lodge everywhere, especially in the spleen, liver, and brain . . . a kind of stroke through the whole body. . . . As the strokelike condition progresses and capillaries in the internal organs become jammed with clots, the hemorrhaging begins: blood leaks out of the capillaries into the surrounding tissues. This blood refuses to coagulate . . . its cells are broken. You are stuffed with clots, and yet you bleed like a hemophiliac who has been in a fistfight. Your skin develops bruises and goes pulpy, and tears easily, and becomes speckled with purple hemorrhages called petechiae, and erupts in a maculopapular rash that has been likened to tapioca pudding. Your intestines may fill up completely with blood. Your eyeballs may also fill with blood. Your eyelids bleed. You vomit a black fluid. . . . In the pre-agonal stage of the disease . . . the patient leaks blood containing huge quantities of virus from the nose, mouth, anus, and eyes, and from rips in the skin. In the agonal stage, death comes from hemorrhage and shock. . . . Ebola seems to crush the immune system.[45]

Unlike Ebola, which struck a few hundred people then vanished back into the jungle until next time, the Black Plague wiped out one-third of Europe's population over a period of 150 years during the late Middle Ages. When the Spanish army entered Mexico carrying small-pox in 1520, 3.5 million Aztec Indians were killed within two years. In Europe smallpox killed some four hundred thousand people every year until English physician Edward Jenner fashioned a cowpox in-noculation that impeded the virus. In 1967 there were still 2 million deaths a year from the disease worldwide.

The most recent incident of plague, caused by the bacterium *Yersinia pestis*, occurred in India at the turn of the twentieth century and killed more than 10 million people. At the same time, a small outbreak among Chinese occurred in San Francisco. Another outbreak of the magnitude of that in India has been postulated for the United States.

The 1918 influenza A pandemic claimed 20 million lives worldwide in less than a year. It is believed to have originated in the United States, went to France, then returned to the U.S. where it spread rapidly from New York to California. Since then there have been serious global influenza outbreaks on at least five occasions.

As bacteria struggle to "cope" with ecological disruption, they are likely to mutate into increasingly sophisticated, virulent forms. One example appears to be the type causing Brazilian purpuric fever, which first emerged in 1984, quietly erupting in a town in Sao Paulo state. "Symptoms included high fever, vomiting, and abdominal pain, followed by the development of purpura and shock owing to vascular collapse." The ten children who were unfortunate enough to contact this bacterial creature died within forty-eight hours of the onset of fever.[46]

With global warming and subsequent droughts, grain becomes subject to toxic fungal infestations like aflatoxin, a threat to both humans and livestock. With more and more dam projects throughout the world, other diseases breed. At the turn of the century, prior to the building of the Sennar dam on the Blue Nile, schistosomiasis did not exist in that area nor was malaria a health problem. By the 1950s, however, following construction of the dam, hundreds of people had died of the severest form of malaria, which is carried by *Plasmodium falciparum*. By 1975, all the villages in the area were malaria endemic.[47] In India, the clearance of forests has transformed several zoophilic species into anthropophilic ones. Vast new areas have been made endemic for malaria. By the year 2000 in India, malariogenic conditions will have increased by 40 percent as a result of massive new irrigation projects that coax seepage from dikes and dams to provide breeding sites for *Anopheles culicifacies* and *Plasmodium falciparum*, mosquitos and malarial parasites.

By the year 2000, there will be some 425 cities each with over a million people, ideal feeding grounds for certain disease-causing organisms. As the aging structures of populations (i.e., in those countries with current fertility replacement) attain a plateau, under high pollution regimes (such as throughout the low fertility, ecological hot spots of Central Europe), mass immunosuppression might invite "opportunistic infections" among large populations too weakened to fight them off. It is now estimated that at least 150 million urban poor are "in a permanently weakened condition because they carry one or more parasites."[48] According to Joshua Lederberg, Robert Shope, and Stanley Oaks, Jr., not only ecological disruption, but "new medical treatments and technologies . . . have created additional openings for opportunistic pathogens."[49]

And then, there is AIDS. Recently, the World Health Organization raised its estimates of global HIV-positive victims to 40 million in the year 2000, 90 percent of them in developing countries. Other reports have projected as many as 100 million victims, particularly in India and Africa. Some health officials have offered their opinion that the AIDS epidemic in Africa will likely peak at 20 to 30 percent of the adult population, which would mean a decline in the continent's growth rate from 3 to 2 percent. If past famines in Africa are any indication of the resulting population resurgence, AIDS too will have a limited long-term impact on the growth rate of the human population overall. However, add famine, a few major wars, a host of new bacterial killers and ecological viruses to the existing AIDS crisis, and who knows? The numbers might start to come down. Or maybe not. If there is a comprehensive law of population averages (measles virus, for example, requires about three hundred thousand human hosts in order to survive), there is no way to divine it because human ecological disorder has engendered so many untraceable parts of the equation.[50]

Not only are projections vulnerable in a world conspicuously lacking in consensus, but—in the ecological sense—even the current growth rates of any given nation are elusive. While it appears that the population percentage-of-increase in the early 1990s was highest in Jordan, Israel, Oman, and Saudi Arabia, these countries attribute much of that growth to immigration; whereas in Rwanda, Kenya, and Swaziland, it is homegrown growth (excess of births over deaths). Italy is said to have the lowest growth in the world, a TFR of 1.32. At the same time, Italy has the highest pollution growth, and Total Population Dynamic, in all of Europe.

Even where the highest TFRs might be expected eventually to come down (i.e., Rwanda from 8.3, Yemen from 7.7) other high TFRs are now going even higher—in Mozambique, Tanzania, and Niger, for example.

And there are a host of other crucial quantitative observations that defy certitude or predetermination. For example, the World Bank estimates that by the year 2000 the developing world will suddenly start declining in growth—just like that. Why should it, coming right after the largest population growth rate in world history? And there are now the largest proportion of childbearing females, ever! How can it be assumed that within one generation, 22 million fewer children will be born every year? The reason, of course, is that it simply "has to be" that way if fertility and mortality are going to even out, or harmonize, by the year 2150. This "harmony" is an arbitrary goal that says more about the human psyche and its perennial faith in the future, than it does about any hard science of numbers. If, as the World Bank Working Papers attest, populations in the developing world between the ages of 0 and 14 will have increased from year 2000 to year 2025 by 188 million, then, if anything—at projected TFRs of between 3.3 and 2.88, then subsequently 2.44 and 2.35—the population momentum for the childbearing sector will continue to more than double. In other words, more women having slightly fewer children, will continue to push the population figures higher and higher. In India, where the TFR is expected to drop to 3.74 by the end of the century, that lurking momentum will be enormous.[51]

In the year 2030, the number of women in South Asia (including India) in their most fertile age period, between fifteen and twenty-nine, is projected to be 173 million, up from 110 million in 1990. Total fertility in all low income countries is currently 3.592. By the year 2030, that figure is expected to drop to 2.397. Will it happen? Only on paper, for certain. A good indication of the inherent vulnerability of such data is the total number of females in their childbearing years (fifteen to forty-four) as calculated in 1995 and again in 2150. In every significant country, there will be many more such women 155 years hence. Unless all other assumptions are fulfilled in the domains of healthcare, education, environmental remediation, and economic equality, these numbers indicate anything but stabilization.

To easily visualize the dilemma, take a region like the Gaza Strip in Israel. In 1995 the number of females between fifteen and forty-four will be something like 135,000. In the year 2150, there will be 504,000 of them. Add to that an additional 84,000 girls in the ten to fourteen age group (who must also be considered as possible childbearing adults) and the total rises to 588,000 women. If each of them has a so-called harmonious TFR of just over two children, then within a single generation following the stabilized year of 2150, the population of Gaza would jump from the projected 2.9 million by over 30 percent! Every woman won't have two children. Some will have five, others none. But the data suggests the built-in uncertainties. In countries like Iran, Iraq, and Pakistan, Egypt and India, Indonesia, Brazil, Mexico, Nigeria, the Philippines, Vietnam, and China, these inherent flaws in the mathematics—inherent volatilities of culture and economics, politics and religion, ecology and health—provide no sure indication whatsoever of the planet's future. Iran, for example, will have 3.7 million young women between the ages of ten and fourteen in 1995. By 2150, there will be an estimated 14 million. Already in that country, the fundamentalist sexual mores are rapidly breaking down. The veil on women is being lifted, skirts are being shortened, prostitution in parts of the country is a fact of life. In China, 1995 will see 44 million girls added to the "teenage" sector, but in the year 2150, 53 million of them will be added. In India, overall there will be 254 million teenage girls in 1995, but in 2150, 370 million of them. In Malaysia, 1995 will see 3.2 million teenage girls, but the year 2150 will see over 8 million of them. In Pakistan, the gap is enormous: from a 1995 figure of 33 million teenage females to nearly 80 million females of childbearing age in 2150. Officials in many of these countries may insist all they like that there are no teen pregnancies, but the fact remains that millions of teenage girls are having babies and in nearly every country such teen pregnancies are on the rise, as they are dramatically in a place like Los Angeles. Many of these young people, in over forty countries, in countless interviews and discussions with me, have expressed their common depression, sense of futility, and fatalism about the planet as they view it, adding fuel to their determination to enjoy life while they can. There are very few good arguments to dissuade them from having sex in a world where tigers and chimpanzees are going extinct. Even the prescriptions for "safe sex" lose some of their cogency in a world so fraught with decay.

Add to these intuitive syndromes one other astonishing fact: worldwide the average age of first menstruation is dropping! This trend corresponds with what must be seen as a global emphasis on earlier sex. Throughout Latin America, birthrates among teen mothers are extremely high—15 percent in El Salvador, for example. By the year 2000, there will be nearly 130 million teenagers in Latin America. Only 10 percent of them use any form of contraception, often erratically and hence, ineffectively. In Africa, 40 percent of all women have their first child before the age of eighteen. In at least one coastal district of Kenya, a country that has until recently placed a premium on female education, girls are getting married as early as the age of nine. In Kenya, 60 percent of all abortions are performed on teenagers, although abortion is still considered a "criminal act" in that country. Every year, 5 percent of all teenagers worldwide contract a sexually transmitted disease. Twenty percent of all those who contract AIDS are in their teens. In Thailand, UNESCO estimates that eight hundred thousand of the 2 million prostitutes are adolescents and children.

Only Brazil, among the fastest growing large nations, will have a slightly lower fertile female count in the twenty-second century than it does now—by a few million. Yet, Brazil's population will have doubled from its 1990 figure of 150 million to over 300 million by then.[52]

Again, according to the "harmony" (stabilization) scenario, China's and India's populations are expected to become stable by the year 2150, at 1.8 and 1.85 billion respectively, while Indonesia is expected to reach 375 million at that time. The Working Papers see Indonesia's population doubling by the year 2110, yet some government officials within Indonesia see it happening by the year 2040. Even in a defined, relatively small area—Shenyang, China, for example—projections are at odds. Shenyang's Environmental Protection Bureau has declared the city's annual population growth rate to be 1.5 percent. The U.N., however, estimates a growth rate of exactly double that![53] Such discrepancies do not recommend demography as anything like the hard science it would like to be.

Recently there were claims that Rwanda's TFR had come down from 8 to 6.2. But within weeks, a World Bank official told me he had received word from other demographers in the country that the figure could not be trusted. A "lack of data comparability" from country to country is hampering any sense of global accuracy.[54] Similar doubts have been expressed by African U.N. officials about Kenya's recent demographic survey, with its promising results.

In sum, future contraceptive technologies and medical breakthroughs, increasing economic disparities, massive urbanization, emerging viruses susceptible to ecological disruption, a fast-growing population base among teenagers, and the unpredictability of fertility rebounds following disasters, all bode of population instability.

In spite of the imprecision and huge discrepancies of demographic projections, one unambiguous fact remains: the current size of the human population has wreaked unprecedented damage on the biosphere, and is going to accelerate that damage. Millions of plant and animal species have been driven to extinction. Hundreds of millions of innocent children, women, and men have been murdered. A billion people are hungry, morning, noon, and night. The ozone layer is thinning, with consequences that are lethal for every living organism. The air, water, and soil across the planet have been fouled. The forests in many countries are gone or nearly gone. And the mammary glands of every mother on Earth are now infiltrated with DDT and other harmful chemicals. These essential facts—truths that distinguish this century from any other in our history—are all the by-product of uncontrolled human fertility and thoughtless behavior. Even if we should somehow manage to merely double our population size by the next century, attaining, in other words, a number of 11 billion, the ecological damage will be catastrophic, unimaginable.

In spite of the foregoing democratic minefields, there is a new sense of ecological consciousness and stewardship throughout the world; a spirit of environmental community that is rapidly trying to mobilize in order to meet this disaster head-on. Whether a similar mobilization to meet the even more difficult crisis of as yet unborn children can be accomplished quickly, humanely, and firmly, will determine whether any kind of qualitative human future is even possible.

# A Global Truce

*Muddling through mists of self-appointed cloud*
*Our hearts are tear ducts, erratic,*
*blind, where light would otherwise*
*this recklessness guide.*
*Congealing 'round a purpose, Earth,*
*To heal and strive.*
*And hence, this hope, our species finds.*

## THE POWER TO HEAL

Even in the heart of demographic madness, besieged by dooms-day graphs, mathematical "regions of instability," and the overpower-ing numbers that they imply, we must recognize that it is never too late to end this war. The question is not whether we can, but whether we will act, in multilateral unison, decisively, and with sufficient dispatch to do what is necessary, region by region, country by country, on a global scale.

There are no pat miracle cures for the crisis of human overpopula-tion, though there are countless temporary anodynes for those with the health, money, or leisure time to pursue them. Overpopulation and its ills, like the world war to which I have likened them, bring out the absolute worst in human nature, though this fact is initially dis-guised by its participants, our most elite corps, our very children, soon to become unwitting warriors. The system of human fertility is the machine that doesn't know it's broken. Confronting this paradox may well exceed the limits and tolerance of rationality, or of genes.

But beyond the paradox itself, the worst has been popular for millen-nia. From the war historian's vantage, the human landscape is an archaeological testament to tyrannical stupidity and cruelty. Each pit of ash and shard, all those plains and depressions of rubble, invoke the vanity and futility of lives taken before their time in the mad stampede of human vengeance or greed. Civilization after civilization has partaken of this mayhem. "And they utterly destroyed all that was in the city, both man and woman, young and old, and ox, and sheep, and ass, with the edge of the sword," wailed the poet of "Joshua" (6: 21).[1] Compare these words with a description of but one night's madness along the Merderet River in Normandy, June 1944: "The slaughter once started could not be stopped. . . . Having slaughtered every German in sight, they ran on into the barns of the French farmhouses where they killed the hogs, cows and sheep. The orgy ended when the last beast was dead."[2] For its part, Aztec civilization will be remembered for its priests having slaughtered something like 20,000 victims a year, an estimated 4 million in all who were tortured, dismembered, burned, and tossed off cliffs. In Columbia in 1947, terrorists ripped fetuses from pregnant women and chopped prisoners into tiny fragments before whole villages.[3] A letter from Winston Churchill to his wife in August 1914 reads, "Everything tends toward catastrophe & collapse. I am interested, geared up & happy. Is it not horrible to be built like that? The preparations have a hideous fascination for me. I pray to God to forgive me for such fearful moods of levity."

Prior to the outbreak of World War I, this same dementia afflicted chief functionaries of five major and two minor powers over a period of ten days. A czar, a chancellor, a premier, and numerous foreign ministers were all vulnerable to the rash of half-truths, misconceptions, adamant pride, and opportunism that, taken together, acted as a blind momentum, the very personality of war, and resulted in events leading to the death of all those millions of soldiers and civilians. Later on, even while the United States was approving the 1929 Pact of Paris in the interest of peace, it was busily building up its naval defenses. Historian Quincey Wright has cited this contradiction of democracies with reference to the League of Nations, which urged universal repeal of war while resisting the obligations to carry it out. One former assistant secretary of Defense described NATO doctrine as that position by which "we fight with tactical nuclear weapons until we are losing, and then we will blow up the world."

The insane conjurations of war ultimately derive from the more subtle and diffuse violence of an overpopulated world. Among those mainstream, clustering numbers, no fingers can be pointed at any particular dictator. Rather it is the global array of relative indifference, a sluggish nervous system (as author Jonathan Schell so aptly described the human response to the build-up of nuclear weapons, in his book *The Fate of the Earth*), a bureaucratic ineptitude, and a concordance of cultural blinders in nearly every country, that is party to the proliferation of people and their subsequent violence. Family planners and ecologists are not equipped with anything like laser-guided missiles or stealth bombers, nor do they enjoy the same level of political clout— even in today's China—necessary to wage effective counterstrategies.

In trying to picture the violence of overpopulation, one is not able to so easily capture its mode of ecological combat, as did, say, painter John Trumbull in *The Battle of Bunker Hill* (1786), or Albrecht Altdorfer in *The Battle of Issus* (ca. 1528), or Pablo Picasso in *Guernica* (1937). In absence of context, we cannot clarify right and wrong on the battlefield, be it ecological or political.

The wars of, and against, overpopulation are even more elusive, colossal, and unprecedented than all previous conflicts. A crowded slum may serve as a convenient icon of peril, but it has not yet made a sufficient impression on our species. It has not been taken seriously.

As social organisms, we have foundered, failing to straighten our path, as ants in moving succession will do, precisely for the opposite reason all other socially moderated organisms have succeeded—because human behavior appears to be guided not by the urgings and imperatives of any collective goal or decision making process, but rather of individual appetites, and of those few others in our lives to whom immediate gratification is tied—a spouse, children, one's parents, even one's rival. Insects do not appear to function according to such personal compulsions. The "tragedy of the commons" is the tragedy of human individualism, sometimes positive, too often negative.

And yet, therein lies a profound hope. Having posited what I take

to be fundamentally distinct about *Homo sapiens*—our individualism—
I am thus led to this crucial observation, the one that must be mustered
in defense of every tomorrow, namely, our capacity to express love. That
is the power to heal, and it contradicts every gross approximation and
generality of the masses.

Take just one example of this capability. Consider the extraordinary
fact that those few survivors of the Buchenwald Nazi concentration
camp, struck no blow, wreaked no revenge, assailed their perverted SS
tormentors with not so much as an insult, when their turn came, upon
liberation from the camp in April of 1945. General Patton and the Third
Army gave the survivors every opportunity to turn the cards on their
captors. But the prisoners who had suffered so abominably had not
become insane and vengeful, but weary with illumination to the point
of compassion.[4]

Such lessons as forgiveness at Buchenwald do not stand alone, nor
are they easily explained by any science. Our innumerous acts of heroic
self-sacrifice, spiritualism, great art, and nonviolence, transcend, and
even outweigh the more noticeable bursts of bedlam and barbarity
throughout our history. So that while it is possible and tempting to condemn our species, anxiously anticipating its self-destruction, there can
be no greater emotion, no finer tears, than that recognition of our
awesome potential. Herein lies the crux of our biological future, our
*new nature*.

Sadly, ponderously, we are too many. This incredible dilemma imposes a near ethical impasse. But, true to our genetic disposition, we
may still reconcile our numbers by focusing on those very traits that
mark our individuality, our strength. Overpopulation, and its impact,
is a Buchenwald to the nth degree. Our survival absolutely hinges upon
those qualities of forgiveness, faith, art, and nonviolence that are
generously distributed throughout our species, though perhaps less
likely to be exhibited in large groups or bureaucracies.

As never before, the gift of our individual humanity must *inform*
the collective. Global electrification, enormous riches, the mass production of latex condoms and pills, new technologies for easy self-
sterilization and abortion, will not necessarily spare the child, or save

the lion and its habitat, or the lamb from being skewered. This challenge must engender personal missions of utmost conviction and urgency, focused upon extraordinary levels of decisiveness based upon empathy; a thorough schooling in human despair, other animals' suffering, biospheric blight, and the untimely destruction of what once lived and was beautiful. There is no economic nirvana to which we can escape such dead reckonings. We can only serve the world through honesty, directly and responsibly; by opening our eyes to the troubles all around us, and working toward change. A democracy provides no assurance of such change. Only individuals can do that.

To paraphrase Greek philosopher Nikos Kazantzakis, we must critically embrace the realization that "the holiest form of theory is action." Familiarity through ardent education breeds responsibility that, in turn, becomes enshrined. Nothing is even remotely safe that is not commonly esteemed, avowed, recognized, and ultimately held sacred. While it has been pointed out that environmental policies do not deal with the certain death of a known individual, but rather with populations at risk, the fact remains that large sums of money will frequently be spent to save a single life (i.e., a little girl trapped in a drainage culvert, or the last remaining California condors) versus comparatively low sums to save "statistical lives."[5] Short of this sacredness—this belief that the world and all of life is a miracle—our children will suffer as never before, while the Earth gets scraped bare, and is abandoned, like some gigantic Gettysburg.

I have argued throughout this book that two forms of action, in tandem, are essential if we are to end the war—namely, family planning at an enlightened, massive level, and an equally dedicated global strategy for preserving and restoring biodiversity and vast regions of wilderness. Both endeavors will take money, lots of it, and the courageous selection of priorities and ecological hot spots from among so many imperatives. Framed differently, we must acknowledge that we have violated three basic biological and physical laws. As social scientist Robert Gillespie, president of the quietly influential Pasadena California–based Population Communication Organization has written, "We need a global mandate to reverse consumption patterns that are not sustainable, to limit family sizes to replacement levels and to achieve a smooth transition from finite fossil fuel to alternative energy systems."

On the surface, these injunctions seem clear enough. But what new terms and conditions of survival will enable countless billions of rough-moving, carnivorous primates, loosely known as human beings, to forever put behind them the violence of nearly five thousand human generations? By what means might *individuals* forge a global truce that supports and sways not merely the poor and the rich alike, indigenous peoples, local lawmakers, taxpayers, even whole nations, but the entire global community of nations? Is the arithmetic even plausible, that a renaissance in individual ethics and efforts should somehow overturn a global genetic propensity, remaking the species? The critical mass, as it is so often referred to, is actually dependent upon the critical *person.* The continuation of our species hinges upon recognition that our individual morals are linked to an ecological bottom line.

These queries are not easily resolved by mere warm feelings, good intentions, and sporadic voluntarism, whose long-term consequences have, to date, shown little ability to impact social norms. Amidst the strident blare of 15 or 20 billion people—the spastic scenario of the coming century, or centuries—such expectations of any moral and ecological renaissance are especially troubling.

As I look back at the anonymous troughs, now covered with grass, into which my great grandparents were dumped by bulldozer in Vilkamir, or at that slaughterhouse in India one late night, ankle deep in blood, or the scoured remains of what was virgin rainforest in the Cote d'Ivoire or prolific estuary along the Yellow Sea, I now see that every personal encounter with anguish—either directly or secondhand—is actually a cogent building block for a new school of thinking that should radically change my person, and by logical inference, a world of persons. This is the ambition of every student who is still relatively unclouded and undiscouraged, to be able to hope clearly, to presume that human nature is malleable, that it can learn from, and be empowered by, its mistakes. As philosopher Robert Nozick has written, echoing the sentiments of Martin Heidegger, "The evolutionary basis of rationality does not doom us to continuing on any previously marked evolutionary track."[6] The cumulative force of rationality is a tantalyzing haiku, an open book of enormous promise. Do we have the power, as a species, to appreciate cause and effect, to do the right thing, to live together harmoniously, to be wise?

There is scarcely time left to document the casualties of this war, though countless reports in thousands of libraries throughout the world lay out the specific problems in millions of printed pages that few people read. But then, detonators washed up on French beaches, or the near extinction of the gorilla, require little reading to understand.

A book is only a personal revelation, guided by a feeble attempt to see oneself in this overriding tragedy as it is played out; to work toward hope, from the depths of private hell. All genes evolved in the light of family and beauty, of a sense of the divine in nature, of altruistic bonds that can best be seen among mothers of every species. But it is more difficult to imagine that those same genetic impulses could have been prepared for the temptation to heal an entire country, or continent, or planet. As humans, our neurology cannot possibly be expected to extend the love of family, or the multispecial biophilia evident at a place like Amboseli, to the whole world. It is a nonrational, though necessary wish, born of a unique awareness—part information glut, part compassion—that has made many half-crazy with the zeal of an enormous ambition, the engendering of this new nature, the realizing of what it is, or can mean, to be truly *human*.

Our response as a species to this particular notion of humanity has been, of course, a mixed affair. Many have sought to insulate themselves; as individuals, to possess *my* land, *my* home, *my* inviolate political boundaries and identities. Such ephemeral possessions may, as His Holiness the Dalai Lama has said, be necessary for the very purposes of effective micromanagement. But we have managed badly, manipulating the rules of survival with the fervor of an escape from something, a tantrum in the name of self-interest and family preservation, a kind of revenge against that very awareness of the larger world. The result has been abundant destruction. Policymakers and the public have often groaned, in so many words, "Stop confusing me with the facts."

It is a form of tuning out, of simply dismissing instinctive fear. Belief systems are hard to alter. As "human exceptionalists" we have always considered ourselves somehow immune to brute nature. How could we have predicted that mass man, who has sought to transcend the harsh vagaries of "primitive" biology, would have, in effect, created the most primitive nature of all, one more red in tooth and claw than anything nature ever devised?

Ironically, the hope for eventual balance is born in the very thickets and Shanghais of the population crisis. With enough aware people on the planet, there will be those who take up the selfless task of attempting to effect a difference. Certainly several thousand NGOs, and their millions of supporters, are trying. As the Chinese and their 50 million family planning volunteers have discovered, the sheer number of present cadres, of individuals taking the reins in modest, but decisive ways, offers the one and only blind antidote to the predicament of ourselves.

The world, by nature, is perfect. It needs no solutions. But if one accepts the premise that biodiversity is necessary, that this planet has invested nearly four billion years in the propagation of wide-ranging life, then it must be conceded that human nature is somehow mysteriously in trouble. Yet we have the power to heal ourselves, and we must do so if the biological world, of which we are a fragile part, is to survive.

By "World War III" I have meant to designate not only the concerted devastation our species is levying upon the whole planet, but the war we must wage to end this state of affairs. In the final sections of this book, I take up four primary arenas wherein that reconstitution must take priority, must, in short, invoke the same level of national and global mobilization that sent U.S. troops into Nazi Germany: the moral, economic, technical, and policy underpinnings of ecological and demographic healing.

## THE ECOLOGICAL CONSCIENCE

There are any number of cultural examples, both contemporary and historical, to suggest qualities of lived (as opposed to merely thought) nonviolence and ecological sustainability.[7] One of the most extraordinary of these is called Jainism. The Jains are an important challenge and inspiration to all environmentalists, political and religious thinkers.[8] While I have spent considerable time with numerous Jain friends, one of my most lasting impressions of this gentle, ethically realistic culture, is of two naked monks in the Indian village of Taranga, in western India. They had been walking side by side for many years. We sat together, briefly, one morning, and when we parted they recited an ancient Jain salutation, *"Khamemi sabbajive, sabbe jiva khamantu me, metti me sabbabhuyesu veram majjha na kenavi,"* which means, "I forgive all beings, may all beings forgive me. I have friendship toward all, malice toward none."[9] Such words are merely shorthand for a deep vein of ecological conscience that runs throughout Jain tradition.

Most of the approximately seven million Jains (predominantly in India, though many having emigrated to East Africa, western Europe, Canada, and the U.S. during the last century) are vegetarian, and engaged solely in those professions and businesses that only minimally harm the environment, and do not profit from any kind of animal product. Not surprisingly, given the recent revolution in "green" business and commerce throughout the world, Jain ethics are proving more and more profitable. In fact, while Jains have restricted themselves to such largely non-impacting trades as law, publishing, mostly natural textiles, diamond cutting, accounting, and medicine, they have become one of the richest communities in India. Much of their money goes back to the culture from which it derived in the form of charity.

While Jainism is arguably one of the oldest religions in the world, its "modern" renaissance was unleashed in the person of Mahavira (599–527 b.c.), an elder contemporary of Buddha. Mahavira spent most of his adult life naked, having renounced all possessions, including his clothing (as would Saint Francis nearly 1700 years later), and walked throughout India discussing and formulating the rudiments of *ahimsa*, or nonviolence. His teachings have formed the body of a copious literature, several million pages of scripture, and subsequent commentary by followers. Mahavira's first sermon was known as *Samavasarana*, meaning in Prakrit, "a congregation of people and all other animals". He reasoned that all living beings are endowed with a soul that must be respected and nurtured. Equally important to Jainism is the revelation that all souls, all biological beings, are interdependent (parasparopagraho jivanam).

According to Jainism, it is humanity's duty to safeguard and shepherd that interdependency. Consequently most Jains, not merely those few monks who go naked or are clad in white robes, have adopted a philosophy of minimalist consumption. In their thinking, they have categorized life forms according to the number of senses with which those creatures are endowed. A human being has five senses, most plants, one. Humans are restricted to consuming only one-sensed beings. And even that consumption must be sustainable. The whole belief system hinges upon the ecological integrity of those living, and those yet to be re-born; those countless souls as yet to be biologically enshrined, or, conversely, liberated. The process is inherently spiritual, reverential, and compassionate, and poses a working model for how humanity might best embrace responsibility for its undue power over other species. A tree does not need to recall its responsibilities. A human being does. From a Western, conservative religious point of view, one might better appreciate Jain tenets by considering that God's creation in all its magnificence is manifestly shared equally throughout the world. This act of unstinting sharing characterizes a form of behavior, call it religious, just, equitable, or pragmatic. But fundamentally it is about ecological trust in the future. That is God's gift: a living, self-perpetuating planet.

By thoroughly examining humanity's relationship with the biological Earth in this exhaustive manner, the Jains have set out to psychoanalyze the roots of violence and to engineer a method of minimizing impact through the prudent determination of least harm, least pain, and least suffering. Jains acknowledge no Western or Eastern god, or gods; only nature. By defusing this traditional image of a deity atop a pyramid of creation—insisting, instead, that god is in every living being—they have already gone a long way toward establishing a common language of understanding, mollifying religious differences, and providing a revelatory access to nature that is open to all, rabbit and prime minister alike. While they pay homage to their many sages, beginning with Adinath, and much later Parsvanath, and then Mahavira, they do not "worship" these individuals, who were, after all, human beings. Worship, in Jain thinking, is a form of interference, and whereas the Jains' fundamental credo is non-interference, absolute tolerance, and total nonviolence. That means Jains are tolerant of other people's attitudes about God, but will neither condone, nor be party to, violence.

In negotiating their way through the maze of destructive behavior that is endemic to most human life, the Jains have managed to a remarkable degree to impede the accumulation of ill-thoughts, ill-deeds, even inadvertent imposition. This, they say, is the only way to achieve peace. At every juncture of human behavior, the Jains have sought to enshrine gentleness, finding a viable path toward love that can be embraced by an entire community. This ecology of the soul (jiva) consists of care taken in all actions and thoughts: to forgive, to be universally friendly, compassionate, and affirmative; to exercise critical self-examination at all times, as well as restraint in all matters; to fast frequently, and meditate on nonviolence for forty-eight minutes at the beginning of every day (samayika); to follow a holy path that ultimately leads to total renunciation.

By definition, these behavioral codes are strictly ecological; and they extend to family planning. While abortion is diametrically opposed to everything they believe in, lay Jains are not at all averse to making practical choices based upon an empathetic concern for the pregnant woman. Jainism is one of the only religions in history that can claim to have granted totally equal status to men and women. The implications of this extend to the classroom, the bedroom, the workplace, and to the legislative domain. One result has been relatively small Jain families.

And in one other, crucial manner, Jainism goes far beyond mere harmonious introspection: said Mahavira, " . . . a wise man should not act sinfully towards earth, nor cause others to act so, nor allow others to act so. . . . "[10] It is this ecological and moral injunction that has endowed Jainism with its activist disposition and urgently contemporary character.

While vegetarianism has always been obvious to Jains, the increasingly modern revulsion to "hamburger colonialism" is one important key to environmental stability on the planet. Meat-eating is obviously plausible under conditions observable, say, at the Serengeti, where the number of carnivores in relation to prey, is still in balance. But in a world of several billion humans, meat eating is disastrous. Human meat eating kills *tens-of-billions* of animals per year under unimaginably cruel circumstances—a system that Jeremy Rifkin has rightly termed "cold evil," evil imposed from a distance and veiled by a vast technological system that has utterly dulled the senses of consumer-conspirators.[11] Killing, under any humanly ordained circumstances, is unacceptable. But our meat eating goes beyond immediate stupidity and the infliction of pain: it also destroys untold millions of hectares of rainforest and pasture land needlessly, while deeply widening the gap between the rich and poor, between food-sufficient and food-deficient communities. The energy-intensive requirements to grow, spray, harvest, process, and transport a pound of meat versus a pound of soybeans, for example, would consume over seven million barrels of oil a day (just in the U.S.). This amount would be saved were ten percent of all existing American meat eaters to join the existing 30 million U.S. vegetarians and 500,000 vegans. That much oil is roughly the equivalent of total U.S. dependency on foreign sources of petroleum. Thus, vegetarianism would accomplish a number of important reforms: it would lift the burden of cruelty that indicts the human species (and the bad karma inhibiting our liberation, which the Jains have analyzed); it would reduce air pollution; and it would curb the global greenhouse gases produced by methane-producing cattle populations raised merely for slaughter. At the same time, vegetarianism would diminish the impending call for further exploitation of such domestic wilderness areas as the Arctic National Wildlife Refuge in northern Alaska, while lessening the burden associated with public taxes and private premiums expended on health problems derivative of meat eating.

Practically speaking, only a global vegetarian cartel could muster the political clout to sway the human habits of animal protein consumption, which reached the staggering level of 346 billion pounds worldwide in 1992, or 63 pounds of ingested animal corpse for every man, woman, and child. Those figures pertain only to primary domestic factory farming—pork, beef, poultry, and mutton. They do not accurately account for the consumed "exotic" species, such as leopard or whale, bear, deer, moose, or wildebeest. In addition, one must add the nearly 200 billion pounds of fish eaten every year, according to the U.N. Food and Agriculture Organization.[12] A vegetarian political party might well influence a presidential ticket. An international union of vegetarians and ethnobotanists could also wield leverage in the largest meat eating nations—China, India, the U.S., Japan, Brazil, Pakistan, Argentina, New Zealand, Nigeria, Indonesia, Australia, Germany, England, Turkey, Mexico, and Canada—through any number of means, including the mechanism of the strike. An NGO of vegans should be able to boycott and pressure the fashion, poultry, and cosmetic industries to eliminate all animal-based products, whether silk, fur, leather, pearl, or feather. Animal rights organizations, such as PETA (People for the Ethical Treatment of Animals), have already demonstrated some limited success in that realm.

There is an important caveat that attaches to the Jain and vegan way of life, namely, that non-food animal products be replaced by natural fibers or biologically safe processes, of which there are many. Because of the violence inherent to oil spills and toxic waste, the perpetuation of the fossil fuel industry through reliance on synthetics such as vinyl, acrylic, polyester, and artificial rubber, is non-viable. Nevertheless, these products continue to comprise a majority of supposedly non-violent alternatives.

Many may doubt the lobbying power of vegetarians, but the animal rights' following is becoming substantial enough to place people in high offices, to enact legislation, and to change the way society thinks and feels. In a 1994 *Los Angeles Times* poll of 1612 Americans, 47 percent agreed with the following statement by Ingrid Newkirk, cofounder and chairwoman of People for the Ethical Treatment of Animals: "Animals are like us in all important things—they feel pain, act with altruism, they talk and suffer fear. They value their lives, even if we don't understand those lives."[13]

On ethical grounds, the mass torment and murder of other species by strangulation, buzz saw, hydroclipper, tanks of boiling water, furnaces, broiling ovens, clubs, baseball bats, electric jolts, machine guns, and carving knives—as well as the actual eating alive of many species, be they snake, turtle, or monkey—do not suggest a bright future for *Homo sapiens*, particularly if one believes in spiritual reincarnation, a concept out-of-favor with many scientists, economists, and politicians, but fundamental to Jainism. At the unveiling of the U.S. Holocaust Memorial Museum in Washington, D.C., President Bill Clinton stated that the Nazi Holocaust set the standard of evil against which all other acts of inhumanity would be judged. Some thirteen million known and loved humans were annihilated by the Nazis in factory-like settings. It was right after World War II that similar "factory farms" for the killing of animals emerged in many countries. What has changed, however, is that instead of millions of victims in one war, there are now billions of victims every year, with most of the human population—not just one nation—willfully and knowingly conspiring in the crimes. These war crimes against animals, in the fallacious name of human need, go on twenty-four hours every day of the year in every town and city, and on most farms throughout the world. So complete is the conspiracy that the murder has been justified by the perpetrators—by presidents and prime ministers, by jurists, philosophers, scientists, preachers, and doctors, by farmers, teachers, and their children—on the grounds that it is dictated by "human nature."

But it is not human nature. Medical research appears almost unanimous with respect to the countless benefits of a nonviolent diet: a lower risk of osteoporosis (loss of bone mass) and coronary artery disease, lower blood pressure and total cholesterol, fewer digestive system disorders like diverticulosis, a better internal maintenance of blood sugar, and a lower risk of lung and breast cancer.[14]

Amazingly, the U.S. Recommended Daily Allowance of 44 and 56 grams of protein, respectively, for women and men, though excessive, is easily met by vegetarians. For example, some whole grain toast in the morning with a ten-ounce glass of soymilk, one serving of split pea soup for lunch, and a snack of three ounces of peanuts at some point during the day would satisfy the RDA. In fact, most plant foods are richer in protein than are animal products. Chicken is on a par with almonds in that regard (18.6%), while garbanzo beans (20.5%) and lentils (24.7%) both exceed beef (20.2%) and lamb (16.8%). At 34.1%, soybeans are among the highest protein sources known. In addition, there are numerous vegetal-derived sources for Vitamin $B^{12}$.[15]

Those who have perceived meat eating versus vegetarianism as an "either-or" argument have failed to recognize that the overwhelming abundance of all food on the planet is comprised of fruits, vegetables, and grains; and that the taste for meat is acquired, usually the result of about thirty seconds of repeated craving. The human organism is sophisticated, evolved. It obviously has the ability to override those thirty-odd seconds of compulsion, an addiction stoked by a few microscopic nerves in the tongue. Evidence for the strength of that capacity can be read into the fact that some vegetarians have even died for their beliefs.

Most consumers have not sampled even a small proportion of known plants, because of the rage for corporate monoculture throughout the Green Revolution countries. While at least an estimated thirty thousand edible plants have so far been identified, less than twenty are utilized by humanity for over 90 percent of its diet. Such important foods as the variety of wax and buffalo gourds, winged beans, amaranths, tree tomatoes, and spirulina, go largely untried. At least three thousand tropical fruits are known to be high in vitamins, easy to grow, and delicious, but the economics of mass production have not favored them. Writes E.O. Wilson, "It is within the power of industry to increase productivity while protecting biological diversity, and to proceed in a way that one leads to the other."[16] At the same time, it is not wishful thinking to believe that every disease known to humankind could be counteracted by the pharmaceutical properties inherent to that same undiscovered superabundance of vegetation. Wild plants have offered cures for fungal infestation and cancer, treatments for Parkinson's disease and glaucoma, malaria, depression, hemorrhaging, and migraine headaches, while producing heart stimulants and anesthetics. And that's only the beginning.

Vegetarianism offers the clearest insight into the Jain emphasis on maintaining biodiversity. Indeed, anyone who is serious about population issues and helping the world cannot—in good faith—consume flesh. Mahavira took great pains to delineate countless types of organisms in all of their minutiae for purposes of appreciating and protecting them. But in making choices regarding what I have termed ecological triage, the Jains are equally concerned with individuals. Hence, any "utilization" theory that would sacrifice some to save others, is strictly out of keeping with Jain practice, though there are times when such choices must be enacted. Moreover, Jain agriculture must be sustainable and strictly organic.[17]

For myself, Jainism has provided a means of understanding the crucial necessity of making those hard choices. Most Jain doctors will not prescribe animal-derived drugs, or support biomedical research. Jain lawyers and judges are firmly opposed to capital punishment. Few Jains have ever entered the military, though lay Jain doctrine is not opposed to self-defense as a final means of preservation. But it is in the economic realm where Jain insistence on the cost/benefits ratio of compassion and ecological integrity have been most persuasive.

Mohandas Gandhi was educated in his youth by a Jain tutor, his closest friend was a Jain, and he was to base his entire philosophy of nonviolence on Jain principles. Consistent with these are the Jain tenets of *aparigraha* or nonacquisition (nonhoarding), *asteya* (not stealing), and *satya* (truth). Such invocations also happened to be the seminal elements that Gandhi discovered in a far-reaching little book on political economy by British art historian and social critic, John Ruskin. Entitled *Unto This Last*, Ruskin's work was—in Gandhi's own words—"the one that brought about an instantaneous and practical transformation in my life."[18] Gandhi read the work in 1904 while traveling across South Africa by train. Captured by its "magic spell," he was to translate it into Gujarati, under the title *Sarvodaya*, meaning "the welfare for all."

At the time, British industrial imperialism, against which Ruskin railed, had literally appropriated 5 million square miles of tropics and, with the other superpowers in Europe, economically assaulted 85 percent of all developing countries. Gandhi recognized, as did Ruskin and

Charles Dickens (whom Ruskin greatly admired) before him, that society, and modern economics, placed no value on pain—pain of the laborer who works his whole life for virtually no share of the global stream of wealth; pain of all nature, which is not counted as a Soul, a living, feeling being, but a resource to be efficiently uprooted and consumed. The system of utilitarian consumption itself had become a monster (Jeremy Bentham and John Stuart Mill its most celebrated philosophers) and Ruskin had elucidated what he perceived as its dishonesty and irresponsibility. Wrote Ruskin, "So far as I know, there is not in history record of anything so disgraceful to the human intellect as the modern idea that the commercial text, 'Buy in the cheapest market and sell in the dearest,' represents, or under any circumstances could represent, an available principle of national economy. Buy in the cheapest market?—yes; but what made your market cheap? Charcoal may be cheap among your roof timbers after a fire, and bricks may be cheap in your streets after an earthquake; but fire and earthquake may not therefore be national benefits."[19] As James Fallows has pointed out, many economists, including William Lazonick, Friedrich List, and Alice Amsden, have shown that nations historically have gotten rich by "rigging their markets" and "cheating."[20] Ruskin understood that money was not power in itself, but power over people, the essence of cheating. To rid India of British occupation, and reinvest the rural Indian with dignity, Gandhi undertook to inspire village self-reliance. In Ruskin's words, a man must "bake his own bread, make his own clothes, plough his own ground, and shepherd his own flocks."[21] That man is in need of no gold. Gandhi was convinced that an economics of creative employment, ecological sustainability, and social justice would naturally result from Indian independence; that when individuals are empowered within their communities, they will coexist harmoniously with nature.

Ruskin had recognized the fact that local conditions of overpopulation had exerted in England, particularly in the slums of London, and in Ireland, an "unmanageable" "pressure of competition," which—with the literary force of ambiguity—he said only proper "forethought and sufficient machinery" might correct. By the early 1920s, India already numbered more than 260 million people, larger than the present U.S. population but with approximately one-third the territory. Gandhi was not inclined toward population control. He loved children. He was neither daunted by the existing number of Indians, nor the internal competition for resources. What he did not fully anticipate was what the Gandhian, Dr. B.D. Sharma, one of India's foremost thinkers on the subject of sustainable development, has recently described as the "centralization of economic and political power," a pattern of control over its own people that India has adopted after the British pattern. He argues that "honest physical labour is an essential condition of human life and happiness" but that technology has "deskilled" the rural laborer. Sharma has called for the dismantling of the system of development, a de-escalation of the costs associated with trade, advertisements, and transport—money wasted—and a re-provisioning of villagers with the right of self-governance.[22]

Unfortunately, in India, as elsewhere, the possibilities for a revolution in rural economic thinking have been diminished as a result of the continuing population bomb, which has taken extraordinarily high tolls on all rural Indian life. (See chapter 3.) For those hundreds of millions of immigrants, environmental exiles, and migrants to the cities, the rural mentality is no longer so welcome.

Further complicating the picture of a pastoral economics is the fact that the centralization of urban power in nearly every nation has already gone a long way toward rendering that global reservoir of village prowess and tribal wisdom "officially" obsolete. Hence, the essential argument: given the trends toward urbanization, and a doubling, or tripling, or quadrupling of human numbers, what hope is there to revivify the message of Ruskin and Gandhi and Sharma? And, is rural life—with its fragmenting effect on biodiversity—even desirable in a world of so many people?

Dr. A.T. Ariyaratne, a Buddhist social activist and former high school teacher from Sri Lanka, certainly believes it is. Something of a living Gandhi, he has started the Sarvodaya movement throughout South Asia. In 1994, nearly thirty thousand villages have now embraced Ariyaratne's Buddhist economics of self-empowerment and compassion, to varying degrees of success. Those millions of neo-Gandhians are testimony to the willfulness of the ideal. Like the Jains, they have argued that an ecological, nonviolent lifestyle, even in a crowded world, is possible and dignified.

## THE ECONOMICS OF A NEW DEMOGRAPHY

"The heart of the problem," wrote George Wald in the early seventies, "as of all human problems, is one of meaning. Our problem isn't one of numbers, but of the quality of human lives. What we need to do is to produce that size of population in which human beings can most fulfill their potentialities."[23]

The statement, while poetically true, fails to quite encompass the challenge the human species is up against, on two counts. First, no long-term experiment in social engineering, particularly with regard to sexual behavior, has ever succeeded, either in China, or Singapore, or in any other country. Where the TFR has stabilized, it came about by free choice, and in places like France, Germany, and Italy, almost inadvertently. That implies that while family planners and ecologists and many informed citizens know what has to be done, they are unable to dictate policies, at least in any democracy. (The counterargument is that laws are devised not merely to protect, but also to curb our behavior. Smoking on planes, drinking in automobiles, shooting guns on city streets, and selling heroin at schools, have been legislated against, and there is no theoretical reason to believe that one-child families might not also eventually be legally enforced.)

Second, our "potentialities" have collectively and inevitably referred to our progress, our economic development, not to the painter or musician in our souls. At the current rate of economic growth, by 2050 the world economy will have increased another four to five times. Fossil fuel use is escalating 20 percent per decade. Rapid extraction of non-renewable resources and the demolition of other life forms—what I have previously described as the Total Population Dynamic—are resulting from the furious attempt to "grow" per capita income consistent with the virtually global aspiration of achieving a Western lifestyle, or—in the most tragic sense of our predicament—simply trying to survive.[24] Nature does not guarantee a "quality of life" package.

Unquestionably, the greatest challenge to a vision of demographic nonviolence is this very economic growth that continues to spurn the restraints that are so fundamental, not only to the Jains and Buddhists and nearly all spiritual traditions but to biological homeostasis on Earth. It has been estimated that if, as predicted, consumption continues globally at a growth rate of 2 to 3 percent over the next fifty years, the so-called "environmental intensity of consumption (environmental impact per unit of consumption)" must fall by a whopping (and unlikely) 93 percent. Otherwise our growth will never be ecologically reconciled.[25] Considering that in 1993 China's growth rate was 13.2 percent,[26] and that NAFTA, the other fourteen APEC countries (along with China), and the post-World Trade Organization (GATT) regimes, if anything, are going to increase the growth of the global economy, it appears most unlikely that consumption, and environmental impact are going to diminish. Based upon the world's record of natural resource depreciation, it is simpleminded to assume that any boost in productivity, however achieved, will necessarily be good for an economy in the long-term.

The flaws of NAFTA make that particularly relevant.[27] Export-driven growth in developing countries, spurred on by trade liberalization, will not necessarily serve the environment, or the masses, unless production is carefully screened for nonviolence and the preservation of biodiversity. Currently, the $1.35 trillion debt of the developing world is forcing many countries to basically sell off their forests, grasslands, mangrove swamps, coral, exotic species, even their soil. In the euphoria of free trade competition, the rage for highly mechanized monoculture and corporate efficiency is likely to additionally sweep over, or usurp, countless rural economies (the sweat and toil of over 3 billion people), further obfuscating Gandhi's vision. The pre-existing proof lies in this fact, alone: after twelve thousand years of agriculture, just during this century an estimated 75 percent of that genetic agricultural diversity has been eradicated. As biotechnology further boosts yields, there is grave concern that the "old time varieties" will totally vanish.[28] At the same time, the new transgenic industry has already topped a billion dollars. An estimated five thousand mutant creatures have been "created," (such as the designer HLA-B27 rats at the University of Texas with electronic bar codes implanted under their skins to identify the fact they carry a genetically manufactured disease, spondyloarthropathies). Harvard University was the first institution to patent an animal in 1988, OncoMouse. Now, a British firm has created ImmortoMouse.[29] Other firms are looking to quietly boost the size of vegetables by incorporating genes from animals. In absence of genetic labeling, and laws to that effect, vegetarians will be dealt a perverse blow.

The economics of agriculture is only part of a troubling picture, however. According to the most conservative scenario, it is estimated that world income will now increase, as a result of the new economic global accord, by $213 billion annually.[30] But U.S. trade representative Mickey Kantor stated enthusiastically that world economies will grow by three times that amount, reaching a $20 trillion world economy within a decade. And the World Resources Institute has referred to a $50 trillion global economy in the next century, pointing out, at the same time, that if the engineering and proliferation of sustainable technology does not precede that financial crescendo, the environmental damage could be "devastating."[31]

Human fertility transition is simply happening far too slowly to enter into this equation. The economic avalanche has a life of its own, given a current population base of over 5.5 billion. In her extraordinary book *Population Politics*, Dr. Virginia Abernethy, an anthropologist and psychiatrist and editor of the journal, *Population and Environment*, effectively shows how prosperity, contrary to the theory of demographic transition, has actually increased the total fertility rate in country after country, whether China, Turkey, Cuba, or Algeria. The wealthiest nation of all, the U.S., adds nearly 78,846 people to its census every week. In 1992, 4.1 million babies were born there. The Urban Institute has projected a U.S. population of 440 million in the year 2090, despite a U.S. Census Bureau scenario of a stabilized United States in 2020 of under 300 million.[32] In research for the Center For Immigration Studies, demographer Leon Bouvier has even cited the possibility of

500 million Americans some time in the next century. Already (as of the 1970s), according to Lester Brown, "the net benefits associated with economic growth in the United States fell below the growth of population, leading to a decline in individual welfare."[33] Both in terms of the environment and of human welfare, 440 million U.S. consumers is thus a difficult concept to assimilate. And even as post-GATT celebrants are looking toward greater and greater riches, the number of those falling below the absolute poverty line is mushrooming. This global economic schizophrenia offers no insurance whatsoever with regard to ecological safeguards, or even the likelihood of increased governmental largesse.

Today, countries are divided into two types: those exploiting their domestic environs in order to meet the survival needs of their expanding populations, such as India, and those impacting their environment strictly to meet their increasing expectations, such as the United States. In the former case, policy decisions are undermined by huge quantities of future people whose numbers can only be inferred, or falteringly predicted (as in the case of Indonesia, as it proceeds with the commercialization of its forests). Because official demographic "guesstimates" will always err on the side of economic aspiration (the lower the number of people, the higher the imaginary per capita income), such countries are predicating their goals and means upon unrealistic assumptions and rosy pictures that continue to foster maximal extraction policies. The same delusions hold true for countries like the United States, Italy, Great Britain, Japan, and Germany: however modest or fervent the population growth rates, higher and higher income demands will ultimately undermine the ecological effectiveness of any population policies.

With the possible exception of the small Himalayan kingdom of Bhutan, a unique country worthy of international admiration and study, neither the short-term truths or long-term sense of what's right, have yet forged realistic policies of national sustainability.

"The world is not now headed toward a sustainable future, but rather toward a variety of potential human and environmental disasters," writes the staff of the *World Resources 1992/93* report.[34] The word "potential" is frequently inserted diplomatically, as if to stave off total despair, or too literal a realization of the disaster already befalling the planet.

But, in fact, there are (in addition to Bhutan) growing murmurs of ethical and economic transformation, and these I should like to emphasize. They include the aforementioned Sarvodaya movements in India and Sri Lanka, the emerging *kernelfarms* network of organic farms and ecomunicipalities throughout Scandinavia, and the Seikatsu Club Consumers' Co-operative (SCCC) in Japan, with its nearly four hundred thousand ecologically-oriented members. According to economist Paul Ekins, investments in the U.S. totaling some $625 billion are now routinely screened according to at least some ethical consideration. In addition, the International Organization of Consumer Unions (IOCU), founded in 1960, now has 170 groups in sixty-five countries, and it has become, in Ekins' words, "an influential force for change, tackling such problems as ozone layer depletion and toxic waste dumping, the marketing of dangerous drugs and pesticides in the Third World, tobacco, baby foods, irradiation and biotechnology." The IOCU's five so-called "consumer responsibilities" include: critical awareness, action, social concern, environmental awareness, and solidarity.[35] The world's network of NGOs (nongovernmental organizations) has also exerted a profound influence on global possibilities. By bypassing much bureaucracy, NGOs have in essence forged a "United Peoples" network (as Ariyaratne calls it), and have reached an estimated 100 million individuals otherwise forgotten or ignored by their various governments.[36] Collectively, NGOs now provide more than 12 percent of all development aid in the world—over $7 billion.[37] At the same time, the sense of green business ethics has already begun to prove itself as Seikatsu's nearly half-a-billion dollar annual revenues from ecological product sales demonstrates.

Many would argue that it is crucial to focus on that potential, those powerful rays of hope penetrating the demographic inferno. That in spite of our obsessive self-interest, we are capable of learning from past mistakes and doing the right thing. I personally do not doubt that we have the power and the foresight to effect positive change. There are currently well over ten thousand population projects in developing countries.[38] These are not arbitrary bankrolls, or bureaucratic business-as-usual, but determined, often inspired, efforts to help humanity conform to evident laws of nature. It is estimated that all of the population programs together have, historically, thus far prevented 400 million people from being born (more than half in China). In the twenty-first century, if such family planning assistance continues and (hopefully) increases, 4 billion fewer people are expected.[39] Contraceptive prevalence has risen dramatically throughout the world—71 percent of married women in China, more than that on average in most rapidly industrializing countries, and 44 percent, on average across much of the developing world. Child and infant mortality have dropped by a third in the developing countries, particularly among those couples availing themselves of contraception. The use of contraception has also had an important impact on higher levels of adult female literacy.[40] If such data reflects even a semblance of what is actually possible, then it should be key to conceptualizing and enforcing all future political and economic priorities. Indeed, the U.N. slogan—"every child a wanted child"—needs to be politically and economically facilitated in every country by the implementation of better contraceptive and health services.[41] That is not political rhetoric but an elegant summons to a true revolution in biological family values.

At the same time, custodial laws pertaining to the more than 10 million orphans in the world (four hundred thousand in the U.S.) need reexamination. In Hong Kong for example, there are some agencies that will (oddly) not even discuss adoption with a parent who is "more than 20 percent overweight."[42] Only thirty-six thousand of the U.S. foster children qualify for adoption under current regulations. Liberalizing those strictures could only be a "win win" situation with respect to human rights and population politics.

According to Shanti Conly and Joseph Speidel of Population Action International in Washington, D.C., 125 million motivated married couples are currently lacking access to contraceptive supplies and information. Even in the U.S., nearly half of all pregnancies are unintended (the cost to taxpayers of teenage parents in the many billions of dollars annually!), arguing for the Scandinavian approach to the sale of oral contraceptives without a prescription. Other drugs have recently been granted over-the-counter approval (such as ibuprofen and hydrocor tisone creams) yet so far the Food and Drug Administration has opposed placing the pill in this easily accessible category.[43] Given the FDA's current interest in RU-486, sale and distribution of the pill will, hopefully, soon be liberalized.

But if the world's population were to be stabilized below ten billion, between 70 and 80 percent of all couples would need to effectively use contraception, not just have access to it. And the clock is ticking: that stabilization figure will follow only if those couples begin effective use by the year 2000! Every delay, by every year, will mean tens of millions of additional newborns. It is a tall order. It means a doubling of users, and better than a tripling of financial commitments to family planning by governments and private donors—both North and South—during the rest of the 1990s. That translates into a yearly contribution of about $15 billion, or the equivalent of $16 per needy couple. At present, global financial commitments to population control are roughly $4.5 billion, the wealthy (Organization for Economic Cooperation and Development) countries contributing less than $1 billion a year of that, or less than $1 per childbearing couple with unmet contraceptive wants. The U.S. is the largest donor (in overall monetary terms—$352 million in 1991—though not in proportion to its GNP), each taxpayer providing $1.38 per year to international family planning efforts.[44] But it is a paltry sum.

UNICEF, which has long argued that family planning is "the single most important technology" known to the human race, stated in its 1992 *State of the World's Children* report, "the responsible planning of births is one of the most effective and least expensive ways of improving the quality of life on earth, both now and in the future, and one of the greatest mistakes of our times is the failure to realize this potential."[45] Part of the challenge in forging this "new demography" is to convince policymakers that family planning should receive a much greater share of overall development assistance. This is basic. But as of 1992, according to UNFPA, the average allocation for population assistance was a mere 1.34 percent of total ODA (Official Development Assistance), a desperately shortsighted expenditure, given the profound gravity of the population dilemma. In 1991 Switzerland provided less than $7 million, Australia $5.3 million, New Zealand $420,000, Japan $63.1 million, the Dutch $39.6 million, and the English, $46.7 million. These are rich countries that should be more generous and farsighted, but have followed the example of each other. Only (whale-killing) Norway, with its 4.3 million people, has taken slightly more dramatic steps by providing (in 1991) $12.46 per Norwegian, or $53.6 million for international population programs. That worked out to 4.55 percent of Norway's total annual ODA for that year, higher than the 4 percent suggested at the International Forum on Population in the Twenty-first Century, sponsored by the U.N. in Amsterdam in 1989. But nearly all other nations have failed to measure up to that so-called "Amsterdam Declaration". In the U.S., population assistance was 3.13 percent of 1991 ODA.[46]

Conly and Speidel's strategic recommendations to international leaders for remedying this parsimony might be summarized as follows: give far more money, and urge small donors, the World Bank and other U.N. agencies, in addition to UNFPA (such as UNICEF and the World Health Organization) to greatly enhance direct bilateral population aid, as well as managing and supporting multidonor (i.e., NGO) family planning assistance.[47] Multinational corporations interested in long-term image and product lines have been especially disinterested in providing grants for population assistance. But they would be wise to do so, particularly in those countries where low fertility has increased per capita income and boosted consumer buying power, as in Singapore, Shanghai, Chile, and South Korea. While the payback may not be as rapid or as exotic as, say, providing funds to save pandas or Siberian white cranes, a company that invests in family welfare (family values) in its market environments abroad is actually doing much to enhance its promotable image.

The argument for drastically enhanced population funding should be clear enough. From a strictly national budgetary perspective, this is especially so. For example, consider that the developed nations collectively spend the equivalent sum of their total population assistance in all of twelve hours on military endeavors. In the U.S. in 1993, $89 billion was spent on brand-new weapons systems, both for domestic use and sales abroad. In fact, a total of $116.3 billion is still allocated to Cold War expenses every year, even at a time when actual NATO spending has declined from over $450 billion spent in 1989 to $350 billion in 1993. Moreover, there are now twenty-four new potential peace partners in Central and Eastern Europe to spread out the per-country cost burden of defense. But the dollar figures attendant upon other American "priorities" are far more staggering than military costs, such as an estimated $674 billion spent each year to fight crime.[48] Logic dictates that any society that can afford, and feels the moral compunction, to spend such outlays, would, and should also cut costs whenever possible, and that means deterring unwanted expenses and unwanted children. In the broadest sense, there is no greater economic, moral, and ecological recovery to be effected than through family planning. But there are many who have harbored strong reasons for resisting it.

First, the supposed financial deconstraints of the post–Cold War era have been tainted by what former Defense Secretary Les Aspin described (on December 7th, 1993) as a "bigger (nuclear) proliferation danger than we've ever faced before." In part due to the nuclear buildup in North Korea, U.S. military planners have now been instructed to step up their overall preparedness for combating nuclear, biological, and chemical warfare, an initiative that does not bode well for the liberation of new dollars for population assistance. In addition, the Western powers will not soon forget that the bill for Desert Storm was approximately $49 billion.

Second, the Vatican, and those influenced by the Holy See—most noticeably the Republican administration in the U.S. for twelve years—dismissed forceful population control on the myopic grounds that the whole issue was "benign," and that family planning was equivalent to abortion, and therefore immoral. During those twelve years, federal funding for family planning plunged by nearly 70 percent. Today, the U.S. has the highest rate of unintended and teen pregnancies of any industrialized nation. But, despite a more savvy Democratic power base in Washington, the fact of a known 1.6 million abortions in the U.S every year, and surveys indicating that between 70 and 90 percent of all U.S. Catholics believe in artificial birth control (fueled by views of such dissenting Catholics as Father Charles E. Curran who wrote, "*Humanae Vitae* [the pope's encyclical against artificial contraception] appears to be a dead letter today"), those earlier pronatalist sentiments have not altogether been reversed. Congress, the Supreme Court, most Americans, and Westerners in general, are still hopelessly unclear about population priorities.[49] And while it has been argued that the total number of immigrants that can be absorbed at replacement level in the U.S. is 160,000 per year, the *1990 Immigration Reform Act*, which Al Gore, as a senator, voted for (and has reaffirmed since becoming vice-president) increases annual legal immigration by 40 percent to more than one million. NAFTA is likely to propel an additional 14 million Hispanics across the border in coming years. At present, seven hundred thousand "illegals" enter the country every year, and another 3.3 million illegals try to get in, but are thwarted.[50]

And there are other distinct biases hampering the flow of sorely needed population assistance. The provisioning of famine relief comes naturally to any civilized people who can afford it. Population relief should be synonymous with that impulse. But there is a problem: it is estimated that of the 117 developing countries, nearly *half* of them will be dependent on foreign foodstuffs by the turn of the century. Thus, many donor nations are pouring larger shares of their development loans and aid into agricultural and food security projects, hoping to preempt famine situations, and convinced that greater prosperity will translate, ultimately, into fewer mouths to feed. There is, indeed, merciful reasoning to this. But there is a second line of thought, as well: developing nations, certain that there is a Santa Claus, monetarily buoyed by a penitent food allowance, or subsidy, from the West, may be less inclined to strengthen their domestic population policies.

If there is a brutal lesson here, it is endemic not merely to the giving of charity, but to the whole debate over what is most likely to achieve a lowering of the global fertility rate. Thomas Malthus argued against charity because, he said, it fostered population growth and hence, human misery. So-called "lifeboat" ethicists have long argued that by feeding starving children today, one is actually perpetuating a syndrome of starving children later on. The logical inference is that chan̄y, compassion, aid, even medical assistance, increase population, which is certainly true if infant mortality rates are reduced. But according to this same logic, by increasing the standard of living in an effort to engender smaller families, one may well be setting the preconditions for greater exploitation of the environment.

If that is the case, why do anything? Why have smaller families if the end result will be the same, in terms of ecological impact? Does a smaller, wealthier family impact nature with the same severity as a larger, poorer one? It is a classic conceptual entrapment, an ethical no-win situation that can only be intelligently addressesd on a case-by-case basis. Per capita $CO_2$ production in the U.S. is much higher than in Jordan, though the Jordanian TFR is nearly double that of the U.S. Poverty in Senegal has translated into an enormous amount of poaching; the Malaysian middle class is living off the spoils of tropical clear-cutting; impoverished Indians recycle nearly everything—except the forests themselves, which poverty is liquidating. In other words, there are no guarantees that can be predicated merely upon family size or economic status. It is the combination of the two, under the particular, and widely varied, circumstances of human geography, that determines the extent of measurable ecological damage.

But the evidence also suggests that economic performance is nearly always linked to environmental destruction. The passage of GATT and the refining of global free trade (cutting import duties overall by 33%) might also be viewed as a form of charity, a stimulus for increased competition and accelerated exploitation of natural resources, all according to the tired rhetoric of profit. Assessing the pitfalls of sudden prosperity, Virginia Abernethy writes, "Ecological release is the term for fortuitous conditions which lift constraints that would otherwise inspire reproduction caution . . . land redistribution in Turkey, foreign aid and oil revenues in Egypt, and economic liberalization in China—are a small subset of possibilities that can create a perceived surplus, that is, more than people have been accustomed to getting and enough to encourage a preference for large family size."[51] The argument leads to a strange circularity: if, as many have suggested, the underdeveloped countries had managed to dramatically curtail their population growth in the sixties, when many of them launched family planning efforts, it is believed their economic activity today would have been 200 to 300 percent more vital. But then, such prosperity might well have accelerated their fertility rates all over again. Hence, another version of the demographic no-win situation.

Further compounding the competition between food aid and higher levels of population assistance are the recognized costs of environmental cleanup in the Third World, a whole other set of demands upon Western coffers. Those costs have been estimated at a minimum of $250 billion every year.

There is no question that modernity can and does have a tremendously destabilizing effect on existing cultural and fertility norms. What I term economically transmitted diseases (ETDs) have already permanently infected most nations, and there is no predicting how reproductive rates or preferred family size will be affected. The evidence, from country to country, is utterly inconclusive. Families in Uruguay respond differently than those in South Korea, while harsh economic times have induced reproductive restraint in regions and countries as varied as Myanmar, Mexico, the Sudan, Brazil, Tibet, and recently, Kenya. In India and China the opposite has proved to be the case. In Thailand, where the economy is booming at an 8 percent growth rate per year, Bangkok's slum dwellers have doubled in population in the last 30 years. Thirty-five percent of that city's residents are now below the poverty line. What no demographer can yet predict, city by city, is whether such hybrid environments of poverty and wealth are greenhouses for fertility, or for reproductive austerity. This unpredictability has the unfortunate effect of militating against increased government spending to shape the world's demographics.

There is yet a final twist that underscores the unwillingness of most developed countries to spend more money on population control; the huge deficit lurking beneath the West's highly touted GNP. The real development aid may not be traveling from North to South, but the other way around. Economist Manfred Max-Neef has stated, "For five hundred years people of the South have been subsidizing in net terms the growth and the development of the North—$400 billion net goes from Latin America to the North, more than anything that goes from the North to Latin America."[52] In fact, the total combined aid to Third World countries from the North is currently about $50 billion a year. It is thus the North that is indebted, a fact in keeping with the staggering $6 trillion dollar total debt of the U.S.

There are countless solutions, some more imaginative than others, that have been proposed for paying this money back.[53] But the most realistic approaches concern future concepts and plans for stabilizing society's expenditures according to determinations of the true GNP, based upon collaborative perceptions of natural resource use and depreciation. "The true measure of depreciation," writes Robert Repetto, "is the capitalized present value of the reduction in future income from an asset because of its decay or obsolescence."[54]

The first step toward appreciating the value of natural sinks (i.e., the Amazon) and resources (i.e., the oceans, or the air we breathe), which we take for granted, lies in recognizing the costs associated with anything like ecological restitution. As early as 1970, Nobel laureate Joshua Lederbert estimated that medical costs arising from genetic defects in a population of 300 million exposed to 60 percent of the U.S. "acceptable" radiation standards, would be about $10 billion per year.[55] And while the congressional *Price–Anderson Act* limited the public liability of a nuclear disaster to $560 million, Brookhaven National Laboratories estimated that the cost should be at least $7 billion. Moreover, it is believed that the costs of cleaning up after four decades of nuclear weapons' manufacturing in the U.S. will exceed $300 billion in coming years.[56] Exxon has spent over $2 billion thus far on its *Valdez* disaster clean-up. Forest and agricultural losses from air pollution have been shown to cost $46 per person per year in Canada, $15 in Holland, and $17 in Germany. And it has been demonstrated that a 1 percent increase in sulphur pollution levels within a given residential area will result in property devaluation of homes of between 0.06 and 0.12 percent, on a graduated scale. In Basel and Toronto, noise pollution has been shown to have even more dramatic impact on property values. Smog costs the City of Los Angeles about $4 billion a year, according to the South Coast Air Quality Management District.[57] Between 1970 and 1987, the U.S. spent some $400 billion to curtail air pollution. Health costs associated with smog are now estimated to be $30 billion a year, according to Lester Lave of Carnegie–Mellon University. If, as James Flanigan of the *Los Angeles Times* has pointed out, a thirty-cent tax were imposed on each of the 70 to 100 billion gallons of gasoline sold in recent years in the U.S., those health mal-effects could be covered—but not the debt to nature. Which is why such compensatory figures are crude. They only measure the anthropogenic, human-perceived values. We are incapable of knowing, in the spiritual, poetic, or truly ecological sense, what nature's true costs are, or to what extent human behavior may have altered original biotic functions of the planet. With regard to the value of individual life, in the United States annual average per capita health care costs are $2800, slightly higher for children.

Because assigning such costs is arbitrary, dependent upon the human definition of market value, it is inconceivable that any economic system by itself will ever come close to *knowing* nature, though over twenty governments are beginning to revise their national accounting systems so as to internalize environmental accountability. However, only people whose livelihood depends upon natural resources can really begin to appreciate true value.

Even that appreciation is vulnerable to global market forces. If, as some have envisioned, genetic diversity in the wild becomes the "oil" of the next century, two opposing scenarios emerge. First, there is the dire possibility that human ignorance of species and ecosystems (for example, urban entrepreneurs co-opting rural sensitivities) will result in strategies of exploitation based upon crude assumptions of value. This may drive other species and taxa to extinction, particularly when sovereignty over "intellectual property rights" means they are competitively seized, traded, and exhausted in the international free market environment like so many commodities, margins, profits and losses. Conversely, the fashioning of new economic indicators for previously unexamined species, and the sophisticated pursuit of markets for those species, might work to ensure the protection of rainforests and other biota according to new thinking about GNP and what truly constitutes the diverse wealth of a nation. Either way, there will be exploitation. Aspirations are inevitable, as is economic growth, whether they be in the North or the South. Economic growth is arguably not a cure, but a dangerous presumption and any theory of zero economic growth, or ZEG, is simply nostalgic, a form of poetic reverie, pale moons, and romantic arcadias. Aggressive prosperity cannot be held back, anywhere. And while there is no question that wealth has destroyed much of the world—its energies and confusion forming the very basis of World War III—it is hopeless to wage a war in defense of the Earth by combating such wealth. A more politically and ethically realistic solution will be to make that wealth work *for* nature; to ecologically temper and redirect existing, and future profits in comprehensively well-planned, sustainable, and equitable fashions. Ecological sustainability may become the primary organizational impetus of the twenty-first century, but only if compelling economic systems support it. Writes biologist Michael E. Soule,

> Rational people have abandoned the goal of stopping human expansion within the next few decades; now we foresee a century or more of growth in numbers of people in the tropics, perhaps 50 years of growth in the United States. If there is hope for nature, it must rest on the assumption that this binge of reproduction will be a transient blip on the graph rather than a surge to a plateau of permanent planetary obesity. A premise of this "blip theory" is that the twentieth century population explosion will be followed by a slow implosion in the last twenty-first or twenty-second centuries. In the meantime, what should we be doing. . . . The "meantime" . . . is full of possibilities.[58]

Economist Paul Streeten, director of the World Development Institute at Boston University, has cautiously offered a number of methods for meeting those possibilities, for accomplishing the transition from greed to prudence, from monetary ignorance to social wisdom. He suggests the formation of a whole new generation of powerful international organizations that would monitor trade, relieve debt, maintain a multilateral investment trust, and creatively put surpluses to use to assist developing nations. Streeten suggests a Global Environment Protection Agency, an international income tax for agreed-upon criteria, and a body composed of aid-recipient countries to monitor one another's compliance with the very stipulations of aid. Violations of human or ecological rights, international laws, and treaties would come increasingly under the aegis of the International Court at the Hague, the United Nations, and the above-cited Global EPA.[59]

In apparent recognition that such a post–Cold War proactive global beneficence is necessary, in late June 1993 Japan announced a nearly 50 percent increase in its bilateral aid over the coming five years. The country will convert $120 of its $126 billion surplus into development assistance and outright grants, 90 percent of which will *not* be tied to projects allowing only Japanese bidders to compete. More than mere ecological trickle down, that money will be systematically targeted at population problems, the environment, health and human welfare, and infrastructure, the proportions yet to be finalized. Such developmental efforts place Japan firmly in the forefront of the G-7 nations (Britain, Canada, France, Germany, Japan, Italy, and the U.S.) and offer real hope that a new generation of ecological economics is going to be embraced; that we will find the means to harness whatever positive links can be discovered between the environment and human development. The greatest challenge will now be left to other Western countries to follow Japan's example; and to policymakers dictating terms of free trade to force the World Trade Organization (formerly, the GATT) to ensure the adherence to the "protection of life and health" clause, under Article XX(b) and (g) of the treaty, with respect to its stated "extraterritorial environmental objectives." For the United States, just one of the countless legal options for ensuring these aims is provided by the *Pelly Amendment*, part of the *U.S. Fisherman's Protective Act* authorizing trade sanctions against countries that have exhibited ecologically villianous behavior.

## TECHNIQUES OF PEACE

With the serious diversion of hundreds of billions of dollars to conservation and population programs, existing technical and engineering know-how will be increasingly put to work.

Japan's Ministry of International Trade and Industry is predicting a complete phasing out of all CFCs by 2010, with a corresponding reduction of fossil fuel use and the rise of new renewable energy sources; the reversal of desertification through biotechnology by 2020; and net gains from global reforestation by 2030. By 2040, the Japanese predict the emergence of nuclear fusion, orbiting solar power plants, and superconductivity.[60]

While the serious downsides of biotechnology have been presented above, there are also plentiful positive aspects to this powerful, and inevitable, tool, such as new biomass fuels, the microbial recovery of metals, and uncanny biological sensors that might work like thermostats to precisely monitor and control chemical effluents.

Three hundred American cities now rely, to varying degrees, on recycling. By 1997 toilet manufacturers throughout the U.S. will be phasing in low-flush apparatuses to reduce by more than a third the amount of wasted water. Since the *Montreal Protocol* of 1987 and the 1990 London follow-up, CFC production is down 46 percent. Innovative new solvents that come out of alternative CFC research will also reform the ecologically impure world of computer technology. As more and more people work at home, on computers, commuter traffic will diminish, and with it, smog, mental depression, and probably the number of divorces and improperly cared for children. Already, bicycle production has outpaced automobile production, worldwide, by 250 percent.[61] At the same time, in precedent setting legislation, 2 percent of all vehicles sold in California must be emission-free by 1998, and 10 percent by 2003. Ford, Chrysler, and GM are all developing lower emission vehicles, whether electric, or hydrogen-solar, or electric-diesel hybrids. Under the former Bush Administration, a four-year, $260 million joint research venture was undertaken with the three largest automakers to formulate new batteries suitable to high-performance electric vehicles. Other recent pollution control innovations have in-cluded the nation's first auction of tradable air pollution quotas (though, so far, this has proved ineffective).[62] At a recent Los Angeles auto show, nearly every manufacturer was present to show off their new low-emission, electric, or nonemission prototypes. A revolution in pollution-free fuels is underway. By the year 2000, the EC is projecting a fleet of 7 million European emission-free electric cars on the road.[63] All of these trends and technologies were unthinkable even ten years ago.

At the same time, several corporations have begun reducing their wastes and emissions, among them Monsanto Chemical, Dow Chemical, Du Pont, and John Deere. All have shown at least a modest savings through pollution control.

The California Energy Commission has proved that wind power properly installed is more cost-effective than either coal or nuclear, assuming parity in the utility financing domains.[64]

In the U.S., energy savings of some $1 trillion have already been accomplished through efficiency, higher fuel and utility prices, and con-sumption management. Other OECD nations are 20 to 33 percent more efficient still than the U.S. in energy intensity.

The EPA has suggested that the widespread use of high-energy fluorescent lighting in the commercial sector would, by itself, save an additional $19 billion and reduce carbon emissions by 4 percent. And that is just one out of hundreds of simple methods of reducing our na-tional destructiveness and saving money that might be applied to other humanitarian and ecological sectors.[65] Recognizing these many trends and possibilities, the Greenpeace organization has advocated a "Super-fund for Workers" in the U.S., to help effect the inevitable transition to a clean energy economy.

Throughout the world new energy and community development cooperatives have been established: a Savings Development Movement in Zimbabwe; slum dweller self-help groups in Bombay; the Kenya Energy and Environment Organization (KENGO) in Kenya; rural mobilization squads in Ghana and Burkina Faso; and in Bangladesh, two highly successful peoples' organizations—the Rural Advancement Committee (BRAC), which organizes rural labor into self-help projects,

and the Grameen Rural Bank, the focus of which is sustainable develop‐
ment and whose average individual loans to the poor are $70 (total‐
ling $10 million per month). The bank enjoys a 98 percent record of
repayment. The Latin American and Caribbean Working Group on
Rural Energy for Sustainable Development combines work by strategists
in seven Latin American nations and has addressed such issues as pum‐
ping water by use of photovoltaic arrays. In much of Asia, the solar box
cooker—four square feet of sunshine converted to useful energy at vir‐
tually no economic or ecological cost—is witnessing a growing accep‐
tance. And in Haiti, new tree planting schemes have increased the rate
of reforestation by 600 percent.

In more than forty countries where I have conducted research, such
organizations, most in their infancy, are already making a difference,
sometimes profoundly so.

Ecotourism is also just at its beginning. There are countless kinks
to work out but, as E.O. Wilson predicts, " . . . familiarity will save
ecosystems, because bioeconomic and aesthetic values grow as each
constituent species is examined in turn—and so will sentiment in favor
of preservation . . . the better an ecosystem is known, the less likely
it will be destroyed. . . . "[66]

The so-called SLOSS problem (single large reserve or several small
reserves) must be philosophically and effectively resolved for every local
ecosystem, wherever there are people in conflict with wildlife. The
comeback of the California sea otter, the preservation of macaws in
southeastern Peru (in and around the Manu Preserve), and the revolu‐
tion in sustainable thinking, all indicate our ability to micromanage
with a global perspective. We need not remain tragic victims of a com‐
mons syndrome.[67] The aforementioned markets for innovative
biodiversity products are there, if only governments will forcefully steer
their trading partners, and those who own the land—whether corpora‐
tions or small farmers—toward such sustainability. That would mean
that consumers among developed nations, and their elected represen‐
tatives, can no longer expect or demand "sustainability" in the Third
World, unless they are willing to pay for it, which they must. (One very
creative indication of this willingness is the recent formation of a

"Home Shoppers' Network" on British television devoted to Third World products.)

Even if Americans have thus far proven too selfish to approve of a four-cent gasoline tax, the willingness to do so is incipient, ultimately, to the process of democracy. For the alternatives, once understood by lawmakers, are far more dire than high levels of taxation (with which Europeans have already demonstrated a willingness to comply). On March 1, 1994, the Clinton Administration put forth new regulations pertaining to chemical plants that will cost an average family all of $1.50 per year, while scrubbing toxins from the air equivalent to removing nearly 25 percent of all automobiles from U.S. highways. Such sweeping new regulations, "a landmark for public health," according to EPA head Carol Browner, were initiated quietly and without fuss, and show what is possible.

And yet, such *possibilities* have scarcely had time to settle in. Consider the fact that nearly every Environmental Protection Agency the world over, including that of the U.S., has emerged in just the past two decades—following the U.N. Conference on the Human Environment in 1972 at Stockholm. With this newfound agitation for a better world, countless international ecological treaties have been recently concluded, where no previous moderation or understanding, or protection under any laws existed: the 1973 *Washington Convention on Trade in Endangered Species* (CITES), signed by more than one hundred countries; the *Paris Convention* of 1974 for controlling land-based pollutants; the 1982 international whaling ban; the Montreal Protocol of 1987; the 1991 U.N. resolution banning the use of driftnets; the *Bamako Convention* of 1991 protecting ten African countries from toxic waste imports; and the 1991 *Protocol to Protect Antarctica*, which overturned the mining provisions of the earlier *Wellington Convention*. And at the Rio Summit in June of 1991, thirty-nine peoples' treaties were signed by the many NGOs.[68] In its acknowledgement of an interdependent world, the U.S. helped draft most sections of the *Rio Declaration*, as well as the 580-page action plan known as *AGENDA 21* and has, under Clinton, corrected its position on biodiversity and intellectual properties.

The U.S. has committed itself to all the right buzzwords, mobilizing international resources for sustainable development and management, as well as an increase of $150 million a year in U.S. bilateral forest assistance for conservation, and an end to domestic clear-cutting on all federal forestlands. It has called for ratification of the *1989 Basel Convention on the Control of Transboundary Movements of Hazardous Wastes and their Disposal*, making it mandatory for exporting countries to reveal their plans and obtain permission from recipient countries prior to shipping any wastes; enhanced the role of international organizations like UNEP; created a clearer language in various marine treaties to ensure healthy maintenance of endangered populations; and advocated the building of indigenous technological capacity throughout the world. The U.S. has proposed a new network of international chemical risk assessments and management plans; and adopted the *U.N. Framework Convention on Climate Change* that would seek to establish a holding pattern based on 1990 emissions levels by the year 2000.[69]

These are rudimentary precedents, and there are gaping holes. For example, neither the 1971 Convention on Wetlands (RAMSAR) nor the 1972 World Heritage Convention control the necessary financial instruments to do what they are mandated to do. And many signatories to these treaties have not acted responsibly, either because of poverty or greed. According to Hilary F. French, Norway, to cite one example, has been found to be derelict in its commitment to at least 12 of the 27 international environmental treaties it has signed.[70] And so far, the World Trade Organization has failed to live up to its ecological pledge, whether having to do with the dolphin–tuna connection, or unprocessed tropical and temperate forest timber, recycled fiber in newsprint, or asbestos products. As free trade enriches the prospects for individual nations, so too the power of sanctions and popular boycott becomes an important tool of ecologically minded trading partners.

On the domestic front, U.S enforcement of environmental offenders was escalated as of 1992. Ninety-one percent of those convicted in that year served jail sentences and the value of enforcement penalties approached $2 billion. The U.S. Senate passed legislation in 1986 mandating the public's right to know what corporations are doing; and such "public right to know" laws are now being replicated throughout much of the Third World. But they entail, conversely, the *responsibility of awareness* on the part of the public. The public's increasing awareness of environmental issues, helped along by the renaissance in telecommunications, may be the most important single cause for optimism. In Indonesia, such rights are being translated into important, precedent-setting environmental lawsuits by individuals against corporations and government officials.[71]

Vice President Gore has proposed taxing environmental negatives, rewarding the positives, and substituting environmentally competitive products for less responsible ones. It is one thing to say it, but quite another to initiate the actual government incentives that will visibly encourage the free market to act in accordance with green principles. In the Netherlands, the 1989 *Dutch National Environmental Policy Plan* suggested 223 "green" policy changes to better ensure a sustainable economy. Canada, a year later, launched its own *Green Plan for a Healthy Environment.*

In a world of scarcity, political and ecological triage will inevitably figure. Even with the recent example of the Japanese, and an imagined $50 trillion annual economy, the increasing number of newborns on the planet will not make decisions any easier. Will one set out to assist Russia, Brazil, Indonesia, or Mali? The school system in Alabama, or teenage mothers in India? And when conflicts arise, which are to become our national priorities in terms of intervention? Nuclear proliferation in Asia, or illegal clear-cutting in Sumatra? Such questions have always weighed upon political leaders, in one form or another, but never before have ecological and population considerations played such dramatic roles in analysis and national determinations. Also like never before, we are confronted by the evident liability of *systems.* Confront individuals with a problem, and they are likely to embark toward a solution quickly. Pose the same problem to a whole community and the solution is deferred. Consensus requires negotation and time, and invites the possibility of irresolution. Take the problem to the level of a nation, and the deferral time may be compounded over years, even decades. Similarly, an individual steers a vehicle with reflex actions, whereas a captain guides a supertanker, or a nation its economy, at a massively slugglish pace. Reaction times are desperately slow. Individuals are in flesh. Systems are on paper. Conversely, individual reaction times can work to change a system for better or worse (e.g., the stock market crash, or the U.S. withdrawal from Vietnam).

Hence, individual men and women are called upon to become policymakers, to think, to make those determinations, to take their lifestyles into greater consideration than ever before. They will not, as described toward the end of chapter 6, have the means of formulating policies of a nation, but by example they can inspire surprisingly huge assemblages of people. Ethical solutions, reasonableness, beauty, and inspiration, all have in their favor the force of silent majorities, the equivalent power of chain letters, the quiet seduction of an ideal. The number of vegetarians and animal rights' advocates are rapidly increasing. And while the contingent is still minute, a growing number of women are preferring a "child-free" existence. In the U.S., 3.3 million, or 5.7 percent of the approximately 58 million women of childbearing age, are declining to have kids.[72]

From my perspective, it would be a very good thing indeed, if even a smattering of the Jain monks' daily vows infiltrated the consciousness of every nation. That would mean thoughtful hesitation, a neurological hallmark of civilized decorum; nonviolence, the basis for any sustainable, compassionate, and equitable human community on earth; universal one-child families, the only way to begin to slow down the population explosion; and an emphasis in our lives upon sharing, that human and humane capacity that best reflects the abundant generosity inherent to the creation.

For nine days in late 1993, I had the opportunity to speak with many preeminent spiritual leaders from two dozen countries. A living Hindu saint, a Buddhist nun, a Jain monk, a Taoist sage, Protestant and Catholic theologians, His Holiness the Dalai Lama, Islamic and Zoroastrian scholars, yogis, rabbis and tribal elders, interfaith evangelists, and two Holocaust survivors. All were convinced that the world, while gripped by adversity, was poised for positive transformation. By sometime in the twenty-first century, everything about human life would be different, better, resolved. Some (including a Hopi elder), on a darker note referred to an apocalyptic cleansing. Many were bathed in an optimism I greatly admired; a mindset that appeals to countless scientists, economists, and politicians who are singularly encouraged by the prospects of the twenty-first century. These are people who fluently cite a human awakening of consciousness, the first reductions in nuclear weapons, the fact that Germany, Italy, and France, Japan, and the U.S. have held to their peaceful alliances, that communism has all but vanished and that even the Israelis and Palestinians have recognized one another at long last, despite recent setbacks. Such optimism must prevail, in spite of the flaunting of nuclear capability in China, Pakistan, North Korea, and the U.S., the outright turmoil in at least 169 other regions of the world—more small wars being simultaneously waged than at any time in human experience; the political and gender-related tyrannies that rule over countless nations and women in Asia, Africa, the Middle East, and parts of Latin America; and the cumulative *biological war* that our species, *by its very nature*, is engaged in against the planet, a brutal picture of the twenty first-century that is not stable, but ecologically and demographically insane. We know these things, by now. Our acknowledgement itself is an act of meditation poised for selfless, even heroic change. But we must remain optimistic if such changes are to enjoy even half a chance. Awareness itself must be nurtured and protected.

Among more and more of the world's religious thinkers, there is a surge of ecologically aware activism. Jews, Catholics, and Anglicans in the U.S., for example, have sponsored a National Religious Partnership for the Environment. Parishes in hundreds of communities are being encouraged to be in the vanguard of local preservation efforts and to reinterpret classical scripture in a way that reveres God's creation in and of itself, irrespective of humanity's own self-centeredness. Others, nondenominational ecologists among them, have taken to worshipping the Earth, not some prime mover inherent to the planet (i.e., God), but the facts and miraculous details themselves. Perhaps "worship" is not quite the word: *respect*, guided by love, and empowered by the full, unstinting command of conscience.

Indeed, it takes greater courage to be an optimist, to be in love, than merely to mourn. There are important spiritual, behavioral, and self-fulfilling reasons for adopting that positive, emotional paradigm—or up to a point. The precise point becomes fuzzy, hampered by the fundamental uncertainties of data and introversion and of the words themselves upon which a communicating collective so relies—"fertility," "multiple-use," "development," "buffer zones," "edge effects," "connectivity," "habitat," and so forth. We can talk ourselves into nearly anything. So why not talk encouragingly, *savoring* the world as much as we might try to save it? Of course, it is also true that when crucial political and economic policies are guided by *too* simple an optimism, or by words whose meanings cannot be fixed or qualified—the expectation of endless "free" market miracles, or supposedly dazzling new techno-fixes, for example—the results may be irresponsible and dangerous. From a historical and ecological point of view, examples of such dangers long-ago enshrined, would include the invention of explosives, of automation and subsequent factory lines, of banks and fertilizer. Military and nuclear paranoia, runaway global colonialism, and the economic insecurity driving large family sizes and dispossessing local self-sustenance, have subsequently followed. Extrapolating into the future, there is no end to the new scientific and engineering "solutions" that will be tested on, and by, the human aggregate, as our problems are compounded by the sheer quantities of human flesh and blood and need. Will those billions of individuals have the frame of mind, the clarity, to focus judiciously? To be loving?

Having delved into the labyrinth of population pain, it seems clear to me that, as a species, we need all the clarity grounded optimism we can muster. Our children need to be informed and inspired, not daunted. Although the planet is held captive by much that defines our personality and behavior, that aggression and its myriad tragedies need not be destiny.

In refashioning global fate beyond the mere designs of human hope, certain sobering truths must be firmly embraced. I quote: "The indispensable strategy for saving our fellow living creatures and ourselves in the long run, is, as the evidence compellingly shows, to reduce the scale of human activities. The task of accomplishing this goal will involve a cooperative worldwide effort unprecedented in history," write Paul Ehrlich and Edward Wilson, the two leading ecologists of this generation.[73] That invocation has similarly been endorsed by 1670 scientists in 71 countries, 102 of them Nobel laureates, who recently signed a population stabilization statement by the Union of Concerned Scientists, which declares, "Fundamental changes are urgent if we are to avoid the collision our present course will bring about . . . if vast human misery is to be avoided and our global home on this planet is not to be irretrievably mutilated."[74]

Throughout human history, hope and dread have always mingled. But never before have the risks been so permanent, so Earth-bound. Paradise is here, now, if only we will own up to it, accept it, and do our part to keep it true. The solutions I have suggested are logical, reasonable, rational. But I don't know if they are possible.

Unlike, say, the participants in the Cuban Missile Crisis, or those on the first Apollo lunar landing, none of us are today in a position to make things instantly right simply by pushing buttons, moving forward, or backing off. While timing is everything, we must all be prepared for a lifetime of service and diligence. The ethical and ecological responsibilities that being human entails will only increase as humanity finally comes of age.

# Epilogue

"We have met the enemy and they are ours . . . "
(Oliver Hazard Perry, dispatch to Gen. William Henry Harrison, announcing his victory at the battle of Lake Erie, 10 Sept., 1813.)

On the fringes of Tijuana recently, I observed a large number of swallows crossing that scorched and poignant border-purgatory into the United States, freely mocking the queuing automobile mobs that heaved sickly along the hot tarmac, and those who travelled by foot past security forces and magnetometers. The flock's aerial acrobatics, a mere fifty feet above, were graceful and—by contrast to the human follies below—truant to the point of commentary. Here were millions of years of avian evolution passing effortlessly overhead the embattled juggernauts and feverish spasms of human immigration. Swallows have their own problems, of course, such as serious nest competition for woodpecker holes. But they do not take their natural grievances out on the world. They wouldn't know how. Nor are their individual numbers and impact enough to register more than a minute blip on the record of biological activity on Earth.

All of our own intraspecies solipsisms and intrigues, in some cosmic (though certainly not biological) sense, may be no more than a blip as well. As of springtime 1994, in anticipation of the upcoming United Nations International Conference on Population and Development (ICPD) being held in Cairo for a week in September of that year, thousands of concerned citizens and representatives of countless NGOs were airing their often-dramatically-polarized views on population politics in a dozen town meetings around the U.S. At one such meeting, the rancor and opposing voices exploded on issues pertaining to immigration ("them versus us") and such contraceptive technologies as Norplant, deemed by some women to be "racist," even genocidal, and an outgrowth of Western imperialism.

In a hundred other countries, social scientists were busily crafting carefully-toned position papers for presentation at Cairo. The so-called U.N. "preparatory committee" had drafted its own eighty-page summary of policy recommendations. These included the reiteration of principles from the earlier *Rio Declaration* that "human beings are at the center of concerns for sustainable development"; an emphasis on a "decent quality of life" for future generations; a determination, in accord with the *Declaration of Alma Ata*, that all countries "should strive to reduce mortality and morbidity . . . [and] should aim to achieve by 2015 a life expectancy at birth greater than seventy-five years . . . " To do so, of course, under conditions of increasing economic growth might well engender the preconditions for an even-more-robust baby boom than is currently being inflicted upon the world. After all, it was medical technology in the first place that exacerbated an historically growing gap between high birthrates and declining mortality rates.

To meet contraceptive needs and a basic package of primary health care around the world by the year 2015, the social scientists estimated that no more than $17 billion annually would be necessary—which translates into only nickels and dimes being spent per capita. Of that sum, only $5.7 billion would be required by the wealthiest of nations, the United States—where several individual citizens alone are worth more than that.

Every aspect of the U.N. conference, from linguistic etiquette to the potentially divisive nature of protocol, had been thoroughly debated and revised. Such descriptors as "overpopulation" or "population control" or even "birth control" were no longer politic. The term "population bomb" and the views of Malthus were declass. In their place were such anaesthetized expressions as "stabilization" and "reproductive health."

Those charged with engineering the U.N. summit had to weigh the many inextricable links compounding family planning and development: female empowerment; gender equality; social justice; cultural diversity; resource redistribution; quality of life; health care; economic equity; democratic decision-making; access to education; corporate responsibility; the right to have a family, to work, and to play; and the right to happiness. And finally, they had to pay lip service to that vague, groping expression "ecological sustainability." It appeared that demographics, family planning, and the very plight of the world were about nothing more than human rights. Many were arguing, in fact, that there was no "population problem" at all, but rather a problem of greed. Runaway fertility, echoed a chorus, was not the issue, but rather persistent inequality. Many demographers were insisting that the data from throughout the world conclusively evidenced trends suggestive of global stability within a matter of decades. Optimism ran high. Cairo was expected to be a new beginning, which was to say, the previous forty years of international family planning had not been all together effective.

As these preparations for the ICPD were going on around the world in early 1994, the U.S. State Department, Justice Department, and Immigration and Naturalization Service were studying the precedent-setting ruling of a federal judge in Virginia who had determined that a Chinese refugee claiming political asylum in the U.S. had made a valid claim: China's one-child policy, the threat of sterilization, and a couple's thwarted desire to have three children were—in the court's opinion—a violation of human rights. While this ruling was nonbinding on the dozens of other fertility-asylum cases pending throughout the United States, a spokesperson for the Federation for American Immigration Reform called it a "disaster" that could trigger a domino effect.

While the Clinton Administration claimed to place population issues high on its roster of imperatives, sufficiently increased nickels and dimes for family planning were still beyond the means of lawmakers or the understanding or willingness of their constituencies. At the same time, opponents of any Federal funding for abortion continued to threaten not only the Administration's proposed health care package; but also the whole issue of a national population policy.

The many human messages and intentions ricocheting across the planet at the end of the millennium seemed to focus predominantly upon narrow human matters of privilege, exclusivity, national boundary, advantage, and special interest. If anything, the public discussion of "rights" (as opposed to "duties" or "natural law" or basic obligations to other life forms) had been elevated to a level of unprecedented arrogance. Humans—some far more than others—believed themselves possessed of inherent rights to do *anything*, as long as it could be described as socially just, politically or economically necessary, "morally" correct, or fundamental to human—defined as one group's or one person's—happiness. Given the insane and abundant suffering unleashed by our kind upon one another, there was certainly ample rationale for desiring a far better state of affairs and struggling to achieve it. But what about the nonhuman world? I recalled those lines from Immanuel Kant's princely work, *Perpetual Peace* (1795): "War requires no particular motive; it appears ingrafted in human nature. . . . With men, the state of nature is not a state of peace, but of war."

I would prefer not to believe that, in the hopes that belief and optimism themselves may become self-fulfilling. As the Scottish philosopher and historian, David Hume, wrote in *A Treatise on Human Rights*, " . . . nothing we imagine is absolutely impossible. We can form the idea of a golden mountain and from thence conclude that such a mountain may actually exist." Somewhere, sometime, somehow.

In the broader view, perhaps this is what the swallows are also thinking, dreaming, as they gaze flittingly down upon the vast complex of human imbroglios and agitations.

# Notes

PROLOGUE

1. Jack Nelson and David Lauter, "Summit Nations Approve Sweeping Tariff Reductions," *Los Angeles Times*, July 7, 1993, p. A7. President Clinton was speaking in Tokyo during the summer of 1993 Uruguay Round of General Agreement on Tariffs and Trade negotiations.

CHAPTER 1
The Balance of Nature in Antarctica

1. The work of Jostein Goksoyr, discussed in E.O. Wilson, *The Diversity of Life*, Cambridge, MA: Belknap Press of Harvard University Press, 1993, p. 144.
2. See Andrew M. Karmack, *The Tropics and Economic Development: A Provocative Inquiry into the Poverty of Nations*, Washington D.C.: A World Bank Publication, 1976, p. 23.
3. Wilson, op. cit., pp. 210–211.
4. See Lynn T. White, Jr., *Medieval Technology and Social Change*, Oxford: Oxford University Press, 1962.
5. See Colin Clark, *Population Growth and Land Use*, New York: St. Martin's Press, 1977.
6. See Charles Dickens, *Hard Times*, London: Macmillan Publishers, 1983.
7. What finally upset the perfect balance achieved by the Pecos Indians were economic decisions taken by the Spanish in Mexico City, prompting one Francisco Vasquez de Coronado to lead his troops in search of the fabled Cibola, City of Gold. Who could forget that between 1500 and 1650 America's eastern seaboards had yielded 181 tons of gold for European monarchies? One invading group after another pressed throughout North America. Religious monks, eager to convert lost souls, joined the military expeditions. Eventually, the Pecos Indians were destroyed. Farther north, Moravian missionaries would shame the Inuit into general prohibitions against masturbation thus prompting a population explosion. Whatever lessons of population stability and simplicity of lifestyle the North American indigenous peoples had to offer were as yet, hundreds of years ahead of their time. Meanwhile, at

Sukwutnu, the Yokut were also being displaced by a constant stream of homesteaders, ranchers, and farmers. By 1876, the last remaining Yokuts were placed in the Tule River Reservation. By 1905, there were 154 of them left. As for the Ohlone, a few hundred of their descendants still live in the Bay Area. In 1986, I spent the afternoon in a tepee with the last known full-blooded Ohlone, a woman in her nineties who had grown up on an Indian reservation near San Mateo. She still remembered all the wild animals that once populated what is today the East Bay, looking across toward San Francisco.

8. Wilson, op. cit., pp. 247–249.

9. See J. Smith's "Man's Impact upon Some New Guinea Mountain Ecosystems," in *Subsistence and Survival: Rural Ecology in the Pacific*, ed. T. Bayliss-Smith and R. Feachem, New York: Academic Press, 1977, pp. 185–214; and Hagen, "Man and Nature: Reflections on Culture and Ecology," *Norwegian Archaeological Review* 172, no. 5(1): pp. 1–22.

10. See Raymond Dart, "The Predatory Transition from Ape to Man," *International Anthropological and Linguistic Review* I, 1953.

11. Wilson, op. cit., p. 169.

12. See Ron Naveen, Colin Monteach, Tui De Roy, and Mark Jones, *Wild Ice—Antarctic Journeys*, Washington D.C.: Smithsonian Institution Press, 1990, p. 37.

13. See author's film, *Antarctica—The Last Continent*, PBS, 1987; see also author's article "The Next Wasteland—Can the Spoiling of Antarctica Be Stopped?" *The Sciences*, published by the New York Academy of Sciences, March/April 1989, pp. 18–25.

14. See Kim A. McDonald, "Penguins in Peril," *The Chronicle of Higher Education*, January 5, 1994, pp. A6, A7, A15.

15. See author's film, *Black Tide*, Discovery Channel, 1990. Five years after the spill, four years after the 1990 *Oil Pollution Act*, and 95 percent of all oil-bearing vessels in the U.S. are single hull.

16. See Catherine L. Albanese, *Nature Religion in America from the Algonkian Indians to the New Age*, Chicago: University of Chicago Press, 1990, p. 67.

17. Touche Ross Management Consultants, "Reducing Consumption of Ozone Depleting Substances in India; Phase 1: The Cost of Complying with the Montreal Protocol," Touche Ross Consultants, London, England, 1990, p. 55.

18. Chen Ya and Xiong Lei, "Seeing Red over Green Issues," *China Daily*, Beijing, Feb. 12, 1993, p.5.

19. Lester R. Brown, et al., *Vital Signs 1992*, New York: W.W. Norton & Co., 1992, p. 60.

20. Field-Marshall Viscount Montgomery of Alamein, *A History of Warfare*, Cleveland and New York: World Publishing, 1968, p. 33.

21. J.H. Fremlin, "How Many People Can the World Support?" in *Population, Evolution, and Birth Control*, ed. Garret Hardin, San Francisco: W.H. Freeman, 1969.

CHAPTER 2
A Paradox of Souls: China

1. John Bongaarts, W. Parker Mauldin, and James F. Phillips, "The Demographic Impact of Family Planning Programs," *Studies in Family Planning*, volume 21, number 6, Nov/Dec 1990, p. 305.

2. See Jodi L. Jacobson, "Coerced Motherhood Increasing," in *Vital Signs*, by Lester Brown, et al., New York: W.W. Norton, 1992, pp. 114–115.

3. There has, in addition, been some speculation about the year 2000, and its significance to the Chinese in terms of fertility. In his essay "Creating New Traditions in Modern Chinese Populations: Aiming for Birth in the Year of the Dragon," *Population and Development Review*, volume 17, number 4, December 1991, pp. 663–686, Daniel M. Goodkin examines the fact that traditionally in many Chinese cultures there have been baby booms during dragon years. "The dragon will rise again in the year 2000, the gateway to the new millennium, and, for a phenomenon so dependent on an upwelling of emotion, the national symbolic significance of this should not be underestimated," he writes. Furthermore, he notes that this particular dragon year is the Metal Dragon (metal sharing the same characters in Chinese as gold). A Golden Dragon Year is a rare and fortuitous time, one that could well invite a baby boom. "With such an alignment of animal, mineral, and numerological forces," says Goodkin, "expectant parents probably could not hope for a more auspicious sign." (p. 680) When Premier Li Peng spoke at a Beijing Family Planning forum in March of 1992, he too referred to a new baby boom brought on by superstition, but it was not that of the Golden Dragon Year. Rather, he said, 1992 would make the family planner's task more difficult, because 1991—the year of the goat/sheep—had been an inauspicious time to give birth, so people had waited for 1992. See *China Population Today*, volume 9, number 2, April 1992, Beijing: China Population Information and Research Center.

4. John S. Aird, *Slaughter of the Innocents—Coercive Birth Control in China*, Washington D.C.: The AEI Press, 1990, p. 20.

5. Paul Kennedy, *Preparing for the Twenty-First Century*, New York: Random House, 1993, p. 175.

6. In Beijing, as of 1989 there were an estimated 1.3 million "floaters" (migrants) having as many children as they wanted. In Shanghai, the number of such migrants was suspected to be over 1.8 million.

7. See Li Jingneng, "Reproduction Worship and Population Growth in China," *Chinese Journal of Population Science*, volume 4, number 1, 1992, pp. 27–32. Citing Ye Yangzhong, "A tentative study of the child-bearing aspirations of contemporary farmers," *Population Studies*, number 4, 1988, p. 29, Li writes, "There is no question now that throughout rural China, the preferred family norm is two or more children. See also Minja Kim Choe and Noriko O. Tsuya, "Why Do Chinese Women Practice Contraception? The Case of Rural Jilin Province," *Studies In Family Planning*, volume 22, number 1, Jan/Feb 1991, pp. 39–51. Eighty-five percent of those rural married women with one child nevertheless continued to practice contraception, though most reported two as their ideal number.

8. See "China's Population and Development," *Country Report*, prepared by the Chinese Delegation to the Fourth Asian and Pacific Population Conference, Bali, Indonesia, August 1992.

9. Others like Professor Li Jingneng, Director of the Institute of Population and Development Research at Nankai University in Tianjin, argue that the current birthrate is closer to 54,000 per day, or 38 per minute.

10. See Shanti R. Conly and Sharon L. Camp, "China's Family Planning Program: Challenging the Myths," *Country Study Series #1*, Washington, D.C.: Population Crisis Committee, 1992, p. 32.

11. See Song Jian, Tuan Chi-Hsien, and Yu Jingyuan, *Population Control in China*, New York: Praeger, 1985, pp. 263–267.

12. Personal conversations with Professor Wu Cang Ping, Institute of Population Research, the People's University of China, Beijing.

13. In Szechuan, the population growth rate plunged from 3.12 percent in 1970 to 0.67 percent in 1979 and has remained low. Tianjin City also curbed its birthrate dramatically early on. See Luo Hanxian's *Economic Changes in Rural China*, trans. Wang Huimin, *China Studies Series*, Beijing: New World Press, 1985.

14. Chinese non-Han minorities, who account for about 7 percent of the total population, are usually granted exclusions from the population control policies, though there is serious debate over the situation in

Tibet, where some observers have cited instances of forced sterilization and abortion by the Chinese, consistent with allegations of country-wide abuse of fertility rights.

15. See Paul Ehrlich and Anne Ehrlich, *The Population Explosion*, New York: Touchstone Books, Simon and Schuster, 1990, p. 141.

16. See Tu Ping, "Birth Spacing Patterns and Correlates in Shaanxi, China," *Studies In Family Planning*, volume 22, number 4, 1991, pp. 255–263.

17. In fact, some statistics from some regions are known. Population planning in the decade of the 1970s had prevented approximately 59 million births, and most of them were through induced abortions. In 1980, the rate of abortions in the country was estimated to be 574 per 1000 births. See H. Yuan Tien, et al., "China's Demographic Dilemmas," *Population Bulletin*, volume 47, number 1, Washington, D.C., June 1992, p. 13.

18. Conly and Camp, op. cit., p. 33.

19. By 1979, China's death rate had been stabilized at between 6 and 8 per thousand, its birthrate had dropped from 33.43 in 1970 to 17.82 per thousand, and its TFR had declined from 5.8 to 2.75.

20. Aird, op. cit., 1990, p. 21.

21. Ibid., p. 24.

22. See H. Yuan Tien, et al., "China's Demographic Dilemmas," *Population Bulletin*, volume 47, number 1, Washington, D.C., June 1992, p. 7.

23. In both China and India, feminism is divided on this point, the fact that women bear the overwhelming burden of birth control. Family planners themselves are quick to point out that intervention in, and manipulation of, the woman's biological cycle is a more effective means of birth control. Nearly all sterilizations in Asia are among women, though twenty years ago the situation was reversed. Today, the unanimous (and obvious) consensus of family planners is that women need to be empowered to control their own bodies, to reinherit their own destiny, to liberate that which men have co-opted over time. But there is a price, which makes increasing rates of contraceptive prevalence use harder to achieve, particularly in poor countries or regions traditionally bound by cultural machismo.

24. H. Yuan Tien, et al., op. cit., p. 9.

25. The government is presently studying social security systems throughout the world and asking government workers to pay 3 percent of their monthly paychecks into security funds. It is still constitutionally incumbent upon all children to eventually look after their parents.

At present, it is alleged that some 25 percent of the aged in China enjoy the "five guarantees" of social security benefits. But the quality of these services varies. In some cities, pensions are quite respectable for one-child parents, between 60 percent and 100 percent of salary. In Changyi County of Shandong Province, "single-daughter households" have become eligible for old-age insurance, which pays out between 80 and 100 yuan per month, a considerable sum considering that the average annual per capita income in Changyi was 825 yuan in 1990 ($150). Some 350 households have pledged to remain single-daughter families; fecund couples who volunteered to cease all future childbearing activities. The village collectively pays the monthly stipend. It is estimated that this old-age insurance incentive has prevented 230,000 births, or 30 percent of the total population in the County over a very short period. Had the insurance concept not been applied, per capita income would thus have dropped from 825 yuan to 500 yuan. As yet, however, the vast majority of rural China's elderly remain unprotected. Ironically, if China were to actually preserve its one-child revolution into the future, some demographers believe that the high percentage of elderly drawing their pension could actually bankrupt the country. Changyi may be an exceptional case, where existing poverty made it ripe for an incentive experiment. See Gao Zhanjun and Zhang Xueming, "Old-Age Insurance for 'Single Daughter Households' in Rural Changyi," *China Population Today*, volume 9, number 2, April 1992, China Population Information and Research Center, Beijing. Countrywide, however, some have observed that less and less is being done for the aged as the economy grows.

26. Sheryl WuDunn, "China Village Prospers But Retains Old Ways," *New York Times*, January 17, 1993, p. 8.

27. See Theresa Kelleher, "Confucianism," in *Women in World Religions*, ed. Arvind Sharma, New York: State University of New York Press, p. 135.

28. See James Thayer Addison, *Chinese Ancestor Worship*, Shanghai: Church Literature Committee of the Chung Hua Sheng Kung Hui, 1927, p. 15.

29. In Shaanxi Province, nearly half of those parents who already had even two daughters, still wanted a third child, hoping it would be male. Whereas, of those who had had two sons, only 11 percent felt they needed a third child. See Wei Jinsheng, "On the Operating Mechanism of Population Control," *Chinese Journal of Population Science*, volume 4, number 1, 1992, p. 56. See also Li Jingneng, pp. 27–32.

30. See Kelleher, op. cit., p. 143.

31. See Marlyn Dalsimer and Laurie Nisonoff, "Collision Course," in *Cultural Survival Quarterly*, volume 16, number 4, Winter 1992, p. 57.

32. This pathetic gender bias is noticeable in nearly every developing country, where the TFR continues to reside at about 4.6, excluding China. See "China—Accessibility of Contraceptives," *Asian Population Studies Series*, Number 103-B, Economic and Social Commission for Asia and the Pacific, Bangkok, United Nations 1991.

33. See *Fatherhood In Chinese Culture*, Hong Kong: University of Hong Kong, 1983, p. 242.

34. See *Chinese Lives: An Oral History of Contemporary China*, ed. W.J.F. Jenner and Delia Davins, quoted on p. 59 of "Collision Course," by Marlyn Dalsimer and Laurie Nisonoff, in *Cultural Survival Quarterly*, volume 16, number 4, Winter 1992.

35. See Seung-kyung Kim, "Industrial Soldiers," *Cultural Survival Quarterly*, volume 16, number 4, Winter 1992, pp. 54–56.

36. *China Population Today*, volume 9, number 2, April 1992, China Population Information and Research Center, Beijing, p. 15.

37. Aird, op. cit., p. 92. The international organization, Asia Watch, has also reported on fertility-related atrocities in Tibet, including forced abortions, even the witholding of food ration cards from children. See also Aird, p. 69.

38. See Zeng Yi, et al., "An Analysis of the Causes and Implications of the Recent Increase in the Sex Ratio at Birth in China," The Institute of Population Research, Peking University, Working Paper, 1992.

39. See Sten Johansson and Ola Nygren, "The Missing Girls of China: a New Demographic Account," *Population and Development Review*, volume 17, number 1, March 1991, pp. 35–51. In addition to those missing, "excess female infant mortality" has been postulated at some 39,000 per year, or four female deaths per thousand born. It means that it is nearly impossible now to estimate existing female infanticide statistics.

40. Aird, op. cit., p. 32.

41. Ibid., p. 44.

42. See Xie Zhenming, "An Evaluation of the Social Effectiveness of One-Child Policy in China," *Population Research*, volume 7, number 1, 1990, pp. 30–36, Population Research Institute, Anhui University, Hefei, Anhui Province, China. Interestingly, Franklin Roosevelt once said, "The nub of the whole question is this. If a farmer in upstate New York or Georgia or Nebraska or Oregon, through bad use of his land, allows

his land to erode, does he have the inalienable right as owner to do this, or has the community, i.e., some form of governmental agency, the right to stop him?" F.D.R. was inclined to side with a community's power to stop a farmer from destroying his own land. See *Louis Bromfield At Malabar—Writings on Farming and Country Life*, ed. Charles E. Little, Baltimore: Johns Hopkins University Press, 1988, p. xix.

43. Aird, op. cit., p. 83.

44. See *China Population Today*, volume 9, number 2, April, 1992, p. 8, China Population Information and Research Center, Beijing.

45. See Peng Xinzhe, "China's Population Control and the Reform in the 1980s," *Population Research*, volume 7, number 3, 1990, pp. 1–17, in which are discussed certain measures adopted on April 13, 1984 to relax the one-child policy, in particular the right for rural families to have a second child when judged necessary. See also Cai Zuofu, *Population Research*, "A Sociological Perspective of and Measures to Overcome Difficulties in Rural Population Control," *Policy Research Office*, volume 7, number 3, 1990, pp. 43–49, Jiangling County Party Committee, Hubei Province.

46. See Madame Peng, "Efforts to strengthen family planning work at the grass-roots level in order to fulfill the population goals of the eighth five-year plan"—*A Report of a National Conference to Exchange Experience in Family Planning at the Grass-roots Level*, Beijing, November 1990.

47. According to Aird, regardless of one's definition, coercion levels continue to increase, even upon Chinese nationals living abroad. He refers to one professional couple who were pressured by the wife's population control officer in her home factory. Though five months pregnant, she was nevertheless told that her delivery of a child (the couple's second) would mean that all 20,000 employees of her factory back home in China would lose bonuses, that the whole factory might have to shut down, that everyone would be punished. She was ordered to have an abortion at once. See Aird, op. cit., pp.74–75. Similarly, Sheryl WuDunn reports that in the village of Pan Shi near Hong Kong, twice a year, authorities come to the town to take eligible women to the hospital for tubal ligation operations. Those who want to have more than two children flee the village on those days. She also notes that more and more unmarried girls have become pregnant. See WuDunn, op. cit., p. 8. In 1989 the so-called Gansu Law—Gansu being one of the very poor provinces—made it legal for local administrators to forceably sterilize those deemed mentally retarded, based upon utterly unscientific beliefs.

48. Personal interview with Madame Peng Peiyun, State Family Planning Commissioner, February 1993.

49. The evolution of slogans is fairly indicative of where Chinese family planning has gone. In the early 1970s the saying was—"wan-xi-shao," literally, "later-longer-fewer." Then the refrain went, "one is not too few, two is good, three is too many." By 1979 the saying was, "one is best, at most two, never a third." See *China's One-Child Family Policy*, ed. Elisabeth Croll, Delia Davin, and Penny Kane, New York: St. Martin's Press, 1985.

50. The saying now is, "To charge nothing is not enough; the supplies must be sent straight into the home." See "China—Accessibility of Contraceptives," pp. 6–7.

51. See Zhang Tianlu, "The Marriage Pattern and Population Reproduction of the National Minorities of China," *Population Research*, volume 7, number 4, 1990, pp. 27–36, Research Institute of Population and Economics, Beijing Economics College, which discusses such institutions as Tibetan polyandry, consanguineous marriages, visiting and cohabiting companions. Tibet, Xingjiang, Guizhou, Henan, Hainan and Ningxia are still the highest fertility regions in China, all containing minority populations in great number.

52. Aird, op. cit., p. 43.

53. Conly and Camp, op. cit., p. 36.

54. It is of some interest to compare the difficulties, today, of lowering TFR, versus those gains accomplished twenty years ago in China. For example, in just one year, 1973–1974, China went from a TFR of 4.5 to 4.2. See "China's Population and Development," *Country Report*, prepared by the Chinese Delegation to the Fourth Asian and Pacific Population Conference, Bali, Indonesia, August 1992, p. 5. See also, "Study of Stainless Steel Ring and Copper T IUD Efficacy, Use, Cost/Benefit and Conversion in China," UNFPA Project CPR/91/P43, *Project Report*, May 21, 1992, Beijing. Dr. Zhang De Wei, Project Leader.

55. A story is often related how President Carter once proposed that Zhao Ziyang, then premier of China, allow more Chinese to travel to the U.S., to which the Premier apparently replied, "No problem. I will allow 200 million to go to tomorrow if you want." The book, pertaining to China's state-sponsored abortions, is Steven W. Mosher, *A Mother's Ordeal*, New York: Harcourt Brace, 1993.

56. As of mid-1993, UNFPA's "model program" in a few specific communities in China had come to a halt amid American allegations that new evidence for coercion had surfaced. UNFPA's efforts had included

the elimination of family planning targets in certain southern areas of the country. The idea had been to test the effectiveness of "voluntary" fertility self-maintenance, in absence of government control.

57. See *Far Eastern Economic Review*, February 11, 1993, p. 60.

58. According to consultants Stephen Shaw and Jonathan Woetzel of McKinsey & Co., as reported by Mark Clifford, "Consuming Passions," *Far Eastern Economic Review*, February 11, 1993, p. 44.

59. See George Orick, "Combatting Poverty in Rural China: Change Comes to Anwang," the *Ford Foundation Report*, Winter 1992, volume 22, number 4, pp. 3–7.

60. Ehrlich and Ehrlich, op. cit., p. 70.

61. See Aditi Kapoor, "Tiger poaching unabated," *Times of India*, February 20, 1993, p. 3.

62. "Tibet's Grasslands Under Threat," *South China Morning Post*, September 26, 1992.

63. See Susan Woldenberg, "Tibet: Aerie or Abattoir?" pp. 8–10, *The Animals' Voice* magazine, volume 6, number 1, 1993.

64. *Tibetan Environment & Development News*, November, 1992, issue #6.

65. See Jeff Wise, "With Six You Get Bear Paw," *Esquire*, May 1993, p. 47 (in a section, ironically titled by the magazine, "Man At His Best—The Enlightened Traveler"). Wise quotes an Agriculture and Fisheries officer in Hong Kong who was dispatched to capture a clouded leopard seen on the streets of densely packed Mongkok, who stated in some surprise, "Very few people eat leopards in Hong Kong."

66. Ehrlich and Ehrlich, op. cit., p. 129.

67. See Hu Angang and Wang Yi, "The Future Conflict Between Population and Grain Output in China," *Chinese Journal of Population Science*, volume 2, number 3, 1990, Research Center of Ecology and Environment of the Chinese Academy of Sciences, Allerton Press, New York, p. 202.

68. Hu and Wang have examined required grain output based upon a per capita grain requirement of 800 jin per year and have determined that China would have to achieve 700 jin per mu (1 mu = .067 hectares = .165 acres) which is equivalent to the highest petroleum-driven levels in Japan and France. This is not feasible in a cash-poor China, or not yet. The highest non-petroleum driven yield from traditional Chinese agriculture was met in 1984, namely, 482 jin per mu. In 1991 the area of grain harvesting decreased globally by an estimated 2 million hectares and half of that was in China.

69. Five hundred years ago, citizens of Beijing maintained their own open-pit garbage dump behind the Forbidden City. Today, that enormous dump is a park, two hundred feet high, covered in grass and trees.

70. If Beijing had won its bid to host the 2000 Olympic Games, its Vice-Mayor had promised to remove 70 percent of all chimneys in the city which have been implicated in the city's air pollution. It takes the promise of the Olympics, not human deaths, to prompt change. *The China Daily*, February 2, 1993, p. 3. Unfortunately, China did not win.

71. See "Some Very Bad News—An Interview with Amory Lovins," *Newsweek*, Asian edition, February 15, 1993, p. 50.

72. See *Energy in Non-OECD Countries: Selected Topics 1988*, International Energy Agency, OECD, Paris, 1988, p. 43.

73. The Three Gorges dam project at Sandouping, 38 kilometres downstream from Yichang and Zigui counties in central Hubei Province, will be 185 metres high, 175 metres deep and submerge 632 square kilometres including 13 county towns, 140 rural townships and 4500 villages. More than 1.6 million villagers will be displaced over the course of the 18 year construction. The dam is expected to eventually produce 18,000 MW of power. The State is to spend a mere $81 million on reclamation projects though the total budget for the dams is $10 billion.

74. See Dr. Qu Geping, *Environmental Management in China*, United Nations Environment Programme, Beijing: China Environmental Science Press, 1991, p. 260.

75. See Lester Brown, et al., *Vital Signs 1992*, New York: W.W. Norton Publishers, p. 38.

76. Dr. Qu Geping, op. cit., p. 27.

77. China's "Population And Development," op. cit., p. 17.

78. The development of residential areas far away from polluting factories has become an abiding principle of Chinese urban design in keeping with "The Law of the People's Republic of China on Environmental Protection." This means that as more and more technology replaces farm labor, migration into cities will increase, particularly along China's coastal regions, while the interior provinces become increasingly singled out and mined for their resources, without environmental constraint. In the northern part of Shaanxi Province, the Dabaodang Coal Field between Yulin City and Shenmu County has been targeted for its nearly 25 billion tons of coal which exist less than 600 metres below the ground. Large-scale mechanical mining will be employed within the coming few years to get at the fuel. Other gas and chemical

plants are also being built in the region. Synthetic ammonia, methanol, and petrochemical projects will all reap profits for the area. The ecological price tag is not even being questioned.

79. See Thomas Homer-Dixon, "'Destruction and Death" *New York Times,* January 31, 1993, p. E17, adapted from an article in the February, 1993 issue of *Scientific American* and based on research at the Project on Environmental Change and Acute Conflict at MIT.

80. Dr. Qu Geping, op. cit., p. 5.

81. See *The China Daily,* February 3, 1993.

82. In the Middle Ages, to take one sequence of events, the Chinese rage for calligraphy denuded countless watersheds along the country's major rivers trees providing the charcoal necessary for the ink—which in turn unleashed soil erosion and cataclysmic floods that wiped out rice crops and engendered massive starvation. For a discussion of the *shan shui* aesthetic, see Michael Tobias, "A History of Imagination in Wilderness," in *The Mountain Spirit,* ed. Michael Tobias and Harold Drasdo, New York: Overlook-Viking Press, 1979. See also, Michael Tobias, *A Vision of Nature—Traces of the Original World,* Kent, OH: Kent State University Press, 1994.

83. Dr. Qu Geping, op. cit., p. 70.

84. Chen Ya and Xiong Lei, op. cit., p. 5.

85. Dr. Qu Geping, op. cit., p. 9.

86. See World Resources Report 1992/1993, The World Resources Institute, with The United Nations Environment Programme and The United Nations Development Programme, New York: Oxford University Press, 1993, pp. 261—263.

87. That amount of forest cover translates into a staggering 10.9 percent of the terrestial planet, or approximately 130 million hectares.

88. In the past decade, the trees have been mostly oleaster, willows and poplars. Now the government and locals are looking to plant a greater variety of fruit trees so as to help cut back pest infestation. Gao has called for needed loans from international institutions to effect these develoments. See "China Plants A New Great Wall," *Sunday Observer,* Jakarta, February 14, 1993, p. 7.

89. Sheryl WuDunn, "China's Consumers Start to Make a Splash," *International Herald Tribune,* Tuesday, February 16, 1993, pp. 1, 12.

90. Dr. Qu Geping, op. cit., p. 134.

91. See "Some Very Bad News—An Interview with Amory Lovins", op. cit., p. 50.

92. See World Resources Report 1992/1993, op. cit., p. 202.

93. Already the Chinese are following the Japanese model of cultivating fastgrowing timber estates, or acquiring primary raw timber, from other countries, such as New Zealand.

94. See Frank Gibney Jr., "Beijing Rising," *Newsweek*, International Edition, February 15, 1993, pp. 8–13. See also "Interview with David Whittall," *The Asian Wall Street Journal*, February 18, 1993, p. 3.

95. See Nicholas D. Kristof, "China's Headlong Sprint Toward Wealth," *International Herald Tribune*, February 15, 1993, p. 1.

96. See Thomas D. Grant, "From the Paulskirche to Tian An Men," *Bangkok Post*, March 6, 1993, p. 4.

97. See Murray Weidenbaum, "Rising Chinese Economy Creates Prime Opportunity for U.S. Investors," *Los Angeles Times*, June 6, 1993, p. D2.

98. This, according to Kenneth J. Conboy, deputy director of the Heritage Foundation's Asian Studies Center. See also Sam Jameson, "Wary Neighbors Watch China Arm," *Los Angeles Times*, June 15, 1993, Section H11.

99. See Charles Wallace, "Arms Race, Round 2," *Los Angeles Times*, March 23, 1993, pp. 1, 5.

CHAPTER 3
The Ecology of Pain: India

1. The Soviets supplied unsafeguarded "heavy water" to India for years thus assisting Indira Gandhi's strategic goals. See Harald Muller, "Prospects for the Fourth Review of the Non-Proliferation Treaty," SIPRI Yearbook 1990, *World Armaments and Disarmament*, Sipri—Stockholm International Peace Research Institute, New York: Oxford University Press, 1990, p. 557.

2. Seven years before that famine of 1970, India's Shastri government had frantically appealed to President Johnson for long-term food aid. But Johnson and the U.S. Congress had lost patience with India's problems and abolished food guarantees, pushing instead for self-help measures, and sporadic short-run food aid extensions, the precursors of later economically imposed austerity programs. But this "short-tether" policy, as it was called, had failed to stabilize India's agricultural sector, pushing Indira Gandhi closer to Moscow, which in turn cemented nuclear assistance from the Soviets. Angered, Johnson then temporarily cut off all aid whatsoever to India in 1966. The politics of food aid, the outgrowth of failed population policies, thus encouraged nuclear proliferation.

3. See Seymour M. Hersh, "On the Nuclear Edge," *The New Yorker*, March 29, 1993, pp. 56–73. See also Jennifer Griffin, "Pakistan, India Negotiate—But Tensions Increase," *Los Angeles Times*, January 16, 1994, p. M2.

4. See *New York Times*, January 31, 1993, p. E16.

5. Hersh, op. cit., p. 56.

6. "Indo-Pak nuclear war probable: CIA," unsigned article in *The Statesman*, New Delhi, Friday, February 26, 1993, p. 16.

7. See Doug Tsuruoka, et al., "Power Plays," *Far Eastern Economic Review*, February 1993, pp. 47–49.

8. Hersh, op. cit., p. 73.

9. According to later Muslim jurists in India, children were blessings [*nemat*] but not obligations. Not having children was no sin, since the acceptance of blessings was deemed optional. The contemporary legalist Shah Saheb quotes from the *Mishkat*, in which the Prophet said, "Apply your intelligence to your actions. If you find that continuation of a particular action is useful, go ahead with it; but if you realize that its consequences will be bad, then restrain yourself." The Islamic prohibition of intercourse during nursing, known as *ghail* in India, coupled with the *Qur'an's* recommendation that the woman suckle the newborn for two years, reaffirms what is considered international family planning conventional wisdom, namely, that the parents space the birth of children by at least two years. Such spiritual ministrations suggest that, at least in theory, Indian Muslims are content with small family sizes.

10. Dr. George P. Cernada and Dr. A. K. Ubaidur Rob, "Pakistan's fertility and family planning: Future directions," *Journal of Family Welfare*, volume 38, number 3, Bombay, September 1992, p. 47.

11. Accompanying Pakistan's pro-natalist position are other stark corollaries: At least 7 million children are known to be slaves in Pakistan. The Hudood ["limitation"] laws that were ordered in 1979 by military dictator, General Zia ul-Haq made unmarried intercourse punishable by death under religious law, or by ten years in jail under civil law. Virtually no men have been prosecuted, only women. Within a decade of passage of the law, over six thousand women were thrown in jail. Even women who were raped were arrested on adultery charges, adultery being defined as unmarried sex. In such cases of rape, women are granted no right of evidence. The only allowable evidence is if four male Muslims attest to the "total" act of penetration. Minorities in Pakistan—over 3 million people—are all treated like women, which means that they have virtually no civil rights.

12. See V.A. Pai Panandiker and P.K. Umashankar, "Fertility Control-Induced Politics of India," New Delhi: Centre for Policy Research and Family Welfare Program, 1991, p. 12.

13. See Donald P. Warwick, *Bitter Pills: Population policies and their implementation in eight developing countries*, Cambridge: Cambridge University Press, 1982, p. 28.

14. Panandiker and Umashankar, op. cit.

15. India is often described according to ten biogeographic zones: Trans-Himalaya, Himalaya, Desert, Semi-Arid, Western Ghats, Deccan, Gangetic Plain, North East India, Islands, and Coasts. Among the various biological sub-zones, "urban slum" is now considered India's latest environment. Taken together, India's landmass and coastal areas constitute not the richest, but the most diverse ecological tapestry in the world.

16. See Banwari, *Pancavati—Indian Approaches to Environment*, trans. from Hindi by Asha Vohra, New Delhi: Shri Vinayaka Publications, 1992, p. 39.

17. Ibid., p. 94.

18. See Michael Tobias, *Life Force—The World of Jainism*, Berkeley: Asian Humanities Press, 1991, p. 68. See also K.C. Lalwani, *Sramana Bhagavan Mahavira—Life and Doctrine*, Calcutta: Minerva Associates Publications PVT. Ltd., 1975, pp. 176-183. For the most complete biography of Mahavira, see Muni Ratna Prabha Vijaya, *Sramana Bhagavan Mahavira—His Life and Teaching*, New Delhi: K.L. Joshi and Parimal Publications, 1948-1950.

19. Christopher von Furer-Haimendorf, *Tribes of India—The Struggle for Survival*, Berkeley: University of California Press, 1982, p. 305.

20. Ibid., p. 306.

21. See Shivanath Jha, "The gory Ganges—The holy river spawns death in Bihar," *Times Of India*, February 28–March 6, 1993, p. 43.

22. See Catherine Dold, "Tropical Forests Found More Valuable for Medicine Than Other Uses," *New York Times*, April 28, 1992, p. B8.

23. In the Aukre region of Para state in Brazil, for example, environmental organizations are working with the Kayapo Indians to find alternative, sustainable sources of wealth in the forest, including new agricultural techniques and the harvesting of medicinals. But so far, the only market for these products has been a single firm in England that sells Brazilian nut oil. Perhaps all the Kayapo will need are a few good marketing firms. But even then, according to Marguerite Holloway, once a "forest product such as latex becomes commercially important it

is inevitably introduced into higher-yielding plantations . . . Or the material is made synthetically. As a result, the price plummets, and small-scale extraction ceases to be profitable." In "Sustaining the Amazon," *Scientific American*, July 1993, p. 94.

24. See John Kurien and T.R. Thankappan Achari, "On Ruining the Commons and the Commoner—The Political Economy of Overfishing," *Centre For Development Studies Working Paper #232*, Tiruvananthapuram, Kerala, 1989, p. 37.

25. Examples of just some of the products over-exploited for profit include mahua flowers, sal seed (an edible oil), tendu leaf, tusar silk cocoons, edible roots, and mushrooms. See H.S. Panwar, et al., "The Wilderness Factor—Protected areas and people—compatible strategies," *Sanctuary*, volume XII, number 5, 1992, p. 77.

26. Population Crisis Committee, Washington D.C., 1992.

27. According to *The Economic Times*, February 27, 1992, 40 percent of India is below the poverty line, a figure adopted officially by the Indian government. But many others note that with an official per capita GNP of 5651 rupees, or between $160–180 dollars, depending on the currency exchange that day, 80 percent of all Indians are impoverished, even by developing country standards. See *Financial Express*, Bombay, February 28, 1993, p. 6, for the "Text Of Finance Minister's Speech."

28. There are estimated to be some 200 million middle and upper class Indians, leaving nearly 700 million poor or destitute.

29. See Rahul Pathak, "Did You Know", *India Today*, April 15, 1993, p. 54. The government is trying to upgrade the life of these half-a-billion or so individuals. The Rural Development Bill of 1991, or, Panchayati Raj, was meant to ensure regular elections to the local panchayats, or community governing boards. Each panchayat was to provide reservations for scheduled tribes in proportion to their population and one third of all seats were to be reserved for women. As of 1993, however, only two states had actually implemented the bill's provisions.

30. See Charlie Pye-Smith, *In Search of Wild India*, New Delhi: UBSPD, 1993, p. 24.

31. The young Alexander was reported by his biographer to have met several naked yogic masters in the city of Taxila and been convinced that nonviolence and meditation were more powerful than conquest.

32. See Stanley Wolpert, *A New History of India*, 2d ed., New York: Oxford University Press, 1982, p. 59.

33. Ibid., pp. 142–146.

34. Pye-Smith, op. cit., p. 42.

35. See Museum painting number 47-110/1919.

36. For example, it was written in the *Ishopanishad*, "This universe is the creation of the Supreme Power meant for the benefit of all his creation. Each individual life form must, therefore, learn to enjoy its benefits by forming a part of the system in close relation with other species. Let not any one species encroach upon the other's rights." In "Environment and Development—India's Approach," New Delhi: Government of India, 1992, pp. 7–8.

37. Yet to this day, after centuries of soil abuse, little more than 5 percent of eroded land in Kerala has ever been treated with conservation measures, and this paucity of restoration is characteristic of the whole country. See the *Kerala Report*, New Delhi: Government of India, 1985. Three breadbasket states have seen some reclamation efforts, namely, Haryana, Punjab, and Uttar Pradesh. There, tubewells, surface water detention, link drainages, the application of soil amendments like gypsum and pyrites, and the addition of green manures has been carried on for about ten years.

38. Pye-Smith, op. cit., p. 49.

39. An animal protection act was instituted in India in 1912, just as the elephant was offered protection by the *Madras Act* of 1873. The rhino was similarly singled out by the *Bengal Rhinoceros Preservation Act* of 1931. Some 100,000 square miles of Indian lands had been granted "sanctuary" status by 1928, and a *United Provinces National Parks Act* passed in 1934, the first official park coming into existence two years later. Pye-Smith, p. 62.

40. Ibid., p. 53.

41. Ibid., p. 64.

42. By one estimate, there are 45,000 plants (7000 of them endemic), and 75,000 animal species, including 2000 species of birds, 850 species of mammals, and 450 species of reptiles in India. Of these, at least 1500 plants and 138 animals are endangered. B.K. Tikader, *Threatened Animals of India*, Calcutta: Zoological Survey of India, 1983. In addition, there are 250 cultivated species, 60 of which are endemic. See S.K. Saksena, *Environmental Planning Policies and Programmes In India*, New Delhi: Shipra Publications, 1993, p. 65. Other government figures show very different data: 340 mammals, 1200 birds—300 of which are overwinterers, 2000 fishes, 140 amphibians, 420 reptiles, 5000 species of algae, 1600 lichens, 20,000 fungi, 2700 bryophytes, 600 pteridophytes, 50,000 insects, 4000 molluscs, and an unknown number of other invertebrates. Until recently, there were estimated to be as many as 50,000 varieties of rice in India but that number is down to

below 300 now. Twenty years ago, if not for the discovery of one existing wild rice strain in India, namely, *Oryza nivara*, the grassy stunt virus would have devastated rice fields from India to Indonesia. It took the testing of 6273 different strains before the one resistant strain was discovered. Today, with biodepletion having gone on at such a high pace, few options exist for testing among multiple strains. See Wilson, op. cit., p. 301. And it is difficult to imagine that a similarly catastrophic attrition of species is not occurring among other plants and animals, far in excess of that conservative number offered by Tikader and the government. Yet, it is not at all surprising that there is no certainty with respect to the flora and fauna inventory in India. They say it will take at least another century in Australia, for example, to determine even a rough count of the biodiversity.

43. Pye-Smith, op. cit., p. 86.

44. According to one report, rice yields have dropped by half, and wheat by as much as 78 percent in many large regions during just the past few years. See P.K. Joshi and Dayanatha Jha, "Environmental Externalities in Surface Irrigation Systems in India," in *Environmental Aspects of Agricultural Development*, International Food Policy Research Institute, Washington, D.C., 1990, pp. 3–4.

45. Eighty-five million hectares (ha), out of an agricultural base of 143 million ha suffer from severe soil degradation across India. Another 40 million ha are completely unproductive.

46. See "Environment and Development—Traditions, Concerns and Efforts In India," *National Report to UNCED*, New Delhi: Ministry of Environment and Forests, Government of India, June 1992, p. 20.

47. See Raj Chengappa, "Family Planning—The Great Hoax," *India Today*, October 31, 1988, p. 39.

48. See V.A. Pai Panandiker, "Dynamics of Population Growth: Implications for Environment and Quality of Life," New Delhi: Centre for Policy Research, February 1992, p. 6.

49. See *The Life of the People of India*, ASI, Government of India, 1993. The document is 46,000 pages in length and represents the coordinated efforts of 3000 researchers studying 776 traits in each of the 4635 larger communities of the country.

50. Mountain ecologists recommend a maximum of 11 percent sown area where forest is encroached upon by agriculture along steep slopes. See *Mountain People*, ed. Michael Tobias, Norman: University of Oklahoma Press, 1986.

51. See K.B. Pathak and F. Ram who ask, "Under the given environment, are we expecting too much from 'BIMARU' [the heavily populated northern] states to achieve during a comparatively shorter period?" in "Pattern of Population Growth and Redistribution in India," *Demography India*, volume 20, number 1, 1991, pp. 7–14.

52. See H. Govind, "Recent Developments in Environmental Protection in India: Pollution Control," *Ambio*, volume 18, number 8, 1989, p. 429.

53. See Soutik Biswas, "Troubled Waters," *India Today*, February 28, 1993, p. 133.

54. See P.K. Surendran, "Delhi most polluted city in India," *Times of India*, February 6, 1993.

55. As a whole, Indian power plants produce some 30 million tons of coal ash per year.

56. See Mayank Chhaya and Nupur Basu, "Key Facts on the Issue in India—Child Labor: A Stubborn Social Problem," *India Abroad*, February 19, 1993, p. 19. See also Olga Nieuwenhuys, *Childrens' Lifeworlds—gender, welfare and labour in the developing world*, London: Routledge, 1994.

57. This is, of course, a crude indicator because it implies among other things that a man six feet tall might somehow survive in a compound of thirty-six square feet, a cell smaller than what the Department of Defense recommends for minimum bomb shelter space per person. This is certainly not a desirable long-term condition.

58. Wilson, op. cit., pp. 207, 221. See also E.O. Wilson, "Is Humanity Suicidal," *New York Times Sunday Magazine*, May 30, 1993, p. 29. It should be pointed out, however, that this theorem appears to be a very generous one. In at least some parts of India where there have been catastrophic biodiversity losses, it appears possible that less than 90 percent intrusion has nevertheless impacted 50 percent or more of the resident species.

59. Maneka Gandhi was the highest ranking woman in the Janata Dal, or People's Front party, whose leader was Vishwanath Pratap Singh, Rajiv Gandhi's former Finance Minister who became Prime Minister, briefly, in 1989. Mrs. Gandhi was subsequently chosen to be Minister of Environment and Forests.

60. The *Forest (Conservation) Act* amendments of 1988 impose up to fifteen days imprisonment for contravention of the provisions of the 1980 Act, which defines "non-forest purposes" as clearing of trees in order to cultivate tea, coffee, spices, rubber, palms, oil-bearing plants, horticul-

tural crops or medicinal plants, all illegal now, unless otherwise directed, under the authority of the regional chief conservator of forests or the central government.

61. Conversation with Kamal Nath at the Ministry of Environment and Forests, February 22, 1993, New Delhi.

62. See Bhattacharyya, "Economic Development with Reference to Rural Economy and the Eco-Systems," in *Environmental Issues in Agricultural Development*, ed. H. Ramachandran, New Delhi: Concept Publishing Company, 1990, p. 49. Adding to that dialectic is the emerging realization that the global commons can no longer afford any more industrialization along the lines of the past. Speaking of India and China, Paul Kennedy writes, "The heart of the problem, is whether the potential for enhanced per capita standards of living will not be overwhelmed by the millions of newborn children each year. The second, related issue involves a cruel conundrum: is it actually wise for countries possessing half a billion to a billion peasants to attempt nowadays to follow 'the stages of industrial growth' first established in the medium-sized nations of Western Europe 150 years ago, or to try to mimic the high-tech revolution emerging from the very different socioeconomic structures of California and Japan? And does that question imply that China and India should not attempt to catch up, a notion likely to be repudiated by politicians and planners for condemning their countries to remain forever behind the West?" See Paul Kennedy, *Preparing for the Twenty-First Century*, New York: Random House, 1993, p. 182.

63. In 1990, that small sector of the Indian economy relying on petroleum had to cut its oil consumption by 25 percent due to global price increases, despite the country's 22 percent energy shortfall. Increasingly dependent on its primary forests for sustenance, India nevertheless has no fuelwood policy; no national fodder policy. See Ravi Sankaran, "Dudhwaa—forest held to ransom," *Sanctuary*, volume X, number 5, Sep/Oct 1990, p. 81.

64. See Mr. Kamal Nath, Minister of Environment and Forests, India, Address to UNCED, June 1992, Government of India, New Delhi, p. 7.

65. Pye-Smith, op. cit., p. 13.

66. Of profound interest, it has been estimated that three-fourths of all forest exploitation by women is dead wood, not live. This can only mean that it is the male opportunism, whether rich or poor, responsible for most of India's forest losses, though in Africa, the massive deprivation of the forest floor of dead wood has posed serious consequences. See Bina Agarwal, "Engendering the Environmental Debate:

Lessons from the Indian Subcontinent", Center for Advanced Study of International Development, Michigan State University, *Distinguished Speaker Series, Paper #8*, Jan. 1991, pp. 22–23; and Anil Agarwal and Swunita Narain, "Strategies for the involvement of the landless and women in afforestation: Five case studies from India," Geneva: ILO, 1990. Nearly a dozen major chemical and timber companies throughout India have been exploring nonwood fiber alternatives to this forest crisis—growing plywoods, block boards, laminates, concrete shuttering and marine plywoods, as well as MDF—medium density fibreboard, first introduced into the U.S. in the 1960s. The largest plywood factory in the country, Sarda Plywood Industries Ltd, is on a plantation in Assam. Such fast-growing enterprises are also being applied in the southern United States. The issues for India, as for Indonesia, are twofold: first, will the impoverished classes have economic access to such alternative woods in time to save the existing forests; and second, will the political leadership ensure that forest plantations are developed on existing wastelands, rather than in primary forest areas?

67. See Prof. H. Simon and Shri B.B.L. Sharma, *Training Modules For Incorporation Of Family Welfare Messages*, New Delhi: National Institute of Health and Family Welfare, 1990, p. 30.

68. In Brazil, which has thus far incited the most fervent global concerns, approximately 12 percent of the country's total forest area has been lost, to date. However, in some princely regions of Brazilian biodiversity, like Rondonia and Mato Grosso, double that amount had already been destroyed as of 1988. Even where selective cutting has been undertaken, it is known that for each tree logged, "27 other trees that are 10 centimeters or more in diameter are severely injured, 40 meters of road are created and 600 square meters of canopy are opened." See Marguerite Holloway, "Sustaining The Amazon," *Scientific American*, July 1993, p. 95. In the most recent NASA satellite survey of the Amazon, the loss of biological diversity was ascertained to be even worse than previously understood. In addition to actual forest depletion each year, edge effects, such as described by Holloway, were seen to be on the increase. So far, a total of 92,000 square miles have been totally devastated. But according to estimates by David Skole and Compton Tucker (of the University of New Hampshire and NASA Goddard Space Center in Maryland) the total edge effects impact regions encompass 235,200 square miles, or more than 14 percent of the total Amazon.

69. See Paul R. and Anne H. Ehrlich, *The Population Explosion*, New York: Touchstone Books, Simon & Schuster, 1990, pp. 70

70. See Asad R. Rahmani, "Save The Grasslands," *Sanctuary Magazine*, volume X, number 4, 1990, July/August, p. 24.

71. Ibid., p. 21.

72. It is estimated, for example, that each cow consumes on average 27 kilograms of fodder per day, which it scavenges from whatever is available. See S.S. Khanka, "Developments In Ghimtal Gadhera Catchment Area," in *Environmental Issues in Agricultural Development*, ed. H. Ramachandran, New Delhi: Concept Publishing Company, 1990, p. 150. In Karnataka State, the livestock population has totally outstripped available fodder, resulting in a 71 percent deficit. This syndrome has effected the starvation and malnutrition of cattle, and the subsequent destruction en masse of forests.

73. The native Indian blackbuck—a gloriously beautiful antelope—has been reduced from an estimated 4 million to something like 40,000. See Pye-Smith, op. cit., p. 17. It is ironic that humanity has dragged other species into its own population dilemma and made them accomplice/victims.

74. Ibid., p. 30.

75. Ibid., pp. 8–9.

76. Ibid., p. 155. In addition to the ecological mayhem suffered upon resident animal species, there is a second, less discussed side-effect of these many dam projects, and the massive clearance of grassland and forest for human development. A number of virulent ecological diseases have spread throughout India. In the Shimoga District of Mysore, a tick-born Japanese encephalitis broke out in the Kyasanur forest in 1956 when locals tried to clear the stands for rice paddies. More recently, other ecologically related diseases include Dracontiasis (a guinea worm infection), Leishmaniasis (transmitted by sandflies), lymphatic filariasis (Mansonia mosquitoes), and a vast number of malaria cases. With a 27-fold increase in high yielding varieties in India over the past 25 years, heavy agrochemical input has meant that the disease vector mosquitoes have developed multiple resistance to the very pesticides used by public health authorities. India's National Malaria Eradication Program has become the largest health program in the world aimed at a single communicable disease. See E. Narayanan, "Dams And Public Health," *Sanctuary Magazine*, volume X, number 1, 1990, Bombay, pp. 67–73.

77. Outside of India, scientists allege that there are 40 tigers left in China, 350 to 400 in Siberia, 400 in Sumatra, 1500 in Thailand and Malaysia.

78. In 1984, the Wildlife Institute of India was commissioned by the government to assess the network of protected areas in the country and recommend any necessary changes. The two-volume report issued in 1988, "Planning a Wildlife Protected Area Network in India," suggested increasing the number of national parks to 148, and the number of wildlife sanctuaries to 503, far above the existing number.

79. See Ashwani Sharma, "Little protection for HP's protected species," *Times of India*, February 25, 1993.

80. See Aditi Kapoor, "Tiger poaching unabated," *Times of India*, February 20, 1993, p. 3.

81. As Madhav Gadgil of the Indian Institute of Science in Bangalore puts it, deforestation comes about because of "the use of state power to systematically undervalue biomass . . . and organize its supply to those in power at highly subsidized values." Pye-Smith, op. cit., p. 172. Traditional markets that are sustainable and value-added have been usurped by mass producers solely concerned with export of raw materials, which governments have increasingly subsidized, thus devaluing the natural resource and the labor force for purposes of quick profit. The syndrome is the same everywhere.

82. Ibid., p. 108.

83. Ushi Rai, "A quarter of Melghat tiger reserve to be denotified," *Indian Express*, Feb. 25, 1993, p. 9.

84. See Ramesh Bedi and Rajesh Bedi, *Indian Wildlife*, New Delhi: Brijbasi Printers Private Ltd., 1984, pp. 78–79.

85. Ibid., pp. 147–157.

86. The cattle produce dung which is sold to the tea and coffee planters in the Nilgiri hills as fertilizer. The local people do not have the fodder reserves that would be necessary to enable them to provide in-place stall-feeding systems for the cattle. The villagers have no other choice but to let their livestock graze on the forests, denude the commons, and create erosion trenches; and since they are not directly dependent upon the forests for revenues, they can afford, for the time being, to turn a blind eye to the damage. The elephants, however, are in trouble. The only hope, it would seem, is a strictly controlled rotational grazing scheme for the cattle, a concerted repasturization of wastelands and, eventually, a government subsidized stall-feeding regime. But such "solutions" for the time being are distant and unforthcoming. See H.S. Panwar, et al., op. cit., p. 81.

87. Bedi and Bedi, op. cit., p. 24.

88. Ibid., p. 196.

89. Ibid., p. 270.

90. Similarly, when Sharon Camp of the Population Crisis Committee in Washington (now called Population Action International) was asked recently whether she believed there was much hope for wildlife world-wide, her answer was an emphatic, "None." Asilomar Conference on Population and Environment, World Affairs Council of Northern California, Pacific Grove, California, April 30–May 2, 1993.

91. To voice concerns about such abuses, write to the Municipal Commissioner, Carmichael Road, Bombay 400026.

92. Yet the *Qur'an* clearly states, "The flesh and blood of sacrificed animals shall never reach God but your devotion and piety will reach him." 22:37.

93. Saksena, op. cit., p. 41.

94. Rajiv Gandhi set up the National Wastelands Development Board, an eerie harbinger of restoration needs for the twenty-first century, with an aim toward implementing social forestry and wastelands development on protected plantations. Success with social forestry at Arabari in West Bengal, and the Van Panchayat democratically elected village management system throughout Uttar Pradesh, led to a government circular under Maneka Gandhi's term in office asking for a countrywide Joint Forest Management model that would involve locals working with forestry officials. The Uttar Pradesh system requires only that one-third of the villagers agree to form a forest management group. Once accomplished, they automatically become the arbiters of all fodder, timber and fuelwood collection. While it sounds plausible, so far, success has been marginal.

95. According to the Biogeographic Project, however, part of the National Wildlife Action Plan of the Wildlife Institute of India, 1000 square kilometers and a population of 500 to 1000 is considered sufficient to ensure the genetic viability of a community. Many biologists would argue such low figures.

96. See Ravi Chellam, "Wanted—Another home for the Asiatic lion," *Sanctuary*, volume XII, number 5, p. 36

97. *National Report to UNCED*, op. cit., p. 13.

98. Simon and Sharma, op. cit., p. 34.

99. Ibid., p. 88

100. Ibid., p. 39.

101. See Elizabeth Bumiller, *May You Be the Mother of a Hundred Sons—A Journey Among the Women of India*, New Delhi: Penguin Books, India, 1991, p. 42.

102. See F. Max Muller, ed., "The Laws of Manu," *The Sacred Books of the East*, volume 25, New Delhi: Motilal Banarsidass, 1967, 9:2-3.

103. Today, some 4.5 million marriages occur every year in India, and most brides are between 15 and 19 years old. See Dr. Mrs. M.R. Chandrakapure, "New Interventions for a Small Family Norm" *A Seminar Organised by Indian Merchants' Chamber*, Bombay, July 10, 1990. See also Raj Chengappa, "Family Planning—The Great Hoax," *India Today*, October 31, 1988, p. 44.

104. See Catherine Thompson, "Women, Fertility and the Worship of Gods in a Hindu Village," in *Women's Religious Experience*, ed. Pat Holden, London and Canberra: Croom Helm Ltd., 1983, pp. 113-131.

105. Bumiller, op. cit., p. 273.

106. See K.P. Bahadur, *Population Crisis In India*, New Delhi: National Publishing House, 1977, p. 31.

107. See Theodora Foster Carroll, *Women, Religion, and Development in the Third World*, New York: Praeger Press, 1983, p. 57.

108. Private conversation at Dr. Bose's home, New Delhi, February 1993.

109. The six effective pillars of Islam are: *Shahadah*, meaning there is no other God but Allah, and Mohammed is his prophet; *Salah*, the requirement to pray five times a day; *zakah*, the giving of alms; *sawm*, fasting during the month of Ramadan; *Hajj*, pilgrimage to Mecca; and *chador*, the cloak and veil of women.

110. Muller, ed., op. cit.

111. The British legalized remarriage in 1856, though there is still great reticence among Hindu widows to ever remarry. See L.S.S. O'Malley, *Popular Hinduism—The Religion of the Masses*, Cambridge: Cambridge University Press, 1935, p. 113. For the Muslim widow, life can be nightmarish. In India, an indigent divorcee, under the 1986 *Muslim Women Act*, means that the woman is essentially made to carry a begging bowl. Her husband can divorce her instantly, without so much as proffering a reason. Furthermore, the man is free from the liability of maintaining her after three months and ten days. The divorcee must seek help first from relatives, and failing that, from the State Wakf Boards, which tend to be bankrupt.

112. See "Equal Rights Must Start in the Cradle," UNICEF News—Third World Women, Number 76(3), July 1973, p. 17. The government's Department of Women and Child Development has summarized this bias by stating, "In a culture that idolizes sons and dreads the birth of a daughter, to be born female comes perilously close to being born less than human."

113. "Year of the Woman," *Sixty Minutes*, CBS, January 24, 1993.

114. See Bumiller, op. cit., pp. 39–40.

115. See Edward A. Gargan, "In 'Bollywood,' Women Are Wronged or Revered," *New York Times*, January 17, 1993, pp. 11–12.

116. Protection was never assured, however. In 1381, ten thousand women apparently committed suicide at Ranthambhore rather than be taken alive by invaders. See Pye-Smith, p. 61.

117. Today, polygyny is permitted among Muslims in India. Yet Hindu males frequently help themselves to the law. In fact, the proportion of Muslims and Hindus who take more than one wife is about the same. There are no statistics on how many Hindu first wives risk their security by complaining about bigamist husbands. Few appear to. In an important Supreme Court ruling [Priya vs. Suresh, 1971, S.C.1153] the wife had to prove that her husband had in fact married a second woman according to the traditional Hindu rites and ceremonies, including the sacred fire, etc., before she could claim any damages. See Alaka Malwade Basu, "Four wives: A demographic canard," *Economic Times*, New Delhi, March 29, 1993.

118. In several rural areas of India, maternal mortality is as high as 2000 per 100,000, versus 6 in Europe, and an official Indian count of 400 to 500 per 100,000. In times past, the maternal death rate must have been even much higher. See Smt. Avabai Wadia, "Population and Family Planning in the 8th Five Year Plan," Part II, *Family Planning Association of India Publications*, Bombay, 1992, p. 6. See also Katherine Mayo, *Mother India*, New York: Harcourt-Brace, 1927. Young girls who choose an abortion are confronted by an equally dire scenario. According to the National Institute of Family Welfare, 15 percent of all those women undergoing medical termination of pregnancy die in India. In the United States, for comparison, 3 women per hundred thousand are known to die from abortion-related complications, or 1 American death for every 500 Indian deaths.

119. Discussed at length in Bumiller. See also F. Max Muller, ed., 5:164.

120. Most noted among them: Parvati, Siva's consort, daughter of the Himalayas; Syama-Kali, dispenser of boons, dispeller of fear; Raksa-Kali, protectress, Smasana-Kali, creator and destroyer; Laksmi, guarantor of continued rule; Sakti-Maya-Devi, Durga, Uma, Sati, Padma, Candi, and countless others. See Heinrich Zimmer, *Philosophies of India*, ed. J. Campbell, Bollingen Series XXVI, Princeton: Princeton University Press, 1969.

121. See Kavitha Shetty, "Cradles of Mercy," *India Today*, February 28, 1993.

122. Discussed in Bumiller, p. 273.

123. Chengappa, op. cit., p. 42.

124. Until 1993, India had never participated in any world surveys on mortality or fertility, nor has the country ever been willing to sign any international right to foreign access to data.

125. "Population, Development and the Environment: An Agenda for the 1990s," *Proceedings of the National Conference of Non-Governmental Organizations,*" April 14–16, 1991, New Delhi Family Planning Association of India, Bombay, p. 19.

126. The World Bank has given $100 million to the GOI (Government of India) for social marketing of low-cost weaning foods, contraceptives, oral rehydration salts, and safe delivery kits. The program was started in 1991.

127. In Bombay, the director of a university devoted to population studies told me, "You cannot predict poor people's behavior. You can only predict rich people's behavior. In China, the poor people got rid of opium overnight, in one month. That was a miracle. Once people understand their rights in a place like Bihar, another miracle will occur. In 1965 during the Indo-Pakistani War, on one day a week Prime Minister Shastri declared, don't eat cereal, and indeed, the people rallied and abstained as requested. A miracle will happen in India."

128. See Rasheeda Bhagat, "Where condoms are buried in the ground," *Indian Express*, Madras, January 8, 1993.

129. Any contraceptive having to do with males has not done well in India since the mid-1970s. Family planning, in the back of the male's mind, is related to castration. One is thus talking about sexual potency. That has killed the male's participation in family planning. His ego is threatened, say Indian fertility counselers.

130. See Abid Aslam, "India to improve sterilization programme," *Population*, published by UNFPA, volume 18, number 1, January 1992, p. 2.

131. See Shanti R. Conly and Sharon L. Camp, "India's Family Planning Challenge: From Rhetoric to Action," *Country Study Series #2*, the Population Crisis Committee, Washington, D.C., 1992, p. 13.

132. Wolpert, op. cit., p. 267. By that time, India had slid economically to one one-hundreth the level of affluence and industrialization of England. See also Kennedy, p. 11. Many Indophiles have effectively reasoned that India's inability to keep pace with the Western industrial revolution was strictly the result of subjugative British rule, the consequences of which have been an ever-downward spiraling paralysis, a syndrome of poverty from which the country appears in-

capable of lifting itself. Others have argued that India's poverty was there long before the British arrived and if anything the East India Company mobilized the Indian labor force and created the infrastructure by which the country today is making, by some estimates, enormous strides. Put differently, England trained India and the proof can be gleaned from India's escalating GNP. Both perceptions are partly correct. The third missing element, however, is the continuing population boom which for most of India's recent history has skewed all comparisons.

133. See Simon and Sharma, op. cit.

134. See Dom Moraes, *A Matter of People*, New York: Praeger Publishers, 1974.

135. Today, the country's crude birthrate averages about 30 to 33 per thousand. But with its deathrate (of those over the age of five) of 11 per thousand, that makes for a net gain of over 20 per thousand per year—only moderately reduced since 1975. To achieve an NRR-1, or fertility replacement level, the precondition for eventual zero population growth, the country would need to achieve a birthrate of no more than 21, and a death rate of 7, a net gain of no more than 14 per thousand. By comparison, the U.S. birthrate all time high was in 1947 at 26.5 per thousand, but promptly shot back down to around 15 per thousand. See "New Interventions for a Small Family Norm," *A Seminar Organized by Indian Merchants' Chamber*, Bombay, July 10, 1990; See also Paul R. Ehrlich, et al., *Ecoscience—Population, Resources, Environment*, San Francisco: W. H. Freeman, 1977, pp. 100–101.

136. See S. Krishnakumar, "Kerala's Pioneering Experiment in Massive Vasectomy Camps," *Studies in Family Planning*, volume 3, number 8, August 1972, pp. 177–192.

137. See V.H. Thakor and Vinod M. Patel, "The Gujarat State Massive Vasectomy Campaign," *Studies in Family Planning*, volume 3, number 8, August 1972, pp. 186–192.

138. See "Shah Commission Third Report," 1978:153, quoted in V.A. Pai Panandiker and P.K. Umashankar, *Fertility Control-Induced Politics Of India*, New Delhi: Centre for Policy Research and Family Welfare Program, 1991, p. 2.

139. Arrests were made under the *Maintenance of Internal Security Act*, the *Defence of India Rules* and the *Indian Penal Code and Criminal Procedure Code*. See Panandiker and Umashankar, pp. 3–4.

140. See Ashish Bose, "India's Quest for Population Stabilisation: Progress, Pitfalls and Policy Options," *Materials on Demogaphic Questions*,

Herausgeber Wiesbaden: Bundesinstitut fur Bevolkerungsforschung, 1991, p. 23.

141. Later on, when I examine global percentile differentials, even a few parts of one percent will be seen to make huge differences over time in population size.

142. See *This Is Earth Calling—A Children's-Eye View of the Earth's Population and Environment Problems*, New Delhi: UNFPA, in association with Konark Publishers PVT LTD., 1993.

143. In Tamil Nadu, if a couple has two girls and then volunteers for one of the spouses to be sterilized, the government will give the two existing daughters each a bond that they will be able to cash when they are twenty. The bonds are between one and two thousand rupees. In truth, however, the money goes to the parents so that they will have sufficient funds, short of bankruptcy, to pay their daughters' onerous dowries. By holding onto the bonds until the girls are twenty, the government hopes to discourage teenage marriages.

144. Chengappa, op. cit., p. 47.

145. Actually, the Tamil Nadu government had offered incentives for birth control as early as 1956. In the early 1980s in Maharashtra State, local officials who succeeded in reaching certain quotas of prescribed contraceptive prevalance within their community were sent on an all-expenses paid group tour, or honored with a turban, a throwback to the days when kings used to pay tribute to their most loyal subjects. Money to recipients of sterilization was also given out, along with green cards carrying numerous benefits—preferential treatment in health and education, or better chances of getting a bank loan.

146. Chengappa, op. cit., p. 40.

147. Conly and Camp, op. cit., p. 10.

148. Ibid., p. 38.

149. Those supplies include: washed and sun-dried cloth or rags, a washed and sun-dried mat, bed sheet and blanket, a match-box and stove, a bucket and mug, a lantern, a new razor blade, clean cotton, linen, scissors, cord ligatures, an enema can, a mucus sucker, two bowls, a torch, a weighing scale, antiseptics, mercurochrome, and methylated spirit. In addition, the dais are advised to boil all water, to wash everything with soap, and to boil all implements in the bucket for twenty minutes. The dai is warned against cutting the umbilical cord on the floor, or applying any ointment, ash, or cowdung on the mother or child. Only mercurochrome or gentian violet is to be applied onto the stump. The baby should be placed at once against the mother's breast

to benefit from the colostrum, the rich creamy milk of a mother which flows for the first few days. Frequently, Indian women are led to believe that colostrum is harmful; that water sweetened with sugar or jaggery is better for the child. Consequently, many babies die when the fluid given to them is aspirated into their air passage. Often the liquid is prepared unsafely, leading to infection. The government suggests that the dai advocate breast feeding 5 to 6 times a day or more. Because breast feeding inhibits ovulation, while providing the best nourishment for the newborn, it is considered the miracle form of contraception in India, as in most other countries. The dais are also encouraged to teach mothers about oral rehydration solutions to counteract widespread diarrhea, a syndrome that kills some 3 million children every year in the country. The recommended solution can be water with dal, tea, lemon, sugar, salt, coconut, rice or jaggery. One glass of water, one [three-finger] pinch of salt, and one [four-finger] scoop of sugar. Mothers are to be cautioned that the solution should taste no saltier than tears. Fresh solutions should never be boiled, and should be made once or twice a day, and one or two mouthfuls given to the child every few minutes. Breast feeding should be continued during this process. See Simon and Sharma.

150. For developing country comparisons and needs, see "Traditional Birth Attendants," a joint WHO/UNFPA/UNICEF statement, Geneva, 1992.

151. According to one respected population scientist in Bombay, "Once you've lived in a hut you wouldn't want to live anywhere else." He was referring to the "sense of community" to be enjoyed in the slums. Of special note, Dharavi's poor have transformed the adjoining Mahim Creek area into a nature park. A few years ago it was a sprawling dumpsite, seeping with toxic wastes and earmarked for industrial development. "Once the slum-dwellers realized that the park would improve their environment, they were very cooperative," says Shanta Chatterji, the Mahim Nature Park Project Coordinator. See Pye-Smith, p. 182. See also Family Planning Association of India, Bombay: Annual Report of the Family Planning Association of India, 1991.

152. Once again, there is some conflicting data unique to India's situation. According to the findings of the recent Anthropological Survey of India, "there is absolutely no correlation between the status of a woman and literacy or earning." Pathak, p. 51. Mrs. Vohra herself points out that unlike other developing countries which perhaps have had the advantage of aspiring to the levels of other developed cultures through the media, India has been culturally stuck, on its own, and not entirely

open. It doesn't take easiliy to other's customs. And that, apparently, includes literacy.

153. Conly and Camp, op. cit., p. 41.

154. Bumiller, op. cit., p. 262.

155. See V.A. Pai Panandiker and Ajay K. Mehra, with P.N. Chaudhuri, *People's Participation in Family Planning*, New Delhi: Center for Policy Research, Uppal Publishing House, 1987, pp. 128–157.

156. One such boomtown is known as Masinagudi, on the southern edge of the Nilgiri Biosphere Reserve, where tigers, elephants, and the Indian bison are endangered by an increasing human and livestock population, as well as planned hydroelectric projects to service the whole Mudamalai plains region. Some farmers, eager to do away with the two thousand or so elephants in the region, have even resorted to planting explosives in the elephant's favorite food, jackfruit. See Pye-Smith, p. 96. Elsewhere, south of Madras along Tamil Nadu's coast, development is increasingly destroying the available nesting sites for the endangered Ridley's turtles, who are apparently confused by all the lights at night. Locals, meanwhile, grab the eggs and sell them for a pittance. Here is another region where family welfare education must be coupled with environmental education.

157. See T.V. Anthony, "The Family Planning Programme Lessons from Tamil Nadu's Experience," a policy paper, New Delhi: Center for Policy Research, March 1992.

158. See A. Steedhara Menon, *Political History Of Modern Kerala*, Madras: S. Viswanathan, 1987.

159. Calcutta, similarly, has the highest urban population, and highest levels of literacy of any city in India, other than Tiruvananthapuram (formerly Trivandrum), in Kerala. Calcutta, too, witnesses extremely infrequent outbreaks of violence, or even petty crimes.

160. Kurien and Achari, op. cit., pp. 6, 12.

161. Ibid., p. 37.

162. See B.A. Prakash, "Educated Unemployment in Kerala: Some Observations Based on a Field Study," *Working Paper #224*, Tiruvananthapuram, Kerala: Center for Development Studies, 1988. See also B.A. Prakash, "Unemployment in Kerala—An Analysis of Economic Causes," *Working Paper #231*, Tiruvananthapuram, Kerala: Center for Development Studies, 1989.

163. P. Sivanandan, D. Narayana, and K. Narayanan Nair, "Land Hunger and Deforestation: A Case Study of the Cardamom Hills in Kerala,"

*Working Paper #212*, Tiruvananthapuram, Kerala: Center for Policy Studies, 1985 p. 15.

164. Pye-Smith, op. cit., pp. 96–97.

165. See E. Narayanan, "The backwaters of Kerala," *Sanctuary*, volume X. number 5, Sep/Oct 1990, p. 69.

166. Ibid., p. 71. See also Sivanandan, et al., p. 2.

167. See Rajaram Das Gupta, "Conditions of Cropping Intensity," *Working Paper #185*, Tiruvananthapuram, Kerala: Center for Development Studies, December 1983, p. 1. See also K.P. Kannan and K. Pushpangadan, "Agricultural Stagnation and Economic Growth in Kerala: An Exploratory Analysis," *Working Paper #227*, Tiruvananthapuram, Kerala: Center for Development Studies, 1988; and K.P. Kannan and K. Pushpangadan, "Dissecting Agricultural Stagnation in Kerala: An Analysis Across Crops, Seasons and Regions," *Working Paper #238*, Tiruvananthapuram, Kerala: Center for Development Studies, 1990.

168. See Monika Nikore and Marian Leahy, "The Oasis Makers," *World Monitor*, February 1993, pp. 46–52.

169. See Inder Sawhney, "Birth Control Drive Slackening," *Times of India News Service*, February 20, 1993.

170. See K. Srinivasan, "India's Population Problems: Can We Afford to Shove Them Under the Carpet?" *International Institute of Population Sciences Compiled Seminar papers*, 1990, p. 19.

171. See V.A. Panandiker, "Dynamics of Population Growth: Implications for Environment and Quality of Life," New Delhi: Center for Policy Research, February, 1992. Presented at the *International Conference on Population and Environment*, University of Michigan, Ann Arbor, October 1990.

172. Ashish Bose, op. cit., pp. 17, 20.

173. See "Quality of Life and Problems of Governance in India," New Delhi: Centre for Policy Research, 1990.

174. See *World Resources 1992–1993*, a report by the World Resources Institute in collaboration with the United Nations Environment Program and the United Nations Development Program, New York: Oxford University Press, 1992, pp. 237, 239.

175. Ibid., pp. 240, 241.

CHAPTER 4
Nature Held Hostage: Indonesia

1. *Supplement to President's Report to Parliament* on August 16, 1991.

2. See Widjojo Nitisastro, *Population Trends In Indonesia*, Ithaca, NY: Cornell

University Press, 1970. See also "Recent Fertility Trends in Indonesia, 1971-1987," *Working Paper #63*, East-West Population Institute, Honolulu: East-West Center, November, 1991.

3. See Leslie B. Curtin, et al., "Indonesia's National Family Planning Program: Ingredients of Success," *Occasional Paper #6*, Report #91-134-136, Population Technical Assistance Project, USAID-Jakarta and Office of Population Bureau for Research and Development, Washington, DC: Agency for International Development, December 11, 1992, p. 3.

4. The *Straits Times*, Singapore, February 2, 1993, p. 10.

5. "Biodiversity Action Plan for Indonesia," volume 1, Final Draft, August 1991, BAPPENAS—National Development Planning Bureau, Jakarta, p. 18.

6. In East Timor, among the Tetum people, "sexual union and cosmic union . . . are analogous." The primeval vagina from which the ancestors first appeared in the creation, climbing out of the womb onto surrounding vines, is today considered to be inviolate. One must never touch the symbolic vaginas, sculpted in stone, covered in vines. See David Hicks, *A Maternal Religion: The Role of Women in Tetum Myth and Ritual*, Center for Southeast Asian Studies, *Special Report #322*, Northern Illinois University, 1984, p. 3.

7. Curtin, et al., op. cit., p. vii. See also Graeme J. Hugo, et al., *The Demographic Dimension in Indonesian Development*, New York: Oxford University Press, 1987.

8. Though according to research by the writer V.S. Naipaul, only a third of the Muslims in Indonesia actually live as Muslims. See V.S. Naipaul, *Among the Believers—An Islamic Journey*, New York: Vintage Books, 1982. Others have argued that, if anything, there is a strong Islamic revival taking place in Indonesia.

9. Until recently, the median age for marriage was only 15.6 for rural women and 17.3 for urban middle-class women; 21.3 and 22.9 respectively for men. The new emerging ideal marriage is of "parentally-approved self-choice" ties at a later age. See Anju Malhotra, "Gender and Changing Generational Relations: Spouse Choice in Indonesia," *Demography*, volume 28, number 4, November, 1991, pp. 549-570. It must be pointed out, however, that as of 1987, 20 percent of all Indonesian women had nevertheless married by the age of 15, and 45 percent by age 18. See James A. Palmore and Masri Singarimbun, "Marriage Patterns and Cumulative Fertility in Indonesia," *Working Paper #64*, East-West Population Institute, Honolulu: East-West Center, November 1991. See also Walter L. Williams, *Javanese Lives—*

Women and Men in Modern Indonesian Society, New Brunswick, NJ: Rutgers University Press, 1991.

10. According to J. Weeks, it is not Islamic religion, per se, but the inferior status of Muslim women, and their poverty, which causes their high fertility rates. But Weeks also points out that the patriarchal nature of Islamic society has limited the choices of women, which in turn dramatically impacts their freedom of contraceptive use. See J. Weeks, "The Demography of Islamic Nations," *Population Bulletin*, volume 43, number 4, December, 1988. Donald P. Warwick writes, "One of the keys to the success of Indonesia's family planning program in rural Java was winning the neutrality and sometimes even the active assistance of local religious leaders. This was done mainly by consulting these leaders about the overall acceptability of family planning in their region and by showing them respect in other ways." See Donald P. Warwick, *Bitter Pills: Population policies and their implementation in eight developing countries*, Cambridge: Cambridge University Press, 1982, p. 183.

11. And thus, the government has worked to enlist the Ministry of Religious Affairs and the leading Islamic organizations into population issues. Local Islamic leaders have accepted family planning, women's *Qur'an* reading groups have been formed, and over 800 Islamic medical facilities have been integrated into the government's own family planning efforts.

12. See Tahir Mahmood, *Family Planning: The Muslim Viewpoint*, New Delhi: Vikas Publishing, n.d., p. 34.

13. B.F. Musallam, *Sex and Society in Islam—Birth control before the nineteenth century*, Cambridge: Cambridge University Press, 1983, p. 38.

14. See *After Patriarchy—Feminist Transformations of the World Religions*, ed. Paula M. Cooey, William R. Eakin, and Jay B. McDaniel, New York: Orbis Books, 1991, p. 238. See also *Women in World Religions*, ed. Arvind Sharma, State University of New York Press, n.d.

15. It should be pointed out, however, that one Companion, Hazrat Ali bin Abi Talib insisted that "unless the seven stages [atwar] were completed by the foetus its abortion would not be [construed as] burying [the fetus] alive." (Brackets my own) He recited verses 12, 13, and 14 of "Sura Mu'minun" of the *Qur'an* to emphasize his point and was himself confirmed by Hazrat Umar, another Companion, who said, "You [Hazrat Ali] stated the right thing. God bless you." See Mahmood, p. 31.

16. For the beginnings of the legal framework guiding the country's birth control efforts, see "Legal Aspects Of Family Planning In Indonesia," The Committee on Legal Aspects of the Indonesian Planned Parent-

hood Association. *Law and Population Monograph Series #4*, Law and Population Programme, The Fletcher School of Law and Diplomacy, 1971.

17. See Sheikh Abdul Rahman al-Khayyir, "The Attitude of Islam towards Abortion and Sterilization," a report of the proceedings of the International Islam Conference held in Morocco, December 1971, published in *Islam and Family Planning* 2: 345–361, Beirut: International Planned Parenthood Federation, 1974, quoted in *After Patriarchy—Feminist Transformations of the World Religions*, ed. Paula M. Cooey, et al., New York: Orbis Books, 1991, pp. 243–244.

18. See "Adding Choice to the Contraceptive Mix: Lessons from Indonesia," *Asia-Pacific Population & Policy*, number 19, Population Institute of the East–West Center, Honolulu, December, 1991.

19. See Siti Pariani, David M. Heer, and Maurice D. Van Arsdol, Jr., "Does Choice Make a Difference to Contraceptive Use? Evidence from East Java," *Studies in Family Planning*, volume 22, number 6, Nov/Dec 1991, pp. 384–390. The authors conclude, "Whether the user was granted her choice of method was found to be a very important determinant of sustained use of contraceptives. The interaction between whether choice was granted and whether there was husband-wife concurrence on method choice was also important."

20. Curtin, et al., op. cit., p. 7.

21. In 1984 dollars, a hybrid coconut was said to be worth about $20 per year in income. Djamaludin Ancok, *Incentive and Disincentive Programs in the Indonesian Family Planning*, Yogyakarta: Population Studies Center, Gadjah Mada University, 1984.

22. The discount program provides benefits at supermarkets, beauty salons, doctor's and midwive's offices, and offers certain products at reduced rates. A chain of 159 movie theatres is linked to the service, as are book clubs, hotels, restaurants, car garages, and private hospitals. For older members to qualify for membership in the club, they must use permanent methods of birth control and have no more than two children. Younger members must have used an IUD for at least three years, or injections, birth control pills or condoms for at least four years. Members with more than two children must have used birth control for the last five years and pledge to have no more children. While the BKKBN insists that Indonesia's family planning program is strictly voluntary, and lacking any disincentives, according to Djamaludin Ancok, as of 1981, married female public servants, and wives of public servants, were "requested" to become acceptors of family planning. Local schools were also asked to encourage parent participation as an actual requirement for their children's admission. In some areas of cen-

tral Java, parents were strongly encouraged to contribute about five dollars to the village administrator if they had a third baby, and more per each additional child. Other disincentives included withholding a married worker's salary until that person joined a family planning program, and also encompassed determinations of eligibility for government services, whether welfare, irrigation, even permission to undertake a small business. See Ancok, p. 9. In 1973 in Java, female targets, or acceptors of fertile age "were commanded to show up by the (military) village headmen to have their IUDs inserted by mobile teams that travelled the countryside." The women were called "dispiralkan" meaning "had a spiral inserted." See Niels Mulder, *Mysticism and Everyday Life in Contemporary Java—Cultural Persistence and Change*, Singapore University Press, 1978, p. 96.

23. It should be pointed out that there is a second, countrywide program which has also, in effect, unintentionally served, in part, to inhibit long-term methods, namely, the government's very own Blue Circle contraceptive subsidies and distribution program, or KB-Mandiri, as it's known. There are plenty of contradictions in this. KB-Mandiri is, in fact, officially linked to the Effective Contraceptive Method of Choice Program (MKET), which explicitly encourages couples to use long-term or permanent contraception; and, to the Blue Circle Social Marketing Campaign (LIBI) which attempts to foster private sector family planning promotion, sales and contraceptive services. See "Recent Information On Blue Circle and Discount Program," National Family Planning Coordinating Board, Jakarta, 1991. Because Blue Circle pays a 1000 rupiah fee for each short-term injectible contraceptive given by an auxiliary nurse-midwife, the A&M cannot make as much income from any long-term contraceptives (IUDs or Norplant) or from sterilization. Long-term contraceptive prevalence rates are dropping in most of Indonesia's heavily populated areas. There have been 1.4 million voluntary sterilizations in Indonesia, or 146,000 a year, up from 9000 a year in 1974. Yet, sterilization still only comprises 6.6 percent of overall contraception in the country. In addition, there is alleged corruption within the government clinics. Government-provided contraceptives are frequently stolen and then sold by the clinicians. There is, as yet, no mechanism within the clinics to link the patients who come in with the products that go out. Such "free market" contradictions will hopefully be superseded by the government's new Gold Circle Program, which will not subsidize short-term contraceptives, but will encourage a wide cafeteria-style approach to the production and marketing of eighteen different contraceptives, both short- and long-term varieties. Because there are very few pharmacies in rural Indonesia, private com-

panies have not been able to profit on the distribution of contraceptives throughout most of the country. Yet, as soon as the government provides free contraceptives to rural clinics, there is the likely potential for money-making graft. Gold Circle will be a private system. But, not too many experts in Indonesia are optimistic about it. Initially, the government will subsidize Gold Circle. But there will be competition from BKKBN's existing Blue Circle program. Some view the whole conflict as a mess. It was unknown as of 1993 what was going to happen.

24. See "Indonesia Country Profile—Annual Survey of Political and Economic Background 92-93," The Economist Intelligence Unit, New York: Business International, 1993. In a widely conflicting figure, however, the World Bank reports that adult literacy in Indonesia (encompassing both rural and urban), as of 1988, was 81 percent. See *Indonesia—Growth, Infrastructure and Human Resources*, World Bank, Report #10470-IND, Country Department III, East Asia and Pacific Regional Office, May 26, 1992.

25. The international nongovernmental family planning organization known as PATHFINDER, led in Indonesia by Dr. Does Sampoerno, has also addressed the problem of Indonesia's rainy weather and inaccessible communities by launching a system of "floating clinics," boats that can reach remote islands and river bank populations in West Kalimantan, South and Central Sulawesi, Riau Province, North and South Sumatra, Lampun, West, Central and East Java, or, in other words, 69 percent of Indonesia.

26. In addition, nearly 700,000 trainers have been instructed in basic nutrition and primary health care, as part of the BKKBN's Posyandu Health Services program. These cadres are viewed as the final link in nutrition information to mothers. Posyandu actually refers to monthly village nutrition meetings to educate rural villagers and distribute vitamin A capsules. See *Country Report*, Indonesia, *4th Asian and Pacific Population Conference*, August 1992, Bali, Indonesia, p. 36. See also *Indonesia—Growth, Infrastructure and Human Resources*.

27. Employment for women, by itself, is no panacea, however. Currently, there are approximately 2 million Indonesian woman in business. Their presence even dominates high-growth industries like food, beverages, tobacco, electronics, pharmaceuticals and plywood. Most of these females are between ages 21 and 30, single, with only a primary school education, living in virtual poverty. They are paid far less than men. Occupational and safety health standards are either low, or absent. They suffer anemia and malnutrition. As Lin Lim has written, "The

demand for female labor by the labor intensive manufacturing industries is based on the very vulnerability of the women workers who, compared to their male counterparts, are more docile, more deferential to authority, less likely to form unions, or complain, more ignorant and helpless." See "Employment Situation and Training Needs of Women Workers in Garments and Food/Tobacco Processing in Surabaya," *Indonesia Report*, submitted to UNDP ILO, part of preparation for the Industrial Skills Development Project, April 1992.

28. See Sheila Ward, *Service Delivery Systems and Quality of Care in the Implementation of NORPLANT in Indonesia*, Population Council of New York, February 1990.

29. J.S. Parsons, "What Makes the Indonesian Family Planning Programme Tick," National Family Planning Coordinating Board, UNFPA, *Populi*, volume 11, number 3, 1984, p. 11.

30. See Robert Repetto, et al., *Wasting Assets: Natural Resources in the National Income Accounts*, Washington, DC: World Resources Institute, 1989, pp. 57–60.

31. See Budi Utomo, et al., *Abortion in Indonesia—A review of the Literature*, Jakarta: Faculty of Public Health, University of Indonesia, May 1982.

32. Curtin, et al., op. cit., pp. ix, 28. See also *Indonesia—Growth, Infrastructure and Human Resources*.

33. *Biodiversity Action Plan For Indonesia*, op. cit., p. 8.

34. E.O. Wilson, *The Diversity of Life*, Cambridge, MA: Belknap Press of Harvard University Press, 1993, p. 45.

35. *Indonesian Country Study on Biological Diversity*, Ministry of State for Population and Environment, KLH, Jakarta, 1992.

36. According to one researcher, the country contains at least 5 million, or between 10 percent and 17 percent of all known species. See Julian Caldecott, "Biodiversity Management in Indonesia," *Tropical Biodiversity*, volume 1, number 1, 1992, pp. 57–62.

37. As of 1988, the country contained 11,761,000 hectares of alleged wilderness (a 4000 square kilometer section of totally undisturbed area), less than 1 percent of the total land area. See *World Resources 1992–93*, a report by the World Resources Institute in collaboration with the United Nations Environment Program and the United Nations Development Program, New York: Oxford University Press, 1992, p. 263.

38. The analogy with Madagascar, however, does not hold in the long-term: Madagascar has a considerably smaller base of species diversity

than Indonesia. See *World Resources 1992–93*, pp. 304–305.

39. *Biodiversity Action Plan For Indonesia*, op. cit., p. 8.

40. See Haerudin R. Sadjudin, "Status and Distribution of the Javan Rhino in Ujung Kulon National Park, West Java," *Tropical Biodiversity*, volume 1, number 1, 1992, pp. 1–10.

41. A second, somewhat expanded ad ran as follows: "Indonesian Forests Forever; 2000 species of fauna PROTECTED. One-sixth of the world's bird species and one-eighth of its mammal species will live here forever. The tropical forests of Indonesia, home to over 1500 species of birds and 500 species of mammals. Beautiful, unique & exotic. Of these inhabitants, some are dangerous, some helpless while some are even endangered. [sic] The Republic of Indonesia has taken steps to protect their natural habitat by designating 46 million acres as national parks, open to the public, and 75 million acres as protection forests, preserved in their natural virgin state and closed to commercial and recreational use. God gave them one of the world's most beautiful homes. We are their watchmen providing security 24 hours a day. Indonesia's forests are classified as parks, protection, production, and conversion areas, or agro-land. Parks—46 million acres; protection—75 million acres; production—159 million acres; conversion—75 million acres."

42. Quoted in Roderick Nash, *Wilderness and the American Mind*, 3d ed., New Haven and London: Yale University Press, 1982, p. 271.

43. Ibid., p. 360.

44. George Pettie, "Petite Palace," 1576, p. 90, quoted from *The Macmillan Book of Proverbs, Maxims, and Famous Phrases*, selected and arranged by Burton Stevenson, New York: Macmillan, 1948, p. 1617.

45. Wahana Lingkungan Hidup Indonesia (WALHI), whose talented young Executive Director is Mr. M.S. Zulkarnaen. WALHI is the umbrella organization for some two hundred environmental NGOs and has successully launched the first environmental lawsuits in the country.

46. At least twenty countries exceed 10 percent protection, while another twenty countries show 0.0 percent protection. The United States shows an official figure of 10.5 percent of its territory as biologically protected. Though it is more than likely that the discrepancies in Indonesia reflect discrepancies elsewhere. See *World Resources 1992–93*, pp. 298–299.

47. *Biodiversity Action Plan For Indonesia*, op. cit., p. 7.

48. *National Strategy for the Management of Biodiversity*, KLH, Jakarta, 1989.

49. In addition, 14.5 percent from mining, 19.5 percent from manufacturing, and 16.3 percent from trade and tourism.

50. "Indonesia Country Profile—Annual Survey of Political and Economic background 92–93," op. cit., pp. 47–49.

51. Biodiversity Management in Indonesia, op. cit., pp. 57–62.

52. Biodiversity Action Plan For Indonesia, op. cit., p. 17.

53. Mistaking Plantations for Indonesia's Tropical Forest: Indonesia's Pulp and Paper Industry, Communities, and Environment, WALHI and YLBHI, Jakarta, 1992, pp. 13, 15. The South-North Project for Sustainable Development in Asia includes WALHI (The Indonesian Forum for the Environment) and YLBHI (The Indonesian Legal Aid Foundation), the Project for Ecological Recovery (Thailand), Sahabat Alam Malaysia, the Philippines Rural Reconstruction Movement, PROSHIKA (Bangladesh), and Novib (Holland).

54. Government of Indonesia Report, Departemen Kehutanan (Department of Forestry), Jakarta, 1991. One good measure of the success of afforestation and reforestation is the relative change in soil erosion. Presently, Javanese annual on-site erosion is estimated at $330 million in depreciation per year, and the off-site burden is estimated to be an additional $25 to $90 million per year, figures higher than ever before. See Indonesia Forests, Land and Water—Issues in Sustainable Development, Report Number 7A22-IND, Country Department #5, Asia Regional Office, Document of the World Bank, June 5, 1989, p. 58. The re-greening programs going on, called INPRES-PENGHIJAUAN, have since 1976 planted an estimated 3.5 million hectares, but the success rate (trees that survive) has been very low, between 12 and 20 percent. This is attributed to the poor quality of seedlings, and to supply problems.

55. Mistaking Plantations For Indonesia's Tropical Forest: Indonesia's Pulp and Paper Industry, Communities, and Environment, op. cit., p. 7.

56. Ibid., p. 33.

57. Ibid., p. 41.

58. Ibid., pp. 42–53.

59. Thus, an editorial in the Jakarta Post of February 19, 1993, remarks that "timber trade should be developed further and not be curbed because without a market or the possibility of making fair profits there is no incentive at all to maintain forest resources, especially for the developing nations which are still struggling with massive poverty." The belief in the maintenance of forest resources is like trusting hunters to conserve the animals they kill. It has never worked. The rapidly declining duck populations in the U.S. is just one example of such fallacious thinking.

60. Repetto, et al., op. cit. Between 1977 and 1984, the depletion of soils was worth 19 percent of the country's GDP, according to Repetto. At a calculated 4 percent annual depreciation value of the soil's fertility—the same amount of estimated gain in farm production—Indonesia is just breaking even. But that, of course, is a perverse illusion: to break even is to annihilate the future since the soil is not being re-appreciated. Robert Repetto, "Accounting for Environmental Assets," *Scientific American*, June 1992, pp. 94–100.

61. In addition, the country's earnings are cut into by losses stemming from disasters, mostly floods and landslides, the result of watershed destruction and unmanaged logging. An economic figure of $625 million was proposed for losses in the last five years. See "The Land Resources of Indonesia: A National Overview," *Regional Physical Planning Programme for Transmigration*, Overseas Development Administration/Ministry of Transmigration, 1990. At the same time, the yield ratio of fertilizer to crop has dropped, by as much as 50 percent in parts of the country. See "Indonesia Country Profile—Annual Survey of Political and Economic Background 92–93," p. 40.

62. Repetto, et al., op. cit., p. 53.

63. See "Construction of Nuclear Power Plant Likely to go Ahead, *Sunday Observer*, Jakarta, February 14, 1993, p. 1. See also "Government Energy Scenario and an NGO Alternative," *Environesia*, volume 6, numbers 4/5, November 1992, pp. 14–17. In neighboring Malaysia, the Prime Minister has rejected the nuclear option in favor of utilizing the country's 59 trillion cubic feet of proven natural gas, which are believed to be sufficient to meet the country's energy needs for the next century. Others in Malaysia argue that those reserves will be exhausted in twenty years, as in Indonesia, given Malaysia's economic and consumerist surge. See Doug Tsuruoka, Suhaini Aznam, and Carl Goldstein, "Power Plays," *Far Eastern Economic Review*, January 28, 1993, p. 48.

64. However, the Ministry of Forestry and the Indonesian police are cooperating to stop smuggling and illegal logging as best they can, for the time being. Gun provisions for forest rangers have been appropriated. At least 18 sawmill owners involved in illegal operations have been raided. Sixty-nine cases of animal smuggling and illegal logging were taken to court in 1992, but this is probably only a minute smattering of illegal operations, given the range of environmental degradation occurring across all sectors. Such insufficiency of the legal arm argues implicitly for integrating local people into the justice system: asking them to protect their own environmental heritage by

arming them with appropriate information, jurisdictional authority, weapons, and options for viable, non-impactful livelihoods.

65. "Donor" nations have caused other problems in Indonesia, as well. The U.S., Japan, South Korea, and Taiwan, among others, have been dumping their used plastics in Indonesia, 41,117 tons in 1991. This was intended to help a poor country recycle free plastic. Now, officials in Jakarta have discovered that only 60 percent is recyclable, 30 percent is not, and 10 percent is toxic, containing polymer ethylene, polymer styrene, polymer vinyl chloride and copolymer vinyl acetate. With recent new aid pledges from Japan, these situations are likely to begin to change, however. See chapter 8.

66. *Review Indonesia*, #44, February 13, 1993, p. 33.

67. See *World Resources 1992-93*, op. cit., p. 48.

68. See James Gustave Speth, "Coming to Terms: Toward a North-South Bargain for the Environment," *Issues and Ideas*, Washington, D.C.: World Resources Institute, 1990, p. 4.

69. See Narendra Sharma and Raymond Rowe, "Managing the World's Forests," *Finance & Development*, Washington, D.C., June 1992, pp. 31-33.

70. The contributing organizations include the MacArthur Foundation, The World Parks Endowment of the World Conservation Union, the World Heritage Convention, the various United Nations Environmental Program Regional Seas Conventions, the Tropical Forest Action Plan and International Tropical Timber Organization, the International Fund for Plant Genetic Resources, the United Nations Development Program, and the Global Environmental Facility of the World Bank, the largest of such endowments, presently providing $1 billion in grants and low-interest loans to restoration projects. See *Caring for the Earth: A Strategy for Sustainable Living*, IUCN, UNEP, WWF, Gland, Switzerland: IUCN, 1991, p. 22.

71. See Michael Wells, Katrina Brandon, and Lee Hannah, *"People and Parks: An Analysis of Projects Linking Protected Area Management with Local Communities,"* draft report, Washington, D.C.: The World Bank, 1990, pp. 44-46.

72. Personal conversation with World Wide Fund for Nature Country Representative, Dr. Russell Betts. See also Ian Craven, "Community Involvement in Management of the Arfak Mountains Nature Reserve," unpublished paper, Jayapura: World Wide Fund for Nature, 1990.

73. See Vincent Lingga, "Indonesia works hard to take care of its tropical forests . . . but is upset by constant international criticism," the *Jakarta Post*, February 18, 1993, p. 5.

74. Wilson, op. cit., p. 259.

75. The law goes on to say that "Each couple can decide on planning and controlling the number of children and the birth spacing based on their own conscience and responsibility to the present as well as to the future generation. . . . Policies . . . pertain to the regulation of size, structure, composition, growth, and dispersion of population through efforts to reduce mortality rate, through birth control, and by directing the mobility of people in conformity with environmental carrying capacity." *Act of the Republic of Indonesia Number 10 of 1992 concerning Population Development and Development of Prosperous Family*, State Ministry of Population and the Environment, The Republic of Indonesia, Jakarta, 1992.

76. Carl Whiting Bishop, Charles Greeley Abbot, and Ales Hrdlicka, *Man From The Farthest Past*, The Smithsonian Series/7, Washington, D.C.: Smithsonian Scientific Series, 1944, pp. 357–358.

77. J. S. Parsons has called this politically effective system of inducements a "vertical and horizontal sense of commitment." See Parsons, op. cit., p. 10.

78. Private conversation with Mr. M.S. Zulkarnaen in Jakarta, February 18, 1993. See also WALHI and YLBHI, pp. 54–58.

79. *Biodiversity Action Plan For Indonesia*, op. cit. See also M.P. Wells, "Can Indonesia's Biological Diversity be Protected by Linking Economic Development with National Park Management? Three case studies from the Outer Islands," a report to World Bank, 1989; K. MacKinnon, *Biological Diversity in Indonesia: A Resource Inventory*, World Wide Fund for Nature, Bogor, 1990; IUCN, Ministry of Forestry/FAO, *Indonesian Tropical Forestry Action Programme*, 3 vol. Jakarta: U.N. Food and Agriculture Organization, 1991; J. and K. MacKinnon, *Review of the Protected Areas System in the Indo-Malayan Realm*, Gland, Switzerland: IUCN, 1986; J.A. Dixon and P.B. Sherman, *Economics of Protected Areas*, Honolulu: East–West Center, 1990; J.H. de Beer, and M.J. McDermott, *The Economic Value of Non-timber Forest Products in Southeast Asia*, Amsterdam: Netherlands Committee for IUCN, 1989.

80. *Biodiversity Action Plan For Indonesia*, op. cit., pp. 48, 56.

CHAPTER 5
The Forgotten Ones: Africa

1. Jean Dorst and Pierre Dandelot, *A Field Guide to the Larger Mammals of Africa*, 10th ed., London: Collins, 1990, p. 184.

2. Whereas the great apes—all vegetarians—have a brain size of approx-

imately 800 cubic centimeters, Homo's cranial topography ranged, even as early as 750,000 years ago, from between 1000 and 2000 cubic centimeters. During gestation, the Homo erectus fetus grew at the rate of 20,000 neurons a minute, 2.2 milligrams a minute during the third trimester. By the end of its third year of life, the Homo erectus child had achieved a mental bulk of more than 1 kilogram, four-fifths the size of an adult.

3. For a discussion of these evolutionary transitions, see the following articles and books: Noel Boaz and Douglas Cramer, "Fossils of the Libyan Sahara," Natural History Magazine, August 1982, pp. 35–40. Melvin Konner, The Tangled Wing: Biological Constraints in the Human Spirit, New York: Holt, Rinehart & Winston, 1982; Glynn Isaac, "Food Sharing and Human Evolution: Archaeological Evidence from the Plio-Pleistocene of East Africa," Journal of Anthropological Research 34, 1978, pp. 311–325; Frank E. Poirier, Fossil Evidence—The Human Evolutionary Journey, 3d ed., St. Louis: C.V. Mosby, 1981; Richard Leakey, The Making of Mankind, New York: E.P. Dutton, 1981; Donald Johanson and Maitland Edey, Lucy: The Beginnings of Mankind, New York: Simon & Schuster, 1981.

4. By that I do not mean to dismiss other primary agents of change, such as climatic perturbations, shifting curtains of rain, that have, periodically, discombobulated probably every biome in Africa, not just the Sahara: oscillating predator-prey populations, the advance and retreat of glaciers, and natural feast and famine. Even within a very specific, well-documented instance of population boom and bust, there are multiple interpretations. One of the most frequently cited cases is that of the mountain lion and mule deer along the Kaibab Plateau, the north rim of the Grand Canyon. By 1930, the deer, which had reached an estimated 100,000 individuals, suddenly began to starve, their number declining within a decade to a few thousand. Many have argued, including naturalist Aldo Leopold, that it was the absence of mountain lions, which had been systematically eradicated by cattlemen, that encouraged the unnatural excess of deer. But more recently, scholars have pointed out that similar population crashes had occurred among the ungulates with, or without, predators. The same is true of the recent tragic and infuriating conflict between human predators,, wolves, and caribou in the over 4000-square-mile region between Daneli National Park and Fairbanks, Alaska. There, at the urging of hunters, the meager collective of no more than 200 Canis lupus have been systematically strangled and shot by Alaska Department of Fish and Game officials and trappers in the misguided belief that the wolves are responsible for the plunging caribou populations, whose diminished

number are eliminating the opportunities for hunting them, a loathsome and expedient illogic. The wolves have little, if anything to do with the fluctuating number of caribou (a population that periodically vacillates between over ten thousand and less than four thousand in the region). What has not been pointed out is the fact that the wolves, while not officially listed as endangered, are, in fact, within the state of Alaska, totally so. The Alaskan state government is intent upon reducing the wolves in that region to a population less than 30. It is regional genocide. But because there is as yet no underlying rule of thumb, no sure predictive formula, that can be applied to every population, the confusion surrounding the biology of the wolf-caribou connection persists, in spite of such elegant explications as Farley Mowat's classic book, *Never Cry Wolf.* See Daniel B. Botkin, "Ecological Theory and Natural Resource Management—Scientific Principles or Cultural Heritage?" in *Ecological Prospects—Scientific, Religious, and Aesthetic Perspectives,* ed. Christopher Key Chapple, Albany, NY: State University of New York Press, 1994, pp. 65–81. For data pertaining to climatic oscillations in Africa, see W.M. Adams, *Green Development, environment and sustainability in the third world,* London: Routledge, 1990, p. 91

5. See Mike Clary, "Exotic Plants, Animals Imperil U.S. Ecosystem," *Los Angeles Times,* December 5, 1993, pp. A1, 14, 17.

6. See Michael Tobias, "Rapsody of Embers," *After Eden—History, Ecology, & Conscience,* San Diego: Avant Books/Slawson Communications, 1985.

7. Patience W. Stephens, Eduard Bos, My T. Vu, and Rodolfo A. Bulatao, *Africa Region Population Projections, 1990–1991 Edition,* working papers, Population and Human Resources Department, Washington, D.C.: The World Bank, February 1991, WPS 598, p. 26.

8. See E.K. Mburugu, *Factors Related to Stock Ownership and Population Movements; and Perception of Land Pressure and Other Environmental Changes among the Rendille in Marsabit District,* University of Nairobi, Integrated Project in Arid Land (IPAL) Technical Report #F-2—Man in the Biosphere Program, n.d., p. 2.

9. *The Wildlife Conservation and Management Act of 1976* and the 1989 law enacting the Kenya Wildlife Service have thus far managed only minimally to ameliorate this marginalization crisis, a syndrome endemic to East Africa. See *Development Policy for Arid and Semi-Arid Lands (ASAL),* September 1992, Nairobi: Government of Kenya, pp. 28–29.

10. *World Population Prospects 1988,* New York: United Nations, 1989.

Population density in Nairobi is currently estimated to be 1911 persons per square kilometer, but this figure could be way off, considering discrepancies in data pertaining to the actual population size of Nairobi—from 1.5 to 2.5 million inhabitants, up from 8000 in the year 1901 and 118,976 in 1948. See Stephens, et al., op. cit., p. xvii.

11. See "Women at Risk," Society, number 32, Oct. 18, Nairobi, 1993, p. 30. See also Hugh O'Haire, "Think Again," Populi, December 1993/January 1994, pp. 8–10. Regarding the findings of a Berlin round-table conference in late September 1993 ("Population Policies and Programmes: The Impact of HIV/AIDS") writes O'Haire, ". . . the experts added that some sub-national regions and a small number of large cities could experience markedly reduced population growth for short periods. But these downswings would be quickly cancelled by immigration from surrounding areas." In fact, says, O'Haire, "The view that AIDS will take care of the population problem is wrong."

12. Cited on page 41 of Thomas J. Goliber, "Africa's Expanding Population: Old Problems, New Policies", Population Bulletin, volume 44, number 3, November 1989, Washington, D.C.: Population Reference Bureau, and derived from John Bongaarts and Frank Odile, "Biological and Behavioral Determinants of Exceptional Fertility Levels in Africa and West Africa and West Asia," African Population Conference, Dakar, 1988, Liege, Belgium: International Union for the Scientific Study of Population, 1988.

13. Goliber, ibid., pp. 41–42.

14. Ibid., p. 42.

15. Mauritius' TFR dropped faster than any other country in history during the 1960s, from 5.9 in 1962 (with a population of about 700,000) to 3.4 in 1972, to 2.2 in 1983. At the same time, during the 1960s Mauritian primary school enrollment became universal. The country obtained independence from Britain in 1968. By 1989, following the expansion of the textile industry and the widespread employment of women, unemployment in the country was virtually zero, GNP and per capita incomes—currently about $1400 per year—suggested an NIC economy, not an African one. Part of Mauritius' success stems from its tiny defense budget, a mere 0.8 percent of its annual government expenditures. Well educated, with a male life expectancy of sixty-four and a female of seventy-two, the country—with its multicultural background (African, Indian, European, and Chinese) and stable democratic political system—is especially remarkable in that its radically declining TFR occurred during a period of economic stagnation, the reverse of demographic expectations. Official Mauritian family plan-

ning efforts began at that time and played what can only be described as a key role by advertising the benefits of family planning and providing counseling and technology for birth control. Within two years of government support, contraceptive use spread rapidly, and a high CPR plateau has never wavered since. At the same time, the economy picked up. Yet, even this rosy picture of economic and fertility planning triumph has not ensured ecological sanity. Amazingly, tiny Mauritius—a mere dot on the African map, 1900 square kilometers—is the third largest exporter of live primates on the whole continent. Of the four known Mauritian species of mammals, other than humans, three are endangered there. Endemic plants constitute a third of all Mauritian flora, or nearly three hundred plant taxa. Yet, 33 percent are "rare" or "threatened." See Anne Babette Wils, "Mauritius, the LLASA Project," *Materials on Demogaphic Questions*, Herausgeber: Bundesinstitut fur Bevolkerungsforschung, Wiesbaden, 1991, pp. 31–54.

16. U.N. Secretariat, Population Policy Data Bank, Population Division, Department of International Economic and Social Affairs, 1989.

17. See *World Population Prospects—1992 Revision*, New York: United Nations Department for Economic and Social Information and Policy Analysis, pp. 150, 293.

18. Stephens, et al., op. cit., p. 6.

19. See *World Resources Institute 1992–93*, a report by the World Resources Institute in collaboration with the United Nations Environment Programme and the United Nations Development Programme, New York: Oxford University Press, 1992, p. 215.

20. *Kenya Demographic and Health Survey 1993—Preliminary Report*, National Council for Population and Development, Ministry of Home Affairs and National Heritage, Central Bureau of Statistics, Office of the Vice President and Ministry of Planning and National Development, prepared with technical assistance from Demographic and Health Surveys, Macro International Inc., Columbia, Maryland, under contract to USAID, September 1993, pp. 13, 21.

21. In Kenya, six units of capital input are required to produce one unit of output. This ratio is considerably lower in faster growing economies such as those of Malaysia and Thailand.

22. "U.S. official outlines policy on debt relief," the *Standard*, Nairobi, October 18, 1993, p. II.

23. See *Report of the Proceedings of the Workshop on Cost-Sharing in Kenya 1989*, ed. J.E.O. Odada and L.O. Odhiambo, Ministry of Planning and National Development and Kenya Economic Association, UNICEF

Nairobi. See also *Socio-Economic Profiles*, Government of Kenya Ministry of Planning and National Development, and UNICEF, June, 1990.

24. See *District Contraceptive Prevalence Differentials Study—A Case Study of Six Districts*, Population Studies and Research Institute, University of Nairobi, January 1992. See also Wariara Mbugua, *Review and Assessment of Population Data, Policy and Research, Family Planning Programs and Population Information, Education and Communication—A Case Study of Kenya*, UNFPA, Nairobi, October, 1988.

25. See *Association for Voluntary Surgical Contraception, Kenya Assessment Program for Permanent and Long-Term Contraception Assessment Report*, National Council for Population and Development and the Ministry of Health, Division of Family Health, and U.S.A.I.D., 1992.

26. *Kenya Demographic and Health Survey, 1993—Preliminary Report*, op. cit.

27. The survey selected and interviewed 7952 women and 2721 men from 7950 households throughout the country. It was the first such survey in four years, and covered the period from 1990 to 1993. The previous survey covered the years 1984–1988.

28. Official family planning slogans in Kenya read, "*Nafasiza kazini haba, Zaa Watoto Wachache*," or, "Jobs are scarce; have few children"; and "*Punguza Mzigo Wako Tumia Njia Ya Kupanga Uzazi*," meaning, "Lessen your struggle, use a family planning method." Yet, many family planning officials themselves do not set the best example, coming from families of ten and twelve siblings.

29. See *Population, Resources, and the Environment—The Criticial Challenges*, New York: UNFPA, 1991, p. 95. See also *Inventory of Population Projects in Developing Countries around the World 1991/92*, New York: UNFPA, 1992, pp. 275–291; Wamucii Njogu, "Trends and Determinants of Contraceptive Use in Kenya," *Demography*, volume 28, number 1, February, 1991, pp. 83–99.

30. Stephens, et al., op. cit., p. 77.

31. *World Population Prospects—1992 Revision*, op. cit., p. 519.

32. "United Nations Population Fund Proposed Projects and Programmes," recommendation by the executive director, Assistance to the Government of Kenya, Support for a comprehensive population program, Fortieth session, May 1992, Governing Council of the United Nations Development Program, Geneva: UNFPA, p. 4.

33. See *Scout Programme Africa Region—Curriculum Guide and Resource Manual for Family Life Education*, World Scout Bureau, Africa Regional Office and the Margaret Sanger Center of Planned Parenthood of New York, 1988.

34. Founded in 1923 by Dr. Clive Irine, a Scottish missionary and doctor; in 1956, management passed from the Church of Scotland to the Presbyterian Church of East Africa. In 1969, Mzee Jomo Kenyatta, the country's first president, led a famous Harambee, or community self-help jamboree, in which money was raised to expand the hospital to its current 290 beds.

35. See Howard I. Goldberg, Malcolm McNeil, and Alison Spitz, "Contraceptive Use and Fertility Decline in Chogoria, Kenya," *Studies in Family Planning*, volume 20, number 1, Jan/Feb 1989, pp. 17-26.

36. See, John J. Dumm, et al., "Helping Services Meet Demand—An Assessment of A.I.D. Assistance to Family Planning in Kenya," Center for Communication Programs, Johns Hopkins University, 1992.

37. See *PCEA Chogoria Hospital Annual Report*, Chogoria, Kenya: 1992, p. 4.

38. See Professor Philip M. Mbithi, "Implication of Socio-Economic and Cultural Factors on Population and Development," *Report on the Second National Leaders' Population Conference*, NCPD, Republic of Kenya, September 1989, pp. 43-44.

39. Moreover, there is a new "rape culture" in Kenya. On July 14, 1991, male students broke into a female dorm and sexually assaulted fifty young girls at the St. Kizito School in Meru, very close to Gaatia. Nineteen of those girls were killed in the process. In the northern Wajir District, seven girls were raped and shot in 1992. In South Nyanza, fifteen girls were raped at the Hawinga School in 1993. Countless other women have been raped in the Garissa District of the northeast, where the Somali refugees have congregated. These are just a few of the highlights of a trend that may well reflect the fact that "discrimination of women under customary law is constitutional." Psychologically, it seems to follow from the polygynous freedoms taken for granted by so many African males. See Muthoni Murage, "The Law Relating to Women in Kenya," *The Nairobi Law Monthly*, #50, Aug/Sep 1993, pp. 32-33. Such crimes against women are also the precursors of even more comprehensive human war waged against the Earth. For a discussion of that link, see Michael Tobias, *Voice of the Planet*, New York: Bantam Books, 1990, chapter 2. See also the film series "Voice of the Planet," Michael Tobias, Atlanta: Turner Broadcasting Service, February, 1991, Episode Two; *Children & Women In Kenya: A Situation Analysis*, Nairobi: Government of Kenya, U.N. Children's Fund, 1992.

40. As a point of partial documentation, it should be noted that the Nairobi Vegetarian Society has fewer than forty members according to its president.

41. J.P. Handby, et al., "Population Changes in Lions and other Predators," in *Serengeti—Dynamics of an Ecosystem*, ed. A.R.E. Sinclair and M. Norton-Griffiths, Chicago: University of Chicago Press, 1979, pp. 249–262, cited by William Newmark, "The Selection and Design of Nature Reserves for the Conservation of Living Resources," *Managing Protected Areas in Africa—Report from a Workshop on Protected Area Management in Africa*, compiled by Walter J. Lusigi, Mweka, Tanzania: UNESCO, World Heritage Fund, 1992, p. 93.

42. Allen Bechky, *Adventuring in East Africa—The Sierra Club Travel Guide to the Great Safaris of Kenya, Tanzania, Rwanda, Eastern Zaire, and Uganda*, San Francisco: Sierra Club Books, 1990, p. 273.

43. Glory Chanda and Ackim Tembo, "The Status of Elephant on the Zambian Bank of the Middle Zambezi Valley," *Pachyderm* magazine, number 16, 1993, p. 50.

44. Carolyn Alexander, "The Brigadier's Shooting Party," *New York Times*, November 13, 1993, p. Y15.

45. See Fox, "Help for Tanzania," *HSUS News* (Humane Society of the United States), Fall 1993, volume 38, number 4, pp. 30–33.

46. See Michael Satchell, "Wildlife's last chance," *U.S. News & World Report*, November 15, 1993, pp. 68–76.

47. *Kenya Wildlife Service—A Policy Framework and Development Program, 1991–1996*, Nairobi: Kenya Wildlife Service, 1990, p. 4.

48. See G. Brown, "The Viewing Value of Elephants," *The Ivory Trade and the Future of the African Elephant*, Ivory Trade Review Group, vol. 2, Technical Reports, 1989.

49. The Masai tribespeople surrounding the reserves have repeatedly voiced three economic concerns: that the financial benefits of tourism have reached few local pockets; that too little has been received directly by those people who are accommodating the wildlife; and that they are opposed to any new conservation areas that might threaten title to "their" land. See "Identification Study for the Conservation and Sustainable Use of the Natural Resources in the Kenyan Portion of the Mara-Serengeti Ecosystem," *European Development Fund of the European Economic Community*, Interim Report, E. Douglas Hamilton & Association, May 1988, p. 45.

50. *Kenya Wildlife Service—A Policy Framework . . .*, op. cit., p. 184.

51. Ibid., p. 187.

52. See Alison Wilson, "Sacred Forests and the Elders," in *The Law of the Mother—Protecting Indigenous Peoples in Protected Areas*, ed. Elizabeth Kemf, San Francisco: Sierra Club Books, 1993, pp. 244–248.

53. See Robert A. Levine, "Maternal Behavior and Child Development in High-Fertility Populations," *Fertility Determinants Research Notes 2*, New York: The Population Council, September 1984, p. 4, cited in Goliber, p. 27.

54. See *Kenya Wildlife Service—A Policy Framework . . .* , op. cit., p. 20.

55. The crisis is most pronounced in Taita Teva near Tsavo, in the Narok and Kajiado districts, and across the Laikipia District. See Joyce H. Poole, "Kenya's Initiatives in Elephant Fertility, Regulation, and Population Control Techniques," *Pachyderm* magazine, #16, 1993, publication of the World Conservation Union (IUCN) & WWF, p. 62.

56. See *Kenya Wildlife Service—A Policy Framework . . .* , op. cit., p. 62.

57. See Poole, op. cit., p. 64.

58. See *Sub-Saharan Africa—From Crisis to Sustainable Growth—A long-term perspective*, Washington, D.C.: World Bank, 1989, p. 1.

59. See Stephens, et al., op. cit., p. 99.

60. Most of Nigeria, however, is not seeing a CPR improvement. "The results of the 1990 Nigeria Demographic and Health Survey point to high fertility, low levels of contraceptive use, and high mortality and undernutrition among children." *DHS—Demographic and Health Surveys Newsletter*, volume 5, number 1, 1992, p. 3.

61. See John Caldwell, I.O. Orubuloye, and Pat Caldwell, "Fertility Decline in Africa: A New Type of Transition?", *Population And Development Review*, 1992, pp. 214, 236. See also Susan H. Cochrane and S.M. Farid, "Fertility in Sub-Saharan Africa: Analysis and Explanation," *World Bank Discussion*, Paper #43, Washington, D.C.: The World Bank, 1989.

62. See A.J. Ahlback, *Industrial Plantation Forestry in Tanzania: Facts, Problems, Challenges*, Ministry of Natural Resources and Tourism, Dar Es Salaam, n.d.

63. *Central Africa: Global Climate Change and Development*, Technical Report, Landover, Maryland: Biodiversity Support Program, a Consortium of the WWF, The Nature Conservancy, World Resources Institute, U.S. Agency for International Development, 1993, p. V.

64. Peter Matthiessen and Mary Frank, *Shadows of Africa*, New York: Harry Abrams Inc., 1992, p. 83.

65. A local magazine, WALIA, is published in the city of Mopti. Well-written and highly accessible, it covers a broad range of health and ecology issues, but is sadly limited in its distribution.

66. See Nigel Cross and Rhiannon Barker, *At The Desert's Edge: Oral Stories from the Sahel*, London: Panos/Sos Shel, 1992. *See also*

Robin Poulton and Michael Harris, *Putting People First: Voluntary Agencies and Third World Development*, London: Macmillan, 1992.

67. See Peter Warshall, *Mali—Biological Diversity Assessment*, Office of Arid Lands Studies, University of Arizona, Contract No. AFR-0467-C-00-8054-00, (Natural Resources Management Support Project A.I.D., Bureau of Africa, Office of Technical Resources, Natural Resources Branch, Project No. 698-0467), Prime Contractor: E/DI, Washington, D.C.: March 1989.

68. Ibid., pp. 43-44.

69. See *Memoire accompagnant la Carte de Vegetation de l'Afrique*, UNESCO/AETFAT/UNSO, traduit de l'anglais par P. Bamps, 1986. See also C. Rossetti and P. Audry, *Observations sur les sols et la vegetation en Mauritanie du Sud-est, et sur la bordure adjacente du Mali 1959 et 1961*, Rome: FAO, n.d.

70. Warshall, op. cit., p. 32.

71. Ibid., p. 10.

72. Ibid., p. 20.

73. Ibid., p. 34.

74. Ibid., p. iii.

75. Ibid., p. ii.

76. Ibid., pp. 26-27.

77. In the national park, only half of the known animals were seen, and in only one out of six observation attempts. *Rapport Annuel 1992*, Bamako: Ministere Du Developpement Rural et de L'Environnement, Direction Nationale Des Eaux et Forets et Operation Parc Nationale de la Boucle Du Baoule, 1993, p. 49.

78. Volume 1, *Summary Report*, Bamako: Ministry of Planning and National Directorate of Planning, under the International Conference of Donors for the Economic Recovery and Development of the Republic of Mali, December 1982.

79. Stephens, et al., op. cit., pp. 86-87.

80. *Inventory of Population Projects in Developing Countries . . .* , op. cit., pp. 325-330.

81. See "Levels and Trends of Contraceptive Use as Assessed in 1988," *Population Studies*, no. 110, New York: United Nations, 1989, Tables 2 and A.11.1.

82. See Annik Thayer-Rozat, *Plants Medicinales du Mali*, Bordeaux: Ocoe Publishing, 1981.

83. Compare the figures from the *Rapport National Sur La Population*, Mali, August 1993, p. 15, with *Inventory of Population Projects in Developing Countries* . . . , p. 325—discrepancies of a decade, or 25 percent!

84. See *Sub-Saharan Africa—From Crisis to Sustainable Growth* . . . , op. cit., p. 4.

85. Because Bamako is 1200 kilometers from the Atlantic, import duties are prohibitive, 33 percent of the total cost of shipping goods. A weak transportation sector aggravates this pricing syndrome. Yet, the country has had to increase its imports due to declining food production and declining exports, consequences of the persistent drought. An increase in the price of petroleum, inadequate storage facilities for export products during periods of major drops in world prices, inflation, credit exhausted simply to finance deficits rather than for increasing production, have all led to the deterioration of the terms of trade, causing a chronic balance of payments deficit in the country. It is almost impossible to free-up sufficient self-financing for small businesses; to renew productive capital; or to obtain new bank credits. Despite known stores of gold and gypsum, and reserves of bauxite totaling an estimated 800 million tons (with high aluminum content, 40-to-45 percent), iron ore reserves exceeding 1 billion tons, and 3.5 million known tons of manganese, foreign private investment has plummeted, while the domestic financing capability of the government is declining rapidly.

86. The tenets of GATT and of free trade contradict this insular approach to the world, at least in boom economies. NAFTA, for example, is expected to drive 14 million Hispanic immigrants into the U.S. in the coming decades, where America's population will double, well beyond U.S. Census pre-NAFTA projections. But in Africa, just the opposite is occurring: an artificial economy, kept minimally afloat by World Bank, IMF and the African Development Bank loans, and by foreign aid, is also doubling its population every sexual generation, not from immigration, but from high TFRs and the continuing population momentum.

87. These include: Save the Children, the Rockefeller Foundation, PATH (Program for Appropriate Technology in Health), the Population Council, the Johns Hopkins University Population Communication Services, Johns Hopkins Program for International Education in Reproductive Health, International Planned Parenthood Federation, Association Moyenne, Family Health International, the Centre for Development and Population Activities, Association for Voluntary Surgical Contraception, USAID, the Government of Germany, the Canadian International Development Agency, and the African regional office of the World Health Organization.

88. UNFPA hopes to persuade the government to strengthen family planning services in the existing 56 urban and health centers in five regions, while more fully integrating family planning into existing "Maternal Child and Health" centers. Support would be given to train traditional birth attendants and medical staff, and to provide essential drugs, equipment, and a contraceptive cafeteria mix. Awareness of STDs and "Safe Motherhood" would be incorporated into the family planning package as part of Information, Education, and Communication (IEC) strategy. Finally, says Isaiah Ebo, UNFPA hopes to encourage greater child spacing "much in the manner of spacing the planting of maize—that is the analogy people can understand." See *Plan D'Action Sur La Strategie National Pour La Planification Familiale*, Bamako: Ministere de la Santa Publique et des Affaires Sociales a la Banque Mondiale/ Sante, Population et Hydraulique Rurale (PDSII), 1990. See also "United Nations Population Fund Proposed Projects And Programmes," recommendation by the Executive Director, Assistance to the Government of Mali, Support for a comprehensive population programme, Fortieth session, June 1993, Governing Council of the United Nations Development Program, New York: UNFPA.

89. See Femi Ajayi, et al., "Tools of the Trade: Do Farmers Have the Right Ones?" *African Farmer, #5*, November 1990, p. 5.

90. "United Nations Population Fund Proposed Projects and Programmes," (Kenya), p. 8.

91. *Kenya Demographic And Health Survey 1993*, op. cit., p. 29.

92. Two other noteworthy Kenyan women have made a profound difference for the country, as well. Ms. Wangari Matthai started out in the Girl Scouts, and eventually initiated a Green Belt movement that has effected the replanting of hundreds of thousands of seedlings. Ms. Matthai has also prevented at least one high-rise from being built in downtown Nairobi on a pre-existing green area. Similarly, forty-three-year-old Sophia Kiarie, with her eleven children, had moved to the arid town of Ruiru, fifteen miles north of Nairobi in the early 1970s only to realize there was no fuel, no wood, few trees left, even then. Uneducated but motivated, she founded a nursery which to-date has distributed over 600,000 seedlings throughout the country. And she has campaigned to conserve forests by the use of energy-efficient stoves designed by engineers with the Bellerive Foundation in Geneva. Today, a few thousand Kenyans use the stoves, while an estimated million households apply the so-called "C2C" ("cooking to conserve") principles of conservation. Together, the stoves and the C2C method have greatly reduced the loss of heat, and maximized the efficiency of cooking, by harvesting and pruning tree species carefully, cutting, splitting, and

drying firewood properly, and controlling cooking fires in such a way as to minimize loss of energy input. These are beginnings, in absence of which, Kenya's critical resource shortages would have been even worse. See "Kenya: Creating Islands of Green," *Environmental Impact—An Occasional Magazine of the Bellerive Foundation*, 1993, p. 7.

93. See Patricia C. Wright, "Ecological Disaster in Madagascar and the Prospects for Recovery," *Ecological Prospects—Scientific, Religious, and Aesthetic Perspectives*, ed. Christpher Key Chapple, Albany, NY: State University of New York Press, 1994, pp. 11–24.

94. Nampaa Sanogho, *Les Elephants du Gourma*, Bamako: Water and Forestry Department (DNEF), 1980. See also Robert Olivier, *The Gourma Elephants of Mali: A Challenge for the Integrated Management of Sahelian Rangeland*, Nairobi: UNEP, 1983.

95. John Boorman, *West African Butterflies and Moths, A West African Nature Handbook*, London: Longman House Ltd., 1978. See also W. Serle, G.J. Morel, and W. Hartig, *A Field Guide to the Birds of West Africa*, London: Collins Publishers, 1986.

96. See *Rapport Final Phase I: projet du Conservation Dans le Delta Interieur du Mali*, IUCN/CDC, 1988. See also *Mali: Conservation de l'Environnement Dans le Delta Interieur Du Fleuve Niger: Document de Synthese*, Rapport Technique No. 3, Bureau of Africa, U.S.A.I.D., June 1987.

97. Warshall, op. cit., p. 87.

CHAPTER 6
The Price of Development

1. De Montalvo was a Spaniard writing about a rugged nation of black Amazonian women who fed all excess men to griffins, keeping only enough to serve as studs. Gold was the sole metal of this land. See Elna S. Bakker, *An Island Called California*, Berkeley and Los Angeles: University of California Press, 1971, p. xi.

2. See Peter Wilson, *The Domestication of the Human Species*, New Haven: Yale University Press, 1988, p. 181.

3. See Joji Okazaki, *Pure Land Buddhist Painting*, translated and adapted by Elizabeth ten Grotenhuis, Tokyo, New York, and San Francisco: Kodansha International Ltd. and Shibundo, 1977; Saburo Ienaga, *Japanese Art: A Cultural Appreciation*, trans. Richard L. Gage, New York, and Tokyo: Weatherhill/Heibonsha, 1979; and Loraine Kuck, *The World of the Japanese Garden—From Chinese Origins to Modern Landscape Art*, New York and Tokyo: Weatherhill, 1968.

4. William R. LaFleur, *Liquid Life—Abortion and Buddhism in Japan*, Princeton, NJ: University Press, 1992, pp. 96–98.

5. Ibid., p. 110.

6. Ibid., p. 135.

7. Eduard Bos, My T. Vu, and Ann Levin, *East Asia and Pacific Region, South Asia Region, Population Projections 1992–93 Edition,* Policy Research Working Papers, Population and Human Resources Department, Washington, D.C.: The World Bank, November 1992, WPS 1032, pp. 96–97.

8. International North Pacific Fisheries Commission, *Final Report of 1990 Observations of the Japanese High Seas Squid Driftnet Fishery of the North Pacific Ocean,* Seattle: Alaska Fisheries Science Center, Seattle, 1991, Table 24, pp. 193–195; quoted in *World Resources 1992/93,* a report by the World Resources Institute in collaboration with the United Nations Environment Programme and the United Nations Development Programme, New York: Oxford University Press: 1992, p. 181.

9. The U.S. Census Bureau predicts a population of 309 million in the year 2040.

10. By "industrial" here I am referring to the OECD nations, as well as the newly industrialized, and rapidly industrializing nations. By themselves, the OECD nations, or twenty-four countries of the Organization for Economic Cooperation and Development, contained in 1989, 16 percent of the world's population, or 849 million people, had a combined GNP of $15 trillion, and a $17,500 per capita income. The seven largest of these countries consumed 43 percent of the world's production of fossil fuels, most of the metal, forest products, and industrial goods. Those goods included roundwood, paper and paperboard, cement and fertilizer, zinc, tin, nickel, lead, copper, aluminum, and steel. Most of these rich nations maintained trade barriers to keep the agricultural products of the Third World out of their countries, while providing, on average, far less than 1 percent of their GNP to help those same impoverished nations. These same G-7 countries produced 68 percent of the world's industrial waste, and were responsible for 38 percent of the global warming inputs. See *World Resources 1992/93,* pp. 17, 19.

11. See Alex Shoumatoff, *The World Is Burning—Murder in the Rain Forest,* Boston: Little, Brown, 1990, p. 76.

12. See Paul Ehrlich and E.O. Wilson, "Biodiversity Studies: Science and Policy," *Science,* volume 253, August 16, 1991, p. 760.

13. *World Resources 1992/93,* op. cit., p. 119.

14. Ibid., p. 78.

15. See Lester R. Brown, et al., *Vital Signs 1992,* New York: W.W. Norton, 1992, pp. 92–93.

16. E.O. Wilson, *The Diversity of Life*, Cambridge, MA: Belknap Press of Harvard University Press, 1993, pp. 230, 242.

17. Ibid., pp. 254–258.

18. See Robin Poulton and Michael Harris, *Putting People First: Voluntary Agencies and Third World Development*, London: Macmillan, 1992; quoted in Robert Lacville, "Wisdom ancient and modern," the *Guardian Weekly*, February 21, 1993, p. 14.

19. Ibid., p. 14.

20. Five countries in Africa, two in North/Central America, ten in Asia, two in Europe (Luxembourg and Malta) and one in Oceania. Syria has no protected land, nor does the United Arab Emirates, or Iraq, or Cambodia, Somalia or Jamaica. *World Resources 1992/93*, pp. 298–299.

21. Neither wealth, nor the size of human population by itself is a necessarily predictable indication of the likely measure of impact on surrounding habitat. For example, Namibia has all of 22 people per 1000 hectares, but zero percentage of its land could be described as wilderness. Mauritania, however, has 20 people per 1000 hectares, nearly on a par with Namibia, yet 70 percent of its total land area is sufficiently undeveloped to classify it as bona fide wilderness. But what is the true meaning of such wilderness, considering that most of Mauritania's wildlife has been killed? Among the developed countries, Canada and Australia are the two regions with remaining sizable wilderness and low populations and population densities. Twenty-nine percent of little Iceland is wilderness, its density a mere 25 people per 1000 hectares. Nepal—despite its vast, wild Himalayas—has a high density, 1399 people per 1000 hectares, and zero wilderness.

22. Published by the Population Crisis Committee (Population Action Internationa) in Washington, D.C., 1992.

23. Conversation with Paul Papanek, Chief, Toxics Epidemiology Program, County of Los Angeles Department of Health Services, January 1993.

24. For information, write the Sea Turtle Restoration Project, Earth Island Institute, 300 Broadway, San Francisco CA 94133. To protest, write President Salinas de Gortari, Palacio Nacional, Mexico City, DF, Mexico.

25. Joseph Brodsky, "Blood, Lies and the Trigger of History," *New York Times*, August 4, 1993.

26. See Stephen L. Carter, "Strife's Dominion," a review of Ronald Dworkin, *Life's Dominion*, New York: Alfred Knopf, 1993, in *The New Yorker*, August 9, 1993, p. 92.

27. David Rothenberg, *Is It Painful To Think? Conversations with Arne Naess*, Minneapolis and London: University of Minnesota Press, 1993, p. 108.

28. Paul R. Ehrlich and Anne H. Ehrlich, *The Population Explosion*, New York: Touchstone Books, Simon & Schuster, 1990, p. 134.

29. Peter A. Morrison, "Testimony before the House Subcommittee on Census and Population," RAND Corporation, May 26, 1992, p. 5.

30. Paul Kennedy, *Preparing for the Twenty-First Century*, New York: Random House, 1993, p. 221.

31. "The poor disadvantaged tend to suffer disproportionately from environmental degradation but it is extremely difficult to generalize about the impact of poverty on environment. Thus, although extreme poverty may preclude investment in environmental improvement that has a long payoff, it is also true that the richest countries pose the main threat to the ozone layer and the global climate. . . . It is increasingly being recognized that most environmental problems are less the result of individual large-scale development projects that have gone wrong, than the combined consequences of millions of relatively small-scale activities, unsustainable agricultural practices, small-scale polluting activities, and individual decisions to exploit tropical forest resources." Jeremy Warford and David Wheeler, "Integrating Environmental Issues into Economic Policy Making," *Environment Bulletin—A Newsletter of the World Bank*, volume 4, number 2, Spring, 1992, p. 6.

32. See Sergio Munoz and Juanita Darling, interview with "Jaime Serra Puche," *Los Angeles Times*, August 1, 1993, p. M3.

33. Kennedy, op. cit., p. 102.

34. In Seveso, north of Milan, I interviewed scores of scientists and locals to document the continuing tragedy of the 1976 dioxin eruption from a manufacturing plant in that city. Nearly two decades later, the cancers and congenital birth defects are appearing, but the government has still managed to elude a local coalition that is suing for reparation. In the center of town, a park-like grassy knoll conceals 80,000 buried dogs, cats, and cows, as well as countless birds that literally dropped from the sky, having ingested the dioxin. Many other birds dropped their dioxin-laced feces into fields, or perished in other parts of Europe, where the legacy of Seveso remains potent, like the radiation from Chernobyl. See Michael Tobias, *Voice of the Planet*, New York: Bantam Books, 1990, and film program "Voice of the Planet," Turner Broadcasting Service, 1990.

35. *Global Assessment of Soil Degradation, World Map on Status of Human-Induced Soil Degradation*, Sheet 2, Europe, Africa, and Western Asia,

United Nations Environment Program International Soil Reference and Information Center, Nairobi, 1990.

36. *World Resources 1992/93*, op. cit., p. 177.

37. See "The Price of Pollution," *Options*, International Institute for Applied Systems Analysis, Laxenburg, Austria, September 1990, p. 5.

38. Brian Gardner, "European Agriculture's Environmental Problems," paper presented at the First Annual Conference of the Hudson Institute, Indianapolis, IN, April 1990, p. 7.

39. See Peter Verney, *Animals In Peril—Man's War Against Wildlife*, Provo, UT: Brigham Young University Press, 1979, pp. 38–53.

40. See Kenneth Brower, *One Earth*, San Francisco: Collins, 1990, pp. 84–89.

41. *World Resources 1992/93*, op. cit., p. 66.

42. Ibid., p. 62.

43. See Kathleen Hunt, "Death and Life in a Company Town," *Los Angeles Times Magazine*, July 11, 1993, p. 14.

44. See Carey Goldberg, "Flaring Siberian Gas: Torches Light Way to Eco-Disaster," *Los Angeles Times*, July 25, 1993, pp. A1, A10.

45. *World Resources 1992/93*, op. cit., p. 51.

46. See Anthony Rowley, et al., "Heart of Darkness," *Far Eastern Economic Review*, January 28, 1993, pp. 44–46.

47. As late as 1970, the former president of Mexico, Luis Echeverria stated, "We need to populate our country . . . we do not want to control our population." However, by 1973, Article 3, Part II of the *General Law of Population* acknowledged the need for family planning, and was implemented the following year. At that time, Mexico, along with Yugoslavia and China, were the only countries in the world to guarantee family planning as a constitutional right. This change came about as a realization that Mexico's economic "miracle" was being undermined by explosive population growth and declining per capita income. But, unfortunately, much ecological damage has already been unleashed, abortion is still illegal in the country, and while Mexico's family planners have made certain impressive inroads, the country continues to grow rapidly. See Richard A.Nuccio, Angelina M. Ornelas, and Ivan Restrepo, "Mexico's Environment and the United States," *In the U.S. Interest—Resources, Growth, and Security in the Developing World*, ed. Janet Welsh Brown, Boulder, CO: A World Resources Institute Book, Westview Press, 1990, pp. 19–58.

48. *The Cousteau Almanac*, Garden City, New York: Dolphin Books/Doubleday, 1981, pp. 623–24.

49. Tobias, *Voice of the Planet*, and "Voices of the Planet," op. cit.

50. Julia Preston, "A Mother's Success: Some Babies Lived," *International Herald Tribune*, February 16, 1993, pp. 1, 7.

51. See Robert Repetto, "Accounting for Environmental Assets," *Scientific American*, June 1992, pp. 94–100.

52. *World Resources 1992/93*, op. cit., pp. 11–12.

53. Nor have the Amish remained absolutely free of the outer world. Some have left their communities, migrating into cities, while others have invested in outside corporations. In the case of the Yanomami, in August of 1993, twenty tribal women and children were massacred by gold miners for being in the way of capitalism.

54. See Heilbroner's review of Paul Kennedy, *Preparing for the Twenty-First Century, New York Times Book Review*, February 14, 1993, pp. 1, 25.

55. *World Resources 1992/93*, op. cit., p. 27.

56. See *Taiwan 2000*, Steering Committee, Taiwan 2000 Study, Taipei: Academia Sinica, 1989, pp. 11–41; for analysis of pesticide use, see Walden Bello and Stephanie Rosenfeld, *Dragons in Distress: Asia's Miracle Economies in Crisis*, San Francisco: Institute for Food and Development Policy, 1990, p. 179.

57. Kennedy, op. cit., p. 198.

58. Chauncey Starr, Milton F. Searl, and Sy Alpert, "Energy Sources: A Realistic Outlook," *Science*, volume 256, May 15, 1992, pp. 961–986.

59. See "Population Projections by Race/Ethnicity for California and its Counties 1990–2040" *Report 93 P-1—Official Population Projections*, Sacramento: Demographic Research Unit, April 1993.

60. "Clinton Revokes Abortion Curbs," *Los Angeles Times*, January 23, 1993, p. A16.

61. Lecture at the Asilomar Population and Ecology Conference, Pacific Grove, California, April 31, 1993.

62. For population data on the Southern California "megacity," whose current population is nearly 19 million, see Jane R. Rubin-Kurtzman, et al., "Demographic and Economic Interactions in Trans-Border Cities: The Southern California-Baja California Mega-City," USC-Los Angeles/El Colegio de la Frontera Norte, Tijuana, BC, 1992, a paper presented at the Population Geography Symposium, Commission on Population Geography, International Geographical Union, Los Angeles, Aug. 4–7, 1992.

63. In 1990, the U.S. government invested a total of $231.2 million in population research. In constant dollars, that funding level has not appreciably increased since 1970. *Inventory and Analysis of Federal Popula-*

tion Research Fiscal Year 1992, U.S. Department of Health and Human Services, NIH, Public Health Service, prepared by the Office of Science Policy and Analysis in cooperation with the Center for Population Research, National Institute of Child Health and Human Development for the Interagency Committee on Population Research, p. 4.

64. Norman St. John-Stevas, *The Agonizing Choice—Birth Control, Religion and the Law*, London: Eyre & Spottiswoode, 1971, pp. 20–21.

65. Ibid., p. 222.

66. See Joseph A. McFalls, Jr., "Population: A Lively Introduction," *Population Bulletin*, volume 46, number 2, October 1991.

67. Sixty-five percent of those women under 20 are unmarried; 7.3 percent of the births were to women under 15. California teen pregnancies have risen by 36 percent since 1985. Over 14,000 teenagers have had repeat births in the state. Two-thirds of all pregnancies reported in a recent teen survey resulted from alleged molestation or rape. Many other births were simply unintended. It has been shown that births delayed until after the mother was 20 would save each taxpayer $7177. *Facts of Life in California*, published by the Planned Parenthood Affiliates of California, 1993.

68. Claudia Morain, "A Freedom Fighter Packs for Washington," *Los Angeles Times*, March 8, 1993.

69. "1992 Population Picks and Pans," *Press Release*, Washington, D.C., 1993, p. 2.

70. *Facts of Life in California*, op. cit.

71. Laurie Becklund, "A Cry for Love," *Los Angeles Times*, March 15, 1993, pp. E1, E2.

72. Lauri Becklund, " 'I Wanted Somebody to Love,' " *Los Angeles Times*, March 14, 1993, p. E5.

73. Mark Arax, "The Child Brides of California," *Los Angeles Times*, May 4, 1993, pp. A1, A22. The only population in recent American history to exceed in fertility that of the Hmongs were the Hutterites of the 1930s, whose average TFR was 12, because of early marriage and the renunciation of contraception and abortion. See McFalls, p. 4. Certain Hasidic (Orthodox Jewish) communities on the East Coast have also exhibited high fertility rates since the devastating losses of the Holocaust in Europe. At present, there are approximately 5.6 million Jews in the U.S. and among the Hasidim, TFRs range from 3 to 8.

74. See Peter A. Morrison, "California's Demographic Outlook: Implica-

tions for Growth Management," Santa Monica, CA: RAND Corporation, 1991, P-7738.

75. Currently, 1.3 million illegal immigrants reside in California, nearly 800,000 of them in Los Angeles County. Forty-six percent of all births over deaths in the state are from illegals, or non-citizens, and together they account for 17 percent of all students, and require $3.6 billion of the state's education budget, in addition to another $918 million in health and welfare costs, and $243 million for imprisonment expenditures for those illegals convicted of one crime or another. See Robert Gillespie, "Summary of Los Angeles and California Immigration Facts and Demographic Studies," *Population Communication*, Pasadena, 1992.

76. *Report 93 P-1—Official Population Projections*, op. cit.

77. There are half a dozen other points on the freeway system past which more than 300,000 cars travel every day. The only other comparable traffic would be the Chicago Van Ryan Expressway, with 260,000 cars per day. See *1991 Traffic Volume on California State Highways*, Los Angeles: Cal Trans Publication, 1992.

78. See "Travel & Related Factors in California, 1991," Los Angeles: *Cal Trans Annual Summary*, 1992.

79. *Southern Cal Edison Corporation 1991, Annual Report*, 1991.

80. See *Regional Growth Management Plan*, Los Angeles: Southern California Association of Governments, February 1989.

81. California 1990 State Death File, Data Collection and Analysis Unit, County of Los Angeles, unpublished data, Department of Health Services.

82. For at least 2800 years, there have been philosophers and natural scientists who questioned the links between human mental and physical health, and the health of the environment. The Ionian, Thales, the Indian, Bhagwan Mahavira, several pre-Socratics, the Japanese, Kukai, the Roman, Lucretius, and scores of European scientists, poets, and philosophers since that time are among them. Yet, despite these diverse, impassioned pleas on behalf of Nature, and other than a few sporadic regulations, beginning with forestry laws and anti-pollution codes in France and England in the late middle ages, it was not until the early 1970s that governments formally recognized the human–environmental connection. See "Airs, Waters, Places," in *Works of Hippocrates*, trans. W.H.S. Stone, Cambridge, MA: Harvard University Press, 1948; Lucretius, *The Nature of the Universe*, trans. R.E. Latham, Harmondsworth: Penguin, 1951; Clarence J. Glacken, *Traces on the Rhodian Shore*, Berkeley: University of California Press, 1967; "Stone Inscription for the Sramana Shodo," trans. Allan Grapard, in *The*

*Mountain Spirit*, ed. Michael Tobias and Harold Drasdo, Woodstock, NY: Overlook Press, 1979, pp. 51–59; Keith Thomas, *Man And The Natural World—A History of the Modern Sensibility*, New York: Pantheon Books, 1983.

83. See P.L. Kinney and O. Ozkaynak, "Associations of daily mortality and air pollution in Los Angeles County," *Environmental Research*, volume 54, issue 2, April 1991, pp. 99–120. In this study, data for Los Angeles County was analyzed from 1970–1979. Pollution variables included total oxidants, sulfur dioxide, nitrogen dioxide, carbon monoxide and KM (the measure of particulate optical reflectance). Significant correlations were deciphered between air pollution and mortality. "The results of this study show that small but significant associations exist in Los Angeles County between daily mortality and three separate environmental factors: temperature, primary motor vehicle-related pollutions (e.g., CO, KM, NO2), and photochemical oxidants." Other studies have been able to relate smoke and haze to daily mortality in London and New York City. See J. Schwartz and A. Marcus, "Mortality and air pollution in London: A time series analysis," *American Journal of Epidemiology*, volume 131, 1990, pp. 185–94.

84. See Paula K. Mills, et al., "Ambient Air Pollution and Cancer in California Seventh-day Adventists," *Archives of Environmental Health*, Sep/Oct 1991, volume 46, number 5, pp. 271–280.

85. Anderson, et al., found that asthma is a risk factor for occupational exposure concentrations of sulfuric acid. Adults exposed to ambient oxidant pollution in the Los Angeles area also showed respiratory irritant effects—decrements in forced expiratory function. Other studies have shown decreased lung function in children from ozone. Research into the effects of carbon monoxide exposure on subjects with coronary artery disease has also been shown to significantly hasten the onset of angina. See E.N. Allred, et al., "Short-Term Effects of Carbon Monoxide Exposure on the Exercise Performance of Subjects with Coronary Artery Disease, *New England Journal of Medicine*, number 321, November 23, 1989, pp. 1426–1432. Studies on nitrogen dioxide have also yielded evidence of increase in respiratory failure. In another Los Angeles study, the same team of researchers found ample epidemiological evidence to suggest "substantial health risk from acidic ambient particulate pollution" but because human laboratory evidence is still ambiguous, the necessary levels of regulation for acidic aerosols have not yet been clarified. See Karen R. Anderson, et al., "Controlled Exposures of Volunteers to Respirable Carbon and Sulfuric Acid Aerosols," *Journal of the Air & Waste Management Association*, volume 42, number 6, June 1992, pp. 770–776.

86. See James S. Cannon, *The Health Costs of Air Pollution—A Survey of Studies Published 1984–1989*, Los Angeles: American Lung Association, 1990.

87. Ironically, that Los Angelino is by no means the worst offender—and this is what confounds policy makers when trying to act judiciously from the global perspective. Relative per capita emissions are actually the highest in the United Arab Emirates, Qatar, Brunei, Bahrain, even Cote d'Ivoire and Luxembourg. See *World Resources 1992/93*, pp. 209–210.

88. Mary D. Nichols and Stanley Young, *The Amazing L.A. Environment—A Handbook For Change*, Los Angeles: The Natural Resources Defense Council/Living Planet Press, 1991, p. 72.

89. Ibid., p. 107.

90. Seventeen thousand industrial plants contribute to the overall sewage problem in Los Angeles. Of these, at least 241 plants are required to pretreat their waste, but more than half routinely violate the regulations set forth by the EPA to control them. Nichols and Young, p. 93.

91. See Deborah B. Jensen, Margaret Torn, and John Harte, *In Our Own Hands: A Strategy for Conserving Biological Diversity in California*, Berkeley: California Policy Seminar Research Report, 1990, p. 101.

92. Ibid., p. 105.

93. Two hundred agricultural chemicals used in California that might be harmful to human reproductive health have been identified under the state's *Birth Defects Prevention Act of 1986*. But, chemical companies have kept the bill in court. When and if it is ever free of litigation, the bill—like all related bills thus far passed in the state—is not expected to call into question the fate of these chemicals in the general environment, only in people. Jensen, et al., pp. 110–111.

94. See L. Tomenius, "50-Hz electromagnetic environment and the incidence of childhood tumors in Stockholm County," *Bioelectromagnetics*, volume 7, issue 2, 1986, pp. 191–207. See also F. Gobbato and E. Carli, "Possible consequences of urban pollution caused by radio frequency," *Giornale Italiano Di Medicina Del Lavoro*, volume 7, issue 5-6, 1985 Sep/Nov, pp. 165–174. In 1979, N. Wertheimer and E. Leeper published data that showed a strong correlation between cancer in children and high current flow (60-Hz) near their homes in Denver. See Werrtheimer and Leeper, "Electrical wiring configurations and childhood cancer," *American Journal of Epidemiology*, volume 109, 1979, pp. 273–284. See also *Science*, August 23, 1991, p. 964, and the August 1991 issue of the *American Journal of Epidemiology*, in which it was reported that EMF-exposed male electricians, welders, electric equip-

ment repairmen, and broadcast workers showed an increased risk of breast cancer by 200 percent.

95. See W. Feichtinger, "Environmental factors and fertility," *Human Reproduction*, volume 6, issue 8, September 1991, pp. 1170–5.

96. See L.A. Wallace, "Comparison of risks from outdoor and indoor exposure to toxic chemicals," *Environmental Health Perspectives*, volume 95, November 1991, pp. 7–13.

97. See G.W. Evans, S.D. Colome, and D.F. Shearer, "Psychological Reactions to Air Pollution," *Environmental Research*, volume 45, issue 1, February 1988, pp. 1–15.

98. See Federico G. Puente-Silva, "Explosion Demografica y Salud Mental," *Salud Mental*, 1982 Summary volume 5, number 2, pp. 4–7. See also M. Karno, et al., "Anxiety disorders among Mexican Americans and non-Hispanic Whites in Los Angeles," *Journal of Nervous and Mental Disease*, volume 177, number 4, April 1989, pp. 202–209. A Zero Population Growth urban mental stress test administered in 1988 to 192 existing American cities with populations over 100,000 indicated that the cities with the best scores averaged 116,000 people.

99. Personal conversation with Oscar Mendiola, Management Systems, Los Angeles County Mental Health, November 1992.

100. In 1991, there were 3503 crimes per 100,000 residents throughout California, or 1,073,613 in all: 3876 murders, 10,090 rapes, and 330,916 other violent crimes. *Crime and Delinquency in California in 1991*, Sacramento: Office of the Attorney General, p. 109. See also Shawn Hubler, "Homicides in 1992 Set Record for L.A. County," *Los Angeles Times*, January 5, 1993, pp. B1, B4. Ten-, eleven- and twelve-year-olds are now taking guns to school and routinely shooting at people. See the *Time* feature story, "Los Angeles—Is the City of Angels Going to Hell?" by Richard Lacayo, April 19, 1993, p. 36. The "cult of violence" was perhaps best typified by the New York publishing auction frenzy for *Monster: The Autobiography of an L.A. Gang Member* by one Monster Kody, who committed his first gang killing at the age of eleven.

101. It will be another decade before the loudest of the aircraft are phased out and replaced by the quieter 104 DB level aircraft, like the 747-400, the 767, the MD80 and the 737-300-to-400 series. Most small aircraft are in the 80-88 DB range, but, because they are closer and seem slower to the human ear, normally, they sound much louder. The only airport in the country to thus far control sound is the John Wayne airport in Orange County. Newport Beach residents took the Board of Supervisors to court to prohibit all Stage 2 aircraft from land-

ing at the airport, preferring a quieter life to the higher revenues for businesses, which have in fact amounted to hundreds of millions of lost dollars. See *Federal Air Traffic Activities Handbook*, Washington, D.C.: Federal Aviation Agency, 1990, pp. 52–55.

102. *Facts & Figures, 1991*, Los Angeles, Los Angeles Department of Water and Power, 1991.

103. Nichols and Young, op. cit., pp. 14–15. Half of the water that would otherwise enter San Francisco Bay from the hills inland is diverted to Los Angeles. L.A.'s energy, like its water, comes from eleven other states and two provinces of Canada. The energy is stepped down by thousands of transformers and thousands of miles of overhead wires, eventually entering houses at the 110 volt level. It is not at all uncommon for a California family to use 100 kilowatt hours (the equivalent of 100 pounds of burned coal, or 400 eight-hour days of intense physical human labor), and 700 gallons of water every day. At least half of that energy is provided by coal-burning power plants outside the state. At the Navajo Generating Plant near Page, Arizona, the consequence of this energy hunger is a landscape destroyed by strip-mining, and thousands of carbon dioxide and sulfur dioxide emissions. See *Facts And Figures, 1991*, p. 69.

104. Ehrlich and Ehrlich, op. cit., p. 88. Other reports have shown agricultural inefficiency or crop losses of between 5 and 30 percent. See John Bongaarts, "Can the Growing Human Population Feed Itself?", *Scientific American*, volume 270, number 3, March 1994, p. 40.

105. "Land's End," by Meredith Grenier, *The Santa Monica Outlook*, March 17, 1993, p. B1.

106. Singapore and Hong Kong present similar levels of ecological contradiction and irony. Both cities are densely packed, yet fringed with gardens and mountainous ecosystems. An hour commuter train ride from downtown Hong Kong takes you to 740 acres of the Mai Po Marshes Nature Reserve where as many as 50,000 migratory birds—as many as 100 species—show up during peak periods in winter and spring at the mouth of the Peal River estuary. This region plays host to one of the greatest concentrations of shore birds in the world in April and May as they make their way back to northern China, and Siberia. See Barbara Basler, "In Shadow of Hong Kong, a Bird Haven," *International Herald Tribune*, February 26, 1993, p. 6. Singapore—the most densely populated city in the world, at 44,639 people per 1000 hectares, (0.05 acres per person)—has the highest "peace and quiet" value of any of the 100 largest cities. Yet, beginning in 1983, former Prime Minister Lee Kwan Yew reversed his population

control stance and advocated three or more children per couple in order to prevent the country's economy from being ruined by a perceived shrinking talent pool. The government has even begun a matchmaking service for professionals. Whether the country can maintain its "peace and quiet" with all the new expected births is doubtful. See "Singapore Government Plays Cupid, and It Works," *Sunday Observer*, Jakarta, February 14, 1993, p. 4.

107. *Sliding Toward Extinction: The State of California's Natural Heritage, 1987*, prepared at the request of the California Senate Committee on Natural Resources and Wildlife, commissioned by the California Nature Conservancy, prepared by Jones & Stokes Associates, Sacramento, November 1987, pp. 19–25.

108. Emily Yoffe, "Silence of the Frogs," *New York Times Sunday Magazine*, December 13, 1992, p. 64.

109. Jones & Stokes Associates, op. cit., pp. 29–32.

110. Ibid., pp. 26, 28.

111. At least 15,211,000 acres of those are given over today to cities and irrigated agricultural lands. Particularly severe impacts have been felt on grasslands, coastal scrub, foothill woodlands, and redwood forests. In addition, over the last 150 years, 94 percent of valley wetlands, 89 percent of valley riparian regions, 80 percent of coastal wetlands, and 56 percent of all vernal pools—a total of 6,881,100 acres—have been destroyed, a figure that is part of the total 17,281,000 original habitat acres that have been lost. Between 1947 and 1967, 255,800 acres, 53 percent of the state's total coastal estuaries and wetlands, were filled or dredged, four times the national average of wetlands loss. As far as freshwater lakes, the Central Valley's largest, Tulare Lake, once 600,000 acres in extent, the largest such lake west of the Mississippi, is now extinct. In its place is a "900-square-mile ecological ruin sandwiched between poisons leached from the soil by massive irrigation and dropped from the sky by crop dusters." Writes Ted Williams, "If there is a hell for people who love wild things and wild places, it is California's Tulare Lake. . . ." See Ted Williams, "Death in a Black Desert," *Audubon*, Jan/Feb 1994, p. 24. Of the 921,000 acres of riparian forests in the Central Valley which existed prior to the Gold Rush, only 53,000 acres remained by 1980. See Jensen, et al., pp. 32, 33, 49, and 53.

112. Ehrlich and Ehrlich, op. cit., p. 29.

113. Jensen, et al., pp. 59, 66.

114. Ibid., pp. 81, 86.

115. See Rimmon C. Fay, E. Michael, J. Vallee, and G. Anderson, *Southern California's Deteriorating Marine Environment—An Evaluation of the Health of the Benthic Marine Biota of Ventura, Los Angeles and Orange Counties*, Claremont, CA: Center for California Public Affairs, 1972.

116. Ibid., p. 36.

117. Others include the bull trout, the unarmored threespine stickelback, the San Mateo mint, the wolverine (of which fifty or fewer are left), Yosemite onion, hanging gardens manzanita, the blunt-nosed leopard lizard, the brown pelican and bald eagle, the least bells vireo, the tricolored blackbird, the tidewater goby, the California leaf-nosed bat and Mohave ground squirrel, the desert tortoise and island fox, and Belkins Dune tabanid fly, the Santa Monica shieldback katydid and Palos Verdes blue butterfly, the Los Angeles sunflower, Santa Catalina Island mountain mahogany, the arroyo toad and western snowy plover, the California least tern and western yellow-billed cuckoo. The list goes on. California Department of Fish and Game, the Resources Agency print-out, Sacramento, December 16, 1992.

118. Stephen Mather, *Outlook Magazine*, January 1917.

119. See Michael Tobias, *After Eden—History, Ecology & Conscience*, San Diego: Avant Books, 1985, p. 112.

120. Jensen, et al., op. cit., p. 35.

121. However, it must be pointed out that the state's economy is actually a shambles. During the summer of 1992, California was bankrupt for nearly two months. The City of Los Angeles today faces a half-a-billion dollar debt, such that even were it to lay off all 9100 city employees, the deficit would not go away. See Marc Cooper, "A City Divided," *Los Angeles Reader*, April 2, 1993, volume 15, number 25, p. 9. While California's economy is larger than Canada's, and one-tenth the size of the entire U.S., in early 1993, municipal bond analysts called California the weakest state economically in the union. Federal emergency money following the earthquake of January 1994 is no magic wand.

122. See Paul Dean, "To Catch A Thief," *Los Angeles Times*, January 10, 1993, p. E2.

123. See Maura Dolan, "Nature At Risk In A Quiet War," *Los Angeles Times*, December 20, 1992, pp. A1, A28.

124. *Annual Report on the Status of California State Listed Threatened and Endangered Animals and Plants*, State of California, Sacramento: The Resources Agency, Department of Fish and Game, 1991, pp. 12–13. To take but one example of how such pressures are exerted, consider the last wild river in Southern California, the Santa Clara. One hundred

miles of the river run through Ventura and Los Angeles counties. It has withstood development since 1769 when the diarist for the first Spanish exploring party, Father Juan Crespi, described the river's adjacent stands of forest, its wild grapes and herbs. Today, there is still some forest and at least four endangered species that make the river environ home—the unarmored threespine stickleback fish, the brown pelican, Bell's vireo, and the California least tern. But nearly a dozen different gravel mining companies want at the river. So do developers who have planned for thousands of new homes and an enormous industrial park. Farmers with bulldozers also want a piece of it. See "Charting Future of a Wild River," *Los Angeles Times*, April 25, 1993, p. 1.

125. *U.S. Environmental Quality—23rd Annual Report*, Washington, D.C.: the Council on Environmental Quality, Office of the President, 1993, p. 22.

126. See Maura Dolan, "Endangered Species Act Battles for Its Own Survival", *Los Angeles Times*, Dec. 21, 1992.

127. *U.S. Environmental Quality—23rd Annual Report*, op. cit., pp. 12–15.

128. Kennedy, op. cit., p. 293.

129. The QEM (Quality Environmental Management) Subcommittee has been trying to demonstrate the viability of Total Quality Management (TQM) as a method for preventing pollution in major industries. In surveying the efforts of a dozen industrial plants throughout the U.S. the subcommittee was unable to even agree on a definition of what pollution prevention means. *Total Quality Management—A Framework For Pollution Prevention*, Washington, D.C.: Quality Environmental Management Subcommittee, January 1993, p. 45. Cf. with the efforts of the Chinese. See chapter 2.

130. See Rudy Abramson, "The Superfund Cleanup: Mired in Its Own Mess," *Los Angeles Times*, May 10, 1993, pp. A1, A10.

131. See Richard C. Paddock, "Iron Mountain Mines Defy Efforts to Stop Toxic Flow," *Los Angeles Times*, April 10, 1993, pp. A1, A22, A23.

132. *U.S. Environmental Quality—23rd Annual Report*, op. cit., p. 406. The spill of the *Braer* on the rocky shores of Quendale Bay in the Shetland Islands off northern Scotland, on January 5, 1993, suggests that little, if anything, has been learned or acted upon since the March 1989 *Exxon Valdez* disaster. Congress passed the *Oil Protection Act* in 1990, vowing to get tough on tanker regulation, including interim safety measures for single-hull ships while the double-hull tankers were phased in by 2015, and extending the same safety standards to all foreign ships plying domestic waters. Ninety-five percent

of those ships are single-hulled, like the *Braer* was. But much of that Act, according to the Natural Resources Defense Council, has not yet been implemented.

133. Ibid., p. 414.

134. Cary Fowler and Pat Mooney, *Shattering: Food, Politics, and the Loss of Genetic Diversity*, Tucson: University of Arizona Press, 1990, p. 19.

135. Peter Singer, "The Peta Guide To Animal Liberation," Washington, D.C.: People for the Ethical Treatment of Animals, 1993, p. 21.

136. Nearly 20 percent of all U.S. citizens, in fact, are directly involved in the killing. There are estimated to be over 30 million who killed freshwater fish, over 8 million who killed saltwater fish, 8 million small game hunters, a staggering 10 million big game hunters, and nearly 3 million waterfowl hunters. *U.S. Environmental Quality—23rd Annual Report*, pp. 381–382. Throughout the U.S., some 15 million hunting licenses are purchased each year. The sportsmen's caucus would pressure Congress to revise the *Endangered Species Act* to benefit hunters; to alter the *Marine Mammals Protection Act* to allow trophy hunting of polar bears. These are just a few of their demands. A spokesperson for the U.S. Fish and Wildlife Service was quoted by the *Los Angeles Times* as saying, " . . . it is important for the (Fish and Wildlife) service, state agencies, and the hunting community as a whole to engage in outreach to promote hunting among the young and those who live in urban areas." According to *Los Angeles Times* writer John Balzar, President Clinton, after killing a Maryland duck, said gun control "doesn't have anything to do with hunting." John Balzar, "A Surprise Bounty for Hunters," *Los Angeles Times*, March 16, 1994, pp. A1, A13.

137. *World Resources 1992/93*, op. cit., pp. 24.

138. During the Bush administration, it was estimated that no more than 30,000 jobs were at stake in the northwest, fewer than the massive reductions in countless industries in this country in the past two years. And while the National Association of Home Builders has called it a "crisis" that now adds $3600 to the price of each new home, it has been shown by economists that lumber prices today are where they were in constant dollars during the late 1970s. Furthermore, long-term interest rates have more than offset any price increases in the housing sector.

139. Ehrlich and Ehrlich, op. cit., p. 131.

140. *World Resources 1992/93*, op. cit., p. 176. See Benjamin L. Peierls, Nina F. Caraco, Michael L. Pace, et al., "Human Influence on River Nitrogen," *Nature*, volume 350, April 4, 1991, p. 386.

141. Kennedy, op. cit., p. 306.

142. See, for example, *Environmental Data Compendium 1991*, Paris: OECD—Organization for Economic Co-operation and Development, 1991, p. 114.

143. John Bongaarts, "Can the Growing Human Population Feed Itself?", *Scientific American*, volume 270, number 3, March 1994, pp. 36–43.

144. Michael Specter, "Climb in Russia's Death Rate Sets Off Population Implosion," *New York Times*, March 6, 1994, p. 1.

145. Peter W. Kelly, *Thinking Green! Essays on Environmentalism, Feminism, and Nonviolence*, Berkeley, CA: Parallax Press, 1994, p. 96.

CHAPTER 7
Demographic Madness

1. Quoted in Paul R. and Anne H. Ehrlich, *The Population Explosion*, New York: A Touchstone Book, Simon & Schuster, 1990, p. 159.

2. *The Divine Comedy*, trans. Arthur Livingston, Norwalk, CT: The Easton Press, 1978, Canto I, 1–3.

3. Joseph A. McFalls, Jr., "Population: A Lively Introduction," *Population Bulletin*, Population Reference Bureau, Inc., volume 46, number 2, October, 1991, p. 34.

4. Quoted in Kirk J. Schneider, *The Paradoxical Self—Toward an Understanding of Our Contradictory Nature*, New York and London: Insight Books, 1990, p. 17.

5. See J. Hardoy and D. Satterthwaite, *Squatter Citizen: Life in the Urban Third World*, London: Earthscan Publications Ltd., 1989.

6. Two works stand out as preeminent in terms of their having set the stage for multi-variable analysis: Donella Meadows, et al., *Limits To Growth*, a Club of Rome report, New York: Universe Books, 1972; and *The Global 2000 Report*, Washington, D.C.: Government Printing Office, 1976.

7. Discussed by Norman Myers in *Population, Resources and the Environment—The Critical Challenges*, New York: UNFPA (United Nations Population Fund), 1991, p. 72. See R.S. Chen, W.H. Bender, R.W. Kates, et al., *The Hunger Report*, Providence, RI: World Hunger Program, Brown University, 1990.

8. See Barry James, "At the Peak of the Litter," *International Herald Tribune*, February 26, 1993, p. 1.

9. E.E. Hunt, "The Depopulation of Yap," *Human Biology* 26:20-51, 1954, quoted in Virginia D. Abernethy, *Population Politics—The Choices that Shape Our Future*, New York: Insight Books, 1993, p. 62.

10. See John Keegan, A History of Warfare, New York: Alfred Knopf, 1993, p. 26

11. Myers, op. cit., p. 49.

12. Ibid., p. 74.

13. See G.C. Daily and P.R. Ehrlich, An Exploratory Model of the Impact of Rapid Climate Change on the World Food Situation, Stanford, CA: the Morrison Institute for Population and Resource Studies, Stanford University, 1990.

14. See P. Harrison, "Too Much Life on Earth?" New Scientist, volume 126, 1990, pp. 28–29.

15. See Martin P. and Naomi H. Golding, "Population Policy: Some Value Issues," in Arethusa, ed. John Peradotto and John J. Mulhern, special issue, Population Policy in Plato & Aristotle, volume Eight, number Two, Buffalo, NY: State University of New York, Fall 1975, p. 355.

16. William H. Austin, Philosophy Waves, Particles, and Paradoxes, monograph for Rice University Studies, volume 53, number 2, Spring 1967.

17. For certain psychological aspects of the population paradox, see Kenwyn K. Smith and David N. Berg, Paradoxes of Group Life— Understanding Conflict, Paralysis, and Movement in Group Dynamics, San Francisco and London: Jossey-Bass, 1987.

18. See Stephen Cook, "The Complexity of Theorem Proving Procedures," proceedings of the 3d Annual ACM Symposium on the Theory of Computing, New York: Association of Computing Machinery, 1971; and Richard Karop, "Reducibility among Combinatorial Problems," eds. R.E. Miller and J.W. Thatcher, Complexity of Computer Computations, New York: Plenum Press, 1972.

19. See Lao Tzu, Tao Te Ching, trans. D.C. Lau, Baltimore: Penguin Books, 1963, Chapter 56. See also James J.Y. Liu, Language-Paradox-Poetics— Chinese Perspective, ed. Richard John Lynn, Princeton, NJ: Princeton University Press, 1988.

20. See R.M. Sainsbury, Paradoxes, Cambridge: Cambridge University Press, 1988; Marvin C. Shaw, The Paradox of Intention—Reaching the Goal by Giving Up the Attempt to Reach It, Atlanta: Scholars Press, 1988; and Herbert Weisinger, Tragedy and the Paradox of the Fortunate Fall, East Lansing: Michigan State College Press, 1952.

21. See F.A. Shamsi, Towards a Definitive Solution of Zeno's Paradoxes, Karachi, Pakistan: Hamdard Academy, 1973. Conversely, Heraclitus said that all nature is instability, so much so that one cannot even arrest his own movement long enough to even conceptualize stability.

See also Howard P. Kainz, *Paradox, Dialectic and System—A Contemporary Reconstruction of the Hegelian Problematic,* University Park and London: The Penn State University Press, 1988, p. 38.

22. Special report, *Christian Science Monitor,* July 8, 1992, pp. 9–16.

23. *World Resources 1992–93,* a report by the World Resources Institute in collaboration with the United Nations Environment Program and the United Nations Development Program, New York: Oxford University Press, 1992, p. 30. See H. Jeffrey Leonard, *Environment and the Poor: Development Strategies for a Common Agenda,* New Brunswick, NJ: Transaction Books, 1989, pp. 5–7, 19.

24. *Demographic Indicators of Countries: Estimators and Projections as Assessed in 1980,* New York, United Nations, 1982, pp. 3–28; *Population Profile of the United States 1983/4, Current Population Reports,* Special Studies, Series P23, no. 145, Washington, D.C.; U.S. Government Printing Office, 1985, pp. 4–6; *World Development Report 1984,* New York: Oxford University Press, 1984.

25. *Long-Range World Population Projections: Two Centuries of Population Growth, 1950–2150,* New York: United Nations, forthcoming.

26. Traditionally, women with more children experienced on average a mere 20 to 30 menstrual cycles in their lifetime. The rest of the time, they were pregnant. Fertility restraint has meant far more cycles, which tend to agitate a woman's hormone levels and age her ovaries, breasts, and lining of the womb the way the sun ages the skin. It has been estimated that every twelve menstrual cycles may increase a woman's chances of breast cancer by 0.25 percent—one of the little-discussed downsides of contraception.

27. Peter J. Donaldson and Amy Ong Tsui, "The International Family Planning Movement," *Population Bulletin,* Population Reference Bureau, Inc., volume 45, number 3, November 1990, p. 6.

28. To place this in its context, noticeable rapid population increase first accompanied the industrial revolution beginning about 1760. Steam and colonialism have been generally advanced as the "triggers." Our first billion was reached around 1800. The second billion by 1930; the third by 1960; the fourth by 1975; the fifth by 1987; the sixth in the late 1990s. McFalls, op. cit., p. 34.

29. The World Bank is careful to point out that none of the Working Papers officially represent the views, findings, interpretations or conclusions of the World Bank, its board of directors, its management, or any of its member countries.

30. See My T. Vu, Eduard Bos, and Ann Levin, *Europe and Central Asia Region, Middle East and North Africa Region—Population Projections, 1992–93 Edition, Policy Research Working Papers, Population, Health, and Nutrition*, Population and Human Resources Department, PS 1016, Washington, D.C.: World Bank, November 1992, p. xii.

31. Mozambique is the poorest, at $80 per capita, followed by Tanzania, Somalia, Nepal, Lao People's Democratic Republic, Guinea-Bissau, Malawi, Bangladesh, Chad, Ethiopia, and Bhutan—all poorer than India.

32. Vu, et al., op. cit., vii.

33. Ibid., p. xxxii.

34. Ibid., p. xxiii.

35. Deborah Guz and John Hobcraft, "Breastfeeding and Fertility: a Comparative Analysis," *Population Studies*, March 1991, volume 45, number 1, pp. 91–108.

36. Lester R. Brown, et al., *Vital Signs 1992*, New York: W.W. Norton, 1992, p. 76.

37. Geneva: WHO, 1992, p. 1.

38. Myers, op. cit., p. 78.

39. See P. Winglee, "Agricultural Trade Policies of Industrial Countries," *Finance and Development*, volume 26, number 1, 1989, pp. 9–12.

40. W. Feichtinger, "Environmental factors and fertility," *Human Reproduction*, volume 6 number 8, 1991, pp. 1170–1175. "The effects of detrimental environmental influences on reproduction appear to have been proved and they are alarming." Another study reported that "such remarkable changes in semen quality and the occurrence of genitourinary abnormalities over a relatively short period is more probably due to environmental rather than genetic factors." WHO International Workshop on impact of the environment on reproductive health, Copenhagen, October 1991, findings presented by Elisabeth Carlsen, et al., in the *British Medical Journal*, volume 305, September 12, 1992, p. 612. Yet, human population increase is occurring faster than ever before.

41. See Lawrence H. Summers, "Investing In All the People," a policy research paper, Washington, D.C.: World Bank, 1992.

42. They are: malaria, schistosomiasis, lymphatic fliariasis, river blindness, Chagas' disease, leishmaniasis, leprosy and African sleeping sickness.

43. Stephen S. Morse, "Stirring Up Trouble—Environmental Disruption Can Divert Animal Viruses into People," *The Sciences*, New York: New York Academy of Sciences, volume 30, Sep/Oct 1990, p. 18. See also

Stephen S. Morse and Ann Schluederberg, "Emerging Viruses: The Evolution of Viruses and Viral Diseases," *The Journal of Infectious Diseases*, 1990, 162:1-7; and *Emerging Viruses*, ed. Stephen S. Morse, New York: Oxford University Press, 1993.

44. *Emerging Infections—Microbial Threats to Health in the United States*, Institute of Medicine, ed. Joshua Lederberg, Robert E. Shope, and Stanley C. Oaks, Jr., Washington, D.C.: National Academy Press, 1992, pp. 72-73.

45. Richard Preston, "Crisis In The Hot Zone," *The New Yorker*, October 26, 1992, p. 59.

46. Lederberg, et al., op. cit., p. 90.

47. E. Narayanan, "Dams And Public Health," *Sanctuary*, Bombay, volume X, number 1, 1990, p. 65.

48. Eugene Linden, "Megacities," *Time Magazine*, January 11, 1993, p. 36.

49. Lederberg, et al., op. cit., p. 52.

50. Ehrlich and Ehrlich, op. cit., p. 151.

51. Vu, et al., op. cit., p. xxx.

52. See My T. Vu, Eduard Bos, and Ann Levin, *Latin America and the Caribbean Region (and Northern America) Population Projections 1992-93 Edition, Policy Research Working Papers: Population, Health, and Nutrition*, Population and Human Resources Department, Washington, D.C.: World Bank, November 1992, WPS 1033.

53. See Cynthia B. Lloyd, "The Contribution of the World Fertility Surveys to an Understanding of The Relationship Between Women's Work and Fertility," *Studies in Family Planning 1991*; volume 22, number 3, pp. 144-161. See also Shenyang Environmnetal Assessment—Phase I, Summary of Existing Information and Recommendations for Action. Project Manager, Jeff Gersh, Denver, CO: a project of the China Environmental Fund, 1994, p. 2.

CHAPTER 8
A Global Truce

1. One thinks back to the Kassites, Hyksos, Semites, Dorians, Achaeans, Etruscans, Celts, Assyrians, Hittites, Teutons, Bantus, Taishans, Mongols, Tibetans, Parthians, Kushans, Tungus, Turks, Aztecs, Vikings, Huns, Ottomans, English, French, Spanish, Portuguese, Moghuls, Germans, Italians, Austrians, Bosnians, Japanese, Chinese, Koreans, Soviets, Nederlanders, Belgians, Danes, Swedes, Iraqis, Iranians, and Americans, to name but a few of the invading waves of aggressors.

2. See, S.L.A. Marshall, *Men Against Fire*, New York: William Morrow, 1947.

3. See Louis Rene Beres, *Apocalypse: Nuclear Catastrophe in World Politics*, Chicago: University of Chicago Press, 1980.

4. Jacques Lusseyran, *And There Was Light*, trans. Elizabeth R. Cameron, New York: Parabola Books, 1991, p. 308.

5. See the debate over the "money value of human life" in *Environmental Policy Benefits: Monetary Valuation*, Paris: Organization for Economic Co-Operation and Development, 1989, p. 31. With reference to the approximate number of deaths from AIDS each year: In the U.S. in 1994, $30,000 per death is being spent in research, multiplied times fifty thousand deaths—a figure that crosses over between pro-active and retroactive expenditures. See Tomas J. Philipson and Richard A. Posner, *Private Choices and Public Health: the AIDS Epidemic in an Economic Perspective*, Cambridge, MA: Harvard University Press, 1994. With respect to the condors, the state of California has invested, to-date, $15 million on the 76 remaining birds (of which 70 are in zoos) or, $197,368.42 per avian, the approximate cost of bone-marrow surgery for a woman with breast cancer, which most health insurers refuse to pay on grounds that such surgery is "experimental." The difference, of course, is that human females are not an endangered species.

6. Robert Nozick, *The Nature of Rationality*, Princeton: Princeton University Press, 1993, p. 177.

7. Among the precedents, I include the examples of the Fore of New Guinea, the early Tahitian tribals, the Tasaday of Mindanao, the Yurock Indians of California, the West Malaysian Semai, the Tanzanian Hadza, the Australian Murngin and Walbiri, the Dani of Irian Jaya, the Bishnoi of Rajasthan (India), the Pantaneiros, Qollahuaya and Runas of Bolivia. In addition, with respect to vegetarian cultures, one must mention the ancient Essenes of Qumran (Israel), the Gnostics, the Manichaean religion founded by Mani in A.D. 242, the Paulicians of Armenia, followers of Paul of Samosata, and the Bulgarian Paterenes of the tenth century, followers of the priest named Bogomil, who, in Italy, were to found the Cathar creed, also known as the Albigensians in France. According to these latter two religious groups, in the manner of the Jains, all animals were endowed with a soul and were thus considered sacred and inviolate. Later Christian monastic orders held similar views, including the Carthusians, Trappists, and Camaldolese. Most Seventh Day Adventists adhere to the same vegetarian philosophy, based upon the passage in *Genesis* 1:29, that reads "And God said, behold, I have given you every herb bearing seed, which is

upon the face of all the earth, and every tree, in which is the fruit of a tree yielding seed; to you it shall be for meat." See Colin Spencer, *The Heretic's Feast—A History of Vegetarian-ism*, London: Fourth Estate, 1993. See also Christopher Key Chapple, *Nonviolence to Animals, Earth, and Self in Asian Traditions*, Albany, NY: State University of New York Press, 1993, pp. 113–114. See also Ashley Montagu, ed., *Learning Non-Aggression—The Experience of Non-Literate Societies*, Oxford: Oxford University Press, 1978; and Michael Tobias, ed., *Mountain People*, Norman: University of Oklahoma Press, 1986.

8. See Michael Tobias, *Life Force—The World of Jainism*, Fremont, CA: Asian Humanities Press, 1991. See also the Michael Tobias film "Ahimsa—Nonviolence," Denver: KRMA/PBS, 1987.

9. Translated by Professor Padmanabh S. Jaini, University of California at Berkeley.

10. *The Akaranga Sutra*, Book I, Lecture I, Lesson 3, in Jaina Sutras, trans. Hermann Jacobi, New Delhi: Motilal Banarsidass, pp. 4–5.

11. See Jeremy Rifkin, *Beyond Beef—The Rise and Fall of the Cattle Culture*, New York: A Dutton Book, 1992, p. 284. See also Michael Tobias, *Rage & Reason*, New Delhi: Rupa & Company, 1993.

12. See Lester R. Brown, Christopher Flavin, Hal Kane, *Vital Signs 1992*, New York: W.W. Norton, 1992, pp. 28, 30.

13. See John Balzar, "Creatures Great and—Equal?" by John Balzar, *Los Angeles Times*, December 25, 1993, pp. A1, A30. See the 67 Factsheets, published by PETA. In a poll concerning the extinction of species conducted by Peter D. Hart Research Associates and Professor Stephen R. Kellert of Yale University, 70 percent of those questioned had never heard of biodiversity loss, and only 50 percent of so-called environmentalists knew anything about it. Yet, once respondents were informed about the basics of biodiversity, 80 percent proclaimed human obligations to protect flora and fauna, and only 30 percent of those interviewed cited economic reasons for doing so. In fact, when informed of basic ecological systems, 46 percent of respondents even marshaled in support of legislation to protect insects. For a list of animal friendly organizations, products, and animal rights activities, write (and join) People for the Ethical Treatment of Animals (PETA), P.O. Box 42516, Washington, D.C. 20015-0516, telephone (301) 770-PETA. In addition, important information can be obtained from the Ark Trust Inc., P.O. Box 8191, Universal City, CA 91608-0191, telephone (818) 786-9990; Last Chance for Animals (LCA), 18653 Ventura Blvd., Suite 356, Tarzana, CA 91356; In Defense of Animals, 816 W. Francisco Blvd., San Rafael, CA 94901; and EarthSave, 706

Frederick St., Santa Cruz, CA 95062. Animal friendly consumer
catalogues include: Heartland Products Ltd., P.O. Box 218, Dakota
City, Iowa 50529, telephone (800) 441-4692; Creatureless Comforts,
702 Page Street, Stoughton, MA 02072, telephone (617) 344-7496;
and Aesop Inc., P.O. Box 315, North Cambridge, MA 02140,
telephone (617) 628-8030.

14. See the *University of California at Berkeley Wellness Letter*, volume 9,
issue 6, March 1993, pp. 4–5. The human taste for meat is totally out
of kilter with a physiology designed for vegetarianism. Our tooth structure (small canines, powerful molars), our enzymes for digestion, and
our twenty-six feet of intestines, are all characteristic of herbivores, not
carnivores. Meat-eating produces intense acids in the stomach. Because
of the length of our intestines, nine days are required to digest a hamburger. During that time, the deoxycholic acid produced by the
stomach to help in digestion is converted by clostridia bacteria in the
intestine into toxins alleged to be carcinogenic. With the meat remaining so long in the intestines, that sustained toxicity, and the animal
fats that stay solid at 98.6 degrees—clog up, and lengthen the duration
necessary to pass through the colon. During that extended transit time,
the bowel walls reabsorb the toxins, which the body has been trying to
expel.

15. See Billy Ray Boyd, *For the Vegetarian in You*, San Francisco: Taterhill
Press, 1993, p. 16. For information on B-12, see John Robbins, *Diet for a
New America*, Walpole, NH: Stillpoint Publishing, 1987, p. 300. As a
point of cultural evolution, both Burger King and McDonald's are selling veggie burgers at a few of their outlets. In Holland, the McDonald's
"Groeteburger" is made of potatoes, peas, corn, carrots, onions, and
spices. Burger King has been selling a Spicy Bean Burger in the U.K.
for several years, made of kidney beans, carrots, onions, peppers, green
chilies, chili powder, mixed herbs, bread crumbs, potato flakes, garlic,
white pepper, salt, wheat flour and corn flour. As of 1993, neither of
these "burgers" had been introduced into the U.S. market, though
countless brands of soy patties can be found in any health food store.
One of the most prominent meat-eating cultures to educate will be
Canada's, a country possessed of the largest wilderness area in the
world (64 percent of the whole country), one of the lowest population
densities (30 per 1000 hectares), but the highest proportion of grain fed
to livestock of any other nation at 79 percent. Canada's 27 million
residents kill 10 million pigs, 12 million cows, 755,000 sheep and
goats, 416,000 horses, and 108 million chickens every year. This animal
hell thus represents the highest *per capita* contribution to methane in

the upper atmosphere of any country. It is ironic, considering that Montreal was the sight of the first international ozone protocol.

16. E.O. Wilson, *The Diversity of Life*, Cambridge, MA: Belknap Press of Harvard University Press, 1993, p. 299.

17. It has been suggested that a lifestyle based upon the killing and consumption of animals will result in a lower fertility rate than that among farmers. The reason offered is that the meat-eating diet (of the nomad) discourages the mushing up of baby food, and that constant travel makes the company of too many infants a difficult proposition. Furthermore, goes the theory, the farmer's cereal production is a sufficient substitute for breastfeeding. With less breastfeeding, a woman ovulates more rapidly following a birth, thus inviting yet another birth. But for every example, there is a contradiction. While it may be true that the TFR among, say, !Kung bushmen, is low, it is also low among the agriculturally settled, essentially vegetarian Bhutanese, as well as the shifting agriculturalists of the Andaman and Nicobar islands. But the point is deflated on other grounds: the fact is, the world is largely settled, its meat-eating completely uncoupled from a nomadic way of life. One other contradiction concerns organic farming. Dennis Avery, Director of the Hudson Institute Center for Global Food Issues has pointed out that "by 2050, there would be billions of organically induced starvation deaths" if organic farming took over. Because organic farming tends to be slightly lower yielding, according to Avery, a 5 to 7 times increase in the amount of land farmed would be necessary to feed the population in the year 2050, or 30 to 40 million square miles. That, he argues, would spell the absolute doom of all wildlife. But, there is a counterargument, which holds that several new generations of biotechnology will likely enhance the efficiency of organic farming, thus nullifying such spatial concerns. Several European governments are now subsidizing farmers who want to convert to organic agriculture. See Nicolas Lampkin, "Organic Farming and Agricultural Policy in Europe," paper presented at the Conference on Sustainable Agriculture and Agricultural Policy, Quebec Ministry of Agriculture, Fisheries, and Food, Quebec City, Canada, November 1990, p. 7.

18. Quoted in Mark Lutz, "Humanistic economics: history and basic principles," in *Real-Life Economics—Understanding Wealth Creation*, ed. Paul Ekins and Manfred Max-Neef, London and New York: Routledge, 1992, p. 100. See John Ruskin, *"Unto This Last:" Four Essays On The First Principles Of Political Economy,"* London: Smith, Elder and Co., 1862.

19. Ruskin, op. cit., pp. 60–61.

20. See James Fallows, "How the World Works," *The Atlantic Monthly*, December 1993, pp. 61–87.

21. Ruskin, op. cit., p. 43.

22. Dr. B.D. Sharma, "On Sustainability—the voice of the disinherited," *Sanctuary*, volume XII, number 3, 1992, pp. 14–25. It is interesting to note how far the economic transformation in the developing world has gone. Writing in 1782, Jean-Jacques Rousseau said, "I have noticed that only in Europe is hospitality put up for sale. Throughout Asia you are lodged free of charge." Jean-Jacques Rousseau, *Reveries of the Solitary Walker*, London: Penguin Classics edition, 1979, pp. 151–152.

23. From George Wald's essay, "The Human Enterprise . . .", *Population, Environment and People*, ed. Noel Hinrichs, Council on Population and Environment, New York: McGraw-Hill, 1971, pp. 220–221.

24. See Samir Amin, *Maldevelopment—Anatomy of a Global Failure*, London and Tokyo: Zed Books and U.N. University Press, 1990. See also Vaclav Smil, *Global Ecology—Environmental change and social flexibility*, London: Routledge, 1993.

25. See Paul Ekins, "Sustainability first," *Real-Life Economics—Understanding Wealth Creation*, ed. Paul Ekins and Manfred Max-Neef, London and New York: Routledge, 1992, p. 418.

26. The Central Intelligence Agency has been putting out warnings for some time that such growth in China is spiraling "out of control," causing enormous cost-of-living increases, and presenting the leadership—even the reformists—with a presentiment of civil war and economic upheaval. Already, in 1992–1993, over two hundred incidents of rural rebellion across China were reported in Hong Kong newspapers. It is thus difficult to predict how long China's boom will last.

27. Consider the fact that Canada's labor force has seen half a million jobs go to the U.S. in three years of a working Canadian-U.S. Free Trade Agreement, all because average wages are approximately a dollar an hour cheaper in the U.S. But in the Maquiladora Zone, salaried Mexican labor is $13 an hour cheaper than in the U.S., or, as low as 20 cents an hour, in some cases, a third of what it is in much of Mexico. Maquiladora workers enjoy no benefits, no security, and the industries are subject to effectively no environmental regulation. Moreover, because of Mexico's more than $100 billion dollar debt, and austerity conditions imposed by the World Bank to force the country to pay back the money, wages are not likely to rise. Said President Salinas, in response to the Zapatista National Liberation Army's war in Chiapas state, "We know that needs and inequalities persist. We know that benefits and opportunities still are not tangible realities for many." The

passage of NAFTA was pressured by lobbyists for the National Association of Manufacturers, which represents the largest corporations in the U.S., with the most to gain by maintaining low wages, and few, if any regulatory constraints. Environmental transgressions are the province of NAFTA's tri-national disputation board. Yet these are not ecologists, but rather business people whose mandate is to maintain the free movement of trade. The issue of low wages and population control is open to mercenary speculation. Do 20-cents-per-hour laborers have fewer children than those earning 60 cents per hour? Would several million 20-cents-per-hour earners consume less CDs, nails, blouses, toothpaste, or suitcases than those taking home 60 cents per hour? Clearly, neither the 20 centers nor the 60 centers will be buying BMWs any time soon. But what about the low-priced goods? For a "green party" point of view on NAFTA, see Mike Castro, "NAFTA?" *The Green Letter*, Santa Fe, NM: The New Mexico Green Party, Winter 1993–94, pp. 2, 10.

28. See William D. Montalbano, "Dangers of Narrowing the Field," *Los Angeles Times*, December 23, 1993, p. 1. Three examples of the biotechnology takeover, cited by Martin Khor (the "Ralph Nader of Malaysia") are: the development of fructose, which has eliminated jobs for tens of thousands of sugar workers in developing countries; an industrial substitute for natural gums, developed in New York, which has wiped out the Sudan's gum arabic market; and vanilla beans, produced artificially by a Texas biotech firm, which has subsequently put 70,000 natural vanilla growers in Madagascar out of business. See Martin Khor, "Development, Trade, and the Environment: A Third World Perspective," *The Future of Progress*, ed. Helena Norberg-Hodge and Peter Goering, Bristol, UK, and Berkeley, CA: The International Society for Ecology and Culture, 1992, p. 33.

29. See Marla Cone, "The Mouse Wars Turn Furious," *Los Angeles Times*, May 9, 1993, pp. A1, A16.

30. Ian Golding, Odin Knudsen, and Domonique van der Mensbrugghe, "Trade Liberalisation, Global Economic Implications," Washington, D.C.: World Bank and OECD Development Center, 1993.

31. *World Resources 1992–93*, a report by the World Resources Institute in collaboration with the United Nations Environment Program and the United Nations Development Program, New York: Oxford University Press, 1992, p. 27. Even before passage of the GATT, demographer John Bongaarts had estimated that per capita global income would rise from its current average of $3000 to $36,000 by the year 2100. John Bongaarts, "Population Growth and Global Warming," *Population and Development Review*, volume 18, number 2, June 1992, p. 306.

32. Virginia D. Abernethy, *Population Politics—The Choices that Shape Our Future*, New York: Insight Books, 1993, pp. 40, 290. Abernethy quotes from Cuban demographers S. Diaz-Briquets and L. Perez, *Cuba: The Demography of Revolution*, Washington, D.C.: Population Reference Bureau, 1981: "The fertility rises in almost every age group suggest that couples viewed the future as more promising and felt they could now afford more children."

33. Lester Brown, "The New World Order," in *State of the World*, New York: W.W. Norton, 1991, p. 10.

34. *World Resources 1992–93*, op. cit., p. 2.

35. Paul Ekins, "Markets, ethics and competition," in *Real-Life Economics—Understanding Wealth Creation*, ed. Paul Ekins and Manfred Max-Neef, London and New York: Routledge, 1992, p. 324.

36. B. Schneider, *The Barefoot Revolution*, London: IT Publications, 1988, p. 22. That 100 million figure, according to Schneider, leaves an additional 1.9 billion others who are among the poorest on the planet and untouched by any kind of social welfare. Reaching them would cost another $13 billion per year, by his estimates.

37. See John Clark, *Democratizing Development: The Role of Voluntary Organizations*, West Hartford, CT: Kumarian Press, 1991, p. 39.

38. *Inventory of Population Projects in Developing Countries Around the World 1986/1987*, editor-in-chief, Jyoti Shankar Singh, New York: UNFPA.

39. John Bongaarts, W. Parker Mauldin, and James F. Phillips, "The Demographic Impact of Family Planning Programs," *Studies in Family Planning*, volume 21, number 6, Nov/Dec 1990, p. 305.

40. *World Resources 1992–93*, op. cit., p. 87.

41. United Nations, *World Fertility Survey—Major Findings and Implications*, Vaarburg, Netherlands: WFS, International Statistical Institute, 1984, quoted by Norman Myers, *Population, Resources and the Environment— The Critical Challenges*, New York: United Nations Population Fund, 1991, p. 111.

42. Kate Bales, "Adoption: The World Baby Boom," *International Herald Tribune*, February 13–14, 1993, p. 16.

43. See Shari Roan, "A Big Change in How Women Get the Pill?" *Los Angeles Times*, August 17, 1993, pp. E1, E6.

44. See Shanti R. Conly and J. Joseph Speidel, *Global Population Assistance—A Report Card on the Major Donor Countries*, Washington, D.C.: Population Action International, 1993.

45. By analogy, UNDP (the United Nations Development Program) has estimated that to prevent a human death through preventative care

costs between $100 and a $1000, versus a spread of between $500 and $5000, on average, for operative cures.

46. In the U.S., some 160 members of Congress who signed the so-called Beilenson Letter (named after Rep. Tony Beilenson, a California Democrat) are now seeking assurances from the House Foreign Operations Subcommittee of Appropriations for a commitment from Washington of $725 million a year for international population policies, in keeping with the Amsterdam Declaration. This would mean increasing the U.S. per capita contribution to roughly $2.86 annually.

47. Conly and Speidel, op. cit., pp. 39–40.

48. "Cost of Crime: $674 Billion," data compiled by Sara Collins from U.S. Departments of Commerce and Justice; Hallcrest Systems; Ted Miller—National Public Services Research Institute; Mark Cohen—Vanderbilt University; Dorothy Rice—University of California at San Francisco, *U.S. News and World Report*, January 17, 1994, p. 40.

49. For example, despite President Clinton's lifting of the Mexico City policy of the Reagan administration, under the existing 1973 *Helms/Hyde Amendments* in the Foreign Assistance Act, government funds are still not eligible for supporting abortion.

50. It is, furthermore, estimated that every migrant who attempts a border crossing eventually gains entry. On this latter point consider the following theoretical trade-offs, the idealistic pros, and realistic cons of an "Immigration Corps of Conservationists." What if the United States adopted a forceful policy of helping rainforest countries to reduce their populations? Recall Emma Lazarus's sonnet, "The New Colossus," inscribed, in part, in the pedestal of the Statue of Liberty:

> Give me your tired, your poor,
> Your huddled masses yearning
> to breathe free,
> The wretched refuse of your
> teeming shore,
> Send these, the homeless,
> tempest-tost, to me:
> I lift my lamp beside the
> golden door.

Suppose, to this effect, the U.S. opened up its borders to by-invitation-only immigration; targeting those countries and regions where the largest concentration of poor migrant workers were wreaking damage upon significant biodiversity hotspots? Might we receive this potentially enormous labor pool into the most ecologically bereft portions of the United States where they might be retrained for purposes of basic ecological clean-up in this country? There are, of course, no guarantees that this sudden exodus would not leave a conspicuous window of opportunity in the marketplace for successive waves of Third World rainforest despoilers, unless some deal were worked out with those country's governments, based upon the agreed-to perception that substantial per capita income gains to the participating countries would result from respectively lowered populations. Many have argued that as a large immigrant population's demand for public services increases, and the funds for such services diminish, racism and violence result, as witnessed in Germany. Nor can the world afford any more U.S. consumers. Furthermore, the job situation would pose a major challenge. Consider that between the issuance of 439,000 temporary work permits, and 390,000 green cards during the first half of 1992, plus all the illegal aliens trying to enter the labor market, there were more foreign workers than new jobs in 1992 in the United States. Some have argued that it is precisely human boundaries that provide inherent management systems for nature. That by keeping immigrants out, hence forcing nations to deal with their own regional problems, humankind can exercise better micromanagement over its ecological dilemmas. But this argument has been tested for several thousand years and it has faltered. The U.S. has the opportunity to test a new model. In a sense, the U.S. *is* that model. Hence, a thorough analysis of population and ecology issues in the Santa Barbara–Ensenada megacity is quite critical to extrapolating trends and formulating alternative scenarios for the future.

In a sense, one terribly simplistic version of a net asset movement-across-borders policy has already been debated for a decade, namely, the charging of a dollar toll from everyone who enters the United States, never mind the intended purpose (which was to pay for more border patrols). Simple enough? Hardly. Government experts have called it untenable, a bureaucratic morasse, an invitation to corruption, gridlock, a violation of countless treaties with Canada and Mexico, and so on. And yet, the vision of an ecological airlift holds promise, given the fact there will be between 70 and 100 million environmental refugees moving across borders no matter what governments do. Transforming those numbers into a meaningful force for nature should be possible in this world; it would truly assert that political and cultural boundaries and priorities are far less crucial to this Earth than ecological ones. Yet, as Virginia Abernathy points out, even were America to welcome 100 million immigrants every year, it would not make a dent in the world population crisis. See Virginia D. Abernethy, *Population Politics—The Choices that Shape Our Future*, New York, Insight Books, 1993, p. 54. Some have argued for a re-orientation of the fifty states in the U.S. according to the geography of watersheds, not state boundaries. One might also reflect upon new political boundaries based upon animal corridors, or green belts; political micromanagement of human communities based upon a new framework for assessing ecological boundaries of the United States. If anything, by comparison with the inevitable problems of an immigrant conservation corps, a domestic one seems, at least, plausible. Franklin Roosevelt, Al Gore, and Jerry Brown have each outlined different frameworks for motivating such a work force.

51. Ibid., p. 45.

52. Satish Kumar, "Ecological Economics," an interview with Manfred Max-Neef, *Resurgence*, number 155, Nov/Dec 1992, p. 9. The UNDP estimates that the total "aid" from the South to the North equals about $500 billion per year.

53. If social engineering were possible, a year of nationally coordinated voluntarism might be a useful beginning. What would it mean? All production capacity would be scaled to modest consumption, not sales. Considering the fact that less than 5 percent of the world's population (Americans) consume 75 percent of all global energy, egalitarian Depression, that is, a nationally mobilized voluntarism, would have enormous ecological benefits. There would thus be no legal speculation, no investments, no profits, only essential work and hand-to-mouth consumption—whether of food or gas or goods. Public utilities would determine what was essential, on a per capita basis, and immediately scale back all household energy budgets. There would be individuals who fared better than others, but at a much reduced rate of disparity. Meanwhile, the average consumption would plummet. All exchanges of goods would be noted, according to value previously enjoyed by such sales. And that cumulative tabulation would be applied as savings against the national debt. What would prevent hoarding and greed? A quota system, again determined the way sugar and gasoline rations in a time of war are determined—at the source of supply. This would inevitably reduce the discrepancies between the highest and lowest wage earners, which, in the United States, exceed all other nations. Such national voluntarism is very much like pure communism, and while its energies and intentions are noble and pure, and certainly account for much good that takes place in local communities, as a national principle it unfortunately suffers from three outstanding deficiencies: 1) it requires a totalitarian political sysem, 2) it fosters a necessarily enormous and vulnerable bureaucracy for tabulating and policing such a system, and 3) the size of an aging population, and in-

creasing demands for social welfare, will out-tax the ability of governments and citizens to pay for the demographic transition. Furthermore, communism was an environmental disaster, while the socialist entitlement schemes in countries like Sweden have only exacerbated recession and high unemployment.

54. Robert Repetto, "Accounting for Environmental Assets," *Scientific American*, June 1992, p. 96. See also Ernst Lutz and Saleh El Serafy, "Environmental and Resource Accounting: An Overview," Environment Department Working Paper number 6, Washington, D.C.: World Bank, June 1988.

55. In Paul Ehrlich's Foreward to Arthur R. Tampli and John W. Gofman, *Population Control Through Nuclear Pollution*, Chicago: Nelson-Hall, 1970, p. ix.

56. See Thomas L. Freidman, "Cold War Without End," *New York Times Magazine*, August 22, 1993, p. 29. See also Pamela Murphy, "Coming Clean: the Department of Energy, the Public, and the Nation's Nuclear Waste Mess," *The League of Women Voters*, March/April 1994, pp. 14–19.

57. *The Environment At Risk—Responding to Growing Dangers*, National Issues Forums, Iowa: Public Agenda Foundation, 1989, p. 13.

58. Michael Soule, "A Vision for the Meantime," *The Wildlands Project, Wild Earth—Special Issue*, Cenozoic Society, p. 7, 1992.

59. See *Real-Life Economics—Understanding Wealth Creation*, ed. Paul Ekins and Manfred Max-Neef, London and New York: Routledge, 1992, pp. 335–339. In a similar vein, Paul Hawken has argued convincingly for such societal overhauls as the replacement of the conventional tax system with enlightened "green fees," and a "shift from electronic to biologic literacy." Writes Hawken, "That an average adult can recognize one thousand brand names and logos but fewer than ten local plants is not a good sign." See Paul Hawken, "Ecology Is a Serious Business," *Resurgence*, number 163, March/April 1994, pp. 16–19.

60. "The New Earth 21, Action Program for the Twenty-First Century," paper presented by the Japanese Ministry of International Trade and Industry at the U.S./Japan Conference on Global Warming, Atlanta, June 3, 1991.

61. Brown, et al., op. cit., p. 72.

62. The EPA and Chicago Board of Trade took in 171 bids from countless electric utilities, brokerage firms and private investors, competing for some 150,000 allowances sold by sealed bid, each allowance providing the holder a one ton of sulfur dioxide emission per year, or the right to retrade the allowance on the open market. The results—$21 million in

sales and the first step toward cutting in half the nation's $SO_2$ annual emissions. Nevertheless, in the first quarter of 1993, with the new Clinton administration on line, pollution control companies lost a glaring 12.5 percent on the stock exchange, while oil and gas exploration showed a 24.7 percent gain, second only to semiconductor stocks. The problem with the auction was that a single power company from North Carolina evidently acquired 57 percent of the 150,010 permits that were sold specifically to delay having to install anti-pollution scrubbers under federal law.

63. *World Resources 1992–93*, op. cit., p. 203.

64. See Sam Rashkin, et al., *Energy Technology Status Report*, Final Report, Table 4, Sacramento: California Energy Commission, 1991, p. 73.

65. *Green Lights: A Bright Investment in the Environment*, Washington D.C.: U.S. Environmental Protection Agency, 1991, p. 1.

66. Wilson, op. cit., p. 320.

67. See Garret Hardin's groundbreaking essay, "The tragedy of the commons," *Science* 162, December 13, 1968, pp. 1243–1248.

68. See Gillian Phillips, "Vox Populi—The Real Rio?" *Populi*, October 1991, pp. 10–11.

69. *U.S. Environmental Quality—23rd Annual Report*, Washington, D.C.: the Council on Environmental Quality, Office of the President, 1993, pp. 140–143. It is worth pointing out, however, that while Prime Minister Gro Harlem Brundlandt of Norway stressed, in her opening remarks at Rio, the inseparability of the issues of population and environment, ironically Agenda 21 mentions no population action plan, and the Rio Declaration itself only says that "states should . . . promote appropriate demographic policies." The Holy See, which still believes that the world's resources can feed 40 billion people, clearly had a hand in the final documents, basing its defiant and unrealistic bias on the adamant proclamations of Pope Pius XI who declared in 1930 that contraception in any form was a "sin against nature . . . shameful and intrinsically vicious." (Pope Pius XI, "Casti Connubii" [On Christian marriage], Encyclical Letter, December 31, 1930.) Ironically, Pope John XXIII in his essay "Mater and Magister," (July 1961) unambiguously raised the spectre of overpopulation in developing countries. To accommodate a baby boom, which church doctrine prevented him from directly confronting with the obvious contraceptive prescriptions, he instead chose to recommend economic reforms that would guarantee more equitable distribution of wealth worldwide. In his 1968 Encyclical outlawing all birth control (*Humanae Vitae*), Pope Paul VI nevertheless acknowledged that the greatest challenge of modern times was the

rapid increase in population. Fully endorsing *Humanae Vitae*, the current Pope John Paul II (Karol Wojtyla, former poet and playwright), today tours the world condemning birth control, telling AIDS-ridden African populations, and Latino baby boomers in Denver, that the use of condoms is a "sin against nature." While it is believed that no more than one in ten Catholics in the developed nations ascribe to the destructive position taken by *Humanae Vitae*, the Pope's influence in the Catholic developing world, such as Brazil, is widespread.

70. Hilary French, *After the Earth Summit: The Future of Environmental Governance*, Worldwatch Paper #107, Washington, D.C.: Worldwatch Institute, March 1992, p. 28.

71. U.S. Senate, Committee on Environment and Public Works, "Title III—Emergency Planning and Community Right to Know," in *Superfund Amendments and Reauthorization Act of 1986* (P.L. 99-499), Washington, D.C.: U.S. Government Printing Office, 1987, pp. 169–201.

72. That figure comes from a 1988 National Survey of Family Growth, sponsored by the National Center for Health Statistics.

73. Paul Ehrlich and E.O. Wilson, "Biodiversity Studies: Science and Policy," *Science*, volume 253, August 16, 1991, p. 761.

74. "World Scientists' Warning to Humanity," Union of Concerned Scientists, Cambridge, MA: April 1993.

# Selected Bibliography

## ANTARCTICA

Naveen, Ron, and Colin Monteath, Tui De Roy, and Mark Jones, *Wild Ice—Antarctic Journeys*, Washington, D.C.: Smithsonian Institution Press, 1990.

Tobias, Michael, "The Next Wasteland—Can the Spoiling of Antarctica Be Stopped?" *The Sciences*, published by the New York Academy of Sciences, Mar/Apr 1989.

## CHINA

Addison, James Thayer, *Chinese Ancestor Worship*, Shanghai: Church Literature Committee of the Chung Hua Sheng Kung Hui, 1927.

Aird, John S., *Slaughter of the Innocents—Coercive Birth Control in China*, Washington, D.C.: The AEI Press, 1990.

Angang, Hu, and Wang Yi, "The Future Conflict between Population and Grain Output in China," *Chinese Journal of Population Science*, volume 2, number 3, Research Center of Ecology and Environment of the Chinese Academy of Sciences, New York: Allerton Press, 1990.

"China—Accessibility of Contraceptives," *Asian Population Studies Series*, number 103-B, Economic and Social Commission for Asia and the Pacific, Bangkok, United Nations, 1991.

*China Population Today*, volume 9, number 2, April 1992, China Population Information and Research Center, Beijing.

"China's Population and Development," *Country Report*, prepared by the Chinese Delegation to the Fourth Asian and Pacific Population Conference, Bali, Indonesia, August 1992.

Choe, Minja Kim, and Noriko O. Tsuya, "Why Do Chinese Women Practice Contraception? The Case of Rural Jilin Province," *Studies In Family Planning*, volume 22, number 1, Jan/Feb 1991.

Conly, Shanti R., and Sharon L. Camp, "China's Family Planning Program: Challenging the Myths," *Country Study Series #1*, the Population Crisis Committee, Washington, D.C., 1992.

Croll, Elisabeth, Delia Davin, and Penny Kane, eds., *China's One-Child Family Policy*, New York: St. Martin's Press, 1985.

Dalsimer, Marlyn and Laurie Nisonoff, "Collision Course," *Cultural Survival Quarterly*, Winter 1992, volume 16, number 4.

Geping, Dr. Qu, *Environmental Management in China*, United Nations Environment Programme, Beijing: China Environmental Science Press, 1991.

Hanxian, Luo, *Economic Changes in Rural China*, trans. Wang Huimin, China Studies Series, Beijing: New World Press, 1985.

Jacobson, Jodi L., "Coerced Motherhood Increasing," *Vital Signs*, Lester Brown, et al., New York: W.W. Norton, 1992.

Jian, Song, Tuan Chi-Hsien, and Yu Jingyuan, *Population Control in China*, New York: Praeger, 1985.

Jingneng, Li, "Reproduction Worship and Population Growth in China," *Chinese Journal of Population Science*, volume 4, number 1, 1992.

Jinsheng, Wei, "On the Operating Mechanism of Population Control," *Chinese Journal of Population Science*, volume 4, number 1, 1992.

Johansson, Sten and Ola Nygren, "The Missing Girls of China: a New Demographic Account," *Population and Development Review*, volume 17, number 1, March 1991.

Mosher, Steven W., *A Mother's Ordeal*, New York: Harcourt Brace, 1993.

Orick, George, "Combating Poverty in Rural China: Change Comes to Anwang," *The Ford Foundation Report*, volume 22, number 4, Winter 1992.

Peiyun, Madame Peng, "Efforts to strengthen family planning work at the grass-roots level in order to fulfill the population goals of the eighth five-year plan"—a report of a National Conference to Exchange Experience in Family Planning at the Grassroots Level, Beijing, November 1990.

Ping, Tu, "Birth Spacing Patterns and Correlates in Shaanxi, China," *Studies In Family Planning*, volume 22, number 4, 1991.

Sharma, Arvind, ed., *Women In World Religions*, Albany, NY: State University of New York Press, 1987.

Tianlu, Zhang, "The Marriage Pattern and Population Reproduction of the National Minorities of China," *Population Research*, volume 7, number 4, 1990.

Tien, H. Yuan, with Zhang Tianlu, Ping Yu, Li Jingneng, and Liang Zhongtang, "China's Demographic Dilemmas," *Population Bulletin*, volume 47, number 1, June 1992, Washington, D.C.

Wei, Dr. Zhang De, Project Leader, "Study of Stainless Steel Ring and Copper T IUD Efficacy, Use, Cost/Benefit and Conversion in China," UNFPA Project CPR/91/P43, *Project Report*, May 21, 1992, Beijing.

Xinzhe, Peng, "China's Population Control and the Reform in the 1980s," Population Research, volume 7, number 3, 1990.

Yangzhong, Ye, "A tentative study of the childbearing aspirations of contemporary farmers," Population Studies, number 4, 1988.

Yi, Zeng, et al., "An Analysis of the Causes and Implications of the Recent Increase in the Sex Ratio at Birth in China," the Institute of Population Research, Peking University, Working Paper, 1992.

Zhanjun, Gao, and Zhang Xueming, "Old-Age Insurance for 'Single Daughter Households' in Rural Changyi," China Population Today, volume 9, number 2, April 1992, China Population Information and Research Centre, Beijing.

Zhenming, Xie, "An Evaluation of the Social Effectiveness of One-Child Policy in China," Population Research, volume 7, number 1, 1990, Population Research Institute, Anhui University, Hefei, Anhui Province, China.

Zuofu, Cai, Population Research, "A Sociological Perspective of and Measures to Overcome Difficulties in Rural Population Control," Policy Research Office, volume 7, number 3, 1990, Jiangling County Party Committee, Hubei Province.

## INDIA/SOUTH ASIA

Anthropological Survey of India, The Life of the People of India, Government of India, 1993.

Antony, T.V., "The Family Planning Programme Lessons From Tamil Nadu's Experience," a policy paper, New Delhi: Center for Policy Research, March 1992.

Aris, Michael, Views of Medieval Bhutan: The Diary and Drawings of Samuel Davis, 1783, Washington D.C.: Smithsonian Institution, 1982.

Bahadur, K.P., Population Crisis in India, New Delhi: National Publishing House, 1977.

Banwari, Pancavati, Indian Approaches to Environment, trans. Asha Vohra, New Delhi: Shri Vinayaka Publications, 1992.

Bedi, Ramesh, and Rajesh Bedi, Indian Wildlife, New Delhi: Brijbasi Printers Private Ltd., 1984.

Bose, Ashish, "India's Quest for Population Stabilisation: Progress, Pitfalls and Policy Options," Materials on Demogaphic Questions, Herausgeber Wiesbaden: Bundesinstitut fur Bevolkerungsforschung, 1991.

Bumiller, Elisabeth, May You Be the Mother of a Hundred Sons—A Journey among the Women Of India, New Delhi: Penguin Books, India, 1991.

Cernada, Dr. George P., and Dr. A.K. Ubaidur Rob, "Pakistan's fertility and family planning: Future directions," *Journal of Family Welfare*, volume 38, number 3, Bombay, September 1992.

Chengappa, Raj, "Family Planning—The Great Hoax," *India Today*, October 31, 1988.

Coburn, Broughton, *Nepali Ama—Portrait of a Nepalese Hill Woman*, Santa Barbara, CA: Ross-Erikson Ltd., 1981.

Conly, Shanti R., and Sharon L. Camp, "India's Family Planning Challenge: From Rhetoric To Action," *Country Study Series #2*, the Population Crisis Committee, Washington, D.C., 1992.

"Equal Rights Must Start in the Cradle," *UNICEF News—Third World Women*, number 76(3), July 1973.

Furer-Haimendorf, Christoph von, *Tribes of India—The Struggle for Survival*, Berkeley: University of California Press, 1982.

Govind, H., "Recent Developments in Environmental Protection in India: Pollution Control," *Ambio*, volume 18, number 8, 1989.

Hersh, Seymour M., "On the Nuclear Edge," *The New Yorker*, March 29, 1993.

Holloway, Marguerite, "Sustaining the Amazon," *Scientific American*, July 1993.

Joshi, P.K., and Dayanatha Jha, "Environmental Externalities in Surface Irrigation Systems in India," *Environmental Aspects of Agricultural Development*, International Food Policy Research Institute, Washington, D.C. 1990.

Kannan, K.P., and K. Pushpangadan, "Agricultural Stagnation and Economic Growth in Kerala: An Exploratory Analysis," *Working Paper #227*, Tiruvananthapuram, Kerala: Centre for Development Studies, 1988.

Khanka, S.S., "Developments in Ghimtal Gadhera Catchment Area," *Environmental Issues in Agricultural Development*, ed. H. Ramachandran, New Delhi: Concept Publishing Company, 1990.

Krishnakumar, S., "Kerala's Pioneering Experiment in Massive Vasectomy Camps," *Studies in Family Planning*, volume 3, number 8, August 1972.

Kurien, John, and T.R. Thankappan Achari, "On Ruining the Commons and the Commoner—The Political Economy of Overfishing," Tiruvananthapuram, Kerala: Centre For Development Studies, *Working Paper #232*, 1989.

Mehra, G.N., *Bhutan—Land of the Peaceful Dragon*, New Delhi: Vikas, 1974.

Menon, A. Sreedhara, *Political History of Modern Kerala*, Madras: S. Viswanathan Publishers, 1987.

Moraes, Dom, A Matter of People, New York: Praeger Publishers, 1974.

Narayanan, E., "The backwaters of Kerala," Sanctuary, volume X, number 5, Sep/Oct 1990.

Nath, Kamal, Address to UNCED, New Delhi: Ministry of Forests and Environment, Government of India, June 1992.

National Report to UNCED, "Environment and Development—Traditions, Concerns and Efforts in India," New Delhi: Ministry of Environment and Forests, Government of India, June 1992.

Nieuwenhuys, Olga, Childrens' Lifeworlds—gender, welfare and labour in the developing world, London: Routledge, 1994.

Nikore, Monika, and Marian Leahy, "The Oasis Makers," World Monitor, February 1993.

O'Malley, L.S.S., Popular Hinduism—The Religion of the Masses, Cambridge: Cambridge University Press, 1935.

Muller, F. Max, ed., "The Laws of Manu," The Sacred Books of the East, volume 25, New Delhi: Motilal Banarsidass, 1967.

Olschak, Banche, C., Ancient Bhutan—A Study on Early Buddhism in the Himalayas, Zurich: Swiss Foundation for Alpine Research, 1979.

Panandiker, V.A. Pai, "Dynamics of Population Growth: Implications for Environment and Quality of Life," New Delhi: Center for Policy Research, February, 1992. Presented at the International Conference on Population and Environment, University of Michigan, Ann Arbor, October 1990.

Panandiker, V.A. Pai, Ajay K. Mehra, with P.N. Chaudhuri, People's Participation In Family Planning, New Delhi: Center for Policy Research, Uppal Publishing House, 1987.

Panandiker, V.A. Pai, and P.K. Umashankar, "Fertility Control-Induced Politics Of India," New Delhi: Center for Policy Research and Family Welfare Program, 1991.

Panwar, H.S., Ruchi Badola, Bitapi Sinha, and Chandrashekhar Silori, "The Wilderness Factor—Protected areas and people—compatible strategies," Sanctuary, volume XII, number 5, 1992.

Pathak, K.B., and F. Ram, "Pattern of Population Growth and Redistribution in India," Demography India, volume 20, number 1, 1991.

"Population, Development and the Environment: An Agenda for the 1990s," proceedings of the National Conference of Non-Governmental Organizations," 14–16 April 1991, New Delhi Family Planning Association of India, Bombay.

Prakash, B.A., "Educated Unemployment in Kerala: Some Observations Based on a Field Study," *Working Paper #224*, Tiruvananthapuram, Kerala: Centre For Development Studies, 1988.

Pye-Smith, Charlie, *In Search of Wild India*, New Delhi: UBSPD, 1993.

"Quality of Life and Problems of Governance in India," New Delhi: Center for Policy Research, 1990, n.a.

Rahmani, Asad R., "Save the Grasslands," *Sanctuary*, volume X, number 4, Jul/Aug 1990.

Saksena, S.K., *Environmental Planning Policies and Programmes in India*, New Delhi: Shipra Publications, 1993.

Sankaran, Ravi, "Dudhwa—a forest held to ransom," *Sanctuary*, volume X, number 5, Sep/Oct 1990.

Sharma, Arvind, ed., *Our Religions*, San Francisco: HarperCollins, 1993.

Shetty, Kavitha, "Cradles of Mercy," *India Today*, February 28, 1993.

Simon, Prof. H., and Shri B.B.L. Sharma, *Training Modules for Incorporation of Family Welfare Messages*, New Delhi: National Institute of Health and Family Welfare, 1990.

Sivanandan, P., D. Narayana, and K. Narayanan Nair, "Land Hunger and Deforestation: A Case Study of the Cardamom Hills in Kerala," *Working Paper #212*, Tiruvananthapuram, Kerala: Center for Policy Studies, 1985.

Thakor, V.H., and Vinod M. Patel, "The Gujarat State Massive Vasectomy Campaign," *Studies in Family Planning*, volume 3, number 8, August 1972.

*This Is Earth Calling—A Children's Eye View of the Earth's Population and Environment Problems*, New Delhi: UNFPA, in association with Konark Publishers PVT LTD., 1993.

Thompson, Catherine, "Women, Fertility and the Worship of Gods in a Hindu Village," *Women's Religious Experience*, ed. Pat Holden, London and Canberra: Croom Helm Ltd., 1983.

Tikader, B.K., *Threatened Animals of India*, Calcutta: Zoological Survey of India, 1983.

Tobias, Michael, *Life Force—The World of Jainism*, Berkeley: Asian Humanities Press, 1991.

——— , ed., *Mountain People*, Norman: University of Oklahoma Press, 1986.

Wadia, Smt. Avabai, "Population And Family Planning in the 8th Five Year Plan," Part II, Bombay: Family Planning Association of India Publications, 1992.

Warwick, Donald P., *Bitter Pills: Population policies and their implementation in eight developing countries*, Cambridge: Cambridge University Press, 1982.

White, J.C., "Journey Through Bhutan," *National Geographic*, Washington, D.C., 1909.

Wolpert, Stanley, *A New History of India*, 2d Ed., New York: Oxford University Press, 1982.

Zimmer, Heinrich, *Philosophies of India*, ed. J. Campbell, Bollingen Series XXVI, Princeton, NJ: Princeton University Press, 1969.

INDONESIA

*Act of the Republic of Indonesia #10 of 1992* concerning Population Development and Development of Prosperous Family, State Ministry of Population and the Environment, The Republic of Indonesia, Jakarta, 1992.

"Adding Choice to the Contraceptive Mix: Lessons from Indonesia," *Asia-Pacific Population & Policy #19*, Honolulu: Population Institute of the East-West Center, December 1991.

al-Khayyir, Sheikh Abdul Rahman, "The Attitude of Islam towards Abortion and Sterilization," a report of the proceedings of the International Islam Conference held in Morocco, December 1971, published in *Islam and Family Planning 2*, Beirut: International Planned Parenthood Federation, 1974.

Ancok, Djamaludin, "Incentive and Disincentive Programs in the Indonesian Family Planning," Yogyakarta: Population Studies Center, Gadjah Mada University, 1984.

"Biodiversity Action Plan For Indonesia," volume 1, Final Draft, BAPPENAS—National Development Planning Bureau, Jakarta, August 1991.

"Biodiversity Management In Indonesia," *Tropical Biodiversity*, volume 1, number 1, 1992.

Bishop, Carl Whiting, with Charles Greeley Abbot and Ales Hrdlicka, *Man from the Farthest Past*, volume Seven of The Smithsonian Series, Washington, D.C.: Smithsonian Scientific Series, 1944.

Caldecott, Julian, "Biodiversity Management in Indonesia," *Tropical Biodiversity*, volume 1, number 1, 1992.

*Caring for the Earth: A Strategy for Sustainable Living*, IUCN, UNEP, WWF, Gland, Switzerland: IUCN, 1991.

Cooey, Paula M., William R. Eakin, and Jay B. McDaniel, eds., *After Patriarchy—Feminist Transformations of the World Religions*, New York: Orbis Books, 1991.

*Country Report*, Indonesia, Fourth Asian and Pacific Population Conference, August 1992, Bali, Indonesia.

Curtin, Leslie B., Charles N. Johnson, Andrew B. Kantner, and Alex Papilaya, "Indonesia's National Family Planning Program: Ingredients of Success," *Occasional Paper #6*, Report #91-134-136, Population Technical Assistance Project, USAID-Jakarta and Office of Population Bureau for Research and Development, Washington, D.C.: Agency for International Development, December 11, 1992.

de Beer, J.H., and M.J. McDermott, *The Economic Value of Non-timber Forest Products in Southeast Asia*, Amsterdam: Netherlands Committee for IUCN, 1989.

Dixon, J.A., and P.B. Sherman, *Economics of Protected Areas*, Honolulu: East-West Center, 1990.

"Government Energy Scenario and an NGO Alternative," *Environesia*, volume 6, numbers 4/5, November 1992.

*Government of Indonesia Report*, Departemen Kehutanan (Department of Forestry), Jakarta, 1991.

Hugo, Graeme J., Terence H. Hull, Valerie J. Hull, and Gavin W. Hones, *The Demographic Dimension in Indonesian Development*, New York: Oxford University Press, 1987.

*Indonesia Country Profile—Annual Survey of Political and Economic Background 92–93*, The Economist Intelligence Unit, New York: Business International, 1993.

*Indonesian Country Study on Biological Diversity*, Ministry of State for Population and Environment, KLH, Jakarta, 1992.

"Indonesia Forests, Land and Water—Issues in Sustainable Development," Report #7A22-IND, Country Department V, Asia Regional Office, Document of the World Bank, June 5, 1989.

"Indonesia-Growth, Infrastructure and Human Resources," Report #10470-IND, Country Department III, East Asia & Pacific Regional Office, Document of the World Bank, May 26, 1992.

IUCN, Ministry of Forestry/FAO, *Indonesian Tropical Forestry Action Programme*, 3 vols., Jakarta: U.N. Food and Agriculture Organization, 1991.

*Legal Aspects of Family Planning in Indonesia*, The Committee on Legal Aspects of the Indonesian Planned Parenthood Association, Law and Population Monograph Series #4, Law and Population Programme, The Fletcher School of Law and Diplomacy, 1971.

Lim, Lin, *Employment Situation and Training Needs of Women Workers in Garments and Food/Tobacco Processing in Surabaya*, Indonesia Report, submitted to UNDP ILO, part of project preparation for the Industrial Skills Development Project, April 1992.

MacKinnon, J., and K. MacKinnon, *Review of the Protected Areas System in the Indo-Malayan Realm*, Gland, Switzerland: IUCN, 1986.

_____ , *Biological Diversity in Indonesia: A Resource Inventory*, World Wide Fund for Nature, Bogor, 1990.

Mahmood, Tahir, *Family Planning: The Muslim Viewpoint*, New Delhi: Vikas Publishing, n.d.

Mulder, Niels, *Mysticism and Everyday Life in Contemporary Java—Cultural Persistence and Change*, Singapore: Singapore University Press, 1978.

Musallam, B.F., *Sex and Society in Isalm—Birth control before the nineteenth century*, Cambridge: Cambridge University Press, 1983.

Naipaul, V.S., *Among the Believers—An Islamic Journey*, New York: Vintage Books, 1982.

Nitisastro, Widjojo, *Population Trends In Indonesia*, Ithaca, NY: Cornell University Press, 1970.

Palmore, James A., and Masri Singarimbun, "Marriage Patterns and Cumulative Fertility in Indonesia," *Working Papers #64*, East-West Population Institute, Honolulu: East-West Center, November 1991.

Pariani, Siti, David M. Heer, and Maurice D. Van Arsdol, Jr., "Does Choice Make a Difference to Contraceptive Use? Evidence from East Java," *Studies in Family Planning*, volume 22, number 6, Nov/Dec 1991.

Parsons, J.S., "What Makes the Indonesian Family Planning Programme Tick," National Family Planning Coordinating Board, UNFPA, *Populi*, volume 11, number 3, 1984.

"Recent Fertility Trends in Indonesia, 1971–1987," *Working Papers #63*, East-West Population Institute, Honolulu: East-West Center, November, 1991.

"Recent Information on Blue Circle and Discount Program," National Family Planning Coordinating Board, Jakarta, 1991.

Repetto, Robert, "Accounting for Environmental Assets," *Scientific American*, June 1992.

Repetto, Robert, et al., *Wasting Assets: Natural Resources in the National Income Accounts*, Washington, D.C.: World Resources Institute, 1989.

Sadjudin, Haerudin R., "Status and Distribution of the Javan Rhino in Ujung Kulon National Park, West Java," *Tropical Biodiversity*, volume 1, number 1, 1992.

Sharma, Narendra, and Raymond Rowe, "Managing the World's Forests," *Finance & Development*, Washington, D.C., June 1992.

Speth, James Gustave, "Coming to Terms: Toward a North-South Bargain for the Environment," *World Resources Institute Issues and Ideas*, Washington, D.C.: World Resources Institute, 1990.

*The Land Resources of Indonesia: A National Overview*, Regional Physical Planning Programme for Transmigration, Overseas Development Administration, Ministry of Transmigration, 1990.

Utomo, Budi, Sujana Jatiputra, and Arjatmo Tjokronegoro, *Abortion in Indonesia—A review of the literature*, Jakarta: Faculty of Public Health, University of Indonesia, May 1982.

WALHI and YLBHI, *Mistaking Plantations for Indonesia's Tropical Forest: Indonesia's Pulp and Paper Industry, Communities, and Environment*, Jakarta, 1992.

Ward, Sheila, "Service Delivery Systems and Quality of Care in the Implementation of NORPLANT in Indonesia," Population Council of New York, February 1990.

Weeks, J., "The Demography of Islamic Nations," *Population Bulletin*, volume 43, number 4, December 1988.

Wells, M.P., *Can Indonesia's Biological Diversity be Protected by Linking Economic Development with National Park Management?* Three case studies from the Outer Islands, a report to the World Bank, 1989.

Wells, M.P., Katrina Brandon, and Lee Hannah, "People and Parks: An Analysis of Projects Linking Protected Area Management with Local Communities," draft report, Washington, D.C.: The World Bank, 1990.

Williams, Walter L., *Javanese Lives—Women and Men in Modern Indonesian Society*, New Brunswick, NJ: Rutgers University Press, 1991.

## KENYA

Brown, G., "The Viewing Value of Elephants," *The Ivory Trade and the Future of the African Elephant*, Ivory Trade Review Group, volume 2, Technical Reports, 1989.

Chanda, Glory, and Ackim Tembo, "The Status of Elephants on the Zambian Bank of the Middle Zambezi Valley," *Pachyderm*, number 16, 1993.

*Children & Women in Kenya: A Situation Analysis*, Nairobi: Government of Kenya, U.N. Children's Fund, 1992.

*Development Policy for Arid and Semi-Arid Lands (ASAL)*, Nairobi: Government of Kenya, September 1992.

*District Contraceptive Prevalence Differentials Study—A Case Study of Six Districts*, Population Studies and Research Institute, University of Nairobi, January 1992.

Dumm, John J., Richard M. Cornelius, Roy Jacobstein, and Barbara Pillsbury, "Helping Services Meet Demand—An Assessment of A.I.D. Assistance to Family Planning In Kenya," Center for Communication Programs, Johns Hopkins University, 1992.

Goldberg, Howard I., Malcolm McNeil, and Alison Spitz, "Contraceptive Use and Fertility Decline in Chogoria, Kenya," *Studies in Family Planning*, volume 20, number 1, Jan/Feb 1989.

Hamilton, E. Douglas & Association, "Identification Study for the Conservation and Sustainable Use of the Natural Resources in the Kenyan Portion of the Mara-Serengeti Ecosystem," European Development Fund of the European Economic Community, *Interim Report*, May 1988.

Handby, J.P., et al., "Population Changes in Lions and other Predators," in A.R.E. Sinclair and M. Norton-Griffiths, eds., *Serengeti—Dynamics of an Ecosystem*, Chicago: University of Chicago Press, 1979.

"Kenya: Creating Islands of Green," *Environmental Impact—An Occasional Magazine of the Bellerive Foundation*, 1993.

*Kenya Demographic and Health Survey 1993—Preliminary Report*, National Council for Population and Development, Ministry of Home Affairs and National Heritage, Central Bureau of Statistics, Office of the Vice President, and Ministry of Planning and National Development, prepared with technical assistance from Demographic and Health Surveys, Macro International Inc., Columbia, Maryland, under contract to USAID, September 1993.

*Kenya Wildlife Service—A Policy Framework and Development Program, 1991–1996*, Nairobi: Kenya Wildlife Service, 1990.

Levine, Robert A., "Maternal Behavior and Child Development in High-Fertility Populations," *Fertility Determinants Research Notes 2*, New York: The Population Council, September 1984.

Mbithi, Professor Philip M., "Implication of Socio-Economic and Cultural Factors on Population and Development," University of Nairobi, *Report on the Second National Leaders' Population Conference*, NCPD, Republic of Kenya, September 1989.

Mbugua, Wariara, *Review and Assessment of Population Data, Policy and Research, Family Planning Programs and Population Information, Education and Communication—A Case Study of Kenya*, Nairobi: UNFPA, October 1988.

Mburugu, E.K., *Factors Related to Stock Ownership and Population Movements; and Perception of Land Pressure and Other Environmental Changes among the Rendille in Marsabit District*, University of Nairobi, Integrated Project in Arid Land (IPAL) Technical Report #F-2—Man in the Biosphere Program, p. 2, n.d.

Murage, Muthoni, "The Law Relating to Women in Kenya," *Nairobi Law Monthly*, #50, Aug/Sep 1993.

Njogu, Wamucii, "Trends and Determinants of Contraceptive Use in Kenya," *Demography*, volume 28, number 1, February 1991.

*PCEA Chogoria Hospital Annual Report*, Chogoria, 1992.

Poole, Joyce H., "Kenya's Initiatives in Elephant Fertility, Regulation, and Population Control Techniques," *Pachyderm*, #16, 1993, publication of the World Conservation Union (IUCN) and WWF.

*Scout Programme Africa Region—Curriculum Guide and Resource Manual for Family Life Education*, World Scout Bureau, Africa Regional Office, and the Margaret Sanger Center of Planned Parenthood of New York, 1988.

"United Nations Population Fund Proposed Projects and Programmes," Recommendation by the executive director, Assistance to the Government of Kenya, support for a comprehensive population programme, Fortieth session, Governing Council of the United Nations Development Programme, Geneva: UNFPA, May 1992.

Wilson, Alison, "Sacred Forests and the Elders," *The Law of the Mother—Protecting Indigenous Peoples in Protected Areas*, ed. Elizabeth Kemf, San Francisco: Sierra Club Books, 1993.

## MALI

Boorman, John, *West African Butterflies and Moths, A West African Nature Handbook*, London: Longman House Ltd., 1978.

*Mali: Conservation de l'Environnement Dans le Delta Interieur Du Fleuve Niger: Document de Synthese*, Rapport Technique #3, Bureau of Africa, USAID, June 1987.

Olivier, Robert, *The Gourma Elephants of Mali: A Challenge for the Integrated Management of Sahelian Rangeland*, Nairobi: UNEP, 1983.

*Plan D'Action Sur La Strategie National Pour La Planification Familiale*, Bamako: Ministere de la Santa Publique et des Affaires Sociales a la Banque Mondiale/ Sante, Population et Hydraulique Rurale (PDSII), 1990.

*Rapport Annuel 1992*, Bamako: Ministere Du Developpement Rural et de L'Environnement, Direction Nationale Des Eaux et Forets et Operation Parc Nationale de la Boucle Du Baoule, 1993.

*Rapport Final Phase I: projet du Conservation Dans le Delta Interieur du Mali,* IUCN/CDC, 1988.

Sanogho, Nampaa, *Les Elephants du Gourma,* Bamako: Water and Forestry Department [DNEF], 1980.

Serle, W., G.J. Morel, and W. Hartig, *A Field Guide to the Birds of West Africa,* London: Collins Publishers, 1986.

*Summary Report,* Bamako: Ministry of Planning and National Directorate of Planning, under the International Conference of Donors for the Economic Recovery and Development of the Republic of Mali, volume 1, December 1982.

Thayer-Rozat, Annik, *Plants Medicinales du Mali,* Bordeaux: Ocoe Publishing, 1981.

"United Nations Population Fund Proposed Projects and Programmes," Recommendation by the executive director, Assistance to the Government of Mali, support for a comprehensive population programme, Fortieth session, Governing Council of the United Nations Development Program, New York: UNFPA, June 1993.

Warshall, Peter, *Mali—Biological Diversity Assessment,* Office of Arid Lands Studies, University of Arizona, Contract #AFR-0467-C-00-8054-00, (Natural Resources Management Support Project AID, Bureau of Africa, Office of Technical Resources, Natural Resources Branch, Project #698-0467), Prime Contractor: E/DI, Washington, D.C.: March 1989.

PAN-AFRICAN

Ahlback, A.J., *Industrial Plantation Forestry in Tanzania: Facts, Problems, Challenges,* Ministry of Natural Resources and Tourism, Dar Es Salaam, n.d.

Ajayi, Femi, et al., "Tools of the Trade: Do Farmers Have the Right Ones?" *African Farmer,* number 5, November 1990.

Armen, Jean-Claude, *Gazelle-Boy,* New York: Universe Books, 1974.

Bechky, Allen, *Adventuring in East Africa—The Sierra Club Travel Guide to the Great Safaris of Kenya, Tanzania, Rwanda, Eastern Zaire, and Uganda,* San Francisco: Sierra Club Books, 1990.

Bongaarts, John, and Frank Odile, "Biological and Behavioral Determinants of Exceptional Fertility Levels in Africa and West Africa and West Asia," *African Population Conference,* Dakar, 1988, Liege, Belgium: International Union for the Scientific Study of Population, 1988.

Caldwell, John, I.O. Orubuloye, and Pat Caldwell, "Fertility Decline in Africa: A New Type of Transition?" *Population and Development Review*, 1992.

*Central Africa: Global Climate Change and Development*, Technical Report, Landover, Maryland: Biodiversity Support Program, a consortium of the WWF, The Nature Conservancy, World Resources Institute, and U.S. Agency for International Development, 1993.

Cochrane, Susan H., and S.M. Farid, "Fertility in Sub-Saharan Africa: Analysis and Explanation," *World Bank Discussion Paper #43*, Washington, D.C.: The World Bank, 1989.

Cross, Nigel, and Rhiannon Barker, *At the Desert's Edge: Oral Stories from the Sahel*, London: Panos/SOS Sahel, 1992.

*DHS—Demogaphic and Health Surveys Newsletter*, volume 5, number 1, 1992.

Dorst, Jean, and Pierre Dandelot, *A Field Guide to the Larger Mammals of Africa*, London: Collins Publishers, 10th ed., 1990.

Goliber, Thomas J., "Africa's Expanding Population: Old Problems, New Policies," *Population Bulletin*, volume 44, number 3, Washington, D.C.: Population Reference Bureau, November 1989.

Isaac, Glynn, "Food Sharing and Human Evolution: Archaeological Evidence from the Plio-Pleistocene of East Africa," *Journal of Anthropological Research 34*, 1978.

Johanson, Donald, and Maitland Edey, *Lucy: The Beginnings of Mankind*, New York: Simon & Schuster, 1981.

Leakey, Richard, *The Making of Mankind*, New York: E.P. Dutton, 1981.

"Levels and Trends of Contraceptive Use As Assessed in 1988," *Population Studies #110*, New York: United Nations, 1989.

Matthiessen, Peter, and Mary Frank, *Shadows of Africa*, New York: Harry Abrams Inc., 1992.

Newmark, William, "The Selection and Design of Nature Reserves for the Conservation of Living Resources," *Managing Protected Areas in Africa—Report from a Workshop on Protected Area Management in Africa*, compiled by Walter J. Lusigi, Mweka, Tanzania: UNESCO, World Heritage Fund, 1992.

O'Haire, Hugh, "Think Again," *Populi*, December 1993/January 1994.

Poirier, Frank E., *Fossil Evidence—The Human Evolutionary Journey*, 3d ed., St. Louis: C.V. Mosby, 1981.

Poulton, Robin, and Michael Harris, *Putting People First: Voluntary Agencies and Third World Development*, London: Macmillan, 1992.

Stephens, Patience W., Eduard Bos, My T. Vu, and Rodolfo A. Bulatao, *Africa Region Population Projections, 1990–1991 Edition*, Working Papers, Population and Human Resources Department, Washington, D.C.: The World Bank, February 1991, WPS 598.

*Sub-Saharan Africa—From Crisis to Sustainable Growth—A long term perspective*, Washington, D.C.: World Bank, 1989.

Wright, Patricia C., "Ecological Disaster in Madagascar and the Prospects for Recovery," in *Ecological Prospects—Scientific, Religious, and Aesthetic Perspectives*, ed. Christpher Key Chapple, Albany, NY: State University of New York Press, 1994, pp. 11–24.

## CALIFORNIA

"Annual Report on the Status of California State Listed Threatened and Endangered Animals and Plants," State of California, Sacramento: The Resources Agency, Department of Fish and Game, 1991.

Bakker, Elna S., *An Island Called California*, Berkeley and Los Angeles: University of California Press, 1971, p. xi.

Barrett, S.A., and E.W. Gifford, *Miwok Material Culture—Indian Life of the Yosemite Region*, Yosemite Natural History Association Inc., Yosemite, 1933.

*Facts & Figures, 1991*, Los Angeles: Los Angeles Department of Water and Power, 1991.

*Facts of Life in California*, published by the Planned Parenthood Affiliates of California, 1993.

Fay, Rimmon C., with E. Michael, J. Vallee, and G. Anderson, *Southern California's Deteriorating Marine Environment—An Evaluation of the Health of the Benthic Marine Biota of Ventura, Los Angeles and Orange Counties*, Claremont, CA: Center for California Public Affairs, 1972.

Gillespie, Robert, "Summary of Los Angeles and California Immigration Facts and Demographic Studies," *Population Communication*, Pasadena, 1992.

Jensen, Deborah B., Margaret Torn, and John Harte, *In Our Own Hands: A Strategy for Conserving Biological Diversity in California*, Berkeley: California Policy Seminar Research Report, 1990.

Kinney, P.L., and H. Ozkaynak, "Associations of daily mortality and air pollution in Los Angeles County," *Environmental Research*, volume 54, issue 2, April 1991.

Mills, Paul K., David Abbey, W. Lawrence Beeson, and Floyd Petersen, "Ambient Air Pollution and Cancer in California Seventh-day Adventists," *Archives of Environmental Health*, volume 46, number 5, Sep/Oct 1991.

Morrison, Peter A., "California's Demographic Outlook: Implications For Growth Management," Santa Monica, CA: RAND Corporation, 1991, P-7738.

Nichols, Mary D., and Stanley Young, *The Amazing L.A. Environment—A Handbook for Change*, Los Angeles: The Natural Resources Defense Council/Living Planet Press, 1991.

"Population Projections by Race/Ethnicity for California and its Counties 1990–2040" *Report 93 P-1—Official Population Projections*, Sacramento: Demographic Research Unit, April 1993.

Rashkin, Sam, et al., *Energy Technology Status Report, Final Report*, Sacramento: California Energy Commission, 1991.

*Regional Growth Management Plan*, Los Angeles: Southern California Association of Governments, February 1989.

Rubin-Kurtzman, Jane R., et al., "Demographic And Economic Interactions in Trans-Border Cities: The Southern California–Baja California Mega-City," paper, USC-Los Angeles/El Colegio de la Frontera Norte, Tijuana, BC, 1992, presented at the Population Geography Symposium, Commission on Population Geography, International Geographical Union, Los Angeles, August 4–7, 1992.

Stokes Associates, Jones &, *Sliding Toward Extinction: The State of California's Natural Heritage, 1987*, prepared at the request of the California Senate Committee on Natural Resources and Wildlife, Commissioned by the California Nature Conservancy, Sacramento, November 1987.

GENERAL

Abbey, Edward, *Desert Solitaire: A Season in the Wilderness*, New York: Ballantine Books, 1977.

Abernethy, Virginia D., *Population Politics—The Choices That Shape Our Future*, New York: Insight Books, 1993.

Adams, W.M., *Green Development, environment and sustainability in the third world*, London: Routledge, 1990.

Amin, Samir, *Maldevelopment—Anatomy of a Global Failure*, London and Tokyo: Zed Books and U.N. University Press, 1990.

Bello, Walden, and Stephanie Rosenfeld, *Dragons in Distress: Asia's Miracle Economies in Crisis*, San Francisco: Institute for Food and Development Policy, 1990.

Beres, Louis Rene, *Apocalypse: Nuclear Catastrophe in World Politics*, Chicago: University of Chicago Press, 1980.

Bongaarts, John, W., Parker Mauldin, and James F. Phillips, "The Demographic Impact of Family Planning Programs," *Studies in Family Planning*, volume 21, number 6, Nov/Dec 1990.

Bos, Eduard, My T. Vu, and Ann Levin, *East Asia and Pacific Region, South Asia Region, Population Projections 1992–93 Edition*, Washington, D.C.: Policy Research Working Papers, Population and Human Resources Department, The World Bank, November 1992, WPS 1032.

Botkin, Daniel B., "Ecological Theory and Natural Resource Management —Scientific Principles or Cultural Heritage?" in *Ecological Prospects— Scientific, Religious, and Aesthetic Perspectives*, ed. Christopher Key Chapple, Albany, NY: State University of New York Press, 1994.

Boyd, Billy Ray, *For the Vegetarian in You*, San Francisco: Taterhill Press, 1993.

Brower, Kenneth, *One Earth*, San Francisco: Collins, 1991.

Brown, Janet Welsh, ed., *In the U.S. Interest—Resources, Growth, and Security in the Developing World*, Boulder, CO: A World Resources Institute Book, Westview Press, 1990.

Brown, Lester, "The New World Order," *State of the World*, New York: W.W. Norton, 1991.

Cannon, James S., *The Health Costs of Air Pollution—A Survey of Studies Published 1984–1989*, Los Angeles: American Lung Association, 1990.

Chapple, Christopher Key, *Nonviolence to Animals, Earth, and Self in Asian Traditions*, Albany: State University of New York Press, 1993.

Chen, R.S., et al., *The Hunger Report*, Providence, RI: World Hunger Program, Brown University, 1990.

Clark, Colin, *Population Growth and Land Use*, New York: St. Martin's Press, 1977.

Clark, John, *Democratizing Development: The Role of Voluntary Organizations*, West Hartford, CT: Kumarian Press, 1991.

Conly, Shanti R., and J. Joseph Speidel, *Global Population Assistance—A Report Card On The Major Donor Countries*, Washington, D.C.: Population Action International, 1993.

Cook, Stephen, "The Complexity of Theorem Proving Procedures," proceedings of the 3rd Annual ACM Symposium on the Theory of Computing, New York: Association of Computing Machinery, 1971.

Cousteau, Jacques-Yves, *The Cousteau Almanac—An Inventory of Life on Our Water Planet*, Garden City, NY: Dolphin Books/Doubleday, 1981.

Daily, G.C., and P.R. Ehrlich, *An Exploratory Model of the Impact of Rapid Climate Change on the World Food Situation*, Stanford, CA: The Morrison Institute for Population and Resource Studies, Stanford University, 1990.

*Demographic Indicators of Countries: Estimators and Projections as Assessed in 1980*, New York: United Nations, 1982.

Donaldson, Peter J., and Amy Ong Tsui, "The International Family Planning Movement," *Population Bulletin*, Population Reference Bureau, Inc., volume 45, number 3, November 1990.

Ehrlich, Paul, Anne Ehrlich, and John Holdren, *Ecoscience: Population, Resources, Environment*, San Francisco, W.H. Freeman, 1977.

Ehrlich, Paul, and Anne Ehrlich, *The Population Explosion*, New York: Touchstone Books, Simon and Schuster, 1990.

Ehrlich, Paul, and E.O. Wilson, "Biodiversity Studies: Science and Policy," *Science*, volume 253, August 16, 1991.

Eibesfeldt, Irenaus Eibl, *The Biology of Peace and War: Men, Animals, and Aggression*, trans. E. Mosbacher, New York: Viking Press, 1979.

Ekins, Paul and Manfred Max-Neef, eds., *Real-Life Economics—Understanding Wealth Creation*, London and New York: Routledge, 1992.

*Ending Hunger—An idea whose time has come*, The Hunger Project, New York: Praeger Publishers, 1985.

*Environmental Data Compendium 1991*, Paris: OECD—Organization for Economic Cooperation and Development, 1991.

*Environmental Policy Benefits: Monetary Valuation*, Paris: Organization for Economic Cooperation and Development, 1989.

Evans, G.W., S.D. Colome, and D.F. Shearer, "Psychological Reactions to Air Pollution," *Environmental Research*, volume 45, issue 1, February 1988.

Feichtinger, W., "Environmental factors and fertility," *Human Reproduction*, volume 6, issue 8, September 1991.

Fowler, Cary, and Pat Mooney, *Shattering: Food, Politics, and the Loss of Genetic Diversity*, Tucson: University of Arizona Press, 1990.

Fremlin, J.H., "How Many People Can the World Support?" in *Population, Evolution, and Birth Control*, ed. Garret Hardin, San Francisco: W.H. Freeman, 1969.

French, Hilary, *After the Earth Summit: The Future of Environmental Governance*, Worldwatch Paper #107, Washington, D.C.: Worldwatch Institute, March 1992.

Gardner, Brian, "European Agriculture's Environmental Problems," paper presented at the First Annual Conference of the Hudson Institute, Indianapolis, Indiana, April 1990.

Glacken, Clarence J., *Traces on the Rhodian Shore: Nature and Culture in Western Thought from Ancient Times to the End of the Eighteenth Century,* 2d ed., Berkeley: University of California Press, 1976.

Goldin, Ian, Odin Knudsen, and Domonique van der Mensbrugghe, "Trade Liberalisation, Global Economic Implications," Washington, D.C.: World Bank and OECD Development Center, 1993.

Golding, Martin P., and Naomi H. Golding, "Population Policy: Some Value Issues," in *Arethusa,* eds. John Peradotto and John J. Mulhern. Special issue, *Population Policy in Plato & Aristotle,* volume Eight, number Two, Buffalo, NY: State University of New York, Fall 1975.

Guz, Deborah, and John Hobcraft, "Breastfeeding and Fertility: a Comparative Analysis," *Population Studies,* volume 45, number 1, March 1991.

Hardin, Garret, "The tragedy of the commons," *Science* 162, December 13, 1968.

_____ , *Living Within The Limits—Ecology, Economics & Population Taboo,* New York: Oxford University Press, 1993.

Hardoy, J., and D. Satterthwaite, *Squatter Citizen: Life in the Urban Third World,* London: Earthscan Publications Ltd., 1989.

Harrison, P., "Too Much Life on Earth?" *New Scientist,* volume 126, 1990.

Hippocrates, "Airs, Waters, Places," in *Works* , trans. W.H.S. Stone, Cambridge, MA: Harvard University Press, 1948.

International North Pacific Fisheries Commission, *Final Report of 1990 Observations of the Japanese High Seas Squid Driftnet Fishery of the North Pacific Ocean,* Seattle, WA: Alaska Fisheries Science Center, 1991.

*Inventory and Analysis of Federal Population Research Fiscal Year 1992,* U.S. Department of Health and Human Services, NIH, Public Health Service, prepared by the Office of Science Policy and Analysis in cooperation with the Center for Population Research, National Institute of Child Health and Human Development for the Interagency Committee on Population Research.

*Inventory of Population Projects in Developing Countries Around the World 1991/92,* New York: UNFPA, 1992.

Kaplan, Robert, "The Coming Anarchy," *Atlantic Monthly,* February 1994.

Karmack, Andrew M., *The Tropics and Economic Development: A provocative inquiry into the Poverty of Nations,* Washington D.C: A World Bank Publication, 1976.

Karop, Richard, "Reducibility among Combinatorial Problems," in R.E. Miller and J.W. Thatcher, eds., *Complexity of Computer Computations*, New York: Plenum Press, 1972.

Keegan, John, *A History of Warfare*, New York: Alfred Knopf, 1993.

Kennedy, Paul, *Preparing for the Twenty-First Century*, New York: Random House, 1993.

Khor, Martin, "Development, Trade, and the Environment: A Third World Perspective," in *The Future of Progress*, eds. Helena Norberg-Hodge and Peter Goering, Bristol, UK, and Berkeley, CA: The International Society for Ecology and Culture, 1992.

Konner, Melvin, *The Tangled Wing: Biological Constraints in the Human Spirit*, New York: Holt, Rinehart & Winston, 1982.

LaFleur, William R., *Liquid Life—Abortion and Buddhism in Japan*, Princeton, NJ: Princeton University Press, 1992.

Lampkin, Nicolas, "Organic Farming and Agricultural Policy in Europe, paper presented at the Conference on Sustainable Agriculture and Agricultural Policy, Quebec Ministry of Agriculture, Fisheries, and Food, Quebec City, Canada, November 1990.

Lederberg, Joshua, Robert E. Shope, and Stanley C. Oaks, Jr., eds., *Emerging Infections—Microbial Threats to Health in the United States*, Institute of Medicine, Washington, D.C.: National Academy Press, 1992.

Leonard, H. Jeffrey, *Environment and the Poor: Development Strategies for a Common Agenda*, New Brunswick, NJ: Transaction Books, 1989.

Linzey, Andrew, and Tom Regan, eds., *Animals and Christianity—A Book of Readings*, New York: Crossroad, 1990.

Little, Charles E., ed., *Louis Bromfield At Malabar—Writings on Farming and Country Life*, Baltimore: Johns Hopkins University Press, 1988.

Liu, James J.Y., *Language-Paradox-Poetics—A Chinese Perspective*, ed. Richard John Lynn, Princeton University Press, 1988.

Lloyd, Cynthia B., "The Contribution of the World Fertility Surveys to an Understanding of the Relationship Between Women's Work and Fertility," *Studies in Family Planning 1991*, volume 22, number 3.

Lucretius, *The Nature of the Universe*, trans. R.E. Latham, Harmondsworth: Penguin, 1951.

Lutz, Ernst, and Salah El Serafy, "Environmental and Resource Accounting: An Overview," *Environment Department Working Paper #6*, Washington, D.C.: World Bank, June 1988.

Mahavira, Sramana Bhagavan, *The Akaranga Sutra*, Book I, Lecture I, Lesson 3, in *Jaina Sutras*, trans. Hermann Jacobi, New Delhi: Motilal Banarsidass.

Manfred, Max-Neef, "Ecological Economics," an interview by Satish Kumar, *Resurgence*, number 155, Nov/Dec 1992.

Mansfield, Susan, *The Gestalts of War: An Inquiry into its Origins and Meanings as a Social Institution*, New York: Dial Press, 1982.

McFalls, Joseph A., Jr., "Population: A Lively Introduction," *Population Bulletin*, volume 46, number 2, October 1991.

Meadows, Donella, et al., *Limits To Growth*, a Club of Rome report, New York: Universe Books, 1972.

Montagu, Ashley, ed., *Learning Non-Aggression—The Experience of Non-Literate Societies*, Oxford: Oxford University Press, 1978.

Morse, Stephen S., "Stirring Up Trouble—Environmental Disruption Can Divert Animal Viruses into People," *The Sciences*, New York: New York Academy of Sciences, volume 30, Sep/Oct 1990.

Morse, Stephen S., and Ann Schluederberg, "Emerging Viruses: The Evolution of Viruses and Viral Diseases," *The Journal of Infectious Diseases*, 1990.

Morse, Stephen S., ed., *Emerging Viruses*, New York: Oxford University Press, 1993.

Musgrave, Ruth S., and Mary Anne Stein, *State Wildlife Laws Handbook*, Center for Wildlife Law at the Institute of Public Laws, University of New Mexico, with Karen Cantrell, Sara Parker, and Miriam Wolok, Rockville, MD: Government Institutes Inc., 1993.

Myers, Norman, *Population, Resources, and the Environment—The Criticial Challenges*, New York: UNFPA, 1991.

Nance, John, *Discovery of the Tasaday, A Photo Novel: The Stone Age Meets the Space Age in the Philippine Rain Forest*, Manila: Vera-Reyes Inc., 1981.

Narayanan, E., "Dams and Public Health," *Sanctuary*, Bombay, volume X, number 1, 1990.

Nash, Roderick, *Wilderness and the American Mind*, 3d ed., New Haven and London: Yale University Press, 1982.

Newkirk, Ingrid, *Free The Animals! The Untold Story of the U.S. Animal Liberation Front & Its Founder, "Valerie,"* Chicago: The Noble Press Inc., 1992.

————, *Save The Animals! 101 Easy Things You Can Do*, New York: Warner Books, 1990.

Nozick, Robert, *The Nature of Rationality*, Princeton, NJ: Princeton University Press, 1993.

Perrin, Noel, *Giving up the Gun: Japan's Reversion to the Sword, 1543–1879*, Boston: David Godine, 1979.

Phillips, Gillian, "Vox Populi—The Real Rio?" *Populi*, October 1991.

Piotrow, Phyllis Tilson, *World Population Crisis—The United States Response*, New York: Praeger, 1973.

Preparatory Committee for the International Conference on Population and Development, Third Session, April 4–22, 1994, Draft Final Document of the Conference, New York: United Nations, 1994.

Preston, Richard, "Crisis in the Hot Zone," *The New Yorker*, October 26, 1992.

Rifkin, Jeremy, *Beyond Beef—The Rise and Fall of the Cattle Culture*, New York: A Dutton Book, 1992.

Robbins, John, *Diet for a New America*, Walpole, NH: Stillpoint Publishing, 1987.

Ruskin, John, *"Unto This Last:" Four Essays on the First Principles of Political Economy*, London: Smith, Elder, and Co., 1862.

Sainsbury, R.M., *Paradoxes*, Cambridge: Cambridge University Press, 1988.

Schell, Jonathan, *The Fate of the Earth*, New York: Avon, 1982.

Schneider, B., *The Barefoot Revolution*, London: IT Publications, 1988.

Schneider, Kirk J., *The Paradoxical Self—Toward an Understanding of Our Contradictory Nature*, New York and London: Insight Books, 1990.

Sharma, Dr. B.D., "On Sustainability—the voice of the disinherited," *Sanctuary*, volume XII, number 3, 1992.

Shaw, Marvin C., *The Paradox of Intention—Reaching the Goal by Giving Up the Attempt to Reach It*, Atlanta: Scholars Press, 1988.

Shepard, Paul, *Nature and Madness*, San Francisco: Sierra Club Publishers, 1982.

Shoumatoff, Alex, *The World Is Burning—Murder in the Rain Forest*, Boston: Little, Brown, 1990.

Singer, Peter, "The Peta Guide To Animal Liberation," Washington, D.C.: People for the Ethical Treatment of Animals, 1993.

Smil, Vaclav, *Global Ecology—Environmental change and social flexibility*, London: Routledge, 1993.

Smith, Kenwyn K., and David N. Berg, *Paradoxes of Group Life—Understanding Conflict, Paralysis, and Movement in Group Dynamics*, San Francisco and London: Jossey-Bass, 1987.

Soule, Michael, "A Vision for the Meantime," the Wildlands Project, *Wild Earth—Special Issue*, Cenozoic Society, 1992.

Spencer, Colin, *The Heretic's Feast—A History of Vegetarianism*, London: Fourth Estate, 1993.

St. John-Stevas, Norman, *The Agonizing Choice—Birth Control, Religion and the Law*, London: Eyre & Spottiswoode, 1971.

Summers, Lawrence H., "Investing in All the People," a policy research paper, Washington, D.C.: World Bank, 1992.

*Taiwan 2000*, Steering Committee, Taiwan 2000 Study, Taipei: Academia Sinica, 1989.

Thacher, Wendy, "Tests That Fail, Drugs That Kill," in *Good Medicine*, from the Physicians Committee for Responsible Medicine, volume II, number 4, Autumn 1993.

*The Environment At Risk—Responding to Growing Dangers*, National Issues Forums, Iowa: Public Agenda Foundation, 1989.

*The Global 2000 Report*, Washington, D.C.: Government Printing Office, 1976.

"The New Earth 21, Action Program for the Twenty-First Century," paper presented by the Japanese Ministry of International Trade and Industry at the U.S./Japan Conference on Global Warming, Atlanta, June 3, 1991.

"The Price of Pollution," *Options*, International Institute for Applied Systems Analysis, Laxenburg, Austria, September 1990.

Thomas, Keith, *Man And The Natural World—A History Of The Modern Sensibility*, New York: Pantheon Books, 1983.

Thoreau, Henry David, *The Journal of Henry D. Thoreau*, eds. Bradford Torrey and Francis H. Allen, 2 vols., New York: Dover, 1962.

Tobias, Michael, ed., *Deep Ecology*, San Diego: Avant Books, 1985.

Tobias, Michael, *After Eden—History, Ecology, & Conscience*, San Diego: Avant Books/Slawson Communications, 1985.

————, *Environmental Meditation*, Freedom, CA: Crossing Press, 1993.

————, "Jainism and Ecology: Views of Nature, Nonviolence, and Vegetarianism," in *Worldviews and Ecology*, eds. Mary Evelyn Tucker and John A. Grim, *Bucknell Review*, Lewisburg, PA: Bucknell University Press, 1993.

————, *A Naked Man*, Fremont, CA: Jain Publishing, 1994.

————, *Rage & Reason*, New Delhi: Rupa & Company, 1994.

————, *Voice of the Planet*, New York: Bantam Books, 1990.

*Total Quality Management—A Framework For Pollution Prevention*, Washington, D.C.: Quality Environmental Management Subcommittee, January 1993.

Toynbee, Arnold, *Mankind and Mother Earth: A Narrative History of the World*, New York: Oxford University Press, 1976.

Tuan, Yi-Fu, *Topophilia: A Study of Environmental Perception, Attitudes, and Values*, Englewood Cliffs, NJ: Prentice-Hall, 1974.

Tunnard, Christopher, *A World with a View: An Inquiry into the Nature of Scenic Values*, New Haven, CT: Yale University Press, 1978.

Tzu, Lao, *Tao Te Ching*, trans. D.C. Lau, Baltimore: Penguin Books, 1963.

United Nations, *World Fertility Survey—Major Findings and Implications*, Vaarburg, Netherlands: WFS, International Statistical Institute, 1984.

U.S. *Environmental Quality—23rd Annual Report*, Washington, D.C.: The Council on Environmental Quality, Office of the President, 1993.

U.S. *National Report on Population*, a report for the U.S. Department of State in preparation for the 1994 International Conference on Population and Development, Washington, D.C.: Population Reference Bureau, October 1993.

U.S. Senate, Committee on Environment and Public Works, "Title III— Emergency Planning and Community Right to Know," in *Superfund Amendments and Reauthorization Act of 1986* (P.L. 99-499), Washington, D.C.: U.S. Government Printing Office, 1987.

Verney, Peter, *Animals In Peril—Man's War Against Wildlife*, Provo, UT: Brigham Young University Press, 1979.

Vijaya, Muni Ratna Prabha, *Sramana Bhagavan Mahavira—His Life and Teaching*, 6 vols., New Delhi: Parimal Publications, 1989.

Vu, My T., Eduard Bos, and Ann Levin, *Europe and Central Asia Region, Middle East and North Africa Region—Population Projections, 1992–93 Edition*, Policy Research Working Papers: Population, Health, and Nutrition, Population and Human Resources Department, PS 1016, Washington, D.C.: World Bank, November 1992.

———, *Latin America and the Caribbean Region (and Northern America) Population Projections 1992–93 Edition*, Policy Research Working Papers: Population, Health, and Nutrition, Population and Human Resources Department, WPS 1033, Washington, D.C.: World Bank, November 1992.

Wald, George, "The Human Enterprise . . .", in *Population, Environment and People*, ed. Noel Hinrichs, Council on Population and Environment, New York: McGraw-Hill, 1971.

Warford, Jeremy, and David Wheeler, "Integrating Environmental Issues into Economic Policy Making," *Environment Bulletin—A Newsletter of the World Bank*, volume 4, number 2, Spring, 1992.

White, Lynn T., Jr., *Medieval Technology and Social Change*, Oxford: Oxford University Press, 1962.

Wilson, Edward O., *Biophilia—The Human Bond with Other Species*, Cambridge, MA: Harvard University Press, 1984.

Wilson, Edward O., "Is Humanity Suicidal," *New York Times Sunday Magazine*, May 30, 1993.

Wilson, Edward O., *The Diversity of Life*, Cambridge, MA: Belknap Press of Harvard University Press, 1993.

Wilson, Peter, *The Domestication of the Human Species*, New Haven: Yale University Press, 1988.

Winglee, P., "Agricultural Trade Policies of Industrial Countries," *Finance and Development*, volume 26, number 1, 1989.

*World Armaments and Disarmament*, Sipri—Stockholm International Peace Research Institute, New York: Oxford University Press, 1990.

*World Population Prospects—1988*, New York: United Nations, 1989.

*World Population Prospects—1992 Revision*, New York: United Nations Department for Economic and Social Information and Policy Analysis, 1992.

*World Resources Report 1992/1993*, The World Resources Institute, with the United Nations Environment Program and the United Nations Development Program, New York: Oxford University Press, 1993.

Zampaglione, Gerardo, *The Idea of Peace in Antiquity*, trans. Richard Dunn, South Bend, IN: University of Notre Dame Press, 1973.

# Index

# About the Author

Michael Tobias obtained his Ph.D. in the History of Consciousness from the University of California-Santa Cruz. A former Assistant Professor of Environmental Affairs at Dartmouth College, and Visiting Associate Professor of the Humanities at California State University-Northridge, he has spent twenty-five years tracking humanity's love-hate relationship with the Earth, research for which has taken him to more than fifty countries. A social scientist, frequent lecturer, best-selling novelist, prolific film director and producer, screenwriter, historian, anthropologist, and ecological humanist, his diverse oeuvre has focused, principally, upon the embattled human spirit and its perennial wellsprings of hope and renewal. Tobias was a former MacNeil-Lehrer NewsHour correspondent, and PBS Executive Producer for the acclaimed television series, *The Power Game* (about politics in Washington). His dozens of award-winning films (garnering major international film festival prizes, and five Emmy and Ace Award nominations), and more than twenty previous books, have been viewed, or read, in over forty countries. They have influenced political decisions affecting base operations in Antarctica and the engineering of double-hulled oil tankers, as well as persuading countless numbers of people to become vegetarian.